Laws of Medicine

Amirala S. Pasha

Editor

Laws of Medicine

Core Legal Aspects for the Healthcare Professional

 Springer

Editor
Amirala S. Pasha
Division of General Internal Medicine
Mayo Clinic
Rochester, MN, USA

ISBN 978-3-031-08161-3   ISBN 978-3-031-08162-0   (eBook)
https://doi.org/10.1007/978-3-031-08162-0

This Springer imprint is published by the registered company Springer Nature Switzerland AG
The registered company address is: Gewerbestrasse 11, 6330 Cham, Switzerland

This book is dedicated to

Emily and Dunkan,

for their love, patience, support, and encouragement.

Preface

Both law and science shape the clinical practice of medicine. As demonstrated by the breadth of this book's Table of Contents, numerous areas of the law influence clinical practice, some of which are even considered beyond the traditional "health law" topics. Therefore, a better understanding of these laws is paramount for any successful healthcare professional. The *Laws of Medicine: Core Legal Aspects for the Healthcare Professional* is a primer on US laws that affect clinical practice for healthcare professionals with no legal background.

Entire volumes have been written on the content of each chapter of this book with deep analysis by subject matter experts, often accompanied by lengthy and insightful editorials on their policy implications. This book is not that type! Rather, this book focuses on providing healthcare professionals, at any stage of their life-long learning, with concise introductory content on US laws that broadly affect clinical practice. It has scant focus or discussion on policy implications. If topics within a given chapter are of interest, the reader can obtain more in-depth knowledge from other widely available sources.

The book comprises 13 parts. The first part briefly introduces the US legal system as well as important legal concepts relevant to topics discussed in later chapters. Therefore, unless the reader already possesses a substantial understanding of the US legal system, reading the first part's chapters is highly recommended. This will allow for a better understanding of subsequent chapters. Although the hope is for the reader to read and understand every chapter, given the demands of clinical practice, readers may not wish to or be able to do so. With that in mind, each chapter is written in a modular fashion, in most cases eliminating the need to know or reference other chapters.

Each chapter starts with a summary of the chapter's content and relevant legal concepts in bullet points before discussing the topics in detail. An application section is provided in many chapters to clarify essential issues by reflecting on clinically relevant case law or clinical vignette(s). Depending on the subject and its relevance to daily clinical practice, some chapters only scratch the surface while others dive deeper.

Finally, no preface in a law book is complete without a disclaimer! So, here it is: Laws change frequently and they can vary slightly or greatly between jurisdictions. Therefore, an introductory book covering such a large swath of laws cannot comprehensively analyze every law and its legal implications for various scenarios.

Consequently, while this book can make readers more familiar with concepts and legalese relevant to a given area of law, it is not meant as a substitute for legal advice and should not be used in such a manner. Readers must seek competent legal counsel regarding legal concerns.

Rochester, MN   Amirala S. Pasha

Contents

Contributors

Jamie Abrams, JD Washington College of Law, American University, Washington, DC, USA

Shawn Bayern, JD College of Law, Florida State University, Tallahassee, FL, USA

Molly Berkery, JD, MPH Healthcare Regulatory Attorney, San Diego, CA, USA

Teneille Brown, JD S.J. Quinney College of Law, Center for Law and the Biomedical Sciences (LABS), Center for Health Ethics, Arts and Humanities (CHeEtAH), University of Utah, Salt Lake City, UT, USA

Nicholas J. Diamond, JD Georgetown University Law Center, Washington, DC, USA

Lauren DiGiovine, JD Littler, Mendelson P.C., Boston, MA, USA

Stephen M. Donweber, JD, MLIS Boston University School of Law, Boston, MA, USA

Jonathan H. Ferry, JD Bradley Arant Boult Cummings LLP, Charlotte, NC, USA

Summer Ghaith Mayo Clinic Alix School of Medicine - Arizona Campus, Phoenix, AZ, USA

Marc D. Ginsberg, JD, LLM University of Illinois—Chicago School of Law, Chicago, IL, USA

Ashleigh Giovannini, JD, LLM, CIPP/US Nixon Gwilt Law, Memphis, TN, USA

Alice Hall-Partyka, JD Crowell & Moring LLP, Los Angeles, CA, USA

Laura D. Hermer, JD, LLM Mitchell Hamline School of Law, Saint Paul, MN, USA

B. Jessie Hill, JD School of Law, Case Western Reserve University, Cleveland, OH, USA

Brian Holoyda, MD, MBA, MPH Martinez Detention Facility, Martinez, CA, USA

Anne P. Hovis, JD The Piorkowski Law Firm, PC, Washington, DC, USA

Edmund G. Howe, JD, MD Department of Psychiatry, Uniformed Services University, Bethesda, MD, USA

Toni Jaeger-Fine, JD Fordham Law School, New York, NY, USA

Kate Johansen, JD Department of Public Affairs, Mayo Clinic, Rochester, MN, USA

Andrew M. Knoll, MD, JD Cohen Compagni Beckman Appler & Knoll, PLLC, Syracuse, NY, USA

Jason J. Krisza, JD, Esq Wilentz, Goldman & Spitzer, P.A., Woodbridge, NJ, USA

Jacqueline Landess, MD, JD Medical College of Wisconsin, Milwaukee, WI, USA

Rachel A. Lindor, MD, JD Department of Emergency Medicine, Mayo Clinic, Phoenix, AZ, USA

Catherine London, JD, MPH London Legal Consulting, LLC, Minneapolis, MN, USA

Jeffrey S. Lubbers, JD Washington College of Law, American University, Washington, DC, USA

Stephen McJohn, JD Suffolk University Law School, Boston, MA, USA

Benjamin J. McMichael, JD, PhD The University of Alabama School of Law, Tuscaloosa, AL, USA

Lyndsay E. Medlin, JD Bradley Arant Boult Cummings LLP, Charlotte, NC, USA

Robert S. Olick, JD, PhD Center for Bioethics and Humanities, State University of New York Upstate Medical University, Syracuse, NY, USA

Department of Internal Medicine, George Washington University School of Medicine and Health Sciences, Washington, DC, USA

Amirala S. Pasha, DO, JD Division of General Internal Medicine, Mayo Clinic, Rochester, MN, USA

Joseph D. Piorkowski Jr, DO, JD, MPH The Piorkowski Law Firm, PC, Georgetown University Law Center, Washington, DC, USA

Cristah Artrip Prost, MD School of Medicine, Uniformed Services University, Bethesda, MD, USA

Lesley Ramey S.J. Quinney College of Law, University of Utah, Salt Lake City, UT, USA

Montrece Ransom, JD, MPH National Coordinating Center for Public Health Training, New Orleans, LA, USA

Anthony Rizzotti, JD Littler, Mendelson P.C., Boston, MA, USA

Ana Santos Rutschman, SJD Charles Widger School of Law, Villanova University, Villanova, PA, USA

Emely Sanchez, JD, MPH The Network for Public Health Law, Mid-States Region, Edina, MN, USA

Richard S. Saver, JD UNC School of Law and UNC School of Medicine, Chapel Hill, NC, USA

Michael F. Schaff, JD, LLM, Esq Wilentz, Goldman & Spitzer, P.A., Woodbridge, NJ, USA

Richard Silbert, MD Division of General Internal Medicine, Mayo Clinic, Rochester, MN, USA

E. John Steren, JD Epstein Becker & Green, P.C., Washington, DC, USA

Stacey A. Tovino, JD, PhD College of Law, The University of Oklahoma, Norman, OK, USA

Ashley VanDercar, MD, JD Case Western Reserve University, Northcoast Behavioral Healthcare, Northfield, OH, USA

Michael Vinluan, MD, JD, FCLM University of Maryland, Baltimore, Baltimore, MD, USA

Medical-Legal Consultant, Washington, DC, USA

Part I

Introduction to US Legal System

U.S. Government

Stephen M. Donweber

Contents

Key Points
- The three branches of government for the United States as a whole and each of the 50 states are the legislative, executive, and judiciary.
- The United States has adopted a strong form of federalism, which describes a system involving shared power between federal and state governments.
- Sources of law in the United States include statutes enacted by the legislature, regulations promulgated by executive agencies, and cases issued by the judiciary.
- Statutes must be read in conjunction with the cases and regulations that interpret and implement them.
- Courts in the United States must have jurisdiction to properly hear a claim and exercise power over the parties. Federal courts must have subject matter jurisdiction as they have the power only to hear and resolve a limited set of claims. All courts must have personal jurisdiction over the parties to the lawsuit.
- The United States is a common law country, which means that its law includes judicial decisions written by judges in the absence of an applicable statute.
- The proper analysis of case law involves the use of precedent; that is, the use of earlier cases with similar facts to inform the decision in the current case.
- Courts in the United States are hierarchical and territorial.

S. M. Donweber (✉)
Boston University School of Law, Boston, MA, USA
e-mail: donweber@bu.edu

© The Author(s), under exclusive license to Springer Nature Switzerland AG 2022
A. S. Pasha (ed.), *Laws of Medicine*,
https://doi.org/10.1007/978-3-031-08162-0_1

Legal Concepts

- The Constitution: The document establishing the structure of the American government.
- Branches of the Government: The major governmental divisions establishing the separate powers governing the United States.
- Executive: The branch of government responsible for enforcing the law. The President is the head of the Executive Branch. At the state level, the Executive Branch is headed by a governor.
- Judiciary: The branch of government responsible for resolving disputes and interpreting the law. Judges make up the judiciary.
- Legislature: The branch of government responsible for enacting law. The federal legislative branch consists of two houses, the House of Representatives and the Senate. Most states have a roughly parallel configuration with two legislative houses.
- Administrative or regulatory law: Laws or regulations promulgated by the executive branch.
- Statutory law: Laws or legislation enacted by Congress or a state legislature.
- Case law: Judicial opinions written by judges.
- Common law: A system where judges make law through writing opinions resolving disputes between two or more parties.
- Federalism: The system of dual sovereignty and shared powers as between the states and the federal government in the United States.
- Jurisdiction: The power of a court to hear a case and make legal judgments binding the parties to the lawsuit.
- Venue: The proper federal district court in which a lawsuit must be brought.
- Precedent: An earlier decided case that furnishes a basis for deciding later cases involving similar facts and issues.

1   Introduction

Everyone is subject to the law. The first step in navigating and adhering to the law is developing an understanding of the system of government in the United States, at both the federal and state levels. The three branches of government at the federal and state levels—legislative, executive, and judiciary—all produce law. The law takes the form of statutes enacted by the legislature, regulations promulgated by executive agencies, and opinions issued by the judiciary. These three sources of law are deeply interrelated and must always be considered together.

Case law, or judicial opinions, forms a vital part of the U.S. legal system. In this regard, the United States is what is known as a common law country, where courts resolve disputes and judicial opinions or cases are more than mere interpretations of statutes, but rather constitute the law themselves even when decided in the absence of a statute. The law in a case or opinion takes the form of precedent, which means that the decisions in earlier cases control the decisions in later cases with the same or similar facts. The more similar the facts, the more controlling the precedent. The power of the precedent derives from the hierarchical and territorial nature of the court system. By territorial, I mean that courts can only decide disputes and create binding precedent within a particular territory, like a state for a state court or a collection of states for a federal appeals court. Additionally, only cases from courts higher in the judicial hierarchy in a particular territorial jurisdiction are binding on lower courts in that jurisdiction. All other cases are simply persuasive.

2   Discussion

2.1   The U.S. Constitution

The Constitution established the structure and powers of the government of the United States (*See* Fig. 1).

Article I created the national legislature or Congress, dividing it into two houses, the House of Representatives and the Senate. Members of the House of Representatives were to be elected every two years, and members of the Senate every six. Article I also set forth the limited, enumerated powers of Congress. These powers include, among others, the powers to lay and collect taxes, regulate commerce, coin money, establish post offices, and declare war. Laws passed by Congress must pass both houses and be presented to the President for signature [1].

Article II created the Executive, the offices of the Presidency and Vice Presidency, declaring that the "executive Power shall be vested in a President of the United States of America." The President and Vice President serve four-year terms [2]. The Executive Branch under the President is vast. It encompasses numerous departments or agencies that are headed by political appointees called "Secretaries." The heads of the major agencies serve as the President's cabinet. Although the Executive Branch is often said to merely "enforce" laws, it actually creates many rules of its own through regulations drafted by the federal agencies in their area of expertise.

Article III created the federal judiciary, establishing that the "judicial power of the United States, shall be vested in one Supreme Court, and in such inferior courts as the Congress may from time to time ordain and establish." Article III also established the limited jurisdiction of the federal courts (federal question and diversity) and set forth the edict that federal judges serve for life [3].

Article IV discussed the relationship between the states as to the "public acts, records, and judicial proceedings of every other state" (they shall be given "full faith and credit") and as to the "privileges and immunities" of the citizens of each state (they should all be the same). Article IV also permitted the admission of new states by the federal government [4].

Article V set forth the process for amending the Constitution, providing that "Congress, whenever two thirds of both houses shall deem it necessary, shall propose amendments to this Constitution, or, on the application of the legislatures of two thirds of the several states, shall call a convention for proposing amendments, which, in either case, shall be valid to all intents and purposes, as part of this Constitution, when ratified by the legislatures of three fourths of the several states [5]."

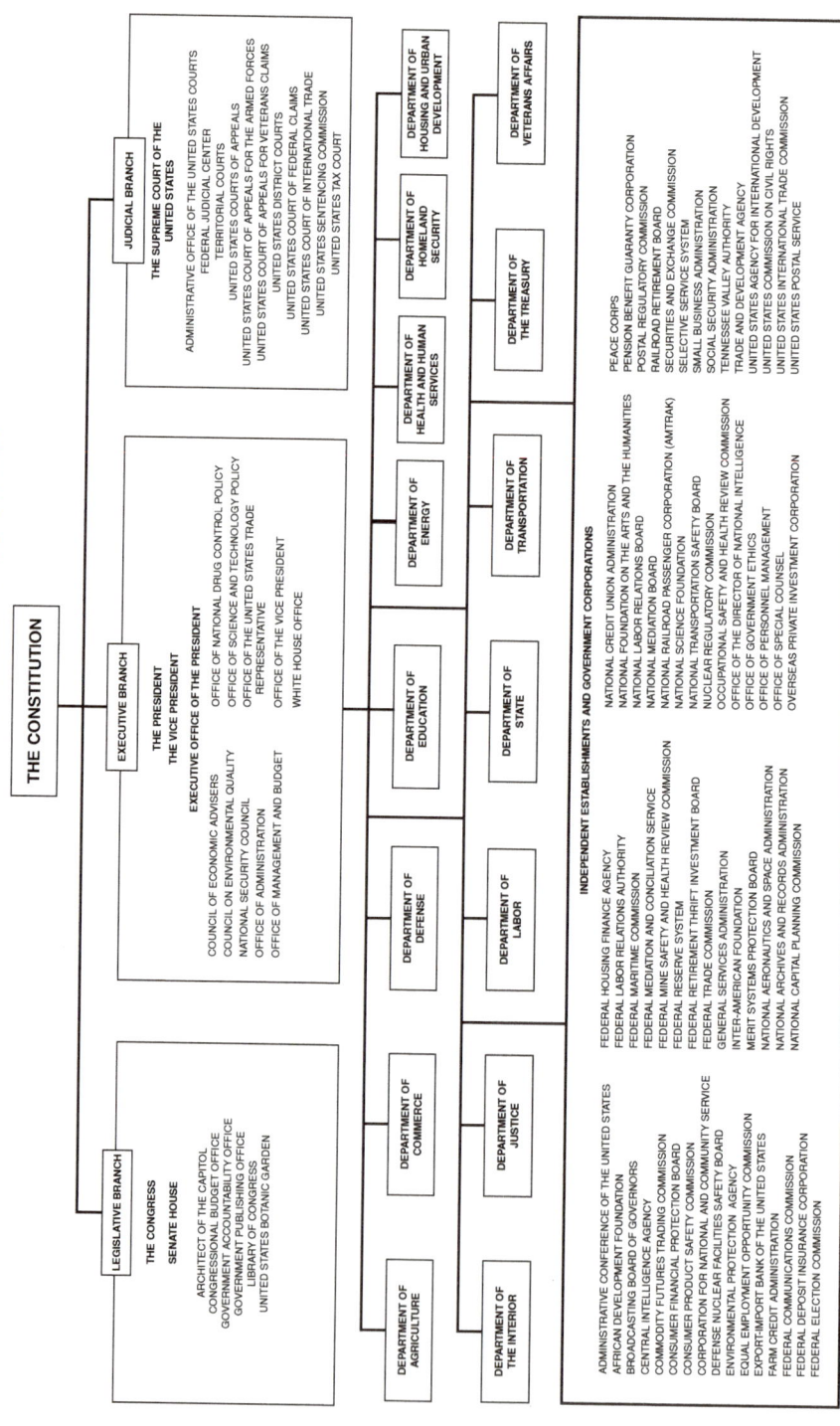

Fig. 1 The Government of the United States [U.S. government source in the public domain, *available at* https://www.usgovernmentmanual.gov/ReadLibraryItem.ashx?SFN=Myz95sTyO4rJRM/nhIRwSw==&SF=VHhnJrOeEAnGaa/rtk/lOg==]

Article VI provided that debts incurred prior to the adoption of the Constitution will be honored under it, that the Constitution and federal laws are the "supreme law of the land", and that public officials "shall be bound by oath or affirmation, to support this Constitution." Article VI also provided that no religious test "shall ever be required as a qualification to any office or public trust under the United States [6]."

Finally, Article VII provided that ratification by nine states shall be sufficient to adopt the Constitution between the states so ratifying [7].

The Constitution also contains 27 amendments, generally focused on the protection of individual rights and liberties from governmental intrusion [8].

States are also directed by the Constitution to form a "republican form" of government. Though not explicitly required to create a three-branch structure, States follow the federal model with an executive branch, headed by a governor, a legislative branch of elected representatives, and a judicial branch. State constitutions further grant local governments powers to regulate such matters as forming a police and fire department, creating municipal courts, etc. [9].

2.2   The Three Branches of Government and What They Do

The U.S. Constitution establishes three separate but equal branches of government, namely, the legislative branch, the executive branch, and the judicial branch (see Fig. 2). Under this system, each branch has its own powers and responsibilities but must also work with other branches for the government to function.

Legislature. The legislature—whether federal or state—is considered the central and most important governing branch. The federal legislature and most of those in the states are *bicameral*, composed of two houses. The two houses at the federal level are the House of Representatives and the Senate. Seats in the House of Representatives are proportionally allocated by state population. At present, there are 435 members of the House of Representatives. There are 100 members of the Senate, two per state [10].

The legislature enacts laws through the production of legislation or statutes. These are laws passed by a legislative body following introduction of a bill, a complex deliberative process, and interaction between and among both houses of the legislature and the executive branch [11]. The process requires that both the House and Senate pass the same form of a bill. When a bill is introduced in either body, it is assigned to a committee appropriate to the bill's subject matter. The committee will research, deliberate, and mark up the bill. The bill then comes to the floor of the relevant body where it is voted on. If approved it is sent to the other body for approval, where it may be amended even further. Until both bodies pass the same bill, it cannot be sent for approval to the executive. If the bodies disagree on amendments to a bill, a conference committee may be established, with members from both bodies, to address the dispute. The conference committee is formally empowered to present a compromise in a conference report to be approved by the House and Senate [12].

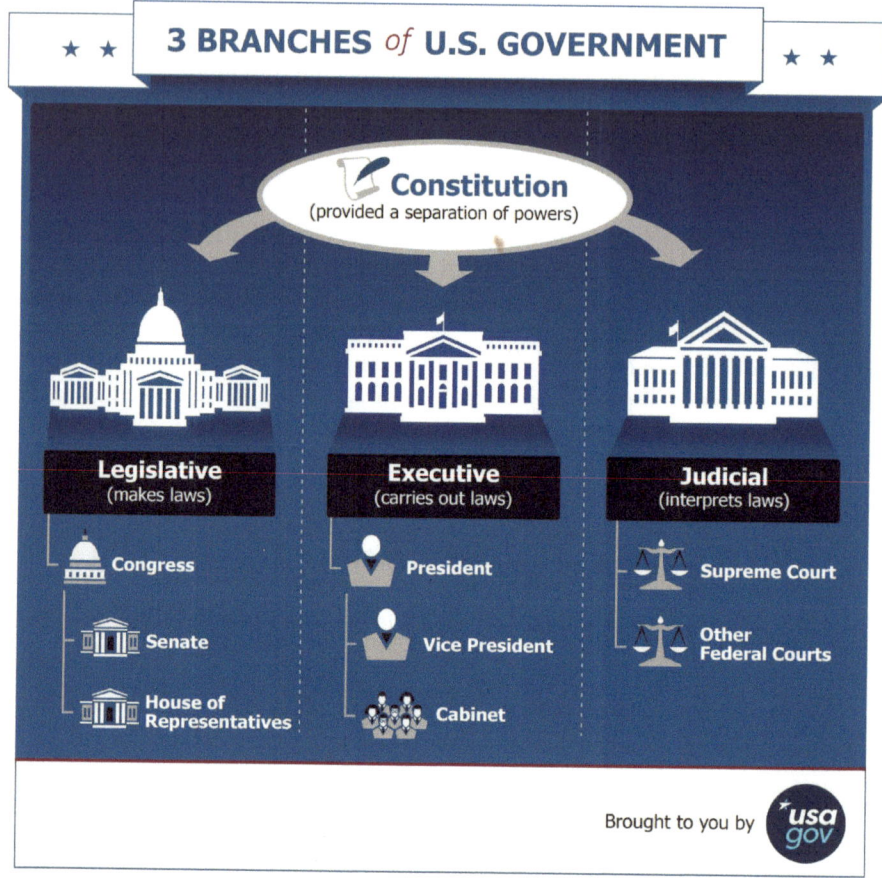

Fig. 2 The 3 Branches of U.S. Government [U.S. government source in the public domain, *available at* https://www.usa.gov/branches-of-government]

Once a majority in both houses (or just one if the state is unicameral) approves the final version of the bill, it can go before the executive, President or Governor, to be signed and enacted into law. The executive, however, can choose to veto the bill, which the legislature may then, with enough votes, override [11]. At the state level the process can vary by state. For instance, Tennessee gives a governor 10 days to veto a bill, or else it automatically becomes law, and only a constitutional majority is needed by the legislature to override a veto [13]. In contrast, Iowa gives the governor only three days to sign or veto a bill before it becomes law, and a two-thirds majority vote from both chambers of the legislature is required to override a veto [14].

Statutes often, but not always, frame their requirements in general language that is subject to interpretation by courts or amplification by executive agencies. This is due to the nature of the legislative process itself, where not every detail of a particular law is included, nor every term defined. The judiciary then will resolve disputes under the statute through statutory interpretation while executive agencies often amplify the requirements of a statute or provide details through the promulgation of regulations, which detail how a broad statute is to be enforced.

Executive. The Constitution only mentions the president in Article II [2]. Over time, this cryptic vesting of power grew into the modern sprawling executive branch. The executive branch generally constitutes most of the federal or state bureaucracy. At the federal level, the executive branch is composed of executive agencies, including the executive agencies whose leaders (secretaries) make up the president's cabinet [10].

These executive agencies—through the federal rulemaking process, which includes notice and an opportunity for public comment—promulgate regulations in the respective subject matter areas for which they are responsible [11]. In one sense, regulations provide the details for the broad policy mandates set forth in statutes. Put another way, regulations are the rules an agency writes to explain how a statute will be implemented. In some situations, the rules relate to a specific statute and subject matter and in other cases to the statute establishing an agency. These rules have the legal force of the statute behind them. In other words, they have the force of law. Rules derive their power to prohibit or regulate actions to the extent that the statute behind them is valid and the rules established are in line with the meaning of the statute. Courts may apply judicial review to such rules, but courts must also afford a certain amount of deference to an agency's determination of how to implement a statute [15].

Judiciary. Article III created the federal judiciary [3]. At the time the Constitution was debated and ratified, state courts had been around for well over 100 years. *Federal* courts for the nation as a whole, however, were something entirely new, and, as a result, the founders were concerned about encroachments on state court authority [16]. For this reason, the grant of federal court jurisdiction in the Constitution is limited to questions arising under federal law and to claims between citizens of different states (there are others, but these are the main ones) [3, 17, 18]. Congress has limited federal court jurisdiction even further through legislation. The Constitution itself only created the Supreme Court. Congress created the lower federal court system via legislation, beginning with the Judiciary Act of 1789 [19].

The judiciary produces case law. Case law comprises the written opinions of judges resolving all or part of a dispute between two or more parties in the course of a civil or criminal proceeding. Case law will often interpret the common law or a statute in the context of an individual dispute. The interpretation creates precedent, which will then shape future interpretations of the common law or the applicable statute in other disputes [20].

2.3   Federalism

The term (and concept) "federalism" describes the relationship between the state and federal governments in the United States of America (see Fig. 3). Specifically, federalism describes a system of dual sovereignty where the federal government

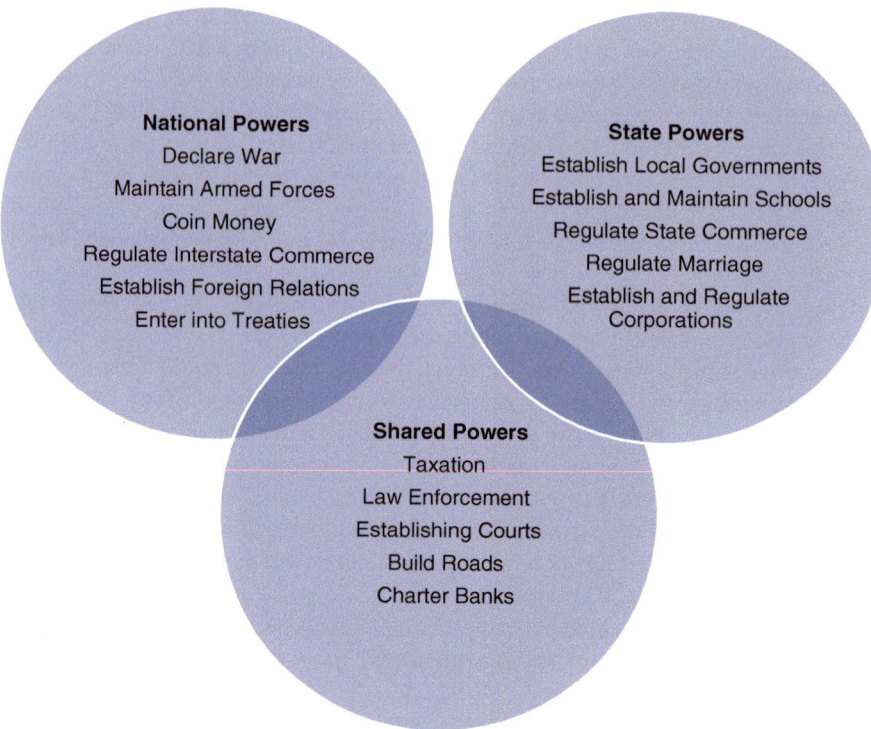

Fig. 3 Federalism in the United States [Created by the Author]

possesses certain limited powers of governance and the states possess others [21]. Americans are subject to and considered citizens of both the federal and state governments. In this mixed government scheme, the powers of the state and federal governments can be exclusive or they can overlap. Even though the U.S. has a strong central government and federal law is supreme, the powers of the central government are limited and the states possess all other powers of governance unless prohibited by the Constitution.

The Constitution gives the federal government certain enumerated powers, including the powers to declare war and maintain the armed forces, to coin money, to ratify treaties, etc. Congress has additionally passed legislation expanding federal powers, including granting the federal government jurisdiction over patent rights [22]. Otherwise, states have the right to create their own form of government and are granted general police powers not given to the federal government. Therefore, states establish their own criminal codes, laws regarding right to property, education systems, etc. In the courts, this system of dual sovereignty means that the federal and state courts exist in parallel, with the federal courts exercising limited jurisdiction (can only hear certain cases) and the state courts exercising general jurisdiction (can hear any case). Some powers of the federal and state governments are concurrent. For example, both state governments and the federal government have the right to levy taxes.

The power to regulate commerce is likely the most important power that the federal government possesses. The Commerce Clause refers to Article 1, Section 8, Clause 3 of the U.S. Constitution, which gives Congress the power "to regulate commerce with foreign nations, and among the several states, and with the Indian Tribes [1]."

The Commerce Clause empowers Congress to regulate activities that seemingly occur solely within one state. While this can sometimes lead to controversy, the Supreme Court has recognized that even local activities can have an impact on commerce between the states, and that these local activities therefore come within the ambit of the Commerce Clause.

Because it is Congress that has the power to regulate interstate commerce, the states may not do so. This implied restriction is known as the Dormant Commerce Clause, wherein the Supreme Court has held that states cannot pass laws that interfere with interstate commerce or that discriminate against out-of-state goods or services [23]. The restriction is self-executing and applies even in the absence of a conflict between federal and state statutes.

The concept of federalism is embodied in the 10th Amendment to the Constitution, which provides that "powers not delegated to the United States by the Constitution, nor prohibited by it to the states, are reserved to the states respectively, or to the people [24]."

2.4   Supremacy Clause

The Supremacy Clause in the U.S. Constitution is deeply interrelated with the concept of federalism.

The Supremacy Clause explains:

"This Constitution, and the laws of the United States which shall be made in pursuance thereof; and all treaties made, or which shall be made, under the authority of the United States, shall be the supreme law of the land; and the judges in every State shall be bound thereby, anything in the Constitution or laws of any State to the contrary notwithstanding [6]."

This means on its face that federal law is supreme to state law. It's more nuanced than that, however. The nuance flows from the fact that the powers of the federal government are limited and specifically enumerated in the Constitution. The enumerated powers include, among others, the power to coin money, to regulate commerce, to declare war, to raise and maintain armed forces, and to establish a Post Office [1]. In these areas, yes, federal law is supreme, and where federal law is supreme, it trumps state law under the doctrine of preemption. This doctrine means that if state laws conflicts with federal law, federal law overrides state law. For example, FDA regulations will preempt state regulation or state court decisions regarding prescription medication. But in many other areas, federal law is not supreme.

For example, there are many powers like establishing schools and local governments and areas of law like contracts, torts, or property law that are within the exclusive province of the states [25]. Importantly, on questions of state law, it is the state supreme court and not the U.S. Supreme Court that has the last word.

2.5   Sources of Law

In addition to the U.S. Constitution, other primary sources of law include statutes, regulations, and case law (see Fig. 4).

Statutes
Statutes are laws passed by legislative bodies, whether federal or state. The federal legislature—the United States Congress—passes laws during legislative sessions. Each session lasts two years.

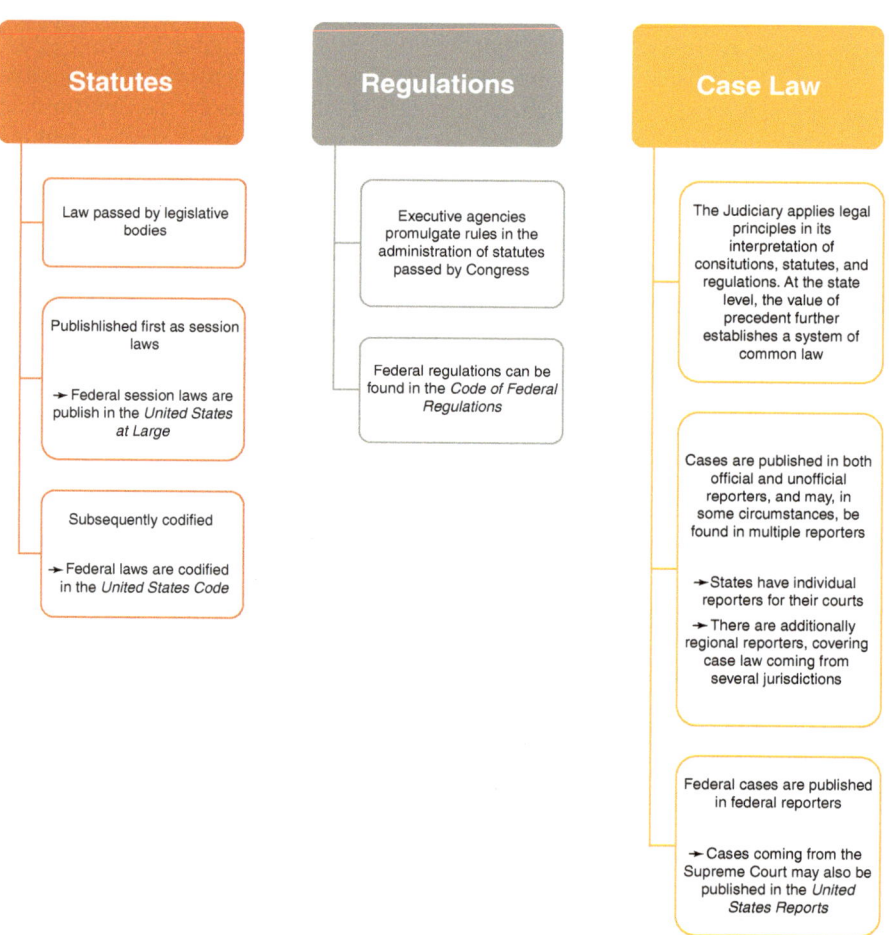

Fig. 4 Sources of Law and where they are published. [Created by Anastassia Korin]

Statutes are generally published in two forms, first as a session law and then as part of a code. The session law is the law or Act exactly as passed by the legislature. Session laws at the federal level are published in chronological order of passage in a set of books called the *United States Statutes at Large* [26]. The federal session laws are called *public laws*, and identified as, for example, Public Law No. 111-148, which refers to the 148th law passed in the 111th session of Congress. Most federal session laws also have a name, like, for example, Health Insurance Portability and Accountability Act of 1996 (HIPPA). This name is known as the session law's popular name. Federal session laws are also identified by where they are published in the Statutes at Large. The session law published on page 637 of volume 31 of the Statutes at Large will be cited as 61 Stat. 637.

Because session laws are published in chronological order and likely contain sections involving different subject areas, they do not lend themselves to successful legal research. For that reason, following passage by the legislature, a session law is *codified* and placed in a *code*. At the federal level, session laws are codified in the *United States Code* (U.S.C.). Codes are organized by subject. At the federal level, the United States Code is organized into 51 titles or subject areas, including Food and Drugs, Hospitals and Asylums, Public Health and Welfare, and more [26].

Regulations

Regulations or rules are the work product of executive agencies. At the federal level, Congress will often empower federal agencies to promulgate regulations that work in conjunction with statutes. The U.S. Government itself defines a regulation as:

> A regulation is a general statement issued by an agency, board, or commission that has the force and effect of law. Congress often grants agencies the authority to issue regulations. Sometimes Congress requires agencies to issue a regulation; sometimes Congress grants agencies the discretion to do so. Many laws passed by Congress give Federal agencies some flexibility in deciding how best to implement those laws. Federal regulations specify the details and requirements necessary to implement and to enforce legislation enacted by Congress [27].

Regulations, therefore, largely determine what statutes mean and the manner in which they are enforced. In addition, federal agency actions and interpretations of statutes, so long as not unreasonable, are accorded great deference by the federal courts, thus lending even greater importance to the actions of federal agencies in relation to legislation [15]. Therefore, regulations have the force of law and the legal power of statutes behind them. Notably, administrative law, which encompasses both law passed by administrative bodies and legal principles governing the power of government agencies, also places limits on the ability for such agencies to act. For instance, delegation of law-making power to administrative agencies by Congress, currently deemed constitutional, still cannot violate the doctrine of the separation of powers and is regulated by the Administrative Procedure Act (APA). Federal regulations are published four times per year in the Code of Federal Regulations or C.F.R. The organization of the C.F.R. mirrors that of the United States Code [28].

Although the details vary, all states have a system parallel to that of the federal government where state executive agencies issue regulations to supplement legislatively enacted statutes.

Case Law

As will be discussed in greater detail below, a case is a written decision issued by a judge or panel of judges at either the trial court, intermediate appellate court, or supreme court level regarding a dispute between two or more parties. The decision generally resolves the entire dispute or single/multiple issue(s) that have arisen during the course of the dispute.

In resolving the dispute, the judge will state and bring to bear applicable legal principles. These principles will for the most part derive from federal or state statutory (or regulatory) sources or the common law. It is the judge's duty to apply these legal principles to the facts of the case before her.

Cases are published in reporters. A reporter is a print set of books, organized by volume number, in which a particular jurisdiction's cases are collected in chronological order [29]. There are separate reporter systems for the federal judiciary and also for each of the states. State cases are also published in regional reporters, which collect cases decided by the courts of several different states.

A case citation will include the named parties and the date of the cited opinion. If the case was published in a reporter, the volume and page number where the case can be found are written in the citation. Unless the reporter cited to only publishes opinions from a single court, the case citation will also identify which court heard the matter. For example, the final opinion for the dispute between Jane Costa, plaintiff, and the Boston Red Sox Baseball Club, decided in 2004, heard before the Appeals Court of Massachusetts, and published in the second series of the North Eastern Reporter in volume 809, starting on page 1090, would be cited as Costa v. Boston Red Sox Baseball Club, 809 N.E.2d 1090 (Mass. App. Ct. 2004). Not all cases will be published in a regional reporter, thus the reporter alone may identify the court in which a case was heard. For example, only cases from the Massachusetts Supreme Judicial Court are published in the Massachusetts state reporter abbreviated as Mass., so a citation to a case published in the state reporter will appear as, for example, Coblyn v. Kennedy's Inc., 359 Mass. 319 (1971).

2.6   Statutes vs. Other Sources of Law

As noted above, statutes are laws passed by the legislature and then signed by the executive. Although ostensibly supreme to other sources of law (e.g., judicial opinions and regulations) they really aren't. Here's why.

Statutes are subject to interpretation. Of course, the words of a statute matter, but unless otherwise defined (and even then), statutory words only take shape into something meaningful and enforceable when interpreted by an executive agency through the promulgation of regulations or by a court through the issuance of judicial opinions.

Regulations. Regulations provide the details that statutes often omit. In fact, it is often up to executive agencies to implement statutes and provide the contours for enforcement. For example., the Clean Water Act provides generally: "The objective of this chapter is to restore and maintain the chemical, physical, and biological integrity of the Nation's waters [30]." Not a lot of guidance in other words. But then, in the very next section the statute states:

> "The Administrator [of the Environmental Protection Agency, an executive agency] shall, after careful investigation, and in cooperation with other Federal agencies, State water pollution control agencies, interstate agencies, and the municipalities and industries involved, prepare or develop comprehensive programs for preventing, reducing, or eliminating the pollution of the navigable waters and ground waters and improving the sanitary condition of surface and underground waters [31]."

It is therefore up to the EPA., not Congress, to determine what "clean water" means and how to achieve it. The statute, thus, cannot be considered without reference to applicable regulations.

Judicial opinions. It is up to the courts, federal and state, to determine what statutes mean. This is the case because generally statutes are only enforceable through the court system. Courts therefore have tremendous power over statutory interpretation. For example, part of the Patient Protection and Affordable Care Act guarantees access to health insurance to those with preexisting conditions [32]. That guaranty is simply not understandable without reference to the cases that interpret it (and also of course not. understandable without reference to the regulations that implement it).

In short, a statute standing alone seldom provides the answer a clinician would need in navigating the law. All available authority, including regulations and cases, must be considered as well.

2.7   The United States is a Common Law Country

Common law countries are those in which judicial decisions resolving disputes are considered "law" in the absence of an applicable statute. The common law tradition derives from practice in the medieval courts of England, where the law was considered "common" to all the king's courts across the country as a whole [33].

The coherence of a common law system depends on application of the idea of "precedent." To adhere to precedent means that a court will follow earlier decisions that resolve similar claims based on similar facts. In so doing, courts imbue the common law system with a level of predictability that benefits courts and litigants alike [20]. In the same manner, with individual judges issuing individual decisions in various jurisdictions, the common law also achieves a dynamism that permits flexibility and application of the law to different, even unique, circumstances [20]. As stated by Oliver Wendell Holmes, Jr.:

> The life of the law has not been logic: it has been experience .... The law embodies the story of a nation's development through many centuries, and it cannot be dealt with as if it contained only the axioms and corollaries of a book of mathematics. In order to know what it is, we must know what it has been, and what it tends to become [34].

The key feature of decisions decided under the common law is the absence of an applicable statute. To be sure, courts in common law countries most certainly interpret legislation, but the pure common law is based on cases alone. Traditional subject areas controlled by the common law include contracts, torts, property law, agency, and wills and estates [25]. Although federal courts can interpret common law principles originating in the states, they are not sources of the common law themselves. Only state courts can formulate common law doctrine [35]. That means, with one exception, there are 49 different (but nevertheless very similar) common law systems in the United States, one for each state. Louisiana has a civil law system [36]. The District of Columbia is a federal territory, not a state; therefore, D.C. courts rely on federal law and interpret laws passed by their legislative body, the Council of the District of Columbia.

2.8   The Judiciary and the Case Law System

A case is a written decision issued by a judge or panel of judges at either the trial court, intermediate appellate court, or supreme court level regarding a dispute between two or more parties. The decision generally resolves the entire dispute or single/multiple issue(s) that have arisen during the course of the dispute.

In resolving the dispute, the judge will state and bring to bear applicable legal principles. These principles will for the most part derive from federal or state statutory (or regulatory) sources or the common law. It is the judge's duty to apply these legal principles to the facts of the case before her.

In a common law system like the United States, much law is not based in statute (that is, enacted by the legislature) but rather in judicial decisions. Any court, at any level, can issue a judicial decision. The decision, or case, is binding on the parties and serves as precedent for future cases with similar facts [33].

The United States has one court system for the country as a whole (the federal system) and then parallel systems in each of the 50 states. Cases may be brought in either of the two systems. State courts generally can hear any case brought before them while federal courts are limited to hearing cases raising a question of federal law [17] or cases involving a dispute between citizens of different states, or between a citizen of a state and a citizen of a foreign country, where the amount in controversy exceeds $75,000 [18].

The organization of the federal and state court systems is exactly parallel.

3   Application

Case #1: <u>Riegel v. Medtronic, Inc., 552 U.S. 312 (2008)</u> Federal law under the Medical Device Amendments of 1976 (MDA) established safety oversight of medical devices. The statute also specified that states could not establish requirements pertaining to the safety or effectiveness of devices that differ from federal law. Federal oversight of medical devices varied based on the class of the device and required that the FDA review a manufacturer's design and labeling of medical devices. Further permission had to be granted by the FDA to make any changes to such devices.

Medtronic, Inc. designed, labeled, and manufactured catheters, identified as a Class III device requiring FDA premarket approval under the MDA. Charles Riegel had a Medtronic catheter that ruptured in his coronary artery during heart surgery. He and his wife brought suit against Medtronic, alleging negligence, breach of implied warranty, and claims of strict liability applicable under New York common law. Plaintiffs' common law claims, however, were based on requirements in the design, manufacture, label, and sale of such devices. The MDA established that the FDA approval process governed such regulation. Therefore, the Supreme Court held that the MDA preempted the state law. Plaintiffs did not claim that Medtronic violated federal law, therefore, the Supreme Court upheld the appeal's court decision to grant summary judgment in favor of defendants.

Case #2: <u>Chevron U.S.A., Inc. v. Natural Resources Defense Council, Inc., 467 U.S. 837 (1984)</u> Amendments to the Clean Air Act of 1977 established certain requirements on States that had not met the Environmental Protection Agency's (EPA) national air quality standards. States had to establish a permit program in which polluters would have to obtain a permit from state regulators to build "new or modified major stationary sources." Permits were only to be granted if the polluters met specific requirements geared towards abating new pollution. The amendments did not specifically define a "stationary source."

The EPA, an executive agency, interpreted the term "stationary source" as used in the Clean Air Act to allow existing plants emitting air pollutants to install or modify pieces of equipment without obtaining an additional permit as long as it did not increase the total pollution emitted from the source. Plaintiff, the Natural Resource Defense Council (NRDC), challenged the EPA's definition, arguing that the term "stationary source" encompassed any individual piece of pollution-emitting equipment. The Court, in reviewing the statute and subsequent administrative agency's interpretation of the statute, determined that the EPA's policy was a reasonable and therefore permissible interpretation of the Clean Air Act. The Supreme Court therefore reversed the lower court's decision to apply plaintiff's definition, holding that courts cannot impose their own interpretations of an ambiguous statute when Congress was otherwise silent on the issue, and the agency's construction of the statute was permissible.

4   Summary

Clinicians need to be aware of applicable law in their respective jurisdictions. This means being aware of both federal law and the law of the state in which they do business or reside. Relevant law generally takes three forms, all related to the branch of government from which the law originates. The legislature enacts statutes, laws generally applicable in the pertinent jurisdiction. The executive promulgates— through rulemaking—regulations, which implement and sometimes interpret statutory law. And the judiciary issues judicial opinions that resolve disputes and bind the parties to the dispute while at the same time serving as precedent for future cases based on similar facts. The power of the precedent derives from the hierarchical and territorial nature of the court system. Only cases from courts higher in the judicial hierarchy in a particular territorial jurisdiction are binding on lower courts in that jurisdiction.

Acknowledgements Many thanks to my amazing research assistant Anastassia Korin for her essential and valuable help in drafting and assembling this chapter.

References

1. U.S. Const. art. I.
2. U.S. Const. art. II.
3. U.S. Const. art. III.
4. U.S. Const. art. IV.
5. U.S. Const. art. V.
6. U.S. Const. art. VI.
7. U.S. Const. art. VII.
8. U.S. History and Historical Documents [Internet]. Washington; USAGov; 2000 Sept 22; [updated 2019 July 8; cited 2021 June 22]. Available from: https://www.usa.gov/history.
9. Our Government: State and Local Government [Internet]. Washington; The White House; [cited 2021 June 22]. Available from: https://www.whitehouse.gov/about-the-white-house/our-government/state-local-government/.
10. Branches of the U.S. Government [Internet]. Washington; USAGov; 2000 Sept 22; [updated 2021 Jan 21; cited 2021 June 22]. Available from: https://www.usa.gov/branches-of-government.
11. How Laws are Made and How to Research Them [Internet]. Washington; USAGov; 2000 Sept 22; [updated 2021 Apr 23; cited 2021 June 22]. Available from: https://www.usa.gov/how-laws-are-made.
12. Rybicki E. Conference Committee and Related Procedures: An Introduction [Internet]. Congressional Research Services. 2021 May 21. Available from: https://crsreports.congress.gov/product/pdf/RL/96-708/18.
13. How a Bill Becomes a Law [Internet]. Nashville (TN): General Assembly; [cited 2021 Oct 24]. Available from: https://www.capitol.tn.gov/about/billtolaw.html.
14. How a Bill Becomes a Law [Internet]. Des Moines (IA): Legislative Service Agency. 2014 Jan 1. Available from: https://www.legis.iowa.gov/DOCS/Resources/HowABillBecomesALaw.pdf.
15. Chevron, U.S.A., Inc. v. Natural Resources Defense Council, Inc., 467 U.S. 837 (1984).
16. Wright CA, Miller AR. Federal Practice and Procedure. 3rd ed. St. Paul (MN): West Group (Thomson Reuters); c2008. (Cooper EH, Freer R. Jurisdiction and Related Matters: The Federal Judicial System; vol. 13, Section 3502, The Constitution).

17. 28 U.S.C. § 1331.
18. 28 U.S.C. § 1332.
19. Judiciary Act of 1789, 1 Stat. 73 (current version codified in scattered sections of 28 U.S.C.).
20. Watford PJ, Chen RC, Basile M. Crafting Precedent [book review]. Harvard Law Review. 2017 Dec;131(2):843-80. Review of: Garner, BA, et al. The Law of Judicial Precedent. St. Paul (MN): Thomson Reuters. 2016.
21. Bond v. U.S., 572 U.S. 844, 854 (2014).
22. Patent Act of 1793, Pub. L. No. 1-7, 1 Stat. 109 (1793) (codified as amended in sections of 35 U.S.C.).
23. Dormant Commerce Power: Overview [Internet]. Washington: Congressional Research Service; [cited 2021 Oct 24]. Available from: https://constitution.congress.gov/browse/essay/artI_S8_C3_1_4_1/.
24. U.S. Const. amend. X.
25. The Federal Court System in the United States: An Introduction for Judges and Judicial Administrators in Other Countries. 4th ed. Washington: Federal Judiciary; 2016. Available from: https://www.uscourts.gov/sites/default/files/federalcourtssystemintheus.pdf.
26. Researching Federal Statutes [Internet]. Washington: The Library of Congress; [updated 2020 Dec 31; cited 2021 June 23]. Available from: https://web.archive.org/web/20210425112323/https://www.loc.gov/law/help/statutes.php.
27. Reginfo.gov: Frequently Asked Questions [Internet]. Washington: U.S. General Services Administration [cited 2021 June 23]. Available from: https://www.reginfo.gov/public/jsp/Utilities/faq.jsp.
28. Code of Federal Regulations [Internet]. Washington; U.S. Government Publishing Office; [updated 2021 Mar 17; cited 2021 June 22]. Available from: https://www.govinfo.gov/help/cfr.
29. Carr E., Legal Research: A Guide to Case Law [Internet]. Washington: Library of Congress; 2019 Sept 9; [updated 2020 Nov 13; cited 2021 June 21]. Available from: https://guides.loc.gov/case-law.
30. 33 U.S.C. § 1251.
31. 33 U.S.C. § 1252.
32. 42 U.S.C. § 18001.
33. Dainow, J. The Civil Law and the Common Law: Some Points of Comparison. The American Journal of Comparative Law. 1966-1967;15(3):419-435.
34. Oliver Wendell Holmes. The Common Law. Mark DeWolfe Howe, editor. Boston (MA): Little, Brown and Company; 1963.
35. Erie R.R. Co. v. Tompkins, 304 U.S. 64, 78 (1938).
36. Algero MG. The Sources of Law and the Value of Precedent: A Comparative and Empirical Study of A Civil Law State in A Common Law Nation. Louisiana Law Review. 2005 Winter;65(2):775-822.

The Courts

Toni Jaeger-Fine

Contents

Key Points

- Courts, as non-political institutions, are charged with interpreting and applying the law.
- As a result of the system of federalism, the U.S. has a federal judicial system which is the product of federal law; and each state has its own court system which is a product of the law of that state. Federal and state systems differ, and state systems vary among themselves.
- The U.S. has an adversarial system of litigation, which means that the parties take primary responsibility for zealously representing their clients' positions, while courts are neutral and primarily reactive.
- Federal courts are courts of limited competence, consistent with federalism. This means that their jurisdiction extends only to cases that meet the requirements for subject matter jurisdiction and justiciability.
- Federal courts principally are courts of general jurisdiction although there are some specialized federal judicial bodies. Federal judges are nominated by the President and are appointed upon confirmation by the Senate. In order to preserve their independence, judges have life tenure, and their salaries may not be reduced while in office.

T. Jaeger-Fine (✉)
Fordham Law School, New York, NY, USA
e-mail: tfine@fordham.edu

- Federal courts are organized on the basis of both hierarchy and geography. There are district court at the first level, followed by courts of appeals, and the Supreme Court at the apex. Geographically, there are district courts and courts of appeals throughout the nation. One geographically based court of appeals plus one or more federal district court comprise a circuit. Circuits are relevant both for judicial administration and for the application of the principle of stare decisis.
- The Supreme Court's jurisdiction is almost entirely discretionary.
- State courts vary among themselves based on the relevant state law, and have different provisions for the selection of judges, the structure of the courts, jurisdictional rules, rules of procedure and evidence, and the like. State judiciaries typically have more specialized courts than are found at the federal level.
- Court selection is based on considerations of subject matter jurisdiction, personal jurisdiction, and venue.
- The applicable standard of review will vary depending on the type of case and the jurisdiction.
- Courts base their rulings on precedent under the principle of stare decisis. The application of stare decisis is a foundational norm but is a concept that is elastic and subject to judicial discretion.
- Courts, as guardians of the constitution, must ensure that both procedural and substantive due process rights are protected.

Legal Concepts
- Federalism: The constitutional power sharing arrangement between the national/federal government and the states.
- Federal courts: Courts created and empowered by the Constitution.
- U.S. district courts: Trial level federal courts.
- U.S. courts of appeals: Intermediate appellate federal courts.
- U.S. Supreme Court: Federal court of last resort.
- Circuit: Geographic area that contains one federal court of appeals and one or more district courts.
- *En banc* review: Discretionary review by a (usually) full court of appeals of a decision of a three-judge panel of that court.
- *Certiorari* review: The exercise by the U.S. Supreme Court of its discretionary appellate jurisdiction.
- State courts: Courts created and empowered by their respective states.
- Competence: The authority of a federal court to review specific cases.
- Subject matter jurisdiction: The power of a court to hear a particular type of case based on whether the Constitution vested that power in the federal judiciary.
- Federal question jurisdiction: Exists when a claim arises under federal law.
- Diversity jurisdiction: Exists when the parties are citizens of different states or countries, and the amount in controversy exceeds $75,000.
- Removal jurisdiction: Allows a defendant to move a case from state court to federal court when it originally could have been brought in federal court.

- Supplemental jurisdiction: Allows a court to hear state law claims related to a federal claim properly before the court.
- Justiciability: The fitness of a particular dispute for judicial resolution based on standing, ripeness, and non-mootness.
- Standing: An element of justiciability, standing requires that the plaintiff has a concrete, genuine, particularized interest in the case and its outcome.
- Ripeness: An element of justiciability, ripeness requires that the harm complained of has matured to a point where it is ready for judicial resolution.
- Non-mootness: An element of justiciability, non-mootness provides that any conflict that has been settled or otherwise rendered moot by subsequent events may not be heard by the federal courts.
- Standard of proof: Level of assurance that must be satisfied by a party seeking to prove a fact
- Personal jurisdiction: The ability of a court to compel the presence of an out of state defendant to respond to a lawsuit.
- Venue: The geographic distribution of a case within a particular jurisdiction.
- Precedent: Earlier-decided cases.
- Binding precedent: Precedent that a court is required to apply to a subsequent apposite case; also known as mandatory or controlling precedent.
- Persuasive precedent: Precedent that a court should consider but need not apply; in the absence of controlling precedent, a court may apply persuasive precedent if it is persuaded that the analysis is correct.
- Stare decisis: The use by courts of principles announced in earlier judicial decisions.
- Common law: Law derived from one or more judicial decisions.
- Due process: Procedural and substantive rights that protect individuals against state deprivation of life, liberty, and property interests.

1   Introduction

This chapter introduces the U.S. courts, exploring the role of the courts in our governmental system as well as the multiple judicial systems—federal and state—that comprise the courts under our system of federalism. The chapter then goes on to explore the federal courts, looking at their hierarchical and geographical organization; the qualifications, selection procedure, and tenure of federal judges; federal courts as courts of general or specialized jurisdiction; the competence of the federal courts; and proceeding through the federal courts. The chapter then explores the state courts, which vary from state to state given that each state court system is the product of its respective state law. It then considers the principal standards of proof used in various cases. This chapter next explores the defining role of stare decisis in our judicial system, under which courts apply principles announced in earlier decided cases. Finally, this chapter explores the way in which courts protect procedural and substantive due process rights.

2   Discussion

2.1   Introduction to the Courts

Under the U.S. system of separation of powers, the courts represent the third branch of government at both the federal and state levels. The legislative branch is charged with *making* the law; the executive branch is charged with *enforcing* or *executing* the law; and the job of the judicial branch is to resolve disputes between and among parties and, as the Supreme Court noted in its landmark decision in *Marbury v. Madison* (1803), to "say what the law is" [1]. The courts do this through largely interpretation and the use of precedent.

The courts provide an important check on the work of the legislative and executive branches, known jointly as the "political" branches because they are chosen through an electoral process and represent their respective constituencies. The courts, however, have a higher call. They are not responsive to any electorate but instead are charged with preserving and protecting constitutional rights, even when they prove unpopular. As guardians of the Constitution [2]—including preservation of the limits on the legislative and executive branches and limits on the power of the national government—the courts by design are anti-majoritarian bodies. Courts safeguard the Constitution through the process of judicial review, under which they have the power to review and invalidate actions of coordinate branches of government.

Litigation before U.S. courts is considered to be *adversarial*. The adversarial system is one in which advocates represent their parties' position before a judge or jury. This system is contrasted with the inquisitorial system found in many civil law countries in which the judge actively investigates the case. In the adversarial system, judges are viewed as impartial arbiters of legal issues, generally reactive to motions made by a party rather than proactive decision-makers. The adversarial system relies on the zealous advocacy of the attorneys as representatives of their clients.

2.2   Dual Federal and State Court Systems

The United States of America is a federalist system under which governmental powers are shared by the national (also known as federal) government and state governments. Because of the federalist nature of the U.S. governmental system, there is a federal judicial system and an independent, sovereign judicial system in each of the U.S. states (and territories). These court systems are independent, and for the most part, they operate in parallel. Accordingly, litigants go through one system and generally do not go from one system to the other. There are exceptions, the most important being that decisions of state courts of last resort may be brought to the Supreme Court when they involve issues of federal law (as discussed below).

The federalist nature of our government is deeply rooted in and informed by our history. Early European settlers came to the U.S. seeking greater personal freedoms, transparency, and accountability. Finding these goals thwarted under British rule, they fought—and won—a bloody and expensive war to gain independence. The framers of our Constitution were very much the product of their time. Having just prevailed over the British in the American Revolution, they were wary of highly centralized governing structures, and sought to find a balance between a decentralized structure and one that would vest power in a national government sufficient to meet the aspirations of the new nation.

Guided by these principles, the founding fathers created a national or federal government that had ample but limited powers—its powers were limited to those that were enumerated in the Constitution. Each branch of the national government was provided with certain powers which it could not exceed.

This system of federalism thus has its roots deep in the history of the founding of our nation and represents a fundamental precept of our governmental and legal systems. The issue of where the boundaries of federal and state power reside remains a deeply divisive issue in law and politics, and the courts have a crucial role in identifying and preserving these boundaries and defining the respective roles of federal and state power.

The U.S. judicial systems are very much the product of our federalist structure. As with the other branches of government, courts exist both at the federal and state levels. Along with their respective legislative and executive branches, state courts are sovereign and independent from the federal courts.

2.3   The Federal Courts

Organization
Article III of the Constitution provides for a Supreme Court and vests in Congress the power to "ordain and establish" courts that fall below the Supreme Court in the judicial hierarchy. Congress, in the Judiciary Act of 1789 [3], provided the structure for the federal courts that is the foundation for the current federal court structure. Broadly speaking, the federal courts are organized hierarchically and geographically.

Hierarchical Organization of the Federal Courts
The hierarchical organization of the federal courts resembles a pyramid with three tiers. At the base of the pyramid are the U.S. district courts, which are trial-level courts. In the middle of the pyramid are the U.S. courts of appeals, the intermediate federal appellate courts. At the apex sits the U.S. Supreme Court, the federal court of last resort.

Geographical Distribution of the Federal Courts
Geographically, federal district courts and courts of appeals are spread across the country.

There are 94 U.S. district courts, organized within state borders. Each state has between 1 and 4 federal district courts, a number set by Congress to reflect the anticipated amount of federal judicial business in the state. In states in which there is a single district court, it is referred to by the name of the state, e.g., the U.S. District Court for the District of New Jersey. In states in which there are more than one federal district court, they will be referred to as the Central, Eastern, Middle, Northern, Southern, or Western district of the state, e.g., the U.S. District Court for the Southern District of New York.

Above the district courts on the federal court hierarchy sit the courts of appeals, sometimes referred to as circuit courts. There are 12 geographically based circuits, each with one court of appeals; 11 of them are numbered, e.g., the U.S. Court of Appeals for the Second Circuit, plus the U.S. Court of Appeals for the D.C. Circuit.

Finally, there is a single Supreme Court with nationwide jurisdiction.

Qualifications, Selection, and Tenure of Federal Judges

Unlike analogous provisions for members of the U.S. Congress or the President, Article III provides no qualifications for federal judges to serve. Presumably that is because the process for selecting federal judges would act to ensure that capable and otherwise appropriate people are selected for these posts. Article III of the Constitution stipulates the process by which federal judges are chosen: they are nominated by the President and appointed upon the "Advice and Consent" of the Senate.

Despite the overtly political process through which federal judges are selected, their independence from the political branches is preserved through a number of Constitutional provisions. Article III section 1 provides that judges serve "during good Behaviour" (which always has been interpreted to confer life tenure) and that a federal judge's compensation may not be reduced during his or her tenure. A sitting federal judge can be removed from her or his post only upon conviction of an impeachable offense.

Federal Courts as Courts of General or Specialized Jurisdiction

Federal Courts as Courts of General Jurisdiction

Article III courts for the most part are courts of general jurisdiction, which means that these courts are not specialized but hear the full range of matters that come to the federal courts. This includes both criminal and civil matters dealing with a wide range of topics.

Specialized Federal Judicial Bodies

In addition to the traditional Article III courts of general jurisdiction described above, there are a number of specialized entities in the federal judicial system, as outlined below. These entities have a limited and defined subject matter jurisdiction and include entities outside of the judicial branch as well as specialized trial and appellate courts.

There are a number of federal adjudicative entities that fall outside of the judicial branch, including military courts at the trial and appellate levels, the Court of

Veteran's Affairs, the U.S. Tax Court, and a range of federal administrative agencies and boards. These entities were not created pursuant to Congress' Article III power; they fall within the executive branch of government, although they perform adjudicative functions pursuant to the Administrative Procedure Act and specific rules applicable to particular agencies. Agency adjudicative procedures generally begin before an administrative law judge (ALJ), subject to review by the full board or commission. Final agency decisions, in turn, are subject to federal court review. Depending on the agency, appeals go to the district court or directly to the courts of appeasl.

In addition, there are several specialized trial-level courts which supplement the district courts of general jurisdiction, such as the bankruptcy courts, the Federal Court of International Trade (an Article III court), and the Court of Federal Claims (also an Article III court). The United States Court of Federal Claims has nationwide jurisdiction over many monetary claims against the United States. It has exclusive jurisdiction over claims in excess of $10,000, and shares jurisdiction with the U.S. district courts for claims $10,000 and below. The docket of the Court of Claims includes bid protests/government contracts claims, claims for vaccine compunction, tax refund suits, civilian and military pay claims, intellectual property claims, and various other statutory claims against the U.S. It does not hear claims arising under the Federal Tort Claims Act (FTCA). In addition, there is one specialized appellate court, the U.S. Court of Appeals for the Federal Circuit which hears certain custom and patent matters.

Limited Competence of the Federal Courts

Broadly speaking, there are two limits on the competence of the federal courts: subject matter jurisdiction and justiciability. Both are based on the language in Article III, section 2 of the Constitution which defines the federal "judicial power" along with "such Regulations as the Congress shall make."

Subject Matter Jurisdiction

The federal courts have authority over only certain types of cases—the states retain exclusive jurisdiction over cases that do not fall within the subject matter jurisdiction of the federal courts. This is a product of the system of dual sovereignty that lies at the heart of our legal system: a federal government of strong but limited powers; and the states, which hold residual governmental power.

There are three jurisdictional possibilities: exclusive state court jurisdiction (when there is no federal court jurisdiction); exclusive federal court jurisdiction (when a statute so provides); and overlapping jurisdiction where the federal and state courts have concurrent jurisdiction.

The principal forms of federal court subject matter are *federal question jurisdiction* and *diversity jurisdiction*. Under federal question or "arising under" jurisdiction, federal courts can hear cases that "arise under" federal constitutional or statutory law [4]. Under diversity jurisdiction, a case can be brought in federal court if the parties are citizens of different states, or of a state and a foreign country; and the amount in controversy is in excess of $75,000, an amount that can be (and in the past has been) adjusted by Congress [5]. By definition, diversity cases arise under state law rather than under federal law. Diversity jurisdiction is a response to the concern that defendants might have that state courts would be biased in favor of their own citizens in cases involving significant amounts of money. Federal courts will have jurisdiction only when there is "complete" diversity; diversity will be destroyed if there is at least one defendant of the same citizenship as any one of the plaintiffs. Citizenship for individuals is based on a person's domicile—physical presence plus present intent to remain there indefinitely. As the Supreme Court established in *Hertz v. Friend* (2010), a corporation is a citizen of the state of incorporation *and* the state in which its principal place of business is located. The principal place of business is the "nerve center," usually the company headquarters [6].

There are also a few derivative forms of federal court subject matter jurisdiction. Under *removal jurisdiction*, if the plaintiff brings an action in state court that could have been brought in federal court, the defendant may "remove" the case to federal court [7]. Removal allows a defendant to transfer a case from state court to the federal court that is in the same geographic area. The defendant, however, may not exercise removal jurisdiction if the case is based on diversity jurisdiction and the case was brought in the defendant's home state court. Under *supplemental jurisdiction*, federal courts may resolve state law matters closely related to a federal question to be litigated in the same case [8].

Because subject matter jurisdiction goes to the competency of a court to hear and decide a case before it, subject matter jurisdiction may not be waived by the parties or ignored by the court. If a federal court decides that it does not have subject matter jurisdiction over a case before it, it will dismiss the case, and the plaintiff will have to refile in state court. Likewise, when a state court determines that it lacks subject matter jurisdiction it must dismiss the case and the plaintiff will have to refile in federal court. Objections to a court's subject matter jurisdiction can be raised at any time in the proceedings, including on appeal, and is one of the few matters that will be raised by the court on its own motion (*sua sponte*).

Justiciability

The justiciability requirement derives from the Constitution's insistence in Article III section 2 that the federal judicial power extends only to "cases" and "controversies." From these terms have emerged several interrelated doctrines that essentially preclude federal courts from giving advisory opinions. These doctrines are standing, ripeness, and non-mootness. The doctrine of standing requires that the plaintiff have a concrete, genuine, particularized interest in the case and its outcome. The ripeness requirement insists that the harm complained of has matured to a point where it is ready to be resolved judicially; pre-enactment review of legislation is not available. The non-mootness doctrine renders non-justiciable any conflict that has been settled or otherwise rendered moot by subsequent events.

Proceeding Through the Federal Courts

Proceeding Through the U.S. District Courts

The federal district courts are courts of first instance in most cases, and they serve primarily as trial level courts whose goal is fact-finding. At the federal district court level, a single judge presides, and the parties, who are generally represented by counsel, are responsible for most elements of the prosecution of the case, including the development of facts, the framing of the legal issues, and the presentation of evidence at trial.

The judge or a jury may be responsible for fact-finding. In many civil and criminal cases in the United States, there is a constitutional right to trial by jury. This right does not extend to preliminary injunctions and other forms of equitable relief. The right to a trial by jury significantly affects trial strategy and trial procedures. The parties may waive the right to a jury trial. When the right to a jury trial is waived, the judge acts as finder of fact in what is called a *bench trial*.

Proceeding Through the U.S. Courts of Appeals

The federal courts of appeals are intermediate appellate courts. A party aggrieved by a final judgment of a district court has the right to appeal to the court of appeals for the circuit in which the district court sits. Federal courts of appeals also hear appeals directly from federal agencies when prescribed by statute.

Courts of appeals cases are heard by panels of three judges selected randomly from among the judges on that court of appeals. On appeal, the facts as determined below (in the district court or agency) are nearly always considered to be dispositive; the appellate court reviews cases to determine whether there has been an error of *law* committed by the lower tribunal that warrants reversal. Appeals also are limited to matters raised and properly preserved below. And a decision will not be reversed for harmless error (i.e., the error identified would not have changed the outcome).

A party dissatisfied with the decision of the court of appeals may seek rehearing *en banc*, under which all or some number of the active judges on the court review the decision of the three-judge panel. Rules about *en banc* review vary somewhat from circuit to circuit. *En banc* review, however, is entirely discretionary with the court of appeals, and in most circuits the grant of rehearing *en banc* is exceptional.

Proceeding Through the U.S. Supreme Court

There is a single federal Supreme Court. It hears cases from the federal courts of appeals. Cases may also come to the Supreme Court from the state courts of last resort if the cases below were decided on an "adequate and independent" federal ground, which means that the decision could not have rested solely on state law grounds. If only state law matters are presented, the federal courts have no special expertise and, under notions of comity, there is no basis upon which the federal courts should hear the matter. For example, in a typical claim for negligence, the issues would be limited to the application of state law and the decision of the state court of last resort would be the final judgment.

The Supreme Court's jurisdiction may be divided into original and appellate jurisdiction. When the Court's original jurisdiction is invoked, a case goes directly to the Supreme Court. These are highly unusual and are limited to cases "affecting Ambassadors, other public Ministers and Consuls," in the language of Article III, section 2; and cases in which a state sues a state. This latter category generally involves boundary disputes, which in the past arose much more commonly than they do at present.

The Court's appellate jurisdiction in turn can be divided into mandatory and discretionary appellate jurisdiction. The Court's mandatory appellate jurisdiction, once a healthy portion of its docket, has been largely eliminated by statute.

The overwhelming majority of the cases that come to the Supreme Court is under its discretionary appellate jurisdiction, also known as *certiorari* jurisdiction, after the device by which parties ask the Court to accept the case for a merits review. As this suggests, the Court's *certiorari* jurisdiction is entirely discretionary with the Court. Of the nearly 10,000 petitions filed each year, in recent years it typically hears only 90–100 of them.

The Court's case selection process demonstrates an important function of the Supreme Court: it is not primarily a court of error correction, as is the case of many national courts of last resort in other countries. It instead is a court primarily concerned with the uniformity of federal law. Indeed, the bases for which the Court has indicated it may grant review involves conflicting interpretations of federal law by lower courts. The Supreme Court as a matter of practice grants the petition for a writ of *certiorari* when 4 of the 9 justices vote in favor of hearing the case. When the Court denies the writ, it does not give a reason for or explain its denial, although very occasionally a justice may write an opinion dissenting from the denial of the writ.

There are 9 justices on the Supreme Court, and they hear cases *en banc* unless one or more of the justices recuses himself or herself. A recusal is where a judge declines to hear a particular case because of some conflict of interest, real or perceived, that might undermine the appearance of impartiality of a judge.

2.4   The State Courts

Under the U.S. legal system, states are independent, sovereign political subdivisions. As such, each state has its own legislative, executive, and judicial branches. State courts operate interpedently of their federal counterparts and of each other.

Just as the federal courts are created by federal law, each state judicial system is the product of the laws of its respective state. Accordingly, state courts vary, and each will have its own structure, jurisdictional rules, method of selecting judges, rules of evidence and procedure, and other details.

There are, however, some commonalities among many state courts. First, most state courts have a large-scale structure that resembles that of the federal courts in that they are comprised of trial-level courts, intermediate courts of appeals, and a court of last resort. The descriptive names of these courts are used here intention-ally, as the specific names of these courts will vary depending on the state. In almost every state, the court of last resort is called the "Supreme Court," but caution is needed: in New York, for instance, the court of last resort is the Court of Appeals, while the lower courts are the Supreme Court Trial Division and the Supreme Court Appellate Division.

State court systems often have more specialized courts than are found at the fed-eral court level. Common in state systems are criminal courts, family courts, traffic courts, juvenile courts, commercial courts, probate courts, housing courts, and small claims courts, among others. State courts also have the authority to create local courts, which many have done to deal with many matters within their jurisdiction.

Depending on the jurisdiction, state court judges are chosen either through an electoral process, which may either be partisan or non-partisan (in the sense that judicial nominees run under a party label and with the support of the party); or appointments, usually by the governor and often with the advice or recommenda-tions of a nominating commission or other body. In some states a combined process of judicial elections and appointments is utilized.

As discussed in the section on *stare decisis* below, the state court of last resort is the ultimate judicial arbiter of questions of state law. All courts—federal or state—interpreting the law of a particular state are to "stand in the shoes" of the state court of last resort of that state and apply the law as it believes that court would do. As with many aspects of the court systems, this is a reflection of our federalist system and the importance of preserving state sovereignty.

2.5   Standards of Proof

The standard of proof describes the level of assurance that must be demonstrated by a party seeking to prove a fact. The applicable burden of proof varies depending on the court and the type of case, particularly whether the case is civil or criminal. The burden of proof is on the party bringing the case (the plaintiff in a civil case or the prosecutor in a criminal case).

In criminal cases, the prosecutor must prove all elements of the alleged crime *beyond a reasonable doubt*. Under this very high standard, the prosecution must convince the finder of fact that there is no other reasonable explanation for the events in question based on the evidence presented. The jury must find that the pros-ecutor proved each element of the crime to a virtual certainty.

In civil cases, the standard of proof is typically a *preponderance of the evidence*, which requires that the plaintiff prove only that each element is more likely than not. This simply requires a tilting of the scale in favor of the plaintiff. In many jurisdic-tions, a standard of *clear and convincing* is used in cases involving fraud, wills, and certain other matters. This standard is more rigorous than a preponderance of the evidence and requires that the evidence be highly and substantially more likely to be true than untrue (i.e., that the thing being proved is highly probable).

2.6   Choosing a Court

There are three considerations in selecting a court in which to litigate: subject matter jurisdiction; personal jurisdiction; and venue.

Subject matter jurisdiction asks whether the case can be brought in federal court or state court, and is discussed in part 2.3.4.1, supra. As noted above, there can be exclusive federal court jurisdiction, exclusive state court jurisdiction, or overlapping/concurrent federal-state court jurisdiction.

Personal jurisdiction asks whether a court can compel the presence of an out of state defendant to answer the litigation in that state. Like subject matter jurisdiction, this raises constitutional issues; personal jurisdiction reflects concern over the due process rights of the defendant. The Supreme Court has held that a defendant can be brought into an out of state court if he purposefully availed himself of the protections of the state so that requiring the defendant to answer suit in the state would not "offend[] fundamental notions of fair play and substantial justice." International Shoe v. Washington (1945). The minimal purposeful contacts test is heavily fact-dependent but can include activities like conducting business within the state and visiting the state. Unlike subject matter jurisdiction, a defendant may waive any objection to personal jurisdiction.

Venue refers to the specific place in which a litigation will occur. Once it is determined whether the case will go to federal or state court (subject matter jurisdiction), and the state or states in which the defendant is liable to suit (personal jurisdiction), venue rules offer the specific court or courts in which litigation is proper. Venue sometimes is prescribed by statute. For example, venue in cases involving real estate is generally where property is located. In other cases, venue may be proper where the cause of action arose; where a particular fact or situation occurred; where the defendant resides or has a place of business; where the plaintiff resides or has a place of business; where the defendant is served with a complaint; or where the seat of government is located.

2.7   Stare Decisis and the Importance of Judicial Precedent

Introduction

Stare decisis is a defining principle in U.S. law. The term means "let it stand"; the "it" refers to principles of law announced in earlier-decided cases. Stare decisis thus stands for the proposition that courts should consider, and oftentimes follow, principles of law announced in earlier decided cases.

Benefits of Stare Decisis

Stare decisis provides for a number of benefits. Above all, the principle is designed to promote the fairness, predictability and integrity of the judicial system. Directing courts to abide by principles of law established in earlier-decided cases leads to more predictable, fairer decisions. Stare decisis helps to ensure that judges are not motivated by bias or other irrelevant factors when making decisions because they must generally explain departures from precedents. These attributes in turn result in the development of stable and consistent legal principles. When there is a change to the judicial personnel, it does not mean that there will be a wholesale reversal of established lines of decisions. To the contrary, stare decisis ensures some level of fidelity to established precedent.

Stare decisis at the same time enhances the integrity of the judicial system and judicial processes. When stakeholders see that courts treat similarly situated parties in a similar way, a level of trust develops in that system that might otherwise not be present. This level of trust trickles down to the citizenry, which develops a sense of confidence in the stability and fairness of the legal and particularly the judicial systems. The application of reasoned principles across the board inspires confidence in the rule of law and the institutions that support and promote it. As Chief Justice Roberts stated in *Citizens United v. Federal Election Comm'n* [9]:

> Fidelity to precedent—the policy of *stare decisis*—is vital to the proper exercise of the judicial function. "*Stare decisis* is the preferred course because it promotes the evenhanded, predictable, and consistent development of legal principles, fosters reliance on judicial decisions, and contributes to the actual and perceived integrity of the judicial process.

The application of stare decisis principles also promotes enormous efficiencies. Stare decisis means that courts do not need to consider anew every legal principle that comes before them but consult earlier decided cases involving the same legal issue for guidance. Additionally, parties whose positions are not supported by legal precedents will be discouraged, and hopefully deterred, from bringing frivolous cases and baseless appeals on well-settled points of law. For these reasons, a system of binding precedents is thought to have the effect of inducing the settlement of disputes and reducing court backlog.

It is also said that the system of stare decisis promotes well-reasoned judicial decision-making. Courts, especially at appellate levels, are well aware that their written decisions will be consulted by generations of lawyers and law students, as well as their fellow judges, which may incentivize them to take special care in formulating their opinions and in articulating adequate, well-reasoned bases for their decisions.

Application of Stare Decisis Principles
The application of stare decisis is complex and more of an art than a science, which embraces the exercise of judicial discretion.

When courts are confronted with legal issues that are not readily disposed of by constitutional or statutory language or some other normative principle, the U.S. common law tradition has been for courts to look at what other courts have done in similar kinds of cases. In this way, the judicial system promotes the values discussed above, and the law develops over time incrementally to respond to specific facts presented.

The common law system, indeed, is characterized by the reliance on precedent that is at the heart of our system of *stare decisis*. Common law refers to law that derives from earlier judicial opinions, and a common law system is one that relies on such precedent. Common law systems are distinguishable from civil law systems, which rely heavily on codified law and general rules of construction that allow them to be adapted to new facts and circumstances.

Scope of the Earlier Decision

Because the entire principle of stare decisis rests on the fundamental similarity among cases, courts must consider whether one or more earlier cases are relevant to the current case. This turns first on a determination of whether the *legal issue(s)* in the precedent and in the current case are the same or analogous. If they are not, the entire bases for the doctrine falls away. The court must also evaluate whether the *facts* in the earlier case(s) are analogous to the facts presented in the current case. Courts generally base their rulings on specific facts presented by the parties. Accordingly, in subsequent cases where different facts are presented, the court must evaluate whether the material facts (i.e., facts that are legally relevant) are substantially the same as in the precedent, or whether the differences are significant enough to warrant a different result. The litigants are important in this process, because, under our adversarial system, they will take on the role of attempting to convince the judge that the facts are highly analogous or readily distinguishable, depending on the party's position.

Consider an example: there is a precedent (Case A) that raises the same or a similar legal issue as is raised in the current case (Case B). Plaintiff wants the court in Case B to apply the holding reached in Case A, because if the court does so, Plaintiff will win. Plaintiff argues to the court in Case B that the facts presented in Case A and Case B are essentially comparable and that any differences in the facts are not legally significant. Defendant does not want the court in Case B to apply the legal principle established in Case A and thus argues to the court that any similarities between the facts are merely superficial; there are significant differences in the facts between the cases, differences which suggest a different outcome. It will be up to the court in Case B to decide which party has presented the better argument.

Finally, a court considering the application of precedent must consider whether it is being asked to apply what is properly considered part of the earlier court's *holding* (the answer to the legal question presented) or the court's *rationale* (the reason or reasons the court reached the conclusion that it did). While it is incompatible with accepted judicial behavior, courts occasionally issue dictum—statements that go beyond the legal issue or facts presented. Because dictum conflicts with the court's duty to avoid advisory opinions and the general expectations of our adversarial system, it does not carry binding precedential effect. What is properly classified as dictum, however, may not be entirely clear and, again, the parties will present arguments to the court as to why it should treat such statements as within the scope of the earlier decision or more appropriately as non-binding dictum. It is ultimately the court that decides.

Whether an Earlier Judicial Decision Is Binding or Persuasive

In addition to ascertaining the scope of an earlier judicial ruling and its applicability to a current case, a court must determine whether the relevant principle of law is binding on it or is merely persuasive. When precedent is binding (also called "controlling" or "mandatory"), the court must follow the applicable principle of law. When persuasive, the court should consider the precedent but remains free to deviate from it. Whether precedent is binding or mandatory depends on the relationship between the court or courts that issued the precedent and the court currently deciding the matter.

In general, the decisions of higher courts bind lower courts within the same system of hierarchy. Here is how this works in the federal court system with respect to questions of federal law:

Decisions of the U.S. Supreme Court: decisions of the Supreme Court of the United States on questions of federal law bind all federal and state courts. Supreme Court decisions can be overruled by the Supreme Court itself, although decisions on statutory questions may be prospectively overruled by legislative action. Supreme Court decisions on constitutional matters may be prospectively overruled by a formal constitutional amendment which, as discussed below, are exceptionally rare.

Decisions of the U.S. Courts of Appeals: decisions of the federal courts of appeals are binding within their circuit—the district court(s) within that circuit and the court of appeals itself. Thus, the first three-judge panel to decide an issue binds the entire circuit, including other panels of that court of appeals. Decisions of a court of appeals do not bind courts of appeals or even district courts in other circuits.

There are a number of ways in which a court of appeals decision can be overruled or modified: by legislative action, in the case of a decision on a matter of statutory law; by a decision of the Supreme Court overruling or reversing the appellate court decision; or by the court of appeals itself sitting *en banc*.

Decisions of the U.S. District Courts: district court decisions do not have binding precedential effect. Courts may look to these decisions but they are never binding, even within the same district court.

This all makes a lot of sense and enables a level of experimentation before a definitive, nationwide decision is reached by the Supreme Court. District courts are free to reach their own conclusions, until the court of appeals in its circuit makes a decision, which then binds that circuit. Courts of appeals may—and often do—disagree with each other, reaching disparate decisions on the same or similar questions of law. This is known as a split between or among circuits. When this happens, the Supreme Court may decide to resolve the question, which decision then binds the entire nation.

When Will the Supreme Court Overrule Itself?

The Supreme Court generally follows its own precedent as a matter of good judicial practice and fidelity to U.S. concepts of the rule of law. But as indicated above, the Supreme Court may overrule itself. When will it do so?

First, the Court is more likely to overrule itself on decisions based on constitutional principles than it is on those based on statutory law. When the Supreme Court makes a decision on the basis of statutory interpretation, there is a relatively easy fix for errors—if Congress believes that the Court's decision was wrong, it can simply modify the statute, which effectively prospectively overrules the Supreme Court's precedent. When it declines to do so, the Court determines that Congress has acqui-esced in the judicial determination, and that overruling the decision would be incompatible with notions of separation of powers. While this is somewhat of a fic-tion because legislatures do not always act in such a deliberative fashion, it is a fic-tion that makes a good deal of sense and puts the burden where it belongs—on Congress—to amend the statute in response to a Court decision that it does not like.

However, when the Supreme Court makes a decision on a matter of constitu-tional law, as a practical matter the only way to correct an error is for the Court to overrule itself. Congress alone cannot change the outcome of a judicial decision based on constitutional precepts. And the process for amending the Constitution is so burdensome and politically divisive that it remains highly improbable; in our 240-year history, there have been only 27 amendments to the Constitution, the last one in 1992. Accordingly, given the challenges associated with formal constitu-tional amendments, the Court must step in and correct itself when it believes it has made an error with regard to a constitutional matter.

Even in such cases, however, the Court does not overrule itself lightly, and gener-ally insists on some special justification before overruling precedent. The justices do not overrule themselves simply because there is a majority on the Court that believe the earlier decision was wrong. The Court has articulated several factors that it considers when making this determination but there seems to be no definitive test that the Court will apply consistently, leaving it open to charges that its decisions on whether to overrule precedent are principled or are outcome driven. These decisions tend to be highly controversial among the public and the members of the Court themselves.

Super Precedent

The stare decisis effect of a legal principle arguably is strongest when it has been affirmed multiple times, a reality that reflects both pragmatism and theory. Super precedent is a term used not by courts but by lawyers and academics to refer to precedent that has been affirmed numerous times and that seems immune from judi-cial overruling. As Professor Michael J. Gerhardt described super precedent:

"Super precedents are the doctrinal, or decisional, foundations for subsequent lines of judi-cial decisions (often but not always in more than one area of constitutional law). Super precedents are those constitutional decisions in which public institutions have heavily invested, repeatedly relied, and consistently supported over a significant period of time. Super precedents are deeply embedded into our law and lives through the subsequent activi-ties of the other branches. Super precedents seep into the public consciousness, and become a fixture of the legal framework. ... Thus, super precedents take on a special status in con-stitutional law as landmark opinions, so encrusted and deeply embedded in constitutional law that they have become practically immune to reconsideration and reversal." [10].

Many scholars disagree as to which decisions constitute super prec-edents and are likely immune from overruling. Some, for instance, argue that *Rose v. Wade* and its progeny protecting the right of a women to have a legal abortion under some circumstances is super precedent, while others believe that these decisions are very much subject to further judicial action. As of this printing, it appears that *Roe* is indeed likely to be overrules.

Stare Decisis and State Law and State Courts

Stare decisis principles in the state courts function more or less as they do in the federal court system, albeit with some variation. The general principle remains that lower courts follow decisions of higher courts within the same system of sovereignty. State courts do not have to follow federal law precedent established by any federal court other than the U.S. Supreme Court. And on matters of state law, a state court of last resort is the ultimate judicial arbiter.

2.8   Due Process

Courts, as guardians of the Constitution, often are called upon to evaluate claims that governmental action violates the due process clause of the 5th or 14th Amendment.

Procedural due process claims require a court to determine first, whether an individual is being threatened with a deprivation of a life, liberty, or property interest by the government; and, if so, what procedure is "due." The inquiries into what procedure is appropriate to the situation involves a three-part inquiry: the private interest that will be affected by the official action; the risk of an erroneous deprivation of such interest through the established procedures; and the government's interest, including the fiscal and administrative burdens that additional procedures would produce. Procedural due process provides for a variability of procedural rights that depends on particular facts presented.

The term "due process" has been interpreted to contain a substantive component as well as a procedural one. Under this notion, the government cannot infringe certain rights unless it demonstrates an adequate justification for the encroachment. The courts apply a sliding-scale approach to such inquiries, balancing the importance of the government insert being served against the right being infringed, and has come up with three tests; the test to be applied depends on the personal interest at issue:

- Strict Scrutiny: strict scrutiny applies when government action threatens a fundamental right (such as interstate travel, voting and ballot access, and privacy matters). Under strict scrutiny, the government must prove that its action is necessary in order to further a compelling government interest.
- Intermediate Scrutiny: in such cases, in order to be upheld, the challenged law must further an important government interest by means that are substantially related to that interest. Intermediate scrutiny applies to certain first amendment matters such as regulation of mass media, adult entertainment, and highway signs.
- Rational Basis Scrutiny: in other cases, government action will be upheld unless the challenger can show that it is not rationally relate to a legitimate government interest.

3   Application

Peter Patient wants to sue Debbie Doctor for alleged malpractice arising out of a procedure by Doctor that he believes she performed in a negligent fashion, leaving him with severe and permanent injuries.

Patient must decide the court in which to sue Doctor. Under the U.S. system of federalism, there are federal courts and state courts. Federal courts have limited jurisdiction and can hear only certain kinds of cases; state court jurisdiction is more expansive.

If Patient and Doctor are citizens of different states and Patient's injuries exceed $75,000, he may file his lawsuit in federal court. State courts have concurrent jurisdiction over many matters, so he would be able to choose between federal and state court. Which specific court he will file in depends on strategic issues, as well as personal jurisdiction and venue considerations.

Regardless of whether Patient sues Doctor in federal or state court, here will be a trial in the absence of a settlement between the partiest. The goal of the trial court is to determine facts in dispute. In most civil and criminal cases, there is a right to a jury trial. When there is a jury trial, the jury is charged with finding facts.

Either party aggrieved by the trial court decision can appeal to the intermediate appellate court. On appeal, the court considers facts decided at the trial court as conclusive and only reviews the case for errors of law. At the appellate level, no new evidence or arguments may be introduced, and the court rests its decision on the record created below.

Patient or Doctor—whichever loses at the intermediate appellate court—can seek review in the relevant court of last resort. In the federal system and most (or all) state court systems, review in the highest court is a matter of judicial discretion.

When the court decides the issue of whether Doctor was negligent, it will be guided by precedent—earlier decisions of courts in the same kinds of cases. Some precedent is binding on the court and other decisions are merely persuasive, meaning that the court can apply their rationale if they are persuaded of their correctness.

4   Summary

This chapter discussed the U.S. judicial systems, exploring the role of the courts in our governmental system as well as the nature of the multiple judicial systems—federal and state—under our system of federalism. The chapter then explored the federal courts, looking at their hierarchical and geographical organization; the qualifications, selection procedure, and tenure of federal judges; federal courts of general or specialized jurisdiction; the competence of the federal courts; and proceeding through the federal courts. The chapter then addressed the state courts, which vary from state to state as each state court system is the product of its respective state law. It introduced the concept of the standard of proof and discussed the major standards that are applied in criminal and civil cases. Finally, this chapter explored the defining role of stare decisis in our judicial system, under which courts consider and often apply principles announced in earlier decided cases.

References

1. *Marbury v. Madison*, 5 U.S. 137 (1803)
2. Constitution of the United States of America
3. Judiciary Act of 1789, officially titled "An Act to Establish the Judicial Courts of the United States"
4. Federal question jurisdiction statute, 28 U.S.C. §1331
5. Federal diversity jurisdiction statute, 28 U.S.C. §1332
6. *Hertz v. Friend*, 559 U.S. 77 (2010)
7. Federal removal jurisdiction statute, 28 U.S.C. §1441
8. Federal supplemental jurisdiction statute, 28 U.S.C. §1367
9. *Citizens United v. Federal Election Comm'n*, 558 U.S. 310 (2010)
10. Michael J. Gerhardt, *Super Precedent*, 90 Minn. L. Rev. 1204 (2006)

Anatomy of a Lawsuit

Joseph D. Piorkowski Jr and Anne P. Hovis

Contents

Key Points

- Federal courts are governed by the Federal Rules of Civil Procedure and Federal Rules of Evidence, whereas state courts may have similar or different rules; in addition, each court has its own particular rules.
- A lawsuit is commenced when the plaintiff files a complaint containing allegations regarding each of the elements of the cause of action, including damages.
- The defendant files an answer to the complaint, including any affirmative defenses and counterclaims.
- Prior to trial, the parties engage in the discovery process, including (1) requiring the other party to respond to interrogatories, requests for admission, and document requests and (2) deposing the other party's fact and expert witnesses.
- The trial itself involves opening statements by each party's counsel, the presentation of evidence (witnesses and physical evidence) by each party (beginning with the plaintiff, followed by the defendant), possible presentation of rebuttal evidence by the plaintiff, and closing statements.

J. D. Piorkowski Jr (✉)
The Piorkowski Law Firm, PC, Georgetown University Law Center, Washington, DC, USA
e-mail: jpiorkowski@lawdoc1.com

A. P. Hovis
The Piorkowski Law Firm, PC, Washington, DC, USA
e-mail: ahovis@lawdoc1.com

- The plaintiff bears the burden of proving its case, and the case will be dismissed if the burden is not met.
- When the case will be decided by a jury, there are numerous other procedures involved specifically related to the jury, such as jury selection and instruction of the jury prior to its deliberations.
- Motions may be made by any party's counsel at various times before, during, and after the trial.
- Once a judgment is rendered, either party may appeal the trial judge or jury's determination to a higher court.
- There are various ways in which lawsuits can be resolved out of court, including settlement, mediation, and arbitration.

Legal Concepts
- Plaintiff: A person or entity who files a case against another in a court.
- Defendant: A person or entity against whom a case is filed in a court.
- Complaint: The initial document filed in court by a plaintiff to commence a legal action, which sets forth the elements of the alleged cause(s) of action.
- Cause of action: The grounds on which a legal action properly may be brought; a set of predefined elements that allow for a legal remedy.
- Damages: The sum of money that law imposes as compensation for some breach of a duty or violation of a right.
- Alleged Damages: The amount of damages claimed by a plaintiff in its lawsuit.
- Awarded Damages: The amount of damages determined to be due to a plaintiff at the conclusion of a lawsuit in which the plaintiff prevails.
- Burden of Proof: The legal requirement of which party must provide evidence and the level of evidence that it must provide in order to prevail on its claim.
- Answer: A written document filed by the defendant in response to a complaint, which sets forth denials, defenses and, potentially, counterclaims.
- Pleadings: The initial formal written documents filed with the court in a legal case, typically the complaint and answer.
- Discovery: A process by which each party to a lawsuit gathers information from the opposing party before trial in preparation for a trial, typically including interrogatories, documents requests and/or depositions.
- Interrogatories: Written questions that each party to a lawsuit may submit to the other as part of the discovery process.
- Document request: Written requests for the provision of documents that each party to a lawsuit may submit to the opposing party as part of the discovery process.
- Deposition: A formal proceeding where, as part of the discovery process, lawyers representing both parties in a lawsuit are permitted to ask questions of a witness who is expected to testify on behalf of a party at trial, and the witness must answer the questions under oath.
- Fact Witness: A witness who is called to testify about factual events and firsthand observations.
- Expert Witness: A witness with scientific, technical or specialized knowledge that may help a judge or jury understand some fact or principle about which the expert has specialized knowledge, skill, experience, training or education; a witness must be qualified to testify as an expert.

- Motion: A written or spoken request to a court for a desired ruling or order.
- Moving party: The party who made (if verbal) or filed (if written) a motion with the court.
- Order: A direction or command by a judge, typically in writing.
- Appeal: A challenge to a legal determination of a court, directed to a higher level court.
- Appellant: A person who files an appeal from a decision of the trial court to a higher court and who can be a plaintiff or a defendant in the original lawsuit.
- Appellee: The non-appealing party in a case that has been appealed; the person defending the lower court's decision.

1   Introduction

People or companies who file lawsuits are called "Plaintiffs." The people or companies who are being sued are called "Defendants." Litigation can be very simple. For example, it can involve one person (the plaintiff) suing another person (the defendant) in Small Claims Court over a relatively small amount of money claimed to be owed, where the plaintiff and defendant appear before a judge and get an immediate resolution. In the alternative, a case may involve hundreds or thousands of people or companies as plaintiffs suing multiple defendants (people and/or companies) and may span multiple states. It may involve both judges and juries, fact witnesses and expert witnesses. Despite these two extremes, there are some general principles and procedures common to most lawsuits, which will be discussed in this chapter.

This chapter discusses these general principals and procedures including the rules governing litigation procedure; the litigation process; costs of trial; appeals; and settlement, mediation and arbitration. The primary focus is on the litigation process itself, with sections on commencing a lawsuit (causes of action and damages, filing a complaint, answering a complaint, and additional pleading), pretrial activities (discovery, motions practice, and orders), and trial (judge versus jury trial, opening statements, presentation of case by plaintiff, presentation of defense by defendant, possible rebuttal by plaintiff, closing statements, verdict, and additional motions practice).

2   Discussion

2.1   Governing Rules

It is important to note at the outset of this discussion that different rules apply to different courts and, thus, to the entire litigation process over which that court has jurisdiction. In all U.S. District Courts, regardless of the state in which they are located, the Federal Rules of Civil Procedure (FRCP) [1] and Federal Rules of Evidence (FRE) [2] apply. However, some District Courts also have Local Rules which supplement the Federal Rules. In addition, state rules (governing state courts) vary widely, even though many states have adopted procedural and evidentiary rules that are similar to the Federal Rules. Finally, all criminal courts and most specific-jurisdiction courts (e.g., federal bankruptcy courts, state family law courts) have their own distinctive rules governing matters before those courts. Therefore, the overview set forth in this chapter should be viewed only as an illustration of procedures that are common, but do not apply in all circumstances.

2.2   Commencing a Lawsuit

Cause of Action and Damages

Before there can be a lawsuit, there must be a basis for the lawsuit, called a "cause of action." For instance, if someone agreed to perform a service and then did not perform that service, that failure to perform (breach of contract) is the cause of action. If a physician misdiagnosed a patient, that misdiagnosis is the basis for a medical malpractice cause of action. In addition to the basis for the suit, there also must be "damages," whether physical or financial. So if, in the first example, someone agreed to perform a service and did not, but was paid for the work, the amount of the payment is the amount of the damages. If the service provider was never paid for the work, there may be disappointment that the work was not performed, but there may not be "damages" that are sufficient to bring a lawsuit. Similarly, in a medical malpractice case, if the misdiagnosis resulted in the need for an additional surgery, the cost of the additional surgery and the pain and suffering associated with the additional surgery would be recoverable as damages. A misdiagnosis, however, might not result in damages if it does not alter the treatment plan or the patient's prognosis. Note that there is a difference between "alleged damages," i.e., the amount of the damages claimed by a plaintiff in its lawsuit, and "awarded damages," i.e., the amount of damages that is determined to be owed to the plaintiff at the conclusion of a lawsuit in which the plaintiff prevails.

Filing a Complaint

Elements of the Cause of Action. A lawsuit is officially commenced by the filing of a "Complaint." This is a legal document that sets forth the cause(s) of action. Although it often does not provide much detail, the complaint must include certain elements regarding the causes of action, without which the case may be dismissed early on in the process. Causes of action recognized under the law (e.g., breach of contract, employment discrimination, libel, medical malpractice) have specific elements, each of which must be proven in order for a plaintiff to prevail. If the plaintiff does not address each element of each of its causes of action in its complaint, the complaint will be dismissed as to those causes of action that do not contain all of the required elements. For example, in a medical malpractice case, the complaint must (1) allege that the defendant owed some duty of care to the plaintiff, (2) specify the standard of care that was owed, (3) allege that the defendant breached the standard of care, (4) allege that the plaintiff was injured, and (5) allege that the breach of the standard of care was the cause of the plaintiff's injury. However, for certain causes of action (such as fraud), the complaint must include a higher level of specificity in the allegations.

Damages. Damages may be sought to the extent that they are caused by the wrongful conduct alleged in the complaint. There are different types of damages that may be requested. The typical damages requested are compensatory damages, intended to compensate a party for its losses. Compensatory damages include economic damages (sometimes called special damages) and non-economic damages (sometimes called general damages). The first of these relate to financial losses such as expenses incurred in repairing or replacing damaged property, lost wages, and medical bills that are not covered by insurance. The second category relates to issues that do not have a specified dollar value, such as pain and suffering, mental anguish, and loss of consortium or companionship. In addition, in rare cases, the court will award punitive damages, intended to punish the offending party and to prevent future similar actions by that party and others. Punitive damages are awarded only where the conduct of the offending party was intentional or especially reprehensible.

Burden of Proof. Although the plaintiff often does not need to provide much detail in the complaint, it is the plaintiff's responsibility during the course of the lawsuit to prove each element of the cause of action that it has alleged. This is called the plaintiff's "burden of proof." It is not the defendant's obligation to prove that the allegations are false, and the plaintiff will lose the case if it cannot prove each and every required element, not just a majority or most of the elements. The extent to which the plaintiff must prove each element depends on the court and the state. However, as a general rule, civil courts require that the plaintiff prove each element based on a "preponderance of the evidence." This means that the allegations are more likely true than false. For some purposes in civil trials, such as to obtain punitive damages (discussed above), there is a higher standard of "clear and convincing evidence." The preponderance of the evidence standard is often described as tipping the scales of justice ever so slightly to one side, whereas, according to the U.S. Supreme Court, "'clear and convincing' means that the evidence is highly and substantially more likely to be true than untrue; the fact finder must be convinced that the contention is highly probable." [3] In a criminal case, the accused typically must be found guilty "beyond a reasonable doubt," which is a much higher standard indicating that there is no substantial doubt about the accused's guilt.

Good Faith Basis. Both parties to a lawsuit and their lawyers have an affirmative obligation to ensure that there is a good faith basis for any document filed with the court (whether complaints, motions, or otherwise). Rule 11 of the FRCP provides that the filing of a document with the court serves as a certification, after reasonable inquiry, that the document is not being filed for an improper purpose (such as to harass or delay), that the allegations are based on existing law or a nonfrivolous argument for a change in the law, and that factual contentions or denials have evidentiary support [4]. Rule 11 provides that courts may sanction parties and/or attorneys for a violation, including monetary penalties (such as the award of attorneys' fees to the opposing party) and non-monetary penalties (such as striking filings, which could result in a dismissal of the case) [5]. Most state courts have similar rules requiring good faith in the submission of documents in a lawsuit.

Service of Process. Each court has laws regarding the delivery of the complaint to the defendant (including type of delivery, who is permitted to deliver it, etc.) to ensure that the defendant is aware of the case and can defend itself in court.

Answering the Complaint

Answer. Once a defendant has received the complaint, the defendant is obligated to file a responsive document called an "Answer." This document informs the court of whether the defendant admits or denies each of the allegations included in the complaint as well as setting forth the defendant's explanations or defenses related to the allegations.

Affirmative Defenses. Sometimes even when the plaintiff is able to prove the elements of its cause of action and to meet its burden of proof, the defendant has a sufficient excuse for its actions so that the defendant will not be liable and the plaintiff will not win the case. These can be defenses based on legal rules (e.g., too much time elapsed between when the injury occurred and the complaint was filed or the defendant was a minor when he made a contractual promise but he was too young to commit himself) or may involve additional facts (e.g., although a defendant's car hit the back of the plaintiff's car at a red light, the defendant's car was safely stopped behind the plaintiff's car until a third car crashed into defendant's car, which was the sole reason that the defendant's car then hit plaintiff's car). Once the plaintiff has met its burden of proof, however, the burden of proof shifts to the defendant to prove its affirmative defense(s).

Counterclaims. An answer also may include any counterclaims that the defendant has against the plaintiff. In essence, counterclaims are what the defendant would have included in a complaint against the plaintiff if the plaintiff had not filed first. The defendant has the burden of proof with respect to any alleged counterclaims.

Third Parties. Sometimes the defendant will bring another party into the case because of that party's role in the dispute between plaintiff and defendant. The new party does not replace the original plaintiff or defendant but is an additional party to the litigation.

Additional Pleadings

Documents filed at the beginning of a lawsuit are called "Pleadings." Sometimes additional pleadings are filed after the complaint and answer. For example, if the plaintiff determines that it has additional causes of action to include, it can amend its complaint (in accordance with specified rule related to timing, etc.), and then the defendant has an opportunity to answer the Amended Complaint. If a defendant includes counterclaims in its original answer, the plaintiff has an opportunity to file an "answer" to the claims included against it in those counterclaims.

2.3   Pre-Trial Activities

In many, if not most, cases, the filing of a complaint ultimately does not result in a trial. In addition to cases that are resolved through pre-trial settlement, arbitration and mediation (which are discussed in more detail below), there are a host of other activities that are conducted after the filing of a complaint and prior to the holding of a trial. These primarily include discovery activities and pre-trial motions practice.

Discovery

As stated above, a complaint sets out the required elements of a cause of action, but often does not contain sufficient detail for a defendant to understand specifically what it is alleged to have done wrong. Discovery is a process during which the defendant learns the specific details of the plaintiff's allegations, and, similarly, the plaintiff learns the details of the defendant's defenses. For example, the complaint may allege that the plaintiff was damaged, but the specific dollar amount of the damages alleged may only become clear though the discovery process. Each party is entitled to engage in discovery with respect to the opposing party.

The discovery process is a formal part of the litigation with rules governing many aspects of each type of discovery [6]. Parties are not free to engage in discovery that does not comply with the applicable rules or to ignore the discovery process, as doing so can result in monetary fines or the disqualification of a party's pleadings, which may result in a judgment for the opposing party. If information obtained during discovery satisfies the applicable rules of evidence, the information may be used as evidence at trial.

There are several forms of discovery, some or all of which may be used in a single case.

Written Discovery. Interrogatories are written questions prepared by the lawyers for each party which the opposing party must answer. Parties also may use requests for admission, which are written inquiries requiring a party to either admit or deny specific factual assertions. The purpose of the interrogatories and requests for admission is to gain information, to learn more about the parties, the plaintiff's allegations, and the defendant's responses and defenses, and also to determine if there are any issues on which the two parties agree (e.g., that the defendant concedes). Although parties must answer these written discovery requests, they also may object to certain aspect of them, such as that they are redundant, that they are worded in a manner that is overly broad, or that it would be excessively burdensome for the party to obtain the information being requested.

Document Requests. Document requests seek to have a party provide written copies of requested documents (except in specific circumstances where original documents must be produced). They often are used together with interrogatories. Parties must produce the requested documents except to the extent that there is a valid objection provided.

Depositions. At the time specified by the applicable rules or the judge, each party must reveal to the other party the witnesses that it intends to call to support its case.

There are two types of witnesses, fact witnesses and expert witnesses, and the applicable rules differ in some ways with respect to each.

Fact Witnesses. Fact witnesses are called to testify about factual events and first-hand observations, such as those necessary for the plaintiffs to establish its cause of action and meet its burden of proof. Likewise, they are used by a defendant to establish deficiencies in the plaintiff's case, bolster the defendant's defense, or to establish facts that are contrary to the plaintiff's allegations. Typically the fact witnesses include the parties themselves, and the non-party fact witnesses are called to testify on behalf of the party whose version of events their testimony best supports. *See also* Sect 2.4.4 below.

Expert Witnesses. Expert witnesses are used to help a judge or jury understand some technical or scientific principles about which the expert has specialized knowledge, training or experience. While mostly anyone with first-hand knowledge can be a fact witness, there are some minimum standards that experts have to meet in order to testify as an expert. Although experts are meant to be unbiased and independent, they generally are retained and paid by one of the parties. Therefore, the opinions of testifying experts are almost always aligned with the views of the party who has retained that expert. *See also* Sect. 2.4.4 below.

In order to know what each party's witnesses and experts are going to say at trial (rather than being surprised during trial in a manner that does not permit each party to prepare adequately to address the testimony), each party is permitted to "depose" the other party and the other side's fact and expert witnesses. A "deposition" is a formal proceeding where lawyers from both sides are permitted to ask questions of a witness who will testify on behalf of a party. Questioning by one of the opposing party's lawyers typically comprises the bulk of the deposition. A deposition usually is not held in a courtroom, but the witness must swear to tell the truth under penalty of perjury, just as if giving testimony before a judge. A written transcript of the deposition is prepared and is available to both parties for a fee, and sometimes the deposition is also recorded on video.

Attorneys use depositions not only to gain factual information, but also to elicit concessions, to identify inconsistencies in testimony, and to evaluate a witness's demeanor and gauge whether the witness will be effective at trial. A witness' deposition testimony may be used to impeach the witness at trial by pointing out how the deposition testimony differs from the same witness's testimony at trial.

Expert Reports. Under the FRCP and the rules of some state courts, expert witnesses who are expected to testify at trial must provide a written report of their opinions, the bases for such opinions, and the facts or data supporting such opinions, typically prior to being deposed [7]. As these experts' opinions often relate to technical or scientific matters, the report permits the opposing party time to prepare to intelligently question the expert in his or her area of expertise. Often, expert testimony is required to prove certain elements of a party's cause of action.

Motions Practice

Throughout the course of a litigation, both parties are permitted to ask the courts to do certain things by making a verbal "motion" or filing a written "motion". The party that is asking (or "moving") the court to take some action is called the "moving party." Most motions practice occurs prior to trial, but motions can be made during trial or even after a verdict is rendered. One example of a pre-trial motion is a motion for summary judgment, where one party asks the judge to decide the case with respect to one or all issues. The party making the motion argues that, even if everything alleged by the other party is true, there is no legal dispute and the judge should rule in the moving party's favor. Another kind of motion is to exclude evidence, perhaps because it is too prejudicial, or to exclude an expert witness' testimony because he or she did not follow an accepted scientific methodology in arriving at his or her opinions.

Once one party files a motion, the opposing party is permitted to file a response, arguing against the motion. Depending on the court, the rules, and the motion, the judge may make his ruling solely on the filings or may hold a hearing (in essence a mini-trial) where the parties make an oral argument to the judge, the judge may ask questions, and in some instances, witnesses may be examined. A motion ultimately results in the judge issuing an "Order" which includes his or her decision on the motion.

Orders

In addition to orders issued in response to motions, a judge may issue pre-trial orders on other matters. For example, in some cases, a judge will issue a scheduling order, setting forth deadlines for when certain elements of discovery are due for each party. A judge may issue an order related to how confidential materials are handled or, in a high-profile case, may order parties not to make public statements about the case until the trial is completed.

2.4   Trial

Judge Versus Jury Trial

One variable that makes a significant difference in how a trial is conducted is whether the findings of fact in the trial will be determined by a judge or a jury. There always will be a judge who presides over the trial and continues to have authority over certain decisions. In all cases, the judge will determine legal questions, will rule on objections and motions, will instruct the jury, and has numerous other responsibilities. However, the "trier of fact" (judge or jury) will determine what it believes to be the facts upon which a verdict will be rendered. For example, in a medical malpractice case, a judge will determine whether an expert is qualified to offer an opinion about the plaintiff's diagnosis, but the trier of fact (judge or jury) will decide which of two experts offering conflicting testimony about the diagnosis to believe (i.e. the factual question of which diagnosis was correct).

The right to criminal trial by jury is a Constitutional right [8]. This right binds state as well as federal governments [9]. Federal and state courts must afford a jury to a criminal defendant for any criminal charge with a potential sentence of greater than six months' incarceration [10]. For civil trials, the Constitution affords the right to a jury in many cases in federal courts [11], but not in state courts [12]. However, most states also have adopted a right to trial by jury in certain types of civil cases [13].

When determining whether to request a jury trial in a civil case, a party may prefer a jury when the circumstances of the case may generate sympathy, in hopes that the jury will decide "with their hearts" rather than through objective application of the facts and the law. By contrast, in a case involving two corporate entities, such as a patent infringement case, the parties may prefer to have a judge serve as the trier of fact, due to the technical nature of the claim. Any party can insist on a jury trial in a case where that right exists. Only if all parties agree can they waive the right to trial by jury.

Jury Selection. In the event that the case involves a jury trial, the first step of the trial will be selection of the jurors to hear the case. A pool of potential jurors may be questioned by the attorneys who attempt to weed out jurors who would be biased against their client and to include those who might be sympathetic to or better understand their client's position. Jury selection is both an art and a science, and parties may hire jury consultants to assist in the jury selection process, particularly in high profile or high dollar value cases.

Procedural Overview

After opening statements by each party's attorneys, the presentation of evidence begins. Because the plaintiff typically bears the burden of proof, the plaintiff presents its case first. The plaintiff presents its fact and expert witnesses and physical evidence. During trial, each witness first testifies on behalf of the party who called the witness, and then the opposing attorney has an opportunity to cross-examine that witness. Once the plaintiff rests its case, the defendant has an opportunity to present its case. The defendant has no obligation to present any case at all and should prevail if the plaintiff has not met its burden of proof. The defendant, however, may wish to provide evidence supporting its explanations or denials and establishing its defenses and counterclaims, if any. If the defendant presents a case, the same process repeats itself, with the defense presenting its fact and expert witnesses and physical evidence, and the plaintiff's attorney cross-examining each defense witness. If the defendant raises any new issues during its case or presents a counterclaim, the plaintiff has another chance to address those issues before the trial concludes. The plaintiff also may present additional evidence at the end of trial to rebut the evidence presented by the defendant.

Opening Statements

At the outset of the trial, each party's attorney will make an opening statement that informs the trier of fact (judge or jury) of what the case is about and provides their client's version of factual events and applicable law. The opening statement is a summary of the case, focusing more on the facts of the case than the law, and the first opportunity to begin convincing the trier of fact of the merits of their client's case.

Presentation of Case by Plaintiff

Once opening statements are completed, it is the plaintiff's responsibility to present evidence and legal arguments that meet its burden of proof.

Fact Witnesses. The first order of business is for the plaintiff to present the facts on which it is relying to make its claims. For example, if the plaintiff is arguing that the defendant breached a contract, the plaintiff must first establish certain elements including that (a) a contract existed between the plaintiff and defendant, (b) that the contract required the defendant to take the action that the plaintiff is claiming that the defendant failed to take, (c) that the defendant benefitted from the contract, and (d) that the defendant failed to take the required action. The plaintiff also must present facts related to injuries or damages sustained and that the damages and injuries resulted from (were caused by) the defendant's breach.

This presentation of evidence is accomplished by the "examination" of witnesses, where the one of the plaintiff's lawyers will ask questions of a witness and the witness will answer. The witness typically sits near the judge at the front of the courtroom and swears that the testimony to be given is true, under penalty of perjury.

A lawyer generally is prohibited from asking his or her own witness "leading" questions (i.e., questions to be answered yes or no), but is supposed to invite the witness to testify in a narrative fashion regarding the information that the witness has to share.

Cross-Examination by the Defense. Once the plaintiff's lawyer has completed his or her examination of a fact witness, the witness is then cross-examined by the defense counsel. The purpose of the cross examination primarily is to weaken the impact of the witness' testimony. This can be done in a number of ways, such as eliciting information that puts facts more into context; attacking the witness' credibility by demonstrating prior inconsistent statements, lack of knowledge, or bias; or completely discrediting the witness, such as revealing situations showing that the witness has lied about something important in the past or is lying on the stand. One method that is used is to point out inconsistencies between the witness' sworn deposition testimony and his sworn testimony in court. On cross-examination, the attorney is permitted to ask pointed, "yes-or-no" questions.

Expert Witnesses. If the case involves matters that are highly technical or scientific (whether that be a medical procedure or arguments related to whether a patent on a car engine is being infringed), the plaintiff next will present evidence through one or more expert witnesses. Unlike fact witnesses, the party first must establish that its expert witnesses have the necessary qualifications to offer the proffered testimony.

Under the Federal Rules of Evidence (which are similar to the evidentiary rules in most states), an expert witness can be qualified by demonstrating "knowledge, skill, experience, training, or education" and, upon demonstrating such requisite qualifications, "may testify in the form of an opinion or otherwise." [14]

Additionally, the Federal Rules of Evidence (and many states) also require that (a) "the expert's scientific, technical, or other specialized knowledge will help the trier of fact to understand the evidence or to determine a fact at issue; (b) the testimony is based on sufficient facts or data; (c) the testimony is the product of reliable principles and methods; and (d) the expert has reliably applied the principles and methods to the facts of the case." [15]

The importance of the role of expert witnesses cannot be overstated; many complex cases have come to be regarded as a "battle of the experts." In many cases, the party who presents the most credible expert witnesses will win the case. As with fact witnesses, expert witnesses may be cross-examined.

Introduction of Physical Evidence. Physical evidence refers to any evidence other than testimony. It may include medical records, photographs, charts and graphs, videos, fingerprints, DNA, or items such as the proverbial "smoking gun." Physical evidence must be introduced through a witness who can provide information regarding the creation or discovery of the evidence. The party who is seeking introduction of the evidence must "lay the foundation" for admission, demonstrating both the relevance of the evidence and that its admission complies with evidentiary rules. The law of evidence is a complex series of rules and legal decisions that range from basic concepts, such as not admitting testimony or evidence that is not relevant to the case, to rules prohibiting "hearsay" testimony (although this general rule has numerous exceptions), to discretionary rules in which the judge is able to keep out evidence that is likely to be "confusing" to the trier of fact or "unduly prejudicial."

While the opposing attorney may cross examine the witness who is used to introduce the physical evidence, the opposing party may not introduce its own evidence until it is presenting its case and introducing its own witnesses.

Presentation of Defense by Defendant

Once the plaintiff has rested its case in chief, the defendant has an opportunity to engage in the same process. The defendant introduces its own fact and expert witnesses, which are subject to cross-examination by the plaintiff's attorneys, and its own physical evidence.

Possible Rebuttal by Plaintiff

In certain cases, once the defense has completed its presentation, the plaintiff may have another opportunity to present additional evidence. This may occur, for example, if the plaintiff has additional "rebuttal" evidence (not used in its case in chief) to impeach the credibility of defense witnesses.

Closing Statements

As with opening statements, each party, beginning with the plaintiff, has an opportunity to summarize its theory of the case and the evidence presented to support it. This is the final opportunity for each party to convince trier of fact (judge or jury) that it should win the case.

Verdict

Following the trial, the trier of fact (judge or jury) will reach a verdict and provide it to the parties. In the case of a jury trial, the judge will "instruct" or "charge" the jury on various matters, such as the applicable burden of proof (i.e., to what degree of certainty does the plaintiff have to prove its allegations). The jury instructions themselves are typically the subject of negotiation by the parties or motions to the judge. The jury will then be dismissed to a separate room to deliberate and to attempt to reach a verdict.

The jury is subject to certain rules during trial. For example, jurors are not supposed to discuss the case with anyone else, including a family member. They are not supposed to have a conversation on any subject with a party, attorney or witness in the case. They are asked to avoid looking at newspaper articles, TV, or other electronic news coverage about the case. In some instances, juries are "sequestered," i.e., housed in a separate location, until jury deliberations are completed.

In federal trials, criminal trials for serious offenses, and some state trials (depending on the state constitution or statute), a jury must reach a unanimous decision. In some states, however, a specified majority vote is sufficient. If a jury is unable to reach a unanimous decision (or a decision of the specified majority), a "mistrial" is declared. A mistrial does not result in a verdict for either the plaintiff or the defendant. Civil cases can be re-tried subsequently, but double jeopardy may prevent retrial of criminal cases in some circumstances [16].

Once a decision is reached, a jury verdict typically is read in court with the parties present. A decision by a judge may be announced in court or may be issued in a written opinion.

Additional Motions Practice

Additional motions may be made by the parties' attorneys throughout the trial. This can occur, for example, after the plaintiff has rested its case, prior to submitting the case to the judge or jury, or even after a verdict is rendered. One type of (rarely granted) motion is a motion for judgment notwithstanding the verdict, which asks the judge to set aside a jury verdict because, applying applicable law, no reasonable jury could have reached the conclusion that the jury reached in the case at hand. Other examples include a motion for a new trial (e.g., where a party discovers new evidence) and for remittitur (lowering the amount of the award, such as where the award is otherwise excessive).

2.5   Costs of Trial

Typically in a trial, each party bears its own costs. These costs include not only attorney's fees, but, among other potential costs: filing fees, the cost of copying documents to be produced to the court or the opposing party, the cost of serving documents on opposing parties (where it is not done electronically), the cost of depositions of the opposing party's fact and expert witnesses, and the cost of hiring experts, jury selection consultants, or others who assist in the trial process.

There are a few exceptions to this rule. In some cases, applicable law may permit the prevailing party to request that the court order the other party to pay some of the costs and/or expenses of the prevailing party. Another exception is where a contract between the parties specifies which party bears the costs of the litigation.

2.6   Appeals

After a judgment is entered, either party can file an appeal to a higher court. An appeal asks the higher court to correct what the party believes was an incorrect judicial ruling or decision by the jury. Appeals must be filed within deadlines specified by the applicable jurisdiction. The party that appeals is called the appellant, regardless of whether that party was the plaintiff or defendant during trial. The non-appealing party is the appellee.

Typically the losing party appeals, but a prevailing party can appeal, for example, if the prevailing party believes that the damages awarded at trial are far less than the prevailing party believes they should be. Additionally, once one party files an appeal, the other party can cross-appeal if it also wishes to appeal issues related to the trial. In this instance, both parties are appealing some aspect of the trial court or jury's decision.

2.7   Settlement, Mediation, and Arbitration

Generally

There has been a movement in the past several years encouraging alternative dispute resolution (i.e. using methods such as arbitration and mediation rather than litigation) to resolve legal conflicts [17]. Alternative dispute resolution has several benefits, including a reduction in the costs borne by the parties; a reduction in the length of time before a matter reaches a conclusion; a reduction in the backlog for courts; and, in some instances, a focus on a resolution mutually acceptable to the parties rather than the outcome dictated by law.

Settlement

Settlement of a case is an agreement reached by the parties on the resolution of the case. It may occur at any time, including before, during or after trial. Settlement discussions may be done exclusively through attorneys, may also involve the parties, or, if the parties have not yet retained counsel, may be conducted directly by the parties. Settlement talks may start and stop, even repeatedly, without affecting the course of the litigation.

If a complaint has been filed in a case that is subsequently settled, there typically is some type of recorded resolution of the matter, either by having the plaintiff dismiss the case or by the judge entering an order of settlement. Alternatively, a case may be dismissed by the judge if a plaintiff files a case and then fails to "prosecute" the case (i.e., take the required steps to conduct the litigation within the deadlines required).

Mediation

Mediation is a process that is intended to lead to settlement, although a facilitator (mediator) is involved to assist in reaching a resolution. Engaging a mediator adds an extra cost, but mediation may allow parties who are at an impasse to reach a resolution without going to trial. As with settlement, the focus of mediation is to reach a resolution that is acceptable to all parties (as opposed to determining who is right or wrong under the law). If mediation is unsuccessful, there is no effect on the litigation, which continues as if the mediation had not occurred.

In certain instances (such as divorce cases), state law or a judge's order may mandate that the parties attempt to resolve their disputes through mediation prior to proceeding with litigation.

Arbitration

Arbitration is an alternative to participating in a trial. In essence, arbitration is a somewhat condensed version of a trial conducted before an arbitrator or a panel of arbitrators. The arbitrators are chosen by the parties and must be paid to conduct the arbitration.

Unlike settlement and mediation, the goal of arbitration is not to find common ground to resolve a dispute. Parties are represented by legal counsel who present evidence to the arbitrator(s). Like judges, the arbitrators' role is to make a determination on the merits of each party's case based on witnesses and evidence presented and on governing law. Often the decision of an arbitrator is final and not appealable.

3   Application (Cases and Examples)

Example 1 A plaintiff brings a lawsuit against a defendant Family Medicine physician alleging medical malpractice based on failure to timely diagnose breast cancer. The plaintiff must provide expert testimony that establishes each element of the cause of action, which includes establishing (1) the existence of a duty owed by defendant to plaintiff, (2) a breach of that duty by defendant, (3) that plaintiff was damaged by defendant's breach of that duty, and (4) that the plaintiff's claimed damages were proximately caused by defendant's breach of that duty. Plaintiff's expert testifies during his deposition that, based on the plaintiff's initial presentation, the relevant standard of care (i.e., owed duty) required the defendant physician to refer plaintiff for a biopsy as soon as possible rather than three months later, as actually occurred. The undisputed facts show that, at the time of diagnosis (three months later than plaintiff's expert said was required by the standard of care), plaintiff was diagnosed with carcinoma in situ, the carcinoma was fully resected, and the biopsied mass showed clean margins. A comprehensive workup showed no evidence of metastatic disease. Plaintiff's expert also admits during his deposition that, had the plaintiff been referred earlier for biopsy, with reasonable medical probability, plaintiff still would have had carcinoma in situ. Defendant moves for summary judgment, which is granted by the court on the ground that, even if, for the sake of argument, defendant had violated the standard of care, the delay in diagnosis did not cause any damage to plaintiff because an earlier biopsy would not have altered the stage of her disease or affected her treatment or prognosis. In other words, the testimony of plaintiff's expert did not establish that the alleged violation of the standard of care was the proximate cause of any injury to the plaintiff.

Example 2 A plaintiff brings a lawsuit against a clinician and a medical facility for medical malpractice. The parties engage in the normal discovery process, including a request by the defendants for all of plaintiff's medical records from all treating clinicians during the relevant time period. After a trial by jury, a judgment is entered in favor of the plaintiff in the amount of $10 million. Shortly after conclusion of the trial, the defendants discover that the plaintiff withheld critical medical records from the defendants, in essence defrauding the court. Defendants file post-trial motions for a new trial (based on the new evidence) and for a judgment notwithstanding the verdict. The court orders the parties to engage in post-trial mediation. Both the plaintiff and the defendants recognize the vulnerabilities in their cases and ultimately settle for $4.6 million. Although the settlement amount is less than the verdict that the jury awarded, from the plaintiff's vantage point, it is still a significant amount of money, and the judgment could be vacated entirely if the court grants the post-trial motion and finds that the court was defrauded. From the defense perspective, this still represents a large payout, but there is a chance that the court will deny the post-trial motion, and the $10 million verdict will stand.

Example 3 The plaintiff, the estate of a patient who died of a cardiac arrest, brings a lawsuit against a pharmaceutical company alleging failure to warn that one of its medications caused a prolonged QT interval, an electrical conduction problem that is sometimes associated with cardiac arrhythmias. During the discovery process, it becomes apparent that the company was aware that incidents of prolonged QT interval had been reported to the company as spontaneous adverse event reports, although none of the controlled clinical trials on the medication showed QT prolongation as a potential adverse effect attributable to the medication. An expert in clinical pharmacology on behalf of the plaintiff offers the opinion that the spontaneous adverse event reports of QT prolongation should have caused a responsible pharmaceutical company to undertake further testing to investigate this potential effect. Alternatively, the expert contends that the company should have modified the product labeling to alert physicians and patients that it had received reports of this potentially serious problem. An expert epidemiologist for the defense opines that the reports received by the company of prolonged QT interval were no greater than what is seen in the background population and that receipt of such reports neither creates a duty to further investigate nor a duty to warn. This question will ultimately need to be decided by the trier of fact based on the relative persuasiveness of the experts.

4   Summary

Although specific rules for conducting a lawsuit vary from jurisdiction to jurisdiction, this chapter presents general principles and procedures governing most civil litigation. It includes a discussion of (1) the litigation process, including commencing a lawsuit (causes of action and damages, filing a complaint, answering a complaint, and additional pleading); pretrial activities (discovery, motions practice, and orders); and trial (judge versus jury trial, procedural overview, opening statements, presentation of case by plaintiff, presentation of defense by defendant, possible rebuttal by plaintiff, closing arguments, verdict, and additional motions practice); (2) costs of trial; (3) appeals; and (4) settlement, mediation and arbitration.

References

1. Fed. R. Civ. P.
2. Fed. R. Evid.
3. Colorado v. New Mexico, 467 U.S. 310 (1984).
4. Fed. R. Civ. P. 11(b).
5. Fed. R. Civ. P. 11(c).
6. *See, e.g.,* Fed. R. Civ. P. 26.
7. *See, e.g.,* Fed. R. Civ. P. 26(a)(2).
8. U.S. Const. art. III, § 2; U.S. Const. amend XI.
9. Duncan v. Louisiana, 319 U.S. 145, 149 (1968), which held that "the Fourteenth Amendment guarantees a right to jury trial in all criminal cases which – were they to be tried in federal court – would come within the Sixth Amendment's guarantee."
10. Baldwin v. New York, 399 U.S. 66 (1970).
11. U.S. Const. amend. XII.
12. Minneapolis & St. Louis R. Co. v. Bombolis, 241 U.S. 211 (1916).
13. *See, e.g.*, Arkansas Const., art. II, § 7; California Const., Art. I, § 16; Kentucky Const., § 7; Maryland Const., Declaration of Rights, art. 5(a)(1)-(2); New Hampshire Const., part I, art. 20.
14. Fed. R. Civ. P. 702.
15. Id.
16. U.S. Const. amend. V. *See, e.g.,* Gamble v. United States, 139 S. Ct. 1960 (2019) (double jeopardy not violated where federal government prosecuted defendant under federal law for same the same incident that he pled guilty to in state court).
17. Alternative Dispute Resolution: How The Growth of This Practice Has Led to A Drop in Litigation, Alterna, F., *Law Student Connection*, May 15, 2012, http://nysbar.com/blogs/lawstudentconnection/2012/05/alternative_dispute_resolution.html.

Contract Law

Shawn Bayern

Contents

Key Points
- Contract law is the law of enforceable promises.
- Promises are normally enforceable when bargained for or relied upon.
- Promises are ordinarily interpreted simply by asking how a reasonable person would interpret them; there is no special expertise required to read most contracts.
- Promises are usually enforced by awarding an amount of money known as "expectation damages," but other remedies are sometimes available.

Legal Concepts
- Contract: a legally enforceable promise.
- Consideration: what one party gives up as the price of another party's promise.
- Goods: tangible, moveable property.
- Uniform Commercial Code: a statute in the vast majority of states governing contracts for the sale of goods.
- *Restatement (Second) of Contracts*: a very influential Contracts treatise produced by the American Law Institute.

S. Bayern (✉)
College of Law, Florida State University, Tallahassee, FL, USA
e-mail: bayern@law.fsu.edu

© The Author(s), under exclusive license to Springer Nature Switzerland AG 2022   59
A. S. Pasha (ed.), *Laws of Medicine*,
https://doi.org/10.1007/978-3-031-08162-0_4

- Expectation damages: an amount of money sufficient to put a contract plaintiff in the situation they would have occupied had the promise been performed.
- Performance: the completion of an act or omission that is legally required under a contract.
- Specific performance: an order to compel the defendant to perform a contract.
- Statute of Frauds: a statute that requires a party who seeks to enforce a contract to provide evidence of a signed writing that embodies or memorializes the contract.

1   Introduction

Contract law is the law of enforceable promises. As the most influential authority on American contract law—a treatise called the *Restatement (Second) of Contracts*, produced by the American Law Institute, a group of leading judges, scholars, and lawyers—puts it, "A contract is a promise or a set of promises for the breach of which the law gives a remedy, or the performance of which the law in some way recognizes as a duty" [1]. The *Restatement* is not itself the law, but it has been widely followed by judges across the US.

The principal questions of contract law are (1) is a particular promise legally enforceable?; (2) how are contracts interpreted, and through what processes are they formed?; (3) what exceptional situations excuse performance?; and (4) how (i.e., by what measure of money or other remedies) are contracts enforced? This chapter surveys these questions and introduces the structure and sources of contract law.

In the United States, the law of contracts is primarily a *common law* subject. That is, its rules largely result from, and are found in, court opinions that apply legal principles to particular facts (caselaw). The court opinions themselves constitute the law; lawyers specialize in reading and applying prior cases.

Contract law is not purely caselaw, however. One particular group of statutes known as the Uniform Commercial Code (UCC) has had a significant impact on contract law. Most importantly, Article 2 of the UCC, which has been adopted in almost all US states, governs the sale of *goods* (physical, moveable items—as distinct from, for example, services, real estate, or cryptocurrency). For example, a contract to sell medical equipment or pharmaceuticals within the US is governed by the UCC, whereas contracts to provide services are governed primarily by caselaw. The rules of Article 2 of the UCC are not radically different from most of the background rules of the common law of contracts, but occasionally the UCC repeals or overrules an old judicial doctrine, so it is important to consider its application to any contract to sell tangible, moveable property.

Ordinary contracts are usually governed by state law in the US, which means that technically the rules of contract law and the dispositions of judges can vary from state to state. Because of the standardized UCC and the influence of the Restatement, however, contract law is similar in most states. In general, contracts are enforceable in state court in a claim called *breach of contract*. Contract cases may be brought in federal court if their value is high enough and the parties reside in different states [2].

2 Discussion

2.1 Which Promises Does the Law Enforce?

In American contract law, promises are not legally binding just because they're promises. Two distinct factors can make promises enforceable: (1) bargains and (2) reliance.

Bargains

First and most importantly, promises are enforceable when they are bargained for—that is, when each party to an exchange gives up something as the price of the other party's promise. The requirement that each party give something up is known as *consideration*, and the main function of consideration doctrine in modern law is to distinguish commercial promises from gift promises. Promises that are bargained for are enforceable on that basis alone, whereas promises of future private gifts (e.g., among friends or family members) require some other factor (like reliance) to be legally binding. Pledges to charities tend to be legally enforceable, however [3].

In recognizing bargain contracts, courts do not ordinarily inquire as to the *value* or *equality* of what each side gives up—as long as the parties are actually negotiating, as compared to casting a promise to give a gift into the mere form of a bargain (e.g., "I promise to sell you my car for $10"). For example, a promise to sell a used piece of equipment for $3000, when that item lists in various reliable price guides for $9000, is very likely to be legally binding (without evidence of fraud, duress, or some other concern).

In cases of extreme disparity between the parties, however, or if a court detects that one of the parties couldn't reasonably have understood the terms or exercised free choice in entering the contract, courts may declare a bargain to be *unconscionable* and refuse to enforce it or modify its terms [4]. The power to refuse to enforce, or to modify, unconscionable contracts is open-ended and functions as an opportunity for courts to prevent extreme abuses of one party by another.

Narrower factors that may prevent contracts from being legally binding include fraud (procuring a contract by lying), duress (compelling someone to enter into a contract by means of force or an improper threat), and the incapacity of one of the parties (because of childhood, mental illness, and so forth). Moreover, contracts to commit crimes are not ordinarily enforced, and some kinds of contracts may be illegal or treated as "against public policy" [5].

Reliance

Second, even promises that are not part of a bargain may be legally binding if they are *relied upon*. For example, a promise to pay for a relative's college tuition is not enforceable on its own. But if the relative has relied on the promise—for example, by enrolling in college preparatory courses at some expense of money and time—courts will tend to enforce the promise [3]. The doctrine that a relied-upon promise is enforceable is sometimes known as *promissory estoppel*.

2.2   Contract Interpretation and Formation

The most frequent dispute in contract law is simply about what the parties to a contract meant with their words and other expressions. In general, the law does not impose specialized or technical meanings on words in contracts; to determine what words and other expressions mean, the law typically looks simply to how a reasonable person would interpret them. To put that differently, in general contract law uses what is known as an *objective standard of interpretation*: words mean what they should reasonably have meant to the recipient. The objective standard of interpretation is dominant but not absolute; the law of interpretation can be complex, but the basic principle of interpretation in contract law depends on what reasonable recipients of language and other expressions would understand.

Written and Oral Contracts

American law imposes no general requirement that contracts be in writing. While putting contracts in writing can be useful for evidentiary reasons, oral contracts are normally just as enforceable as written ones. However, a statute known as the Statute of Frauds, which dates to 1677 in England but has been adopted in modern form by American state legislatures, does identify certain types of contracts that require a *signed writing* to be enforced [6]. The Statute of Frauds can be extremely technical, and its details vary from state to state, but the main relevant types of modern contracts for which a signed writing is needed are contracts for real estate, contracts for goods sold for more than $500, and contracts that cannot be completed within one year. Note that the Statute of Frauds does not ordinarily require that "the contract" be in writing, just that the party seeking to enforce the contract produce *some* writing signed by the party against whom enforcement is sought [6]. That writing might be a note of confirmation that follows up on an oral contract.

In modern law, the notion of a *writing* includes electronic records, and *signed* is interpreted very broadly. For example, sending something from an authenticated email account probably counts as a "signed writing" [6]. Courts in general have not favored striking down oral contracts under the Statute of Frauds.

When the parties do adopt written contracts, however—which is of course commonplace in sophisticated commercial environments—the law applies special rules to interpret them. The two most significant rules that contract law applies to written documents are the *parol evidence rule* and the *plain meaning rule*.

The parol evidence rule is extremely complex, but its main principle is that written agreements that appear complete will *discharge*, or cancel, any prior related agreements between the parties. For example, if someone agrees orally to hire an IT company to perform services and the parties agree on terms over the phone, and then the IT company follows up with a written contract that the other party accepts, the written contract is likely to supplant or discharge the prior oral agreement.

The plain meaning rule is simply a doctrine that when a document is clear, it means what it says. Of course, there can be significant dispute over whether a document is clear. The strength and application of the plain meaning rule vary from state to state, but generally courts will try to interpret a written contract in a way that is at least consistent with the written text of the contract. The main effect of the plain meaning rule, particularly in states with strong versions of the rule, is to exclude other sources of evidence (i.e., apart from the writing) about what the parties intended. For example, the plain meaning rule can cut off the admissibility and use of testimony by agents of either party about what led to the contract or what the parties said to each other. The rule is never absolute, however; it will not, for example, prohibit evidence that a contract came about because of duress.

The upshot of these two rules is that writings can be quite powerful in contract law, particularly when they are clear and appear to be complete. That said, modern contract law does not focus exclusively on the text of written agreements. Courts interpret the text in view of many practical and contextual factors, including *course of performance* (how the parties have acted on the contract in the past), *course of dealing* (how the parties negotiated past contracts), and *trade usage* (e.g., what words mean in a particular industry). For example, a written contract might require that a chemical solution being sold be "free from contaminants," but courts will pay attention to evidence that in the parties' industry, "free from contaminants" doesn't mean that the solution needs to be *entirely* free from contaminants but instead that contaminants must be below a certain threshold [7]. In part because of the way all these factors interact, there is a longstanding tension in contract law between judges who want to pay attention to the literal text of agreements and those who want to figure out the substantive deal that the parties intended.

Contract Formation

It is commonly said that all contracts require an *offer* and an *acceptance*. An offer is simply a communication (by an *offeror*) that a reasonable recipient (the *offeree*) would interpret to invite definite and legally binding acceptance. An acceptance is simply a communication that indicates agreement to the terms of the offer. By default, an offer remains open for a reasonable amount of time, but it may also be terminated if the offeree *rejects* it or if the offeror *revokes* it. An old common law rule holds that any purported acceptance by the offeree that doesn't match the terms of the offer exactly is treated as a *counteroffer*, and thus as a rejection of the offer. Courts and commentators have questioned and limited the force of this rule, and the UCC repeals it for contracts within its scope [8].

Ordinarily, all legally significant communications (offers, rejections, revocations, and so on) are legally effective when they're received, including at a corporate mail room or similar location. However, for technical reasons under what is known as the *mailbox rule*, acceptances are effective as soon as they're mailed or otherwise sent. This rule tends to accelerate the formation of a contract, permitting at least one party to begin performing sooner.

2.3   Excuses to Contract Performance

When a contract's duties are completed, they are said to be *performed* and thus extinguished. Normally, if a party does not perform its side of the contract, it is in breach of contract and may be subject to legal liability. Even when a bargain contract is validly formed and would otherwise have been enforceable, however, exceptional circumstances may excuse performance.

One set of possible excuses to performance arises under a doctrine known as *mistake*. Clearly, a party cannot avoid obligations under a contract merely by showing that it was a bad idea for them to enter the contract—or else no contracts would be enforceable—but the law gives relief for some limited types of mistake. One such type, under a doctrine known as *unilateral mistake*, covers what usually amount to typos, slips of the hand, errors in adding numbers, and so forth. The general rule has been that mechanical errors (like typos) can excuse performance when they are, or should have been, obvious to the other side. For example, if a website posts a price of $8.00 instead of $800 for a laptop computer, the owner of the website might not legally be required to sell the laptop for $8.00.

A different type of mistake is governed by a doctrine called *mutual mistake*. This doctrine mainly applies to mistakes of *fact* that were fundamental to the contract and that were so unlikely or unusual that the parties never even considered them. For example, if two people agree to the sale of a car, but neither knows that at the time they formed the contract, the car has been stolen and then completely destroyed in a crash, performance of the contract is likely to be excused on both sides: the buyer has no obligation to pay the purchase price, and the seller is not in breach of contract for failing to deliver a car.

Relatedly, performance may be excused for *unexpected circumstances* that arise after the contract is formed. For example, suppose someone promises to buy a car but then, before it is to be delivered, the car catches fire. In terms of formal legal doctrine, contracts are said to be excused when performance is *impossible* (physically unachievable) or commercially *impracticable* (oppressively expensive) or if the purpose of the contract has been *frustrated* (no longer desired or appropriate). For example, a clinician who promises to perform a procedure but then is prevented from doing so because of illness or injury is probably not in breach of contract.

2.4   Legal Remedies in Contract Law

In general, courts enforce contracts through awards of money (*damages*) or through orders compelling a party to do what was promised (*specific performance*). Damages are the typical remedy in American contract law.

Ordinarily, the law awards a measure of damages known as *expectation damages* when enforcing a bargain contract. Under this measure, a court calculates the sum of money that is necessary to make the plaintiff indifferent, as nearly as possible, between (1) breach plus damages and (2) performance [9, 10]. For example, if a buyer promises to buy goods for $80,000, the seller breaches by failing to deliver the goods, and the buyer now needs to spend $90,000 instead to procure the goods, the buyer's damages against the original seller will be $10,000—the amount necessary to put the buyer back into a situation where they get what they were promised (goods at a net cost of $80,000). Only those damages *foreseeable* at the time the contract was formed are available to the plaintiff, and the plaintiff must prove damages with adequate *certainty*.

Occasionally, courts determine that money is not adequate to put the plaintiff into the situation they would have occupied if the contract had been performed. For example, a buyer promises to buy a rare painting, and the seller fails to deliver. Money could go a long way towards compensating the buyer, but (depending on the buyer's needs) it may not let the buyer replace the rare painting. When damages can't restore the plaintiff to the situation they would have occupied if the promise had been performed, courts may issue orders that compel the defendant to perform—on threat of *contempt of court*, which in principle can put the defendant in jail until they agree to perform. Because such a remedy can be intrusive and coercive, courts (1) do not tend to use it when damages are sufficient, (2) still evaluate the policy considerations, or costs and benefits, of issuing orders in each individual case, and (3) avoid using it entirely (or almost entirely) in some areas, like employment law, in order not to force people into close legal relationships. Specific performance is more easily available in cases where the seller of real estate refuses to deliver title to property.

Other remedies are available. For example, when enforcing gift promises because they have been relied upon, many courts have used a measure known as *reliance damages*: the amount of money needed to restore a plaintiff to the situation they were in before any promise was made [11]. This measure differs from expectation damages because it tries to reverse the effects of the promise, rather than trying to give the plaintiff the benefit of the promise. For example, suppose Aunt promises Nephew that she will buy him a car costing up to $20,000 and, in reliance on this promise, Nephew buys a car for $18,000. Aunt then informs Nephew she will not pay for the car. Nephew, otherwise unable to afford it, sells it in the used-car market for $15,000. Nephew's expectation damages might have been $18,000, but his reliance damages are only $3,000 (the amount of money he lost because of the promise, as compared to his starting point before the promise was made).

3   Application

Vignette #1 An orthopedic clinic contracts to buy a CT scanner from a distributor for $220,000. The distributor fails to deliver the scanner, so the clinic scrambles to purchase another one from a different supplier for $240,000, in the meantime losing 17 days of billings for CT scans. The contract is covered by the Statute of Frauds because the CT scanner is a good being sold for more than $500, so if there is no writing signed by the distributor, the clinic may have difficulty enforcing the contract. If there is such a signed writing, as there is likely to be, the contract will be enforceable as a bargain contract with consideration on both sides. In that case, the clinic may sue the distributor for breach of contract and recover expectation damages. Here, those damages would be at least $20,000, the difference in cost between what the clinic contracted to pay ($220,000) and what it had to pay for a replacement ($240,000). Under the circumstances, it was probably reasonable for the clinic to find the best replacement they could find quickly, even if it was a little more expensive. Other costs that the clinic faced in finding a replacement, including lost profits, may be recoverable too if those costs are judged to be foreseeable at the time the contract was made. Here, the lost profit during the delay is probably foreseeable: what else would the distributor expect the clinic to use a CT scanner for except to diagnose patients and bill for its use?

Furthermore, because the contract is for the sale of a good, it is governed by the statutory UCC rather than the background rules of common law.

Vignette #2 A few days after visiting a potential new office space at an open-house viewing, the manager of a clinical practice emails the landlord to inquire whether the space is still available for lease. (This email is not yet a contractual offer because there is nothing for the recipient to accept; it is commonly called an *invitation to deal*.) The landlord replies that the space is indeed immediately available at a rate of $8600/month for a three-year lease and attaches a signed lease agreement with a space for the manager's signature. The manager signs and scans the lease agreement, but only after crossing out the number "8600" and replacing it with "7600"; the manager then attaches the document to an email reply to the landlord that reads "This is appealing overall, but only at $7600/month. If that works for you, we're ready to go." So far, the landlord has made an offer and the manager has made a counteroffer. The counteroffer refuses and terminates the original offer. The landlord replies, "All right, we have a deal. Contact my assistant for logistical details regarding keys and access codes." That email from the landlord forms a contract by accepting the manager's counteroffer. There is some debate about whether the "mailbox rule" applies to email, although in practice it is not likely to matter much. Had the negotiation been conducted by USPS Priority Mail instead of by email, the contract would have been formed and legally binding as soon as the acceptance was mailed, regardless of when (or even whether) it was received.

4 Summary

Contract law determines how promises are to be enforced. The law does not enforce all promises; the two principal types of contracts that the law enforces are those that are bargained for and those that are relied upon. Most contracts need not be in writing to be enforced, but when the parties do write down their contracts, their writings are likely to have special significance. Even after they are validly formed, performance of contracts may be excused under doctrines concerning mistake and unexpected circumstances. Ordinarily, however, contracts are simply given the meaning that reasonable people would understand them to have, and they are enforced by default by calculating expectation damages, or the amount of money necessary to give the plaintiff the benefit of the promise. Other remedies, such as orders to perform a contract, are sometimes available.

References

1. Restatement (Second) of Contracts § 1 (1981).
2. 28 U.S.C. § 1332
3. Restatement (Second) of Contracts § 90 (1981).
4. Restatement (Second) of Contracts § 208 (1981).
5. Restatement (Second) of Contracts § 178 (1981).
6. Fuller, LL, Eisenberg, MA, Gergen MP. Basic Contract Law. 10th ed. St. Paul: West Academic; 2018. Appendix A, The Statute of Frauds, p. 1089–1142.
7. Restatement (Second) of Contracts § 222 (1981).
8. Uniform Commercial Code § 2-207.
9. Restatement (Second) of Contracts § 344 (1981).
10. Bayern, SJ, Eisenberg MA, The expectation measure and its discontents. Mich. St. L. Rev. 2013;2013(1):1–37.
11. Fuller LL, Perdue WR, The reliance interest in contract damages: Yale L.J. 1937; (46)3, 373–420.

Part II

Public Health Law

Public Health Law

Montrece Ransom, Emely Sanchez, and Molly Berkery

Contents

Key Points
- Public health law is a tool for creating and enforcing public health interventions and is a social determinant of health. The law is a social determinant of health as existing laws and policies impact the fundamental drivers of health inequities.
- Public health authority stems from various legal instruments at various levels of government, including the U.S. Constitution, federal and state statutes, emergency orders, administrative law and regulations, and local ordinances.
- Federalism and limits on public health authority, such as due process and preemption, create barriers and challenges to enacting broad public health laws.
- The Doctrine of Preemption stems from the Supremacy Clause of the U.S. Constitution. Preemption can be expressed or implied and create ceilings or

M. Ransom (✉)
National Coordinating Center for Public Health Training, New Orleans, LA, USA
e-mail: mransom@nnphi.org

E. Sanchez
The Network for Public Health Law, Mid-States Region, Edina, MN, USA
e-mail: esanchez@networkforphl.org

M. Berkery
Healthcare Regulatory Attorney, San Diego, CA, USA

© The Author(s), under exclusive license to Springer Nature Switzerland AG 2022
A. S. Pasha (ed.), *Laws of Medicine*,
https://doi.org/10.1007/978-3-031-08162-0_5

floors. Allowance of preemption may be determined by the doctrines of Home Rule or Dillon's Rule, depending on state classification [1].

- Interventional public health law can be applied to a variety of subjects, including chronic disease control and injury prevention.
- Global health law includes the study of legal norms, processes, and the participation of institutions to create conditions for people across the world that will increase their potential for optimal health [2]. Global health law and policy go beyond national public health law to regulate interstate behavior to reduce the morbidity and mortality of global norms and infectious diseases.

Legal Concepts

- Administrative Law: Branch of law that governs the creation and operation of administrative agencies (federal and state) and gives powers to those administrative agencies to create substantive rules and regulations, including those that form the legal relationships between such agencies, other government bodies, and the public at large [3].
- Dillion's Rule: Doctrine that gives localities a limited ability to adopt public health policies on a narrow range of subjects based on state law [4].
- Federalism: A system of government that divides authority across states and the federal government [1].
- Home Rule: Doctrine that gives localities the broadest possible powers of self-government or autonomy so that localities may adopt initiatives without looking to state law for specific authorization [4].
- Police Powers: Refers to the power of the state, and where so delegated, local, government to exercise reasonable control over the people and property within its jurisdiction to protect the general welfare [5].
- Preemption Doctrine: The Supremacy Clause of the U.S. Constitution states that a higher authority of law will displace a lower authority of law when the two authorities come into conflict. For example, when a state law conflicts with a federal law, the federal law displaces or preempts the state law. Preemption applies to all conflicting laws (e.g., constitutions, statutes, regulations, executive orders, and case law) [1].
  - Ceiling Preemption: Prevents local governments from establishing more stringent ordinances than those established by state law [6].
  - Conflict Preemption: Subcategory of implied preemption that occurs when simultaneous compliance with both federal and state regulations is impossible ("impossibility preemption") or when state law poses an obstacle to the accomplishment of federal goals ("obstacle preemption") [1].
  - Express Preemption: Preemption that is expressly stated or expressed in a clause [1].
  - Field Preemption: A subcategory of implied preemption where federal regulation implicitly precludes supplementary state regulation or where states attempt to regulate a field in which there is dominant federal interest. Applying these principles, SCOTUS has held that federal law occupies several

regulatory fields, including alien registration, nuclear safety regulation, and the regulation of locomotive equipment [1].
- Floor Preemption: Allows local governments to enact ordinances so long as the added ordinances do not go below the minimum level of standards set by the state law [7].
- Implied Preemption: Preemptive authority that is not expressed but rather implied [1].
- Sovereignty: Refers to paramount control of the Constitution and frame of government and its administration; the self-sufficient source of political power from which all specific political powers are derived [8].
- State: A sovereign political body of people that occupies a clearly defined territory [9].
- Tenth Amendment: Amendment to the U.S. Constitution that grants police powers to the states: "The powers not delegated to the United States by the Constitution, nor prohibited by it to the States, are reserved to the States respectively, or to the people" [10]. It specifies that if a power is not explicitly provided to the federal government, it belongs to the states.
- Tribal Sovereignty: The authority of Tribes to exercise jurisdiction over their land and govern their people; as distinct nations, Tribal governments and their laws are unique and reflective of their histories and cultures [11].

1   Introduction

"In the realm of public health ... it's the law that really does the work. That's been demonstrated time and time again in areas ranging from mandating vaccinations to requiring automobile seatbelts; improving workplace safety; the inspections of meat products; and fluoridation of water. Public health succeeds by making healthy choices the norm." [12]

There is no public health without law. Laws regulate all public health authorities, from FDA vaccine approval to the establishment of a health department and mandatory disease reporting required by a state statute, to a local ordinance that requires individuals to wear a mask to prevent the spread of an infectious disease. The law creates the framework and regulates population health [13].

There are at least two distinct categories of public health laws: foundational and interventional. Foundational public health laws create the legal authority for government regulation of public health. This domain of public health law has also been called infrastructural law. Interventional public health law refers to regulations implemented and enforced by the government, public health practitioners, and others in the health system to promote and protect the public's health [14].

The law is also a significant social determinant of health. Law profoundly shapes the context in which we live, learn, play, work, and worship. It necessarily shapes our neighborhood, our housing, and our social environment, and it has the power to distribute resources—to make the things people need to live healthful lives more accessible.

In this chapter, we focus on defining and characterizing public health law by discussing the sources of law and the limitations on government power. By understanding the origins of public health laws and the legal framework for the practice of public health, readers will have a better perspective on the effect of the application of public health laws on health outcomes. We begin by examining foundational public health law. We define and characterize the major sources of this law and explore basic constitutional concepts and legal principles framing U.S. public health practice across relevant jurisdictions. Then, we address interventional public health laws as applied to disease and injury prevention and control. Finally, we end with a discussion of law as a social determinant of health and global public health law.

2   Public Health Law

2.1   Foundational Public Health Laws

Public health law can be defined as the legal powers and responsibilities of the government and its partners across the health system to ensure conditions allow people to be healthy and the limitations on the power of the government to constrain legally protected interests of individuals for the common good [15]. This definition highlights the fundamental role of laws in daily public health practice. Law defines the authority and responsibilities of public health practitioners while also restraining the actions they take in the name of public health. As discussed below, the major sources of law necessary for public health practice include constitutions, statutes, regulations, executive orders, local ordinances, and case law. Primary limitations to that authority include due process and preemption.

2.2   Sources of Public Health Authority

The U.S. Constitution
The U.S. Constitution distributes power between the different government levels: federal, state, and local. This power distribution is called Federalism. The concept of Federalism is particularly important. It is defined simply as the separation of powers between different government levels. The federal government has only those powers listed in the U.S. Constitution. Specifically, under the U.S. Constitution, the federal government has limited, and enumerated powers defined under Article 1, Section 8. Some of these powers are enumerated and explicitly mentioned in the U.S. Constitution, whereas others are implied but necessary to exercise the expressed powers. The federal government's powers can be exclusive (i.e., only the federal government can execute them), or they can be concurrent (i.e., they are shared with the states). Authority outside the scope of the U.S. Constitution, which includes most core public health powers, is reserved for the states. This is because, under the Tenth Amendment of the U.S. Constitution, states have all the powers not reserved for the federal government. General police powers, which include public health

powers, are not one of the limited and enumerated powers of the federal government. Therefore, they are reserved for the states. This means that the U.S. Constitution gives states primary power over public health.

Courts generally allow the exercise of police powers if it is to promote the general health and well-being of the community. States can further delegate these powers to localities via state constitutions or statutes. In sum, in the U.S., the state and local governments have the most power to pass laws to protect the public's health.

Case law on state and local governments' authority over the federal government's authority to protect the health, safety, and welfare of the community is plentiful. In *U.S. v. Lopez,* a high school student bringing an unloaded gun to school was charged with violating federal law, the Gun-Free School Zones Act of 1990 [16]. The U.S. Supreme Court held that the federal legislature exceeded its congressional authority to regulate guns on school grounds because state police powers have control of such local issues. Ultimately, an attempt to reduce gun violence, a public health issue, at the federal level was limited due to principles of federalism.

In *U.S. v Morrison*, the Violence Against Women Act's (VAWA) federal remedy authority was challenged through the Commerce Clause and Fourteenth Amendment's Equal Protection Clause. The U.S. Supreme Court granted certiorari because the lower court had invalidated the federal statute on constitutional grounds, denying victims of gender-based violence from suing in federal court. The U.S. Supreme Court held that the Commerce Clause was intended for economic consequences and that the Fourteenth Amendment's Equal Protection Clause did not apply to state actions, only individuals. Ultimately, the Court concluded that the Act could not provide a federal civil remedy, but rather, it was to be provided by the State. *US v. Morrison* further narrowed the scope of regulated activity by Congress and the federal government [17].

Both cases represent examples of how in the United States, the federal government is limited in regulating public health, which is primarily governed by state laws and local ordinances. State and local governments have a broad and flexible public health legal authority.

Nonetheless, the federal government does have some authority to affect public health. Examples of federal authorities that impact public health include those levying taxes for the common welfare (e.g., school lunch programs, education, Medicare, Medicaid) and regulating interstate commerce (e.g., responding to communicable disease outbreaks across state and national borders).

The federal government may also use its enumerated powers to shape public health indirectly. Federal regulations can incentivize local action, most commonly through funding, by setting standards for federal programs that might have a trickle-down effect. The minimum drinking age in the U.S. is an example of this. In 1984, Congress passed the National Minimum Drinking Age Act, which withheld 10% of federal highway funding from states that did not maintain a minimum drinking age of 21 [18]. In *South Dakota v. Dole*, the Act was challenged because Congress did not have the authority to set a nationwide minimum drinking age and in doing so, infringed on the authority reserved to the states. The U.S. Supreme Court upheld the constitutionality of the Act and ruled that Congress had not violated the Tenth Amendment because it exercised its right to control federal spending [19].

The U.S. Constitution also provides that federal laws are the "supreme law of the land," meaning that federal laws can sometimes prohibit or supersede state or local action. This is called preemption and is discussed later in the chapter.

Statutes

Statutes are laws passed by the federal and state legislatures and enacted via a constitutionally defined process. Congress enacts federal statutes. State legislatures pass state statutes. Ordinances are local laws passed by local or municipal governments within a state (e.g., city councils, county boards of supervisors, or other forms of municipal governance). Ordinances are legally binding and usually apply only to the local jurisdiction to which they pertain [20].

Statutes and ordinances have several functions. First, they express the intent of the legislative body. Second, as illustrated in Table 1, they create and empower agencies and provide the authority to promulgate regulations and oversee regulated activities. Statutes also appropriate money and define certain activities as criminal.

Regulations

As illustrated in Table 1, administrative agencies are created by enabling statutes. Enabling statutes allows administrative agencies (also known as regulatory agencies) to promulgate, interpret, administer, and enforce regulations.

Regulations are meant to implement the intent of statutes. They provide standards to which corporations and individuals must adhere to follow legal standards while engaging in certain activities. They are more detailed and technical than statutes. Table 2 provides an example of an enabling state statute, which provides the general framework, and agency promulgated regulations, which provide specific requirements and provisions related to the statute.

Executive Orders

An executive order is a legally binding directive issued by the President, governor, or other leaders of an executive branch. Executive orders manage operations, mandate agencies' requirements, and have the force of law. They do not require any action by Congress or the state legislature to take effect, and the legislature cannot directly overturn them. However, legislatures can pass subsequent laws to counteract the effect of these executive orders [21].

Table 1 Example of a State Statute Establishing and Empowering an Agency

Ga. Code Ann. § 31-2A-4
The Department of Public Health shall safeguard and promote the health of the people of this state and is empowered to employ all legal means appropriate to that end.
O.R.S. § 431.131
The Oregon Health Authority…shall establish by rule the foundational capabilities necessary to protect and improve the health of the residents of this state and to achieve effective and equitable health outcomes for the residents of this state.

Table 2 Examples of Authorizing Statutes and Resulting Regulations

Authorizing Statute	Resulting Regulation
GA. Code Ann. § 20-2-771 (b) No child shall be admitted to or attend any school or facility in this state unless the child shall first have submitted a certificate of immunization to the responsible official of the school or facility... (c) The Department of Public Health shall promulgate rules and regulations specifying those diseases against which immunization is required and the standards for such immunizations...	**Ga. Comp. R & Regs. 290-5.4-.02** ...Immunization...against the following named diseases, shall be required for entrance into any school or facility operating in the state: (a) Diphtheria; (b) Pertussis... (2) For any child attending any school or facility in the state of Georgia for the first time, a parent or guardian must submit a valid certificate of immunization...
Florida Code Section 112.0455 ...adopt additional rules to support this law... using criteria established by the U.S. Department of Health and Human Services as general guidelines for modeling drug-free workplace laboratories...	**FL Admin Code 59A-24 et al.** Regulates issues such as the following: Drugs to be Tested/Body Specimens, Collection Site and Specimen Collection Procedures, Drug Testing Laboratories—Standards and Licensure, etc.
O.R.S. § 433.272 The Oregon Health Authority shall adopt rules...which shall include, but need not be limited to...(2) The required immunization against diseases...(5) The procedures and time schedule whereby children may be excluded from attendance in schools or children's facilities...	**OAR 333-050-0040** (1) The statement initially documenting evidence of immunization ... must be on a Certificate of Immunization status form or a form approved by the Public Health Division and must include... (2) [A] minimum of one dose each of the following vaccines must be received for new enterers prior to attendance: Polio, Measles, Mumps, Rubella...Varicella...and Diphtheria/Tetanus/Pertussis containing vaccine...

On the federal level, executive orders are based on powers granted to the President in the U.S. Constitution or federal law and must be consistent with those authorities. Article II of the Constitution assigns the President the roles of commander in chief, head of state, chief law enforcement officer, and head of the executive branch. The President has the sole constitutional obligation to "take care that the laws be faithfully executed" and is granted broad discretion over federal law enforcement decisions [22]. Sometimes, the President can issue executive orders directly related to health and healthcare. For instance, President Barack Obama issued executive orders establishing the White House Office of Health Reform and the National Prevention, Health Promotion, and Public Health Council [23].

State governors can also issue executive orders that directly impact public health. These can include orders that establish investigatory commissions to study health problems, create public health programs, direct state public health

Table 3 Examples of State Executive Orders

Public Health Emergency Declaration
On March 9, 2020, Governor DeWine of Ohio signed an Executive Order declaring a State of Emergency, and on March 12, Ohio became the first state in the U.S. to close schools in effort to stop the spread of SARS-CoV-2 (commonly known as COVID-19). States across the country quickly followed suit as the impact of the virus became dire. DeWine and other governors used public health laws to balance the needs of individuals (e.g., employers, employees, students) against the urgent need to protect the health of the community, particularly the states' most vulnerable citizens [24].

agencies, establish funding, or (as illustrated in Table 3) declare a public health emergency and make proclamations to increase awareness about a public health issue [25].

Case Law

Case law is decided by federal and state courts when a challenge or lawsuit is brought that requires the court to interpret constitutions, statutes, regulations, and local ordinances as applicable to the set of facts. Rulings "attach" to statutes and regulations that are interpreted and ruled upon. Previous rulings impact current disputes that involve the interpretation of the same statutes and regulations. If case law comes from the same jurisdiction and is from a higher court, the rulings are binding on the lower court.

There are numerous examples of case law, both federal and state, that address a broad range of public health issues, including but not limited to pollution, water safety, public nuisance laws, food safety, child labor, maternity and paternity leave, narcotics, vaccines, and quarantine [26].

Jacobson v. Massachusetts (1905) was a landmark U.S. Supreme Court case that upheld the authority of states to enforce compulsory vaccination laws. In 1902 there was a smallpox outbreak in Cambridge, Massachusetts. The defendant, Jacobson, refused to be vaccinated, as the law required. He was fined five U.S. dollars, which would be the equivalent of roughly 130 U.S dollars today. He refused to pay the fine and challenged the constitutionality of the vaccination law in court. The U.S. Supreme Court upheld the law, stating: "there are manifold restraints to which every person is necessarily subject for the common good". The Court stated that the government may enact "reasonable regulations" to protect public health and safety. These "reasonable regulations," however, must balance and recognize the individual rights established and protected under the U.S. Constitution [27].

This precedential case remains the most significant public health law case because it limited individual liberties to protect the public's health. It also marked the beginning of the application of modern constitutional analysis to disease control law. Even though the Supreme Court did not extend the ruling in *Jacobson* beyond the facts of the case, it articulated the principles and authority behind state and local contemporary public health powers, which also applies to areas such as chronic disease and injury prevention and control:

"We are not inclined to hold that the statute establishes the absolute rule that an adult must be vaccinated if it be apparent or can be shown with reasonable certainty that he is not at the time a fit subject of vaccination or that vaccination, by reason of his then condition, would seriously impair his health or probably cause his death" [27].

The important takeaway is that *Jacobson* represents the balancing of collective actions or public health interventions for the common good against the rights of individuals; it also indicates how case law serves as a source of authority for public health action.

Jacobson has been invoked in numerous other Supreme Court cases as an example of the baseline exercise of police power. Most recently, the COVID-19 pandemic has brought the ruling in Jacobson back into the limelight as stay-at-home orders, or government mandates regarding vaccination and wearing face masks have been challenged as unconstitutional or as infringements upon individual liberties.

Tribal Sovereignty

Native American Tribes have a special status. Federally-recognized Tribes are sovereign nations that maintain a nation-to-nation relationship with the United States. Tribal sovereignty refers to the right of American Indians and Alaska Natives to govern themselves [11]. With few exceptions, they have the same powers as federal and state governments to regulate their internal affairs. Sovereignty for Tribes means that on their Tribal lands, officially known as Indian Country, Tribes have the power to do what's necessary to control their internal affairs and preserve their self-government. Tribal sovereignty includes the authority to establish their form of government, determine membership requirements, establish law enforcement and court systems, and create their laws and health regulations to protect the health, safety, and welfare of their communities [28].

2.3   Limitations on Public Health Authority

There are two main limitations to consider: due process and preemption. Each is explored below in turn.

Due Process

To determine the scope of the limits on government authority, we can turn to the Bill of Rights. In 1791, shortly after ratifying the Constitution, Congress amended the new Constitution with the Bill of Rights, consisting of the first ten amendments that set forth our liberties [29]. The purpose was to protect individuals from government overreach and undue control. What results is a tension that pulls public health practitioners in two directions: on the one hand, toward the protection of the common good and, on the other, toward the protection of individuals [13].

The individual liberties established by the Bill of Rights include the right to be free of unreasonable searches and seizures under the Fourth Amendment; the right not to be deprived of life, liberty, or property without due process of the law under the Fifth Amendment; and the right not to be denied equal protection of the laws, also under the Fifth Amendment [29]. Both amendments are particularly important to the everyday practice of public health [13].

The Bill of Rights, though part of the federal constitution, also applies to the states through the Fourteenth Amendment [30]. While the Fifth Amendment prohibits the federal government from depriving an individual of life, liberty, or property without due process of law [31], the Fourteenth Amendment prohibits states from the same [30]. Ultimately, the due process clause is meant to limit the ability of the government to pass and enforce laws unless they are fair and not arbitrary. There are two forms of due process: procedural and substantive. Procedural due process means the government must follow certain procedures to protect our rights and looks at how the law is implemented and enforced. When considering procedural due process, the key question is whether the government has allowed the right to fair and impartial legal proceedings before depriving someone of life, liberty, or property [32]. For example, if someone is receiving a benefit from the government (e.g., disability benefits), the government cannot just terminate that benefit when and however it wants to. Before termination is allowed, the government is required to provide written notice explaining that the benefit will be terminated and why, an opportunity to challenge the decision to terminate, and a hearing to review that decision. These are all procedural requirements to protect individual rights.

In contrast, the substantive due process looks at whether the government has just cause (e.g., public safety) for depriving an individual of life, liberty, or property. To decide if the government action is justified, a court will evaluate the relative importance of the individual versus governmental interests at stake [32]. The criteria required of the government become more stringent as the individual interest at stake becomes more significant.

Preemption

The Doctrine of Preemption states that a higher authority of law will displace a lower authority when the two authorities conflict. For example, when a state law conflicts with a federal law, the federal law displaces or preempts the state law due to the Supremacy Clause of the Constitution. Preemption applies regardless of whether the conflicting laws come from legislatures, courts, administrative agencies, or constitutions. Preemption is generally either expressed or implied. Express preemption is expressly stated within a preemptive clause or language stating that it would preempt a lower-level law [1]. For example, some states during the COVID-19 pandemic expressly preempted local level governments from enacting social distancing measures, with some going further and threatening litigation for violations [29]. Implied preemption is much more subtle and can be complicated; it is confirmed when a court finds that there was preemption, although there was no explicit language.

Preemption can provide a ceiling or a floor. Ceiling preemption prevents local governments from enforcing a law or policy that is more stringent than the state's law, creating a ceiling or maximum level of enforcement [7]. This is also true for federal laws and policies that can create ceilings for state government action. One example is tobacco products, where ceiling preemption from state legislatures has been used to prevent local levels and communities from enacting stronger laws such as local smoke-free air laws [33]. On the other hand, floor preemption creates a minimum or floor level, allowing state or local governments to create laws and policies that are more stringent than federal or state laws [7].

Floor preemption limits local control the least. For example, Maine uses floor preemption to regulate pesticide use. The state regulation includes specific language reaffirming local authority to pass more stringent regulations and notes, "These regulations are minimum standards and are not meant to preempt any local ordinances which may be more stringent" [34]. The minimum standards created by the Environmental Protection Agency under the Clean Water Act (CWA) provide an example of floor preemption at the federal level. While states "may develop water quality standards more stringent than required by this regulation," they are preempted from adopting standards below the minimum set in the act [35].

When determining whether a local jurisdiction has the authority to enact laws independent of the state government, states are classified as Home Rule and Dillon's Rule states. Home Rule states give local governments the most authority and limit the amount of interference the state has, allowing localities to adopt laws and policies without needing to refer to the state. Dillon's Rule is a more restrictive doctrine, allowing local governments to have only explicit authority based on the state's law. Dillon's Rule allows for the presumption that local authority is limited, and if the state does not expressly give that authority, the locality's authority on the subject is untenable [36]. Home Rule states are more common than Dillon's Rule states, allowing for more local-level innovation.

A lawsuit filed by Georgia Governor Brian Kemp against Atlanta Mayor Keisha Lance Bottoms regarding a COVID-19 mask policy provides a salient example of preemption in action [37]. While Governor Kemp strongly encouraged the use of masks, neither he nor the Georgia legislature issued any related mandates. About a year into the pandemic, Mayor Bottoms, seeking to control the spread of the virus, ordered a rollback of the re-opening of the city's businesses and mandated a mask policy that required residents to wear face coverings when they were outside or in commercial buildings. Georgia is a Home Rule state, and the Atlanta Charter has specific provisions for local authority to promote safety, health, and general welfare [38]. As such, these actions were presumably within her authority. However, after Mayor Bottoms put the mask mandate in place for the city of Atlanta, Governor Kemp issued an Executive Order that expressly prohibited local jurisdictions from passing and enforcing laws stricter than the State's, effectively creating a regulatory ceiling and preempting Atlanta's mask mandate. Ultimately the lawsuit was dropped, but this dispute highlighted the significant preemption challenges throughout the country during the COVID-19 pandemic. Some state Governors were issuing stay-at-home orders as floors allowing localities to do more, while others created ceilings to prevent local enforcement of more stringent restrictions [39].

3   Law as a Social Determinant of Health

The COVID-19 pandemic has raised a host of foundational public health law-related constitutional questions about the distribution of power amongst federal, state, and local governments, including how state and local governments can use their police powers to implement and enforce social distancing laws. Other legal issues that have arisen include the use of quarantine and isolation, civil liberties and human rights, data sharing and privacy, vaccination requirements, etc. These interventional laws impact public health practitioners, who are also first responders, and all individuals affected by the worst pandemic in history.

The pandemic has also highlighted significant health inequities in the U.S. and across the globe. Minorities and underserved groups (e.g., racial minorities, low-income families, immigrants) have been disproportionately impacted by the virus. These groups are particularly vulnerable because of existing laws and policies that impact the fundamental drivers of health inequities. The law, our legal systems, and legal structures heavily influence health through a structure that perpetuates inequities across the social determinants of health: education, food, housing, income, employment, sanitation, and healthcare access [40].

There is a growing recognition that systemic, institutional, and other forms of racism drive health inequities across the social determinants of health. As Will Jawando, a councilmember in Montgomery County, Maryland, said, "[W]hether it is police-involved killings or disparate health outcomes where [Black-American] patients can't get treatment because they are not seen as being sick, or financial redlining in certain ZIP codes, food deserts, or people of color getting hit by cars more often because their communities aren't walkable—it's all ultimately due to racism" [40]. This recognition has led hundreds of jurisdictions, including Michigan, as discussed in Table 3, to use executive orders to declare racism a public health crisis and emergency. These proclamations acknowledge that state and local governments are central to efforts to eradicate the impacts of racism to create urgency around the fight to end racism across our local policies, practices, and procedures; reduce the racial health equity gap; and truly achieve the conditions that create optimal health for all.

4   Public Health and Legislation "By the People"

Public health law and public health policy are vulnerable to changes in public health authority through legislative decisions. Public health, although seemingly apolitical, is political as it balances individual liberties with safeguarding population health and behavior. As such, the electorate and the elected officials may shape public health authority and laws designed to curb chronic and infectious diseases. Politically divisive issues such as mask mandates, soda taxes, needle exchanges, and universal healthcare are a few examples of how political interference and political capital have been pivotal or detrimental in creating effective public health policy [41].

Further, the law exists in the public domain as elected judges play a role in the judicial system, and elected officials represent constituents when creating legislation [42]. Public health authority, which usually refers to the role that state and local public health officials exercise, has been challenged and curtailed throughout the COVID-19 pandemic. More than 20 states have passed laws limiting the power of public health authorities to enact protective measures such as mask-wearing, quarantines, and vaccine mandates—often shifting final authority to an elected official [43]. This shift in authority, although possibly reflective of the public's polarizing opinions, may have unintended consequences during the next public health emergency. It is important to recognize that unbiased and evidence-based law and policy are pivotal tools in public health and that strategic approaches to political sentiments must be part of any public health response.

5   A Look at Global Public Health Law

Disease itself does not know borders. Global health law includes the study of legal norms and processes as well as the participation of institutions to create conditions for people across the world that will increase their potential for optimal health. Global health law and policy go beyond national public health law to regulate transnational behavior to reduce the morbidity and mortality of global norms and infectious diseases. International bodies, national actors, international legal instruments, and human rights law, compose most of the field today. The process of international lawmaking relies heavily on treaties. While there is no supranational compliance or enforcement, nations must implement treatise through their domestic policy, and international organizations can foster cooperation. However, because of increased globalization, nations often turn to international cooperation to protect their domestic health, which includes wider use of international legal instruments, such as the World Health Organization's (WHO) International Health Regulations (IHR) [44]. The IHRs were substantially revised in 2005 after the outbreak of severe acute respiratory syndrome (SARS) to heighten the surveillance of international health threats and contain diseases within national borders [45].

WHO also uses international conventions and negotiates treatises, where all member countries must be present. The first and only adopted WHO international convention was the Framework Convention on Tobacco Control (FCTC), adopted in 2003 [45]. Although the U.S. did not ratify the FCTC, it still successfully created binding norms to reduce the demand for tobacco products by influencing taxing, pricing, the creation of smoke-free environments, warning labels on packaging, marketing bans, and treatment programs. As of October 2015, 180 countries have ratified the FCTC, and it is one of the most widely subscribed-to treaties in the history of the UN [44].

Although global health law is used as a mechanism to create a healthier world through the cooperation of states, human rights law projects a framework for protecting the rights of individuals. WHO's Constitution was the first international declaration of a right to health and has been found in the core instruments of international human rights law, including the International Bill of Rights, which consists of the Universal Declaration of Human Rights (1948), the International Covenant on Economic, Social and Cultural Rights (1966) (ICESCR), and the International Covenant on Civil and Political Rights (1966). The right to health focuses on individual rights versus the usual focus on community or public health. Tension also existed between the right to health and the exercise of other human rights, such as liberty or privacy. Health and human rights have expanded under the UN and other international organizations. Legal instruments have been used to address several health concerns, such as HIV/AIDS, reproductive freedom, maternal health, refugee health, access to medications, and others [44].

5.1   Interventional Public Health Law: Two Selected Examples

Chronic Disease Prevention and the Law

Application: Hypothetical Case Study

Dr. Amos is a primary care clinician practicing medicine at a hospital in Wisconsin. Many of his patients suffer from diabetes, high blood pressure, and other endemic chronic health problems. As part of his practice, Dr. Amos asks his patients about their lifestyle (e.g., diet, exercise, sleep habits, etc.). Dr. Amos was alarmed to learn how many of his patients drink sugar-sweetened beverages (e.g., soda, juice) daily and provided his patients with information on the negative effects that the consumption of sugar-sweetened beverages can have on their overall health. He included in his patients' records and patient summary notes his recommendation to stop or reduce their daily consumption of sugar-sweetened beverages.

Dr. Amos knew that providing his patients with educational materials and recommending reduced consumption would only have a limited impact on his patients' overall health. Health is a combination of genetics, environment, and behavior, and while Dr. Amos can make recommendations, his patients have the free will to make their own choices. He knows his advice will only impact patients who choose to limit their consumption of sugar-sweetened beverages.

Dr. Amos then began to contemplate what else would aid his patients in making healthier choices. Could their environment and the availability of sugar-sweetened beverages change? He recalls having read an article about a New York City tax law on sugar-sweetened beverages and wonders:

- Could a state law or local ordinance be used as a primary intervention tool to prevent the overconsumption of sugar-sweetened beverages?
- How else could the law be used as a tool to reduce overconsumption of sugar-sweetened beverages and positively influence the beverages people choose to consume?
- Could law be used as a secondary intervention (e.g., mandatory diabetes screening) to address chronic diseases like those of the patients Dr. Amos sees in his practice?

Legal Framework

At every intervention level (primary, secondary, and tertiary), a legal authority can be used to prevent, control, and treat chronic diseases [46]. Legislation, statutes, regulations, and local ordinances have been used in various ways to intervene and reduce the impact of several chronic diseases. One example is how laws that have enforced smoke-free environments have reduced heart attack hospitalizations by 8–17% [47]. Another example is type 2 diabetes, which can be controlled by various methods ranging from the clinical or tertiary prevention level to the primary or societal level.

At the primary level, public health laws such as legislative and regulatory actions to prevent the use of sugar-sweetened beverages and reduce high caloric intake have been used at the local, state, and national levels (e.g., New York's soda tax and New York City's nutritional labeling of menu items). In New York, to reduce the consumption of sugary beverages, the state introduced legislation to include a sugary beverage tax in the budget for 2009 and 2010. However, due to aggressive lobbying efforts, this ultimately failed and was not included in the budget. The New York City Board of Health also passed a regulation in 2008 to require that all food establishments with more than fifteen locations nationwide publish the caloric information of all food products on their menus. Studies have shown this to decrease overall consumption; since then, over 19 other jurisdictions have adopted these laws [48].

Examples of other intervention levels include the secondary intervention level comprising preventative screening and the tertiary level comprising disease management via individual care, such as weight management and clinical care. Lastly, access to health insurance and affordable healthcare, such as expanding access to Medicare and Medicaid and the passage of the Affordable Care Act, have been additional legal interventions to assist those with chronic health conditions at the primary intervention level.

Law and Injury Prevention and Control

Application: Hypothetical Case Study

Dr. Baylee is a clinician who works in the emergency room of a Miami hospital. She recently treated a patient with severe bruising. As part of their training, all clinicians are educated on how to evaluate or treat a patient who is a victim of domestic violence. As such, Dr. Baylee knows her state law reporting obligations for suspected physical abuse. She is also aware of the atypical federal law enacted to prevent and address violence against women.

But as more and more patients present with signs of injury, she begins to wonder:

- How might those laws help women like her patient?
- What is the legal framework for addressing intentional and unintentional injuries in the U.S.?
- How has the law been used as a prevention tool for other injuries (e.g., motor vehicle safety, drug overdoses)?

Legal Framework

Clinicians like Dr. Baylee play an important role in maximizing patient awareness and education of injury prevention laws, some of which are federal but most of which vary considerably among the states.

Injuries, both unintentional (e.g., drug overdose) and intentional (e.g., interpersonal violence), contribute significantly to morbidity and mortality. Millions of nonfatal injuries and hundreds of thousands of deaths cause medical expenses and loss of productivity that totaled $1.8 trillion in 2013, and most of those injuries were preventable [48, 49].

Using public health data to assess, surveil, and develop policy and legal mechanisms to prevent injuries has been an area of interventional public health law. Several state health departments have full authority to surveil and develop injury prevention strategies and programs. For example, the North Carolina legislature has authorized that:

> "[The] Department [of Health and Human Services] shall establish and administer a comprehensive statewide injury prevention program. The Department shall designate the Division of Public Health as the lead agency for injury prevention activities. The Division of Public Health shall: (1) Develop a comprehensive State plan for injury prevention; (2) Maintain an injury prevention program that includes data collection, surveillance, and education and promotes injury control activities; and (3) Develop collaborative relationships with other State agencies and private and community organizations to establish programs promoting injury prevention" [50].

Through public health data and intervention strategies, public health laws and policies have been enacted to reduce the incidence of these injuries, and they have shown to be effective. One example of unintentional injuries and effective public health laws are laws regarding motor vehicle safety. Reducing motor vehicle injuries and deaths was one of the great achievements of public health in the 20th century, and the law was used as a tool to achieve this. Further, federal, state, and local governments have enforced laws regulating highway traffic, licensing, and vehicle inspections [51]. Currently, traffic laws are still important to prevent injury from driving under the influence of alcohol or speeding, as one in three fatal crashes in the U.S. involved drunk driving and almost one in three involved speeding [52].

Another example of an unintentional injury that has been addressed through law is opioid use disorder and overdose. Opioid use disorder involves addiction and misuse of prescription and non-prescription opioids [53]. The opioid epidemic was identified in the early 2000s after an increase in opioid prescriptions given in the 1990s. This was followed by a rise in deaths from heroin and fentanyl beginning in 2010 [54]. An increasing number of states have passed laws to limit the number of opioids medical professionals can prescribe, increasing from 10 states in 2016 to 39 states in 2019 [55]. Because of the increase in overdose deaths, it has become increasingly important for naloxone, an emergency treatment used to reverse the overdose, to be accessible and used promptly. Although originally only administered by first responders, because of its importance and ease of use, by 2017, all 50 states and the District of Columbia passed legislation to facilitate naloxone access by a layperson [56].

Additionally, overdose Good Samaritan laws allow bystanders to engage in emergency overdose assistance. By December 2020, 47 states and the District of Columbia enacted these laws, reducing deaths by approximately 15% [56]. Criminal or civil litigation has been used to hold the opioid market responsible and potentially redistribute funds to invest in opioid control and public health interventions [57].

Public health studies have shown that intentional injuries, including violence, are linked with risk factors and environmental, biological, and systemic stressors. When considering domestic violence, one in three women and one in four men will experience sexual assault, physical violence, or stalking by an intimate partner. Additionally, victims of domestic violence and/or rape are three times as likely to have a mental health condition [58].

At the federal level, the main federal laws addressing violence against women are the Violence Against Women Act (1994) and the Violence Against Women Reauthorization Act (VAWA) (2013), which includes free rape exams, protections for victims who are evicted or displaced from their homes, legal aid and civil protection for survivors, and much more [59]. At the state level, there is variation in the definition of domestic violence and the protections offered to victims of domestic violence. Approximately 38 states place domestic violence penalties within criminal codes and define it within the social services codes [60]. The effect of these laws on the incidence of violence against women and men has been noted. After the passage of VAWA in 1994, violence against females declined by 53% between 1993 and 2008, according to the Bureau of Justice Statistics [61, 62]. It is important to continue to create legal interventions to reduce the incidence of domestic violence against women and men.

6   Summary

As explored in this chapter, the law is both a foundational tool for public health practice and an interventional tool for addressing emerging public health concerns and improving health outcomes.

It is also important to note that public health law is not and cannot be just a domain for lawyers. The reality is that "public health laws are commonly conceived, promoted, administered, and evaluated by public health professionals and others without JD degrees" [63]. Today's public health challenges are daunting: an increasingly aging population; the burden of the social determinants of health leading to increased mortality and morbidity from chronic and other non-communicable diseases; national disasters; the opioid epidemic; the health impacts of climate change and environmental pollution; and the rapid transfer of infectious pathogens that create the potential for more and deadlier global pandemics. If we are to address these and other emerging challenges effectively, public health practitioners must understand the vital role of law in advancing health outcomes. For the modern public health practitioner, the law must be considered as fundamental to practice as surveillance, epidemiology, and other tools in the public health toolbox.

References

1. Federal preemption: A legal primer - fas [Internet]. [cited 2022 Jan 4]. Available from: https://sgp.fas.org/crs/misc/R45825.pdf.
2. Taylor A. Global health law: International law and public health policy [Internet]. Quah SR, editor. International Encyclopedia of Public Health. U.S. National Library of Medicine; 2017 [cited 2022 Jan 4]. Available from: https://www.ncbi.nlm.nih.gov/pmc/articles/PMC7150305/.
3. Administrative law. Legal Information Institute. Legal Information Institute; Available from: https://www.law.cornell.edu/wex/administrative_law.
4. Public Health Law Center [Internet]. Public Health Law Center | Public Health Law Center. [cited 2022 Jan 4]. Available from: https://www.publichealthlawcenter.org/sites/default/files/resources/Dillons-Rule-Home-Rule-Preemption.pdf.
5. Police Power [Internet]. Definition. [cited 2022 Jan 4]. Available from: https://www.merriam-webster.com/dictionary/police%20power.
6. Pertschuk M., Pomeranz J., Aoki J., Larkin M., Paloma M. [PDF] Assessing the impact of federal and state preemption in Public Health: A Framework for decision makers.: Semantic scholar [Internet]. undefined. 1970 [cited 2022 Jan 4]. Available from: https://www.semantic-scholar.org/paper/Assessing-the-impact-of-federal-and-state-in-public-Pertschuk-Pomeranz/4b929e8585c0b54a9d7d93502527f0efcdc0f30f.
7. Preemption issue brief - centers for Disease Control and ... [Internet]. [cited 2022 Jan 4]. Available from: https://www.cdc.gov/phlp/docs/preemption-issue-brief.pdf.
8. What is sovereignty? definition of sovereignty (Black's Law Dictionary) [Internet]. The Law Dictionary. 2011 [cited 2022 Jan 5]. Available from: https://thelawdictionary.org/sovereignty/.
9. State Legal Information Institute [Internet]. Legal Information Institute; Available from: https://www.law.cornell.edu/wex/state
10. U.S. Const. amend. X.
11. National Congress of American Indians [Internet]. Tribal Governance. Available from: https://www.ncai.org/policy-issues/tribal-governance.
12. Former New York City Mayor Michael Bloomberg in an address at the Harvard School of Public Health, 2009; Mayor Michael Bloomberg Receives Award - YouTube
13. Ransom, M. Introduction. Public Health Law: Concepts and Case Studies. Springer Publishing. 2021.
14. Moulton A., Mercer S., Popovic T., Briss P., Goodman R., Thombley M., Hahn R., Fox D. The Scientific Basis for Law as a Public Health Tool. *American Journal of Public Health.* 2009;99:17–24. [PMC free article] [PubMed] [Google Scholar]
15. Gostin L. Public health law in a new century. *Journal of the American Medical Association.* 2000;283:2837–2841. [PubMed] [Google Scholar].
16. *United States v. Lopez.* (93-1260), 514 U.S. 549 (1995).
17. *United States v. Morrison.* 529 U.S. 598 (2000).
18. 1984 National Minimum Drinking Age Act, [23 U.S.C. § 158].
19. *South Dakota v. Dole.* [483 U.S. 203 (1987)].
20. Laws, Policies and Regulations: Key Terms & Concepts. Tobacco Control Legal Consortium. Available from: tclc-fs-laws-policies-regs-commonterms-2015.pdf (publichealthlaw-center.org)
21. American Bar Association [Internet]. What is an Executive Order? Available from: https://www.americanbar.org/groups/public_education/publications/teaching-legal-docs/what-is-an-executive-order-/.
22. U.S. Const. Article II, Section 3.
23. Executive Order 13507, 2009; Executive Order 13544, 2010.
24. Ohio Executive Order 2020-01D [Internet]. Available from: https://governor.ohio.gov/wps/portal/gov/governor/media/executive-orders/executive-order-2020-01-d#:~:text=Executive%20Order%202020-01D%20March%2009,%202020%20Declaring%20a,and%20can%20easily%20spread%20from%20person%20to%20person.

25. Gakh M., Vernick J., Rutkow L. Using Gubernatorial Executive Orders to Advance Public Health. *Public Health Rep.* 2013;128(2):127-130.

26. Public Health Law 101 [Internet]. Centers for Disease Control and Prevention. Available From: https://www.cdc.gov/phlp/docs/phl101/PHL101-Unit-1-16Jan09-Secure.pdf

27. *Jacobson v.* Massachusetts. 197 U.S. 11 (1905).

28. Hoss A., Castagne M. Public Health Law and American Indians and Alaska Natives, Public Health Law: Concepts and Case Studies. 2021.

29. U.S. History.org [Internet]. The Bill of Rights. Available from: https://www.ushistory.org/us/18a.asp.

30. U.S. Const. amend. XIV.

31. U.S. Const. amend. V.

32. The Due Process Clause [Internet]. Available from: https://constitutioncenter.org/interactive-constitution/interpretation/amendment-xiv/clauses/701.

33. Kansas Archives - American Nonsmokers' Rights Foundation [Internet]. no. 2021. Available from: https://no-smoke.org/tag/kansas/.

34. Maine Pesticide Regulation. 01-026 C.M.R. ch.24, § 8 (2014).

35. The Clean Water Act (CWA). 33 U.S.C. §1251 et seq. (1972).

36. Public Health Law Center [Internet]. [cited 2022 Jan 5]. Available from: https://www.publichealthlawcenter.org/sites/default/files/resources/phlc-fs-state-local-reg-authority-publichealth-2015_0.pdf

37. Kemp v. Bottoms, No. 2020CV338387, (Ga. Super. Ct. July 16, 2020).

38. Kemp v. Bottoms unmasked: Emergency powers and state preemption [Internet]. Network for Public Health Law. 2020 [cited 2022 Jan 5]. Available from: https://www.networkforphl.org/news-insights/kemp-v-bottoms-unmasked-emergency-powers-and-state-preemption/

39. Preemption: Executive order by the ... - the reading room [Internet]. Available from: https://readingroom.law.gsu.edu/cgi/viewcontent.cgi?article=3051&context=gsulr

40. Vestal, C., Racism is a Public Health Crisis, Say Cities and Counties, June 15, 2020. Stateline Article. Pew Trusts. Available from: https://www.pewtrusts.org/en/research-and-analysis/blogs/stateline/2020/06/15/racism-is-a-public-health-crisis-say-cities-and-counties.

41. Bias T. Public health is always political: Think global health. Available from: https://www.thinkglobalhealth.org/article/public-health-always-political

42. Martin R. Law, and public health policy. International Encyclopedia of Public Health. 2008;30-39. Available from: https://www.ncbi.nlm.nih.gov/pmc/articles/PMC7150113/

43. Vestal C. New state laws hamstring public health officials. Available from: https://www.pewtrusts.org/en/research-and-analysis/blogs/stateline/2021/07/29/new-state-laws-hamstring-public-health-officials

44. Gostin L., Taylor A. Global health law: a definition and grand challenges. *Public Health Ethics.* 2008;1(1):53–63. Available from: https://www.ncbi.nlm.nih.gov/pmc/articles/PMC7150305/#bib2

45. Gostin L., Sridhar D. Global health and the Law. The New England Journal of Medicine. 2014;370:1732-1740; Available from: https://www.nejm.org/doi/full/10.1056/nejmra1314094.

46. Law as a tool for preventing chronic diseases: Expanding ... [Internet]. Available from: https://www.cdc.gov/phlp/docs/Prev_Chronic_Disease.pdf

47. Prevention [Internet]. Available from: https://www.cdc.gov/pictureofamerica/pdfs/picture_of_america_prevention.pdf

48. Kansagra S., Kennelly M., Nonas C., Curtis C., Van Wye G., Goodman A., Farley T. Reducing sugary drink consumption: New York City's approach [Internet]. American journal of public health. American Public Health Association; 2015 [cited 2022 Jan 5]. Available from: https://www.ncbi.nlm.nih.gov/pmc/articles/PMC4358191/

49. Lagasse, J. Non-fatal injuries cost healthcare $1.8 trillion annually, report shows [Internet]. Available from: https://www.healthcarefinancenews.com/news/non-fatal-injuries-cost-healthcare-18-trillion-annually-report-shows

50. G.S. § 130A-224. Department to establish program. - ncleg.gov [Internet]. [cited 2022 Jan 5]. Available from: https://www.ncleg.gov/EnactedLegislation/Statutes/PDF/BySection/Chapter_130A/GS_130A-224.pdf

51. Achievements in public health, 1900-1999 motor-vehicle safety: A 20th century public health achievement [Internet]. Centers for Disease Control and Prevention. Centers for Disease Control and Prevention; [cited 2022 Jan 5]. Available from: https://www.cdc.gov/mmwr/preview/mmwrhtml/mm4818a1.htm

52. Motor vehicle crash deaths [Internet]. Centers for Disease Control and Prevention. Centers for Disease Control and Prevention; 2016 [cited 2022 Jan 5]. Available from: https://www.cdc.gov/vitalsigns/motor-vehicle-safety/index.html]

53. Opioid use disorder [Internet]. [cited 2022 Jan 5]. Available from: https://www.psychiatry.org/patients-families/addiction/opioid-use-disorder/opioid-use-disorder

54. Understanding the epidemic [Internet]. Centers for Disease Control and Prevention. Centers for Disease Control and Prevention; 2021 [cited 2022 Jan 5]. Available from: https://www.cdc.gov/opioids/basics/epidemic.html

55. Lieberman A., Davis C. Laws limiting the prescribing or dispensing of opioids [Internet]. Network for Public Health Law. 2021 [cited 2022 Jan 5]. Available from: https://www.networkforphl.org/resources/laws-limiting-the-prescribing-or-dispensing-of-opioids/

56. (50 state survey) legal interventions to reduce overdose ... [Internet]. Available from: https://www.networkforphl.org/wp-content/uploads/2021/05/NAL-Final-4-29.pdf

57. Litigation against opioid manufacturers: Lessons from the ... [Internet]. [cited 2022 Jan 5]. Available from: https://www.networkforphl.org/wp-content/uploads/2020/01/3_15_2018-Litigation-Against-Opioid-Manufacturers.pdf

58. Domestic violence: A public health priority [Internet]. Grantmakers In Health. 2019 [cited 2022 Jan 5]. Available from: https://www.gih.org/publication/domestic-violence-a-public-health-priority/#_ENREF_6]

59. Laws on violence against women [Internet]. Laws on violence against women | Office on Women's Health. [cited 2022 Jan 5]. Available from: https://www.womenshealth.gov/relationships-and-safety/get-help/laws-violence-against-women#references

60. McCann M. [Internet]. Domestic violence/domestic abuse definitions and relationships. [cited 2022 Jan 5]. Available from: https://www.ncsl.org/research/human-services/domestic-violence-domestic-abuse-definitions-and-relationships.aspx

61. Modi M., Palmer S., Armstrong A. The role of violence against women act in addressing intimate partner violence: A public health issue [Internet]. Journal of women's health (2002). Mary Ann Liebert, Inc.; 2014 [cited 2022 Jan 5]. Available from: https://www.ncbi.nlm.nih.gov/pmc/articles/PMC3952594/#B20.

62. Catalano S., Smith E., Synder H., Rand M. Female Victims of Violence. In: Statistics UDoJBoJ, ed2009

63. Burris S., Ashe M., Levin D., Penn M., Larkin M. A transdisciplinary approach to public health law: The emerging practice of legal epidemiology [Internet]. Annual Reviews. [cited 2022 Jan 5]. Available from: https://www.annualreviews.org/doi/full/10.1146/annurev-publhealth-032315-021841

Part III

Access to Care

Emergency Medical Treatment and Labor Act (EMTALA)

Rachel A. Lindor and Summer Ghaith

Contents

Key Points

- EMTALA obligations apply to all Medicare-participating hospitals that have a dedicated emergency department.
- EMTALA prevents hospitals from refusing care to patients with emergency medical conditions.
- EMTALA protections apply to all individuals presenting to an emergency department for care, regardless of their insurance coverage status or ability to pay.
- EMTALA requires that hospitals provide all individuals presenting to an emergency department with (i) a medical screening examination to determine if an emergency medical condition exists, (ii) treatment to stabilize the patient, and (iii) transfer to another facility to stabilize the patient if the original hospital does not have the capability to do so.

R. A. Lindor (✉)
Department of Emergency Medicine, Mayo Clinic, Phoenix, AZ, USA
e-mail: Lindor.Rachel@mayo.edu

S. Ghaith
Mayo Clinic Alix School of Medicine - Arizona Campus, Phoenix, AZ, USA
e-mail: Ghaith.Summer@mayo.edu

- Enforcement of EMTALA is overseen by the U.S. Department of Health and Human Services through the imposition of civil monetary penalties and termination of hospital participation in the Medicare program; EMTALA also permits civil lawsuits of individuals harmed by violations of EMTALA.

Legal Concepts
- Emergency. Medical Treatment and Labor Act (EMTALA): A federal statute requiring Medicare-participating hospitals to screen any individual who presents to a hospital's dedicated emergency department for emergency medical conditions, stabilize the patient, if necessary, and transfer the patient if stabilizing treatment cannot be provided.
- Medicare-participating hospital: A hospital or critical access hospital that has entered into a Medicare provider agreement and accepts payment from the Centers for Medicare and Medicaid Services (CMS) for Medicare beneficiaries.
- Hospital: An institution that is primarily engaged in providing diagnostic, therapeutic, or rehabilitation services *to inpatients* by or under the supervision of physicians.
- Dedicated emergency department: Any hospital department or facility that meets any of the following conditions: (1) is licensed by the state as an emergency department, (2) is held out to the public (by name, posted signs, advertising, or other means) as a place that provides care for emergency medical conditions on an urgent basis without requiring a previously scheduled appointment, or (3) during a retrospective review of the previous calendar year, at least one third of outpatient visits were for treatment of emergency medical conditions on an urgent basis without requiring a previously scheduled appointment.
- Medical screening examination: An examination designed to determine if an emergency medical condition exists.
- Emergency medical condition: a condition which presents with acute symptoms of sufficient severity such that the absence of medical attention could be expected to place the health of the individual or an unborn child in serious jeopardy
- Stabilize: to provide medical treatment, with respect to an emergency medical condition, necessary to assure, within reasonable medical probability, there will be no material deterioration of the patient's condition that will occur from or during a transfer or discharge

1   Introduction

The Emergency Medical Treatment and Labor Act (EMTALA), also known as the patient "anti-dumping act", is a federal statute that was enacted in 1986 after several highly publicized incidents in which indigent individuals, including a number of pregnant women, were turned away from emergency departments (EDs) because they could not afford medical care. In short, the law requires Medicare-participating hospitals to provide care to any individual with an emergency medical condition who presents to the hospital's ED. While the legality and constitutionality of requiring hospitals and physicians to provide these services without compensation has been challenged multiple times, courts have repeatedly noted that EMTALA applies only to Medicare-participating hospitals, highlighting the fact that EMTALA is a voluntary program, just as Medicare participation is voluntary.

Hospital's EMTALA obligations are myriad, but the major tenets of the law can be broken down into three main requirements: First, EMTALA requires that hospitals provide an appropriate medical screening exam to any individual presenting to an ED. Second, if an emergency medical condition is found, hospitals must stabilize the patient prior to discharge or transfer. Finally, if the hospital does not have the capability to stabilize the patient, the hospital must arrange for transfer of the patient to a hospital that has both the capacity and capability to do so [1]. This chapter will provide further details on EMTALA's requirements, discuss the enforcement of EMTALA, and provide case examples.

2   Discussion

EMTALA has been augmented by regulations, interpretive guidelines, and case law that have added important nuance to the language included in the original law. Again, EMTALA requires that, at all *Medicare-participating hospitals*, *any individual* who *comes to* the *emergency department* be given *a medical screening examination*, and if an *emergency medical condition* is found, that person must be *stabilized* or, if the hospital does not have the *capability* to stabilize, the patient must be *transferred* [1]. The following section breaks down the language of the statute to better understand the scope of each of its individual terms.

2.1   "Medicare-Participating Hospitals"

EMTALA only applies to Medicare-participating hospitals, defined as those hospitals that have enrolled in a provider agreement and accept payment from the Centers for Medicare and Medicaid Services (CMS) for Medicare beneficiaries [2]. This may exclude some Veterans Administration hospitals, Native American reservation hospitals, and private psychiatric hospitals. Further, EMTALA applies only to hospitals, specifically those with an ED [2]. Hospitals are defined as institutions that are primarily engaged in providing diagnostic, therapeutic, or rehabilitation services *to inpatients* by or under the supervision of physicians. This definition excludes most sites that are devoted primarily to outpatient care, such a private physician offices or ambulatory surgical centers. "Emergency department" is defined more fully below.

2.2   "Emergency Department"

EMTALA applies to hospitals with a "dedicated emergency department," which (counterintuitively) is a broader term than "emergency department." A dedicated

ED includes one that meets any of the following definitions: (1) is licensed by the state as an ED, (2) is held out to the public as a place that provides healthcare for emergency medical conditions on an urgent basis and does not require an appointment, and (3) during a retrospective, representative review of the previous calendar year, at least one third of outpatient visits were for treatment of emergency medical conditions without requiring an appointment [2, 3]. Regarding the second criteria, being "held out" to the public as an ED means that it has been marketed by name, has posted signs, or has engaged in other advertising that indicates it provides emergency medical care. By this definition, any hospital that has an urgent care center, psychiatric intake, a labor and delivery department, or an ambulatory care clinic can be subject to EMTALA, sometimes unexpectedly.

2.3   "Any Individual"

EMTALA states that any individual, not just a Medicare beneficiary, who presents to the ED must be evaluated. This language applies to all people, regardless of age, insurance status, immigrant status, national origin, disability, race or ethnicity, and includes minors, intoxicated individuals, psychiatric patients, and prisoners [4]. Notably, EMTALA mandates that minors be given a medical screening exam and stabilizing treatment even in the absence of guardian consent [4]. Further, according to the Born-Alive Infants Protection Act, if an infant is born alive on hospital property, EMTALA requires the hospital to assess and stabilize that infant [5].

2.4   "Comes To"

EMTALA applies only to individuals who have "come to" an ED. Again, the meaning of this phrase has been expanded considerably since enactment of the statute. By current law, an individual "comes to" an ED and triggers EMTALA if any of the following four scenarios are met:

- Individuals present for care on hospital property within 250 yards of the main building, ambulances, or anywhere that provides acute care (e.g., labor and delivery unit) and hospital staff are informed of their presence;
- Individuals present to the ED and request care, *or* a prudent layperson would have reason to believe they may need an emergency examination;
- Individuals are in a hospital-owned and -operated ambulance (ground or air), even if the ambulance is not on hospital property; or
- Individuals are in a non-hospital-owned ambulance (ground or air) but have arrived on hospital property [2, 4, 6].

Many EMTALA violations have resulted from hospitals not appreciating the expanded definition of "comes to" as outlined above. Refer to the application section to see a detailed example.

2.5   "Medical Screening Examination"

EMTALA mandates that a dedicated ED perform a medical screening examination (MSE) that is aimed at determining with reasonable confidence if a patient has an emergency medical condition [4]. An MSE is not a uniform procedure applied to every patient, but rather includes the entire care process needed to determine whether an emergency medical condition exists. For some patients, the MSE requires only a brief history and exam, while for others it may require a long evaluation and a host of ancillary tests. For example, for a patient with ear pain, the MSE may take several seconds to determine that no emergency medical condition exists. For a patient with chest pain, the MSE may take several hours in the ED and admission to the hospital for serial monitoring or additional tests. In general, EMTALA is satisfied if the screening exam provided is comparable for similar patients and does not discriminate based on ability to pay or other factors. While the MSE requirements for an individual patient are difficult to define, delays in screening patients have repeatedly been held out as EMTALA violations, as the delay itself has been deemed a deviation from the standard MSE required by EMTALA. Similarly, departures from written guidelines and protocols in the ED have been used to buttress claims of EMTALA violations, as these are often held out as synonymous with departures from MSEs. The statute also requires for the MSE to be performed by a healthcare provider who is trained to deal with emergency situations. [7] While this provider does not need to be a physician, EMTALA deems that only persons who are specifically designated for this role by hospital policies as qualified.

2.6   "Emergency medical condition"

EMTALA defines an emergency medical condition as one in which the absence of immediate medical attention could be expected to place the health of the individual or an unborn child in serious jeopardy or result in serious bodily dysfunction [8]. Patients expressing suicidal or homicidal thoughts qualify as having an emergency medical condition under EMTALA [3].

2.7   "Stabilize"

The requirement to stabilize a patient is activated once an emergency medical condition has been found. If no emergency medical condition is identified, stabilization is not required and EMTALA has been satisfied [3]. "To stabilize" is defined under EMTALA as an obligation to provide medical care so that, within reasonable probability, there will be no material deterioration of the patient's condition that will occur from or during a transfer or discharge [9]. Regarding labor, stabilization requires delivery of both the child and the placenta, or the cessation of labor [10]. Patients can be stabilized either in the emergency department to which they initially presented, through admission to the original hospital, or—if the original hospital lacks the capability or capacity, or the patient requests it—by transfer to an outside hospital.

If a physician finds a patient stable for discharge or transfer, and the patient deteriorates afterwards, this is not interpreted as a *de facto* EMTALA violation. As long as the initial determination of stability was reasonable and in good faith, as evidenced by the provider's documentation and the patient's clinical course, this scenario is unlikely to be considered an EMTALA violation. Furthermore, once a patient is admitted to the hospital, EMTALA no longer applies as long as there is no evidence that the admission was in bad faith (e.g., in an effort to avoid EMTALA obligations). Thus, claims that an inpatient was discharged without being stabilized do not fall under EMTALA [11].

2.8   Capability

EMTALA provides that hospitals that do not have the capability to stabilize a patient can transfer that patient to another hospital. CMS defines the capabilities of a hospital based on the ancillary services available at the hospital. Therefore, hospitals cannot elect to make certain services such as obstetrics, ophthalmology, or interventional radiology unavailable to the ED. If the services are available to inpatients, they must be available to the ED as well.

2.9   Transfers

The EMTALA statute defines transfers as any movement of an individual outside of a hospital's facilities to another facility [3, 12]. An unstable patient may be transferred only if the hospital does not have the capacity or capability to adequately stabilize a patient, and the benefits of being treated at the receiving hospital outweigh the risks of transfer. An unstable individual may also be transferred if they request to be transferred even after being informed of the hospital's obligation under EMTALA and the risks associated with the transfer. This conversation and risk discussion should be clearly documented in the patient's chart [12, 13].

To conduct an appropriate transfer, the transferring hospital and receiving hospital must abide by the following outlined obligations: First, the transferring hospital must weigh the risks and benefits of transfer and only transfer if necessary. Second, the transferring hospital must find a receiving hospital that has the special capabilities needed for the patient. Third, the transferring hospital must provide medical treatment to decrease risks associated with transfer. Fourth, they need to send the receiving hospital the patient's medical record including test results as they become available [10]. Lastly, the transferring hospital must fill out the EMTALA transfer certificate which requires information regarding the patient condition, the reason for transfer, the benefits and risks of transfer, the receiving facility and physician information, the mode of transportation, and the patient consent [10].

The receiving hospital must accept a transfer if they have the physical capacity, special capabilities, and qualified personnel needed to treat the patient. The receiving hospital does not need to have a dedicated ED to accept a transfer, but only needs to have the services required to treat the patient [14]. The receiving hospital must document either an acceptance or a rejection certificate as appropriate. Once a transfer has been accepted, the receiving hospital has an obligation to provide appropriate medical assistance to the patient [15]. Once a patient has arrived at the receiving hospital, the EMTALA obligations are no longer on the transferring hospital and instead are placed on the receiving hospital. However, if a receiving hospital refuses to accept an appropriate transfer despite having the necessary resources for the patient, they have violated EMTALA and can be cited for such actions. This action is referred to as "reverse dumping" [16].

2.10   On-call Physicians

Medicare-participating hospitals are required to have a list of on-call physicians to provide necessary treatment to stabilize patients with emergency medical conditions [20]. Obligations of on-call physicians include responding to a call in a timely manner and providing a medical screening exam and stabilizing treatment. If an on-call physician is requested by an emergency physician to appear at the hospital and fails or refuses to respond in a reasonable amount of time, the on-call physician can be held liable for violating EMTALA [22]. In other words, under EMTALA, an on-call physician who never sees or examines a patient can be held liable for failing to appear and provide necessary stabilizing treatment. EMTALA does not define what is considered a reasonable amount of time for an on-call physician response and instead leaves this to the hospital's discretion. These timelines are typically set forth in hospital policies and may differ depending on the urgency of the medical condition or the setting of the hospital.

2.11   Enforcement Mechanisms

The U.S. Department of Health and Human Services (HHS) is responsible for the enforcement of EMTALA [17]. EMTALA obligations are part of the Medicare program voluntary requirements that are required for Medicare-participating hospitals. Enforcement is provided by three different avenues. First, CMS, the agency responsible for the federal Medicare program, can terminate provider agreements with the Medicare program [18]. Second, the Office of Inspector General (OIG) may administer civil monetary penalties [19]. Third, enforcement can be provided by civil lawsuits brought by individuals harmed by violations of EMTALA [20]. On first offense, a hospital will receive a civil monetary penalty. Then, if such action is flagrant or repeated, their participation in the Medicare program can be terminated [4].

Enforcement of EMTALA and discovery of violations is a heavily complaint driven process. Any individual, including a patient, may initiate a complaint.

Alternatively, hospitals (often those that received an improper transfer) may initiate a complaint. Specifically, hospitals that receive an inappropriate transfer are required to report to CMS or the State survey agency within 72 h of the incident. Failure to report improper transfers may lead to termination of the receiving hospital's provider agreement. [4] Finally, hospital staff may self-report and state surveyors performing licensure or recertification surveys may file a complaint. EMTALA does provide whistle-blower protection for reporters and allows the complainant to remain anonymous through the process.

Once a complaint has been filed against a hospital, an investigation begins to assess whether a violation has occurred. Investigations vary regionally, as each of the ten CMS regional offices are responsible for investigating complaints in their region, and they often delegate these activities to local state survey agencies or quality improvement organizations. Regardless of who carries out the investigation, their actions are guided by the CMS State Operations Manual. In brief, investigators meet with hospital personnel to make them aware of the investigation and request access to any relevant documentation, which may include patients logs for up to 12 months, patient medical records, ED policies, transfer consent forms, staffing schedules, ambulance trip reports, provider credentials, on-call schedules, and meeting minutes from the past year. Investigators may also interview facility staff, or anyone determined to be a potential witness[4].

Once the investigation is completed, investigators may recommend a number of potential actions. They may find the complaint unsubstantiated and recommend that no action be taken. Similarly, they may find that the hospital was out of compliance, but the issue giving rise to the non-compliance has already been corrected. In this case, no corrective action is required but the case can be referred to the Office of the Inspector General for consideration of civil monetary penalties. Finally, investigators may request additional review of cases from a physician before making a determination, and they may highlight suspected cases of discrimination [21]. On average, CMS conducts around 500 EMTALA investigations per year, and approximately half of these result in findings of a violation [22].

Two possible administrative penalties exist for EMTALA violations—assessment of civil monetary penalties and termination from Medicare participation. Notably, both hospitals and physicians can face these penalties. Historically, less than 10% of EMTALA violations were accompanied by a civil monetary penalty [23]. The maximum penalty allowed has changed in recent years and is currently just over $100,000 per patient for hospitals with more than 100 beds and half of that for smaller hospitals. The largest civil monetary penalty, approximately $1.3 million, was assessed in 2017 to a South Carolina hospital chain for inappropriate boarding of psychiatric patients at 36 different locations. Assessment of civil monetary penalties against physicians is not common, with only 8 published cases from 2002 to 2015 [23].

When terminating a provider agreement, CMS may follow a 23-day track or a 90-day track depending on the severity of the violation. In both cases, termination can be forestalled by implementing changes that address the violation. Notably, hospitals and physicians have no pre-termination appeals rights under EMTALA; they must either attempt to demonstrate that corrective action has been taken or proceed with termination. Once terminated, the hospital or physician can appeal, but they cannot receive any money from CMS during the appeals process, which often takes multiple years.

Finally, individuals harmed by an EMTALA violation can bring civil lawsuits, such as personal injury claims, against hospitals but not physicians. The above-outlined limits on monetary penalties do not apply in these cases, potentially exposing hospitals to much higher penalties. These civil lawsuits are often brought in conjunction with medical malpractice claims.

2.12   Public Emergencies

In public emergencies, hospitals may not refuse to treat a patient that may be suffering from a condition just because it could potentially place others at risk [24]. During the COVID-19 pandemic, CMS released clarifications and changes to EMTALA guidelines specific to public emergencies to inform EDs on how to manage their patients. First, CMS clarified that the use of telehealth to screen individuals that did not physically appear in the ED does not activate EMTALA or create any obligations under EMTALA [25]. Second, patients who present to the ED solely for the purpose of COVID-19 testing do not activate EMTALA [25]. Third, medical screening examinations can be provided to patients while they remain in their vehicles [25]. These guidelines highlight the adaptive nature of the EMTALA requirements, and the need for ongoing awareness of the law's scope.

3   Application

Case Example: "Medicare-Participating" A man presented in respiratory distress to an emergency department on a Native American reservation, was refused care and redirected to a hospital 10 miles down the road, and subsequently died before he could reach the neighboring hospital. When his family sued under EMTALA, the courts agreed that EMTALA did not apply as that particular emergency department did not participate in Medicare [26].

Case Example: "Hospital" A man presented to a private physician clinic with complaints of chest pain. An EKG was performed, which was interpreted as normal, and the patient was diagnosed with a gastrointestinal issue and sent home. He died two days later due to a myocardial rupture from a missed myocardial infarction. His family tried to bring a case against the clinic under EMTALA, as they felt the screening exam was inadequate. The courts ruled, however, that the patient was not cared for in a hospital, as defined by the statute, and therefore EMTALA did not apply [27].

Case Example: "Any Individual" A woman presented to the hospital in labor and gave birth to an infant. The doctor examined the infant briefly at birth but then did not evaluate the infant again until an hour and a half later, at which point the baby was in severe distress from birth trauma and an undiagnosed anemia and subsequently died. The parents brought a lawsuit under EMTALA for failure to provide a medical screening exam and failure to appropriately transfer the infant to a higher level of care. The hospital was ultimately found at fault under EMTALA, in accordance with the Born-Alive Infants Protection Act [28].

Case Example: "Comes To" In one well-publicized case, a 15-year-old boy who had been shot in the abdomen attempted to reach the ED with the help of his friends. About 35 feet from the ED entrance, he collapsed to the ground, and his friends ran inside to get help for him. The hospital staff refused to retrieve the patient. After some delay, he was ultimately escorted inside by police officers and subsequently died. When CMS evaluated this case for an EMTALA violation, the hospital argued that the boy did not ever "come to" the ED because he never made it inside, and thus its EMTALA obligations were not trigged. CMS rejected this interpretation of the statute and subsequently issued regulations to eliminate this and similar types of cases in which the letter, but not the intent, of the law was being followed. Changes based on this, and similar cases are reflected in the expanded definition of "comes to" the ED, as explained above [29].

Case Example: "Emergency Department" A woman developed burning epigastric pain and right arm pain. She presented to an urgent care that was part of a local hospital, was evaluated by a physician, and ultimately discharged home with a diagnosis of gastric reflux. The next day, she was found dead at home, and an autopsy identified the cause of death as myocardial infarction. In a subsequent EMTALA claim, the hospital tried to argue that EMTALA did not apply, as the patient presented to an urgent care rather than an ED. However, the court disagreed, noting that the choice of the word "urgent" in the name of the clinic suggested that the clinic was trying to attract patients with urgent *and* emergent needs. Second, staff members of the urgent care clinic who were interviewed acknowledged that they knew there was always a risk a patient with a true emergency would present to them. The urgent care clinic again tried to defend itself by pointing out that its website said specifically that the clinic is intended for patients with any "non-emergency needs" and for people who "don't need emergency room care." However, the court rejected this defense and pointed out that the website was not readily available to people driving by and seeing the signage for an "urgent care clinic" and therefore the clinic was still holding itself out in a way that met EMTALA's definition of a "dedicated emergency department" [30].

Case Examples: "Medical Screening Examination" A 65-year-old female presented to an ED with complaints of nausea, vomiting, and chest pain. The ED had a written protocol in place calling for rapidly assessing and assigning these patients to a physician. However, the patient waited for several hours without evaluation. Eventually, her family opted to take her to a nearby physician's office, where she died. The family alleged an EMTALA violation due to failure of the hospital to provide a medical screening exam, and the court agreed, calling the hospital's delay in assessment "constructive dumping" of the patient. In this case, the failure to follow the internal protocols provided ready evidence for the family's claims that the patient did not receive a screening exam comparable to other patients. Along with an EMTALA violation, the hospital was found guilty of malpractice and a jury awarded $700,000 in damages [31, 32].

In a different case, a young man presented to the ED after being involved in a motorcycle accident. After extensive laboratory and radiographic evaluation, he was admitted to the hospital's intensive care unit. At some point after admission, he was diagnosed with a pneumothorax as well as a renal artery laceration that resulted in loss of one of his kidneys. The patient brought an EMTALA claim against the ED, alleging that it failed to administer an appropriate screening examination, and failed to identify his emergency medical condition. However, there was no evidence to suggest that the screening examination was unreasonable or any different than the screening examinations provided to others, and therefore, the failure to identify the emergency medical condition was not an EMTALA violation (though could still be brought forward as a malpractice claim) [33].

Case Example: "Emergency Medical Condition"   A man with a history of bipolar disorder and a previous suicide attempt was brought to the ED by his wife due to concern that his depression was worsening and he required inpatient treatment. The emergency physician disagreed and discharged the patient within several hours of his arrival. The patient went home and committed suicide that same day. His wife subsequently filed an EMTALA claim, alleging failure to address his emergency medical condition. While EMTALA did not originally address mental illness as emergency medical condition, it now specifically includes both suicidality and homicidality as EMCs requiring stabilization, largely due to cases similar to this [34].

Case Example: "Stabilize"   A woman with longstanding asthma presented to an ED due to shortness of breath. She was slightly tachycardic but had otherwise normal vital signs, so she was triaged to the fast-track area of the ED, which was managed by a nurse practitioner. The nurse practitioner ordered nebulizers and steroids but left his shift prior to the completion of the patient's treatments and pre-signed her discharge instructions. The patient was discharged after receiving her treatments, with no re-evaluation documented or vital signs prior to discharge. Despite reportedly being visibly short of breath, she was taken home by her fiancé, where she continued to worsen and suffered a respiratory arrest within 2 h of arriving back at home. She was transported to another ED where resuscitation efforts resulted in her being left in a permanent vegetative state. Although she received an initial exam by the nurse practitioner and multiple medications, the court agreed with her family's argument that the hospital failed to meet its EMTALA obligations for stabilization. The patient's quick demise, along with the lack of documentation by the nurse practitioner, supported this accusation, and the hospital was held accountable for a verdict of more than $2 million as well as an EMTALA violation [35].

Case Examples: Transfers A man was diagnosed with internal bleeding at a rural ED. The rural hospital did not have any blood available to transfuse him and estimated that it would take them at least 12 h to acquire any. The hospital made the decision to transfer the patient to a hospital with additional capabilities, but unfortunately, the patient died in route from hemorrhagic shock. The hospital and physician argued—successfully—in court that the risks of the transfer, which would take less than 12 h, outweighed the risks of staying in the hospital and waiting 12 h to receive blood. Despite the patient's death during transfer, the hospital did not violate EMTALA [36].

In a separate case, a man was transported to a local hospital after being involved in a car accident. The emergency physician determined that he had fractures of his spine and ribs as well as paralysis of his legs and an aortic injury. However, the original hospital did not have the resources to treat this major trauma, so the physician arranged for the patient to be transferred to the nearby university hospital. After the transport helicopter arrived, the university hospital informed him that they did not have the capacity to take care of the patient as all the operating rooms were currently in use. The emergency physician then attempted to transfer to a different hospital, but the on-call thoracic vascular surgeon refused the transfer despite having the capabilities and capacity for the patient. Eventually, the patient was transferred to a third neighboring hospital, where he deteriorated despite aggressive care and subsequently died. The Inspector General asserted that the thoracic surgeon violated EMTALA by refusing to accept a medically necessary and appropriate transfer despite his hospital's specialized capabilities and capacity, thus engaging in "reverse dumping." This case exemplifies the role of the receiving hospital in a transfer under EMTALA [37].

Case Examples: On-call Physicians A general surgeon was scheduled to be on-call for a local ED. However, he wanted to attend an event in a neighboring town, so he arranged for an orthopedic surgeon to cover his call, with no notice to the hospital. A patient was brought to the ED that day with multiple injuries from a car accident. The orthopedic surgeon evaluated the patient, determined that she needed surgery that required a general surgeon, and recommended transfer to another hospital, resulting in a 4-hour delay of her care. The state court opined that an on-call physician must be available to treat patients within a reasonable time period, notify the hospital if they are unavailable beforehand, or supply a replacement and upheld the EMTALA violation based on the on-call physician's lack of responsiveness in this case [38].

4   Summary

In conclusion, EMTALA requires Medicare-participating hospitals to provide a medical screening examination to any individual who presents to a dedicated ED, provide stabilizing treatment if an emergency medical condition is found, and provide a transfer if the hospital does not have the capability to stabilize the patient. Hospitals and physicians can avoid liability under EMTALA both by understanding the law's requirements and intent, but also by giving patients in need the benefit of the doubt when it comes to providing care.

References

1. 42 U.S.C. § 1395dd (2020).
2. 42 C.F.R. § 489.24(b) (2013).
3. Medicare and Medicaid Programs; Hospital Conditions of Participation; Provider Agreements and Supplier Approval, 62 Fed. Reg. 66,726, 66,741-42 (December 19, 1997).
4. U.S. Dep't HHS, CMS, State Operations Manual, App. V. Emergency Medical Treatment and Labor Act (EMTALA) Interpretive Guidelines, Part II, Tag A-2406 (revised 7/16/2010).
5. Born-Alive Infants Protection Act (BAIPA) of 2002, 1 U.S.C. § 8 (Supp. III 2003)).
6. 42 C.F.R. § 413.65(a) (2017).
7. 42 C.F.R. § 489.24(a)(i) (2013).
8. 42 C.F.R. § 1395dd(e)(1)(B).
9. 42 U.S.C. § 1395dd(e)(3)(B).
10. U.S. Dep't HHS, CMS, State Operations Manual, App. V, Emergency Medical Treatment and Labor Act (EMTALA) Interpretive Guidelines, Part II, Tag A-2407/C-2407 (revised 5/29/2009).
11. 42 C.F.R. § 489.24(a)(ii) (2013).
12. U.S. Dep't HHS, CMS, State Operations Manual, App. V, Emergency Medical Treatment and Labor Act (EMTALA) Interpretive Guidelines, Part II, Tag A-2411/C-2411 (revised 5/29/2009).
13. 42 U.S.C. § 1395dd(c) (2020).
14. 42 U.S.C. § 1395dd(c)(2)(B).
15. 42 U.S.C. § 1395dd(g).
16. 42 C.F.R. § 489.20(r)(2) (2018).
17. 42 C.F.R. § 479.24(g).
18. 42 U.S.C. § 1395dd(d)(1)(A).
19. 42 U.S.C. § 1395dd(d)(2).
20. 42 U.S.C. § 1395dd(d)(1)(B).
21. U.S. Dep't HHS, CMS, State Operations Manual, Chapter 5, Complaint Procedures, Sect. 5450.
22. U.S. Gov't Accountability Office, GAO-01-747, Emergency Care – EMTALA Implementation and Enforcement Issues (June 2001).
23. Zuabi N, Weiss LD, & Langdorf MI. Emergency Medical Treatment and Labor Act (EMTALA) 2002-15: Review of the Office of Inspector General Patient Dumping Settlements. West J Emerg Med. 2016; 17(3):245-51.
24. Thomas Hamilton: Emergency Medical Treatment and Labor Act (EMTALA) Requirements and Implications Related to Ebola Virus Disease (Ebola) (2014), https://www.cms.gov/Medicare/Provider-Enrollment-and-Certification/SurveyCertificationGenInfo/Downloads/Survey-and-Cert-Letter-15-10.pdf (last visited Sep 17, 2021).

25. Emergency Medical Treatment and Labor Act (EMTALA) Requirements and Implications Related to COVID-19 (Revised) | CMS, Cms.gov (2021), https://www.cms.gov/medicareprovider-enrollment-and-certificationsurveycertificationgeninfopolicy-and-memos-states-and/emergency-medical-treatment-and-labor-act-emtala-requirements-and-implications-related-covid19 (last visited Sep 18, 2021).
26. Williams v. United States. 242 F. 3d 169 (4th Cir. 2001).
27. King v. Ahrens. 16. F. 3d. 265, 270-1 (8th Cir. 1994).
28. Allen v. Clinton Hma, LLC, CIV-19-0527-HE, 2019 WL 10091138, at *3 (W.D. Okla. Sept. 16, 2019).
29. Sercye-McCollum v. Ravenswood Hosp. Med. Ctr., 140 F. Supp. 2d 944 (N.D. Ill. 2001).
30. Friedrich et al v. South County Hospital Healthcare System et al, No. 1:2014cv00353 - Document 59 (D.R.I. 2016)
31. GONZALEZ, ESTATE OF v. HOSPITAL SAN FRANCISCO, JVR No. 164534
32. Correa v. Hosp. San Francisco, 69 F.3d 1184 (1st Cir. 1995).
33. Green v. Reddy. 918 F. Supp. 329 (D. Kan. 1996).
34. Estate of Siemen v. Huron Med. Ctr., 11-11249-BC, 2012 WL 909820 (E.D. Mich. Mar. 16, 2012).
35. McGhee vs. Confidential, 45 Trials Digest 8th 9, 2005 WL 3030929.
36. Thompson v. Southwest Health Systems, Inc. et al, case number 1:11-cv-00947
37. The Inspector Gen. v. St. Anthony Hospital, DAB CR620 (1999) (Oct. 5, 1999).
38. Millard v. Corrado, 14 S.W.3d 42, 44 (Mo. App. E. Dist. 1999).

Healthcare Reform

Kate Johansen

Contents

Key Points
- Health law is governed by federal and state statutes and regulations. This federalist structure makes reform efforts more complex.
- Most health reform efforts seek to expand coverage, improve access to care, make coverage or care more affordable, and/or drive innovation.
- State health reforms often lay the foundation for federal reforms or reforms in other states. State reforms are typically achieved through state statutory changes and/or waivers coordinated with federal agencies.
- The Affordable Care Act (ACA) is the largest health reform law in U.S. history. It has survived multiple legal challenges and has reshaped the healthcare landscape since its passage in 2010. The ACA's main components are the individual mandate, the establishment of health insurance exchanges, a variety of consumer protections, and Medicaid expansion.
- Health reform is legally and politically complex. Broadscale health reform is regarded as one of the most challenging legislative endeavors. The challenge

K. Johansen (✉)
Department of Public Affairs, Mayo Clinic, Rochester, MN, USA
e-mail: Johansen.katherine@mayo.edu

arises both from the complexity of healthcare and the multiple stakeholders involved, often with competing interests, who seek to shape reform efforts.
- Broad, comprehensive health reform is rarely achieved. More often, health reform occurs incrementally.

Legal Concepts
- Historically, health insurance in the United States was regulated at the state level. The federal government explicitly reserved this power to the states in the 1945 McCarran-Ferguson Act. Subsequent health reforms have expanded federal authority over insurance coverage and/or augmented state oversight with federal requirements.
- Health coverage is a tax-exempt employee benefit. As a result, health coverage in the United States is strongly linked to employment and is governed in large part by federal employee benefits law.
- Medicare and Medicaid were established to provide coverage to the elderly and low-income Americans respectively, neither group of which had strong access to employer-sponsored insurance.
- The ACA rests on a legal framework that requires nearly all people to obtain coverage under the individual mandate but also provides access to quality, affordable coverage through Medicaid expansion and government-run health insurance exchanges.
- To satisfy the ACA's individual mandate, a person must obtain minimum essential coverage, which must meet certain benefit and affordability requirements. Public program coverage, employer-sponsored coverage, or plans sold through ACA exchanges all meet this requirement.
- The ACA has faced several legal challenges but remains in effect. The most significant legal challenge was *NFIB v. Sebelius* in 2012, in which the U.S. Supreme Court upheld the individual mandate—and therefore the rest of the law—but reshaped the Medicaid expansion requirement.
- The ACA is a "floor, not a ceiling." States may exceed the law's requirements but cannot decline to meet them. State Innovation Waivers (or 1332 waivers) are a key tool for state reforms.

1   Introduction

Health reform is a massive topic, rife with complexity and contradiction. It spans federal, state, and local jurisdictions. It derives from thousands of pages of statutes, hundreds of thousands of pages of regulations, and countless volumes of case law. It touches multiple public and private stakeholders operating in an industry that comprises just under 20% of the U.S. economy [1]. And, of course, it represents a deeply personal issue for hundreds of millions of Americans—healthcare.

For the purposes of this chapter, health reform is defined as the process by which federal and/or state governments create policies that impact healthcare by establishing statutes, regulations, or case law. To better understand health reform and its underlying legal concepts, this chapter provides a brief history of the health reform efforts that created the current U.S. healthcare system. It then explores the Affordable Care Act (ACA), the largest and most recent U.S. health reform, discussing its passage, policy framework, implementation, and challenges to the law. Relevant case studies and major legal challenges are also included at the end of the chapter.

2   Discussion

2.1   A Brief History of U.S. Health Reform

Any discussion of health reform requires an historical foundation—how did the U.S. healthcare system evolve to its current form? The U.S. Constitution does not mention healthcare. Early state and federal laws did not focus on healthcare. Indeed, American healthcare as known today did not begin to take shape until the 20th century. Therefore, not surprisingly, U.S. health law from the 18th and 19th century is somewhat limited. As European countries began to transition to public, nationally centralized health systems in the 1800s, the United States maintained through the early 1900s that healthcare was a private issue to be overseen mostly at the local or possibly state level.

As a result, the U.S. healthcare system built a strong private sector foundation prior to the passage of any major national public health reforms. That foundation reflects the convergence of two historical shifts in the early 1900s. First, medical and technological advancements in the 1800s made healthcare more effective and increasingly required procedures and/or hospitalization. This led to professionalization of medicine and a proliferation of care facilities—it also began to increase healthcare costs [2]. Second, physical well-being was closely linked to economic security, as most workers held jobs in factories or agriculture that demanded intense physical labor. This led to the establishment of workers' compensation coverage, a precursor to modern health insurance [3]. Around the same time, Blue Cross and Blue Shield began developing pre-paid healthcare plans for hospital and physician services, respectively. Thus, in the early 20th century, the American healthcare system developed both private care facilities and financing mechanisms, creating stakeholders and a structure that would shape health reform efforts over the next century through today.

Attempts to establish a public national health system ran in parallel as the private health system emerged. For example, in 1912, former President Teddy Roosevelt endorsed health insurance access as part of his platform [4]. Two decades later, President Franklin D. Roosevelt advanced proposals to establish national health insurance within New Deal legislation. But, citing Congressional opposition, he dropped the provision from the Social Security Act to avoid risking the Act's passage. Proposals to establish national health insurance continued throughout the 1930s and 1940s, underscored by the Great Depression, but none would survive the legislative process due to high costs and stakeholder pushback.

In the 1940s, two major legal events occurred that dramatically reshaped all U.S. health reform discussions that followed. First, the National War Labor Board determined that wage freezes—implemented to control inflation during World War II—did not apply to employer benefits, including health insurance [5]. Consequently, employers, who were competing for scarce workforce during the war, began offering health insurance more routinely as employment benefits. Subsequent federal actions further solidified and accelerated this change. In the late 1940s, the National Labor Relations Board (NLRB) ruled that health insurance could be included in collective bargaining, and in 1954 the Internal Revenue Service (IRS) determined that health insurance premiums paid by employers were tax exempt [6]. These actions had two major impacts for future health reform: First, they established employers as a main source of health coverage, linking health insurance to employment. Second, they rapidly increased the number of Americans with health insurance. In 1940, roughly 10% of Americans had some form of health insurance; by 1960, more than two-thirds of Americans had health insurance [7]. This expansion of industry and coverage created new stakeholders—including providers, payers, employers, and unions—who all advocated aggressively on subsequent health reform proposals.

The second major event was passage of the McCarran-Ferguson Act in 1945 [8]. The Act was a response to *United States v. South-Eastern Underwriters Association* (1944), a U.S. Supreme Court opinion from the previous year that held insurance companies could be regulated by federal law under the U.S. Constitution's Commerce Clause [9]. While the case was unrelated to health insurance, its outcome would have permitted federal regulation of health insurance, which was previously overseen by state governments. A variety of stakeholders, including insurance companies and states, lobbied Congress to pass a law to reverse the Court's decision. The result was passage of the McCarran-Ferguson Act, which explicitly establishes states' exclusive authority to regulate insurance. The Act codified a growing patchwork of insurance laws and regulations, preserving 51 separate health coverage systems rather than facilitating consolidated health reform within one federal system.

Because a critical mass of Americans now received health insurance through employment, health reform discussions in the mid-20th century focused increasingly on extending coverage to populations without access to employer-sponsored insurance (ESI). The largest groups without such coverage were elderly and lower-income people who were not employed or whose jobs did not offer benefits. To address this gap, the federal government passed the Medicare and Medicaid Act of 1965. The legislation established Medicare to provide hospital and medical insurance to Americans over age 65 and Medicaid to extend coverage to lower-income Americans. Medicare was a federal program to be overseen by federal agencies. Medicaid was to be jointly administered and financed by the federal government with the states. States could choose whether to participate in Medicaid, as it was structured as—and remains—a voluntary program.

The creation of Medicare and Medicaid is a seminal moment in U.S. health reform. Passed with bipartisan support in both the House of Representatives and the

Senate, the programs extended coverage to millions of Americans. Because of their size, the programs also established government as a major payer in healthcare and therefore a major market leader and stakeholder in health reforms. Rooted in federal agencies, the programs also shifted more control over healthcare to the federal government. Perhaps most importantly, the programs are often viewed as a possible foundation for a national health system or public option.

Indeed, efforts to establish a national health system ran in parallel with other 20th century health reforms. Both Democrats and Republicans put forward universal coverage plans, but none survived. In particular, organized medicine and organized labor often opposed the proposals, expressing concern that a national health system would be financially infeasible or undermine hard-fought gains in employer-based coverage garnered through collective bargaining. Major federal health reform initiatives could not overcome opposition from key stakeholders across the political spectrum.

Consequently, health reform in the late 20th century was often incremental and/ or led by states. The federal government passed laws to regulate ESI (or "self-insured" coverage), develop health maintenance organizations (HMOs), expand coverage for children, make prescription drugs more affordable, and manage health information oversight [10–14]. At the same time, more and more states opted to participate in Medicaid, continued to regulate health coverage not provided via ESI, and began to further expand coverage through state programs [15].

Despite these reform efforts, U.S. healthcare continued to struggle with issues of access, affordability, and equity. The ACA sought a more comprehensive approach to addressing these issues.

2.2   Passing the ACA

The ACA's passage was far from a foregone conclusion. Indeed, the history of U.S. health reform is fraught with failure. The country had been unable to pass any federal major reforms until the 1960s. Since then, the most significant attempt at broadscale reform was spearheaded by then First Lady Hillary Clinton in the early 1990s. The proposal, colloquially referred to as "Hillarycare," notoriously flopped. Equally important, the failed attempt solidified conventional wisdom that achieving major national health reform is a daunting challenge due to the issue's scale, complexity, political pitfalls, and competing stakeholders.

Nevertheless, when President Barack Obama took office in 2008, he was focused on achieving comprehensive health reform. In his favor, he had a strong electoral mandate, and his party, the Democratic party, controlled both houses of Congress. Still, President Obama's advisors, many of whom had experienced other attempts at failed health reform efforts, were skeptical that any major reform could pass, even in a Democratically controlled Congress. But the Obama Administration proceeded and entered what became a grueling two-year legislative process [16].

One of the first tasks was to develop the proposal itself. Like most legislative initiatives, the ACA derives from years of policy research and experimentation. For example, the ACA famously based its health insurance exchanges on proposals put forward years earlier by the Heritage Foundation, a conservative think tank [17]. The truest predecessor to the ACA, though, was the Massachusetts health reform law of 2006.

As noted previously, states had become pioneers of health reform in the late 20th and early 21st century as federal reforms stalled. In 2006, Massachusetts became the first state to pass health reform to establish universal coverage [18]. Signed into law by then-Massachusetts Governor Mitt Romney (a Republican), "Romneycare" required that nearly all Massachusetts residents obtain a minimum level of insurance coverage or pay a penalty. The law also provided free and subsidized coverage to those making less than 150% and 300% of the federal poverty level (FPL) respectively. Further, it established the Massachusetts Health Connector, an independent public authority that set premium subsidies, offered insurance for those without employer benefits, and acted as a broker to offer the insurance, including through an online portal. The state legislature augmented Romney's initial proposal expanding existing state Medicaid programs and imposing penalties on employers of 11 or more workers that did not offer "fair and reasonable" coverage. Romney vetoed this employer penalty provision as well as provisions to create certain new public program benefits and eligibility, but the law's core elements prevailed.

The Massachusetts health reform effort dramatically and quickly improved coverage in the state [19]. Within two years of its passage in 2006, the state had cut its uninsured rate from 7.7% to 3%. This was a timely proof point as the ACA's official legislative journey began in 2008. Not surprisingly, elements of the Massachusetts health reform discussion—such as an individual mandate, health insurance exchanges, premium subsidies, public program expansion, and employer mandates—all found their way into the ACA.

Legislative proposals rarely pass as introduced. The ACA underwent two years of iteration as foundational questions were negotiated, such as whether to include a public option. Key stakeholders pivoted with each new draft, supporting, then opposing the bill. Despite its inclusion of ideas such as health insurance exchanges—which had been proposed by conservative thinkers—the bill found little traction with Republicans. Because Democrats controlled both the House and Senate, the bill remained viable despite the lack of bipartisan support. Because it was a large bill and a presidential priority, the ACA also became a vehicle for many smaller health reform measures. As the bill progressed over two years, it grew to include dozens of new provisions as policymakers capitalized on the opportunity to advance other reforms that had stalled in previous years.

The bill's complexity, strong opposition, and a variety of uncanny political events slowed its progress, but the ACA eventually passed in two parts: the Patient Protection and Affordable Care Act, which was signed into law on March 23, 2010, and the Health Care and Education Reconciliation Act, signed into law on March 30, 2010. The latter bill reflects the use of a budget process known as "reconciliation," which effectively eliminates the need for a filibuster-proof majority and has now become a common path to pass legislation. This intricate legislative history of the ACA's passage is only the first part of understanding the reforms the law generated. The next step is understanding the ACA's policy framework itself.

2.3 Understanding the ACA

When it was passed, the ACA was the largest health reform law—and one of the largest laws passed ever—in U.S. history. Despite many iterations, its overall design still reflected a structure similar to the original proposal and Massachusetts' 2006 reforms. The law also reflected an integrated response to the structural challenges of U.S. healthcare created by a patchwork of state systems, a hybrid public-private coverage model, and decades of incremental reforms. Millions of Americans lacked coverage, having been denied access due to preexisting conditions or simply unable to afford it. For those who had insurance, coverage varied greatly by state and insurance policy, often carried limits that left huge gaps for patients, and was increasingly expensive. Care costs were also escalating, driven by public health and demographic trends. Healthcare as a sector lacked competition and presented many barriers to implementing innovative care delivery and payment approaches. These were among the issues the ACA sought to tackle.

To address these challenges, the ACA asserted three main goals: (1) to provide affordable insurance for more people; (2) to expand Medicaid; and (3) to spur innovative care delivery to lower healthcare costs [20].

To achieve these goals, the ACA proposed a broad structural framework built on the following:

- **Individual mandate**: All Americans (with a few exceptions) would be required to purchase minimum essential coverage.
- **Guaranteed issue**: Insurers would no longer be permitted to deny coverage to an applicant with a preexisting condition.
- **Affordability and market stabilization**: The federal government would set a cap on how much any household would spend on healthcare and administer premium subsidies accordingly. It would also implement market stabilization programs to shield markets from disruption resulting from reforms.
- **Health insurance exchanges**: The federal government and/or states would establish marketplaces where Americans could purchase health insurance that guaranteed certain levels of coverage.
- **Medicaid expansion**: The federal government would expand Medicaid eligibility to those making up to 138% of the federal poverty level.

More than anything, this framework was designed to achieve universal coverage—the outcome that had driven, and eluded, most previous health reform efforts. In short, the ACA requires people to obtain health insurance coverage, and makes coverage more accessible and affordable by expanding public program eligibility and private market subsidies. It also makes coverage more uniform and comprehensive by implementing basic standards and uniformity requirements. Delving further into how the ACA approached coverage provides helpful context into why these changes are key health reforms.

2.4   Access to Coverage: The Individual Mandate, Guaranteed Issue, and Beyond

As noted, the overarching goal of the ACA was to create universal coverage [21]. One of the main ways in which the ACA approached this goal was to establish the individual mandate—the requirement that nearly all Americans obtain minimum essential coverage or pay a penalty. In other words, the individual mandate not only required people to obtain insurance but also to obtain a certain level of insurance. The ACA offered several paths to demonstrate minimum essential coverage to satisfy the individual mandate, including enrollment in ESI, Medicare, or Medicaid or a purchasing a qualified health plan (QHP). QHPs were a new concept under the ACA—they referred to coverage that met the ACA criteria, specifically regarding required benefits and cost structure.

The ACA lays out several criteria an insurance product must satisfy to be a QHP. First, a QHP must cover essential health benefits (EHBs). The ACA originally proposed a national benefit set to standardize coverage across the patchwork of state mandated benefits. It codified 10 basic benefits, such as primary and preventive care, but delegated to federal agencies the task of identifying remaining mandated benefits that minimum essential coverage must include. Rather than make the unpopular decision of what a single EHB included, federal agencies then pushed the decision to states, giving them 10 options based on existing coverage plans to choose among for their respective EHBs. Thus, the ACA provided greater uniformity across state coverage requirements by securing 10 essential benefits in federal law but codifying different mandated benefits sets across states, meaning where a person lives still determines what exactly their insurance must cover.

Second, a QHP must provide a certain actuarial value, meaning it must cover a certain percentage of the overall cost of care provided under the policy. The ACA achieved this by establishing metal levels to organize coverage around corresponding actuarial values. The metal levels include bronze, silver, gold, and platinum, under which a policy must cover approximately 60%, 70%, 80%, and 90% of an enrollee's care costs, respectively. The remaining costs would be paid out of pocket by an enrollee in the form of premiums (a monthly payment to the insurer for coverage), co-pays (a set amount paid by the enrollee for services or products), or co-insurance (a set percentage of overall care costs that is paid by the enrollee).

Last, a QHP must comply with limits on consumer cost-sharing. Most notably, the ACA prohibited policies from capping the annual or lifetime benefits enrollees received. It also secured enrollees' rights to renew their policy each year. All of these QHP elements were part of an overall design to make coverage across states more uniform, reliable, affordable, and consumer-friendly.

While QHPs provided quality assurance, other reforms were required to ensure universal access to insurance. This led to one of the most popular ACA reforms, guaranteed issue—the requirement that insurers provide coverage to applicants even if they have preexisting conditions. Allowing those with preexisting conditions to purchase coverage was essential to reducing the rate of uninsured. It also presented major challenges. For example, if someone was entitled to purchase insurance at any time, a logical person would only purchase insurance when they needed care. If only sick people purchased insurance, the market would enter a "death spiral," where premiums would not cover care costs as healthy people abandoned coverage, knowing they could pick it up as needed. The ACA attempted to solve for this issue by establishing open enrollment periods (OEPs)—a set period at the end of each calendar year when everybody enrolled in health insurance. If a person missed this window, they would remain uninsured until the following OEP unless an intervening circumstance, such as a new job or marriage, otherwise led to coverage.

Along with guaranteed issue, the ACA also eliminated underwriting—the process by which policies were priced according to a person's health status. Under the ACA, insurers could only price coverage according to three factors: geography, age, and tobacco use. Allowing ratings based on these factors, known as community rating, permits states to divide their boundaries into separate rating regions. It has also revealed much higher premium costs in rural areas or those with older residents, where dwindling populations lack volumes to distribute costs. Absent differentiation in one of these areas, though, the ACA sought to equalize premiums irrespective of health status.

The above reforms posed major disruption—and possible threat of collapse—to state insurance markets. For example, guaranteed issue promised a flood of new entrants, many of whom had significant care needs after being denied coverage for years due to preexisting conditions. Further, while the individual mandate carried a penalty, it was unclear if the penalty was sufficient incentive to seek coverage. Insurers also worried that they may wind up with a disproportionate number of enrollees with the highest cost medical needs and be unable to sustain claims payment.

To protect and stabilize markets during the transition to implement its coverage reforms, the ACA established three programs: reinsurance, risk corridors, and risk adjustment. Reinsurance and risk corridors were both temporary programs from 2014 to 2017. Reinsurance provided payments to plans that happened to enroll higher cost individuals, protecting against high premium increases. Risk corridors limited the potential losses or gains of individual plans to a specified range, helping to stabilize premiums and encourage appropriate pricing. Risk adjustment, the only permanent program of the three, redistributes funds from plans with lower cost enrollees to those with higher cost enrollees, preventing against adverse selection [22].[1]

---

[1] "Adverse selection" refers to the concept that if insurance guaranteed to be made available whenever needed, consumers will only purchase coverage when they have health needs or if they are consistently in need of health care. This structure disrupts the insurance market by allowing healthy entrants to opt in and out of coverage, reducing the number of overall enrollees over which health care costs are distributed.

Many were also concerned that the new coverage requirement might lead employers to abandon coverage, relying on employees to seek it independently. For this reason, the ACA also included employer shared responsibility (ESR) provisions, more commonly known as the employer mandate. Under these provisions, employers with more than 50 full-time equivalent employees, a calculation based on complex formulas, were required to provide minimum essential coverage for employees or pay a penalty. While initially delayed, the ESR provisions remain in place. ESI coverage has not significantly eroded since the ACA's passage [23].

Taken altogether, the above policy framework boils down to a series of related steps: require nearly everybody get insurance (individual mandate) and entitle everybody to purchase insurance (guaranteed issue); ensure the coverage is comprehensive, at least somewhat uniform, and high quality (QHPs, EHBs, metal levels); update coverage processes to system gaming and promote pricing equality (OEPs, community rating); and implement guard rails to prevent the market from collapsing, particularly during transition (reinsurance, risk corridors, risk adjustment, ESR).

Underpinning this entire framework was also the key issue of affordability. The ACA originally defined "affordability" by asserting a household should not spend more than roughly 9% of its income on health insurance premiums. To ensure people could access insurance without spending more than 9% of their income, the ACA provided subsidies to those earning 100%–400% FPL. These subsidies, known as advanced premium tax credits (APTCs) would only be available for QHPs sold through ACA exchanges [24].

2.5   Exchanges

Health insurance exchanges are a marketplace in which consumers can compare and purchase insurance. Exchanges established under the ACA may only sell products that meet the law's coverage requirements. Individuals enrolled in ESI, Medicare, or Medicaid would not purchase coverage through an ACA exchange, so exchanges largely affect the fully-insured market, consisting of individuals and small businesses. Because ACA exchanges are the only place where consumers can access subsidies, they have a strong competitive advantage over private exchanges.

The ACA offered three options to establish exchanges: First, it establishes a federal exchange, known as Healthcare.gov. Second, it allowed states to create their own exchanges. Last, it permitted states to pursue a hybrid state-federal exchange. If a state did not establish an exchange, it defaulted to participation in the federal exchange. This is important because, while the ACA offered financial incentives to states to establish exchanges, the majority of states did not—likely due to the politicization of the ACA, pending legal challenges to the ACA's constitutionality, or the high complexity of building and operating exchanges. Further complicating the choice was the question of whether Congress intended to provide subsidies through both federal and state exchanges or just state exchanges. This question culminated in a major legal challenge to the ACA, *King v. Burwell*, which was resolved by the U.S. Supreme Court and is discussed later in the chapter.

States choosing to operate their own exchanges can pursue greater regulatory control over their market. In addition to maintaining more local control over operations, state exchanges can govern which QHPs they permit on exchange, a process called "active purchasing." Still, most states use the federal exchange. Today, consumers can access subsidies through state or federal exchanges, and 15 states have their own exchanges while the rest use the federal exchange or a hybrid exchange [25]. As of 2021, these exchanges connect more than 11 million Americans to coverage [26].

2.6   Medicaid Expansion

Much of the ACA focused on expanding coverage and improving affordability within the fully-insured market. But one of the major reforms of the ACA was Medicaid expansion. Established federally in the 1960s and jointly overseen by the federal and state governments, Medicaid programs varied widely across states, particularly regarding eligibility. The ACA sought to establish one income threshold—138% FPL—across all states. To do so, it gave states a choice: raise their Medicaid eligibility levels to the new ACA standard or forfeit their federal Medicaid funding. The ACA also offered financial support for the transition, providing 95% of the state's cost of Medicaid expansion initially and then tapering federal support for the expansion population to 90% after a few years.

Despite these incentives, many states did not choose to expand their Medicaid programs. A major factor in non-expansion was the U.S. Supreme Court's holding on Medicaid expansion in *NFIB v. Sebelius*, the landmark case upholding the ACA discussed at the end of the chapter. In the case, the Court held that the ACA could not require states to increase Medicaid eligibility or lose all funding but could require the states to choose between ACA Medicaid expansion or maintain their pre-ACA funding. This holding, combined with the politicization of the ACA, led many Republican-controlled states to resist Medicaid expansion. But the benefits of expansion are significant, and more states continue to expand Medicaid. As of 2021, 39 states (including Washington D.C.) have expanded Medicaid under the ACA[27], extending coverage to an additional 14 million Americans [26].

2.7   Additional Provisions and Financing

While the previous framework represents the core of the ACA, the law advanced countless more reforms. The bill's provisions included, but were not limited to, measures to constrain healthcare costs, expand Medicare drug coverage, improve care quality, incent value-based payments, expedite drug and device approval pathways, implement new reimbursement methodologies, update nutritional labeling requirements, ban healthcare coverage discrimination, and create entirely new government agencies.

One particularly notable provision provides for State Innovation Waivers, also known as 1332 waivers. Under this provision, the ACA allows states to apply for an exemption from ACA requirements to pursue innovative reforms at the state level. The federal government may grant waivers from the requirements provided the state's proposal: (1) offers coverage equal to or more than a QHP; (2) does not increase cost-sharing; (3) does not decrease coverage; and (4) does not increase the federal deficit. Several states have utilized these waivers to establish state reinsurance programs to further lower premiums in their individual markets. Some states have also used 1332 waivers in conjunction with 1115 waivers, which relate specifically to Medicaid programs, to attempt to expand public coverage in their states by incorporating measures, such as work requirements, though such requirements continue to face ongoing federal regulatory and judicial scrutiny. A few, such as Washington and Colorado, are using them to establish public options.

Given its size and scope, the ACA was an expensive law, with a cost originally estimated at nearly $1 trillion by the Congressional Budget Office. Thus, it also included several new taxes and fees to finance its provisions [28]. The bill created a new tax on health insurance companies, expanded Medicare taxes on high earners, implemented an excise tax (known as the "Cadillac tax") on coverage providing more than 90% of care costs, imposed a 2.3% tax on medical devices, and a 10% tax on indoor tanning. Many of these taxes were repealed following their enactment, without repealing the corresponding expenses, leaving future leaders to resolve budget gaps.

2.8   Implementing the Affordable Care Act

The ACA's implementation has been beset by politics. Immediately after its passage, Republican leaders declared their intent to repeal the ACA. During the Obama Administration, they voted several times to repeal the law. These votes were largely symbolic because President Obama was unlikely to repeal his signature legislative achievement. The votes took on greater significance in 2016 when Republican assumed control of both the House and Senate and the presidency following the election of Donald Trump, who had campaigned on repealing the ACA.

With the political majorities needed to repeal the laws, Republicans faced a challenge. Many of the ACA reforms, such as guaranteed issue and coverage expansions, had proven popular. The party's approach to health reform shifted from "repeal the ACA" to "repeal and replace the ACA." Congressional leaders put together a few reforms that effectively scaled back the ACA by eliminating the individual mandate, reducing subsidy eligibility, and transitioning Medicaid to block or per capita grants. In the summer of 2017, these efforts failed when the Senate declined to pass the "repeal and replace" bill. Congressional attempts at health reform largely ceased, with the only major subsequent change being reduction of the individual mandate penalty to $0. This change spurred additional litigation, discussed later in the chapter, over whether the provision remained constitutional.

As Congressional efforts faltered, health reform fell to the executive branch. The Obama Administration proliferated regulations to implement the massive law between 2010 and 2016. Then, the first day President Donald Trump took office in 2016, he issued an executive order requiring agencies to "minimize the economic impacts" of the ACA [29]. He also issued subsequent executive orders permitting insurance alternatives that weakened ACA coverage requirements [30]. These actions weakened the ACA but could not eliminate it.

States also played a key role in ACA implementation, as the law required them to make major choices such as whether to expand Medicaid or establish state exchanges while preserving their regulatory role in overseeing the fully-insured market, albeit to ensure ACA compliance. Democrat-led states embraced the law, fully implementing its reforms and often codifying them in their own state statutes. Republican-led states strongly resisted ACA implementation, declining to adopt any ACA reforms for several years. Some later began to implement reforms, particularly Medicaid expansion, as public opinion became more favorable towards the law and financial incentives to implement the ACA remained strong.

The ACA delegated much of its health reform activity to federal agencies. The law envisioned a decade-long implementation period. Thus far, the law has generated tens of thousands of pages of regulation and countless pages of sub-regulatory guidance. ACA regulations also represent major health reforms. For example, regulations under section 1557 of the ACA extended anti-discrimination protections to LGBTQ patients. Thus, much of health reform has and will continue to take place through executive rulemaking.

2.9   Challenges to the ACA

In addition to adverse legislative and executive actions, ACA opponents have mounted several legal challenges against the law. Generally, the courts have upheld the ACA, with a few modifications. The most important U.S. Supreme Court cases considering the ACA are below.

2.10   Constitutionality: *NFIB v. Sebelius* and *California v. Texas*

The first and most important challenge to the ACA was in *National Federation of Independent Business (NFIB) v. Sebelius* [31]. The plaintiffs in this case challenged the constitutionality of the individual mandate. They also argued that the mandate was foundational to the ACA's structure. Therefore, if the mandate was held to be unconstitutional, the plaintiffs contended, the full law should also be struck down. Accordingly, the constitutionality of the individual mandate could determine the fate of the entire ACA. Further, plaintiffs challenged the legality of the Medicaid expansion.

In a 2012 opinion, the Court upheld the ACA in a slim 5-4 majority. Republican-appointee Chief Justice John Roberts joined with more liberal justices to uphold the mandate, although he found the provision's authority in the U.S. Constitution's Taxing and Spending Clause, while the other jurists held the mandate was also authorized by the Commerce Clause. Because the Court upheld the individual mandate, it did not reach the question of "severability," leaving the entire law intact. The Court also held that Medicaid expansion under the ACA was "coercive" and determined that states should be able to choose between the ACA expansion or preserving their pre-ACA Medicaid funding, rather than losing Medicaid funding altogether. This holding undermined the greatest financial incentives for Medicaid expansion—loss of federal funding.

Even after *NFIB v. Sebelius*, legal challenges to the ACA continued. After the individual mandate penalty was reduced to $0, a new challenge was brought forward in *California v. Texas*. Plaintiffs argued that the individual mandate was no longer authorized by the Taxing and Spending Clause because a $0 penalty is not a tax. If the individual mandate could not be authorized under this clause, it lacked a majority to uphold the provision and would not have survived its initial challenge. Further, the Court shifted ideologically to the right after 2012, and its new composition was more likely to overturn the ACA. But the ACA survived this challenge, too. In 2021, the Court held that plaintiffs lacked standing (the legal right to bring a case) and dismissed the challenge [32].

2.11   Exchanges: *King v. Burwell*

Legal challenges addressed not only the ACA's constitutionality but also its intent. In *King v. Burwell*, plaintiffs argued that the ACA's statutory language permitted only state—and not federal—exchanges to administer subsidies [33]. If the courts upheld this interpretation, the federal exchange would no longer be able to administer subsidies, dealing a crippling blow to the ACA's ability to offer affordable coverage. The U.S. Supreme Court took the case and in 2015 held in a 6-3 decision that the ACA permitted both the federal and state exchanges to administer subsidies. The Court reasoned that legislative history reflected that the intent of the law was to make subsidies available to all exchange purchasers. Thus, while other cases upheld the ACA's constitutionality, *King* upheld its functionality.

2.12   Coverage Requirements: *Hobby Lobby*

One of the most well-known challenges to the ACA was *Burwell v. Hobby Lobby Stores, Inc.* [34] In this case, Hobby Lobby, a retail store chain, challenged the ACA's requirement to make certain contraception coverage available to its employees or pay a penalty for not providing required coverage. Hobby Lobby, which was a company privately held by owners who opposed these types of contraception per their opposition to abortion, argued the coverage requirement violated their religious freedoms. The U.S. Supreme Court agreed in a narrow 5-4 holding. In the opinion, the Court held that privately owned companies were not required to provide certain contraception coverages in contradiction to the religious beliefs of the companies' owners.

The Court did not strike down the concept of essential benefits but rather limited its decision to privately held companies and a specific type of benefit. To reach this outcome, the Court held that it must extend rights established by religious freedom laws to Hobby Lobby as a corporation to effectively protect the religious freedoms of the company's owners [35]. Notably, the dissenting Justices in the 5-4 opinion read the Court's holding as an expansion of corporate personhood, a legal doctrine under which corporations are deemed to hold the same or similar rights granted to individuals under federal law [35]. Read either way, it is clear that the case not only impacts ACA coverage requirements but also underscores the importance of ACA litigation as a source of precedent for a wide variety of issues beyond health care.

Overall, these legal challenges reflect that courts have largely upheld the ACA though they occasionally modify its structure and effects. The Biden Administration has indicated it intends to build on the ACA as its foundation for future health reform. At the same time, the U.S. Supreme Court has shifted ideologically and, if presented with a case challenging the law's constitutionality, could reconsider its previous holdings and upend the law. For now, the legislative, executive, and judicial branches are all likely aligned to preserve the ACA, and it remains the starting place for future U.S. health reform.

3   Application

The ACA affected nearly every part of U.S. healthcare. Below are a few scenarios demonstrating the impact of this landmark reform on providers, patients, and the broader healthcare system:

Example 1: Access to Coverage—Guaranteed Issue and EHB Maya is leaving her job, which provides health insurance through ESI, to become a self-employed freelance consultant. She does not qualify for Medicare or Medicaid based on her age and income (300% FPL). She is pregnant and wants to purchase health insurance on the individual market. What are her options for health insurance?

Before ACA—Before the ACA, an insurer could have denied Maya coverage because her pregnancy was considered a preexisting condition. Depending on the state she lived in, she may have had access to a "high risk pool" that offered coverage at a much higher premium, but otherwise she would have had to change her job or reduce income to obtain ESI or qualify for Medicaid. She could also go uninsured.

After ACA—Following the ACA, Maya would be legally entitled to purchase coverage per guaranteed issue. If she purchased through an ACA exchange, she would be eligible for a subsidy because her income is 300% FPL. In accordance with the holding in *King v. Burwell*, she is eligible for the subsidy whether purchasing on a state or federal exchange. Further, because maternity and newborn care is a required EHB benefit category under the ACA, the plan she purchased on exchange would be guaranteed to cover her pregnancy and care for her new child.

Example 2: Affordability—Medicaid Expansion Lee is a farmer whose income is 120% FPL. They do not have access to ESI and are not old enough to qualify for Medicare. What are Lee's options for health insurance coverage?

Before ACA—Before the ACA, Lee's options would depend on the state in which they live. Their income of 120% FPL would have made them eligible for Medicaid in some states and not others. They also would have faced an asset test that might have disqualified them for Medicaid in combination with their income. If they did not qualify for Medicaid, they could try to purchase insurance on the individual market but may find the costs prohibitive. They could also go uninsured.

After ACA—After the ACA, the state in which Lee lives would still play a role, but access to coverage would be greater. If Lee lived in a state that had expanded Medicaid, they would be eligible for the program because their income is under 138% FPL and the asset test was eliminated. But per *NFIB v. Burwell*, Lee's state may not have expanded Medicaid and retained pre-ACA funding and eligibility levels that did not reach 120% FPL. In that case, Lee could choose to purchase coverage on an ACA exchange and receive a subsidy to cover the difference between the premium's cost and whatever 8.5% of Lee's income is. Lee would need to purchase the coverage, though, during the OEP at the end of the year. Lee could also still choose to go uninsured and pay the individual mandate penalty, now $0 as upheld under *California v. Texas*.

Example 3: Continued Innovation—State Innovation Waivers A state wants to establish a public option health plan. The program would allow any resident to enroll in the state's Medicaid program, paying an income-based sliding scale premium. How would the state work with the federal government to establish this program?

Before ACA—Before the ACA, the state would seek a section 1115 waiver to allow expanded enrollment in Medicaid and permit premiums for new enrollees. The state alone would have to pay for the program expansion.

After ACA—The state would likely still submit a section 1115 waiver but also likely seek an ACA-based section 1332 waiver. The 1332 waiver would allow the state to use funds for Medicaid expansion and exchange subsidies to cover the costs of the program. It would also require that the state verify that the proposed public option not decrease coverage or benefits or increase cost-sharing or the federal deficit.

4   Summary

Health reform is a big, complicated, and challenging. U.S. health reform is shaped by many factors:

- A well-established private sector that predated major national reforms
- A federalist structure that divides healthcare oversight between state and federal government

- The role of Medicare and Medicaid as industry leaders and large government payers
- The role of employers as a primary link to coverage
- A record of incremental reform, often resulting from complex policy and politics
- A complex regulatory structure subject to input from multiple competing stakeholders
- A high degree of politicization, particularly around the ACA, and significant political, policy and economic barriers to large scale reform efforts

Many consider health reform synonymous with the ACA. It reshaped the U.S. healthcare landscape, extending coverage to tens of millions of Americans, moving the country towards universal coverage. Congressional attempts to repeal the law or strike it down in the courts have failed. Thus, the ACA remains and will likely be the basis for health reform in coming decades.

As part of those reforms, healthcare will face new challenges, particularly as healthcare innovation continues to outpace the legal structures that govern it. Future reforms will need to address not only access and affordability but also emerging issues such as how to integrate new technologies for care delivery and how best to protect patient privacy in an increasingly digital world. Thus, the need for health reform exists perpetually as U.S. healthcare moves forward.

Acknowledgments None

References

1. Centers for Medicare and Medicaid Services [Internet]. National Health Expenditure Data, NHE Fact Sheet. 2020. [cited 2021 Jul 24]. Available from: https://cms.gov/Research-Statistics-Data-and-Systems/Statistics-Trends-and-Reports/NationalHealthExpendData/NHE-Fact-Sheet.
2. Emanuel, E. Reinventing American Health Care. 1st ed. New York: Public Affairs; 2014. p. 18-33.
3. Guyton, G. A brief history of worker's compensation [Internet]. Iowa Orthop. J. 1999; 19: 106-110. [cited 2022 Jun 4]. Available from: https://www.ncbi.nlm.nih.gov/pmc/articles/PMC1888620/.
4. Kaiser Family Foundation [Internet]. Timeline: History of Health Reform in the U.S. 2011. [cited 2021 Jul 25]. Available from: https://www.kff.org/health-reform/timeline/history-of-health-reform-efforts-in-the-united-states/.
5. Executive Order 9250. Providing for the Stabilizing of the National Economy. 7 FR 7871. 1942. Established pursuant to The Stabilization Act of 1942, 50a U.S.C. § 961. 1942.
6. National Bureau of Economic Research [Internet]. Employer-Sponsored Health Insurance and Health Reform. Bulletin on Aging and Health No. 2. 2009. [cited 2021 Jul 25]. Available from: https://www.nber.org/bah/2009no2/employer-sponsored-health-insurance-and-health-reform.
7. Carroll A. The Real Reason the U.S. Has Employer-Sponsored Health Insurance [Internet]. New York Times. 5 Sept. 2017. [cited 2021 Jul 25]. Available from: https://www.nytimes.com/2017/09/05/upshot/the-real-reason-the-us-has-employer-sponsored-health-insurance.html.
8. McCarran-Ferguson Act. 15 U.S.C. §§ 1011-1015 (1945).

9. U.S. v. South-Eastern Underwriters Ass'n. 322 U.S. 533 (1944).

10. Employee Retirement Income Security Act of 1974. 29 U.S.C. § 18 (1974). Establishing federal oversight of employee benefits, including ESI.

11. Health Maintenance Organization Act of 1973. 42 U.S.C. § 300e (1973). Establishing a federal trial program to support the development of HMOs.

12. Centers for Medicare and Medicaid Services [Internet]. Medicaid.gov. Program History. [cited 2021 Jul 25]. Available from: https://www.medicaid.gov/about-us/program-history/index.html.

13. Medicare Modernization Act of 2003. 42 U.S.C. § 1395w, 101-154 (2021).

14. Health Insurance Portability and Accountability Act of 1996 (HIPAA). 110 Stat. 1936. Pub. L. 104-191 (1996).

15. For example, in the early 1990s, Minnesota established MinnesotaCare, a public program offering coverage to residents who were not eligible for Medicaid but did not have sufficient income to support private insurance coverage. Minn. Stat. Chap. 257L. (2021).

16. PBS. Frontline. Obama's Deal (documentary). [cited 2021 Jul 25]. Available from: www.pbs.org/wgbh/frontline/film/obamasdeal/. Providing an overview of the legislative history of the ACA.

17. Jacobson L. Obama says Heritage Foundation is source of health exchange idea [Internet]. Politifact. 1 April 2010. [cited 2021 Jul 25]. Available from: https://www.politifact.com/factchecks/2010/apr/01/barack-obama/obama-says-heritage-foundation-source-health-excha/.

18. An Act Providing Access to Affordable, Quality, Accountable Health Care (Massachusetts Health Reform Act of 2006) [Internet]. Chapter 58 of the Acts of 2006, Massachusetts General Court. [cited 2021 Jul 25]. Available from: https://malegislature.gov/Laws/SessionLaws/Acts/2006/Chapter58.

19. Long S, Skopec L, Shelto, A, Nordahl K, Kenney Walsh, K. Massachusetts Health Reform at Ten Years: Great Progress But Coverage Gaps Remain [Internet]. Health Affairs. 35:9. Sept. 2016. [cited 2021 Jul 25]. Available from: https://www.healthaffairs.org/doi/10.1377/hlthaff.2016.0354.

20. Affordable Care Act Glossary [Internet]. Healthcare.gov. [cited 2021 Jul 25]. Available from: www.healthcare.gov/glossary/affordable-care-act.

21. Kaiser Family Foundation [Internet]. Summary of the Affordable Care Act. March 2017. This document is based on a more detailed summary of the law's provisions, available at https://www.kff.org/health-reform/fact-sheet/summary-of-the-affordable-care-act/, which was accessed 25 July 2021.

22. Cox C, Semanskee A, Claxton G, Levitt L. Kaiser Family Foundation [Internet]. Explaining Health Care Reform: Risk Adjustment, Reinsurance, and Risk Corridors. Aug. 2016. https://www.kff.org/health-reform/issue-brief/explaining-health-care-reform-risk-adjustment-reinsurance-and-risk-corridors/

23. Claxton G, Rae M, Damico A, Young G, McDermott D, Whitmore H. Health Benefits in 2019: Premiums Inch Higher, Employers Respond to Federal Policy [Internet]. Health Affairs. 2019; 38:1. [cited 2022 Jun 4]. Available from: https://doi.org/10.1377/hlthaff.2019.01026

24. In 2021, the American Rescue Plan Act of 2021, passed to address the COVID-19 pandemic, amended these provisions and lowered the definition of affordability to roughly 8.5% and temporarily expanded APTC subsidies to accommodate the new definition. McDermott D, Cox C, Amin K. Kaiser Family Foundation [Internet]. Impact of Key Provisions of the American Rescue Plan Act of 2021 COVID-19 Relief on Marketplace Premiums. 15 Mar. 2021. [cited 2021 Jul 5]. Available from: https://www.kff.org/health-reform/issue-brief/impact-of-key-provisions-of-the-american-rescue-plan-act-of-2021-covid-19-relief-on-marketplace-premiums/.

25. Kaiser Family Foundation [Internet]. State Health insurance Marketplace Types, 2021. [cited 2021 Jul 25]. Available from: https://www.kff.org/health-reform/state-indicator/

state-health-insurance-marketplace-types/?currentTimeframe=0&sortModel=%7B%22colId%22:%22Location%22,%22sort%22:%22asc%22%7D.

26. HHS.gov [Internet]. New HHS Data Show More Americans than Ever Have Health Coverage through the Affordable Care Act. 5 June 2021. [cited 2021 Jul 25]. Available from: https://www.hhs.gov/about/news/2021/06/05/new-hhs-data-show-more-americans-than-ever-have-health-coverage-through-affordable-care-act.html.

27. Kaiser Family Foundation [Internet]. Status of State Medicaid Expansion Decisions: Interactive Map. 23 July 2021. [cited 2021 Jul 25]. Available from: https://www.kff.org/medicaid/issue-brief/status-of-state-medicaid-expansion-decisions-interactive-map/.

28. IRS.gov. [Internet]. Affordable Care Act Tax Provisions. [cited 2021 Sept 17]. https://www.irs.gov/affordable-care-act/affordable-care-act-tax-provisions.

29. Minimizing the Economic Burden of the Patient Protection and Affordable Care Act Pending Repeal. Exec. Order 13765. 20 Jan. 2017.

30. Promoting Healthcare Choice and Competition Across the United States. Exec. Order 13813, 12 Oct. 2017.

31. Nat'l Fed. Ind. Bus. v. Sebelius. 567 U.S. 519 (2012).

32. California v. Texas. 593 U.S. __ (2021). WL 2459255.

33. King v. Burwell. 576 U.S. 473 (2015).

34. Burwell v. Hobby Lobby Stores, Inc. 573 U.S. 682 (2014). Stating that "protecting the free-exercise rights of corporations like Hobby Lobby…protects the religious liberty of the humans who own and control those companies."

35. Burwell v. Hobby Lobby Stores, Inc. 573 U.S. 682 (2014). Dissent noting that "the determination that [federal religious freedom law] extends to for-profit corporations is bound to have untoward effects" and that "the Court's expansive notion of corporate personhood" will contribute to future lawsuits seeking religious exemptions for other regulations.

Part IV

Tort Law

Malpractice

Benjamin J. McMichael

Contents

Key Points

- Tort law provides an avenue for patients injured by the acts and omissions of clinicians to hold those clinicians liable for monetary damages.
- Tort law operates independently of the criminal system and the disciplinary system maintained by various licensing and regulatory boards.
- Different torts provide recourse in different circumstances, with medical malpractice and informed consent claims covering situations in which a clinician acts negligently and medical battery covering situations in which a clinician acts with intent.
- Understanding how individual torts work and how different individuals and entities may be held liable can help clinicians avoid this liability.

Legal Concepts

- Medical malpractice occurs when a clinician deviates from the standard of care and causes harm to a patient.
- Informed consent claims provide patients legal recourse when their clinicians fail to provide all material information about a proposed course of treatment.

B. J. McMichael (✉)
The University of Alabama School of Law, Tuscaloosa, AL, USA
e-mail: bmcmichael@law.ua.edu

© The Author(s), under exclusive license to Springer Nature Switzerland AG 2022   129
A. S. Pasha (ed.), *Laws of Medicine*,
https://doi.org/10.1007/978-3-031-08162-0_8

- Medical battery occurs when a clinician touches a patient without his or her consent or when a clinician performs a substantially different procedure than the one the patient originally consented to.
- Vicarious liability allows plaintiffs to hold those who can control clinicians liable for the acts and omissions of those clinicians.

1   Introduction

Tort law exists to provide compensation to individuals harmed by the actions of others. A tort is "a civil wrong ... for which a remedy may be obtained, usually in the form of [monetary] damages" [1]. In other words, a tort is a harm inflicted by one person on another that the law recognizes as deserving of compensation. The law of torts governs many everyday activities, from driving—where a negligent driver may be required to compensate someone harmed by that driving—to housekeeping—where a homeowner may be required to compensate a guest following a slip and fall on the homeowner's property. Tort law also provides a basic framework for social interactions by providing recourse to those injured by the intentional conduct of others.

In the healthcare context, a patient harmed intentionally or negligently by a clinician may turn to tort law for compensation. As civil wrongs, any torts committed by clinicians stand wholly apart from the more familiar crimes which governments may prosecute under criminal law. Tort law is also independent of the disciplinary system operated by medical, nursing, and other boards or agencies that license and regulate clinicians. While there may be some overlap, tort liability operates outside the disciplinary system, making statements like "I don't want to risk my license" irrelevant in the context of tort law.

Tort law provides a framework for clinician-patient interactions and governs how clinicians fulfill their duties to their patients. Two broad classes of torts are germane to healthcare. First, the law of negligence in general, and medical malpractice in particular, requires that clinicians act reasonably in delivering care to their patients. Such reasonable care may include taking certain precautions against infections, diagnosing patients correctly, or performing surgeries consistent with accepted standards of care. Reasonability may also include providing patients with sufficient information on various courses of treatment so that they can make fully informed choices about their care. Second, the law of intentional torts prohibits clinicians from taking intentional actions that may harm a patient without the patient's consent. And consent is a key point within intentional torts, as consent transforms a vicious attack into a routine surgery or sexual harassment into a routine breast examination.

Common law—law made by judges relying on precedent over time—forms the foundation of tort law. Recently, however, states have increasingly passed statutes that codify, and sometimes override, the common law of torts. Thus, in some states, the law of medical malpractice can be found almost entirely within judicial opinions and in others, it can be found in both opinions and statutes passed by the state legislature. Importantly, states bear primary (and often exclusive) responsibility for tort law, meaning that federal law has a relatively limited role to play in the law of torts. An important exception to this general rule occurs when clinicians practice in federal facilities, which is discussed in this chapter. The primacy of state law also means that tort law varies across more than fifty different jurisdictions in the United States. This chapter provides a general overview of tort law, but it is important to note that the law of individual states may depart from the general description and specific examples provided here.

2   Discussion

While tort law is a broad area of law that covers many different types of harms, three torts have particular relevance to the delivery of healthcare by individual clinicians. First, medical malpractice, which is essentially a specialized form of negligence, provides a legal remedy for patients who are injured by the unreasonable acts or omissions of clinicians. Second, battery, which is an intentional tort, provides a legal remedy for patients who are treated without their consent or in ways that exceed the consent they have given their clinicians. Third, informed consent, which may be either a type of negligence or a type of intentional tort depending on the state, provides a remedy for patients who were not provided with enough information to make an informed decision about their course of treatment.

2.1   Medical Malpractice

Beginning with what is likely the most familiar tort to clinicians, medical malpractice is essentially a specialized form of negligence applied to a healthcare setting. Taking a step back from that specialized setting for a moment, negligence has four separate elements: duty, breach, causation, and harm. A plaintiff—the person who initiates the lawsuit—has the burden to prove each element by a preponderance of the evidence. For example, consider a simple car crash. If Abby collides with Bill's car while driving and Bill wishes to assert a negligence claim against Abby, Bill must prove that he suffered a harm, that Abby owed him a duty of reasonable care, that she breached that duty, and that her breach of her duty was the cause of his harm. In this case, Abby owes Bill a duty because they were driving on the road together, so it is foreseeable that if she acts unreasonably, Bill could be injured.

Whether Abby breached this duty depends on her behavior at the time of incident. A court will compare her actions to that of a reasonable person. If Abby's behavior was consistent with what a reasonable person would have done, the court will conclude she did not breach her duty and that Bill's claim therefore fails. If Abby's behavior fell below that of a reasonable person, then the court will conclude that she has breached her duty. For example, if Abby was texting while driving, had become drunk before driving, or taken any other action inconsistent with reasonable behavior, a court would conclude that she had breached her duty.

If Abby breached her duty, the court would next consider whether the breach of her duty caused Bill's harm. Causation in tort law has two parts: factual causation and proximate causation. The consideration of factual causation (often called "but-for" causation) involves a simple counterfactual: If Abby had not behaved unreasonably, would Bill's harm still have occurred? If the answer is no, then Abby was a factual or "but-for" cause of Bill's harm. If the answer is yes, then Abby's actions were not a factual cause of Bill's harm. Following this inquiry, the court will next consider whether Abby's actions were the proximate cause of Bill's harm. Proximate causation serves as a mechanism to cut off liability when, although someone's actions are a factual cause of the relevant harm, those actions were too remote to the harm to justify imposing liability. Different states approach proximate causation differently, but one way to evaluate the proximate cause is to consider the scope of the risk associated with the defendant's actions. In this example, a court will ask whether Bill's harm was within the scope of the risk created by Abby's breach of her duty. This inquiry involves asking if the type of harm suffered by Bill was the type of harm that results from the risks that made Abby's conduct tortious. Usually, if the harm was a foreseeable result of the relevant conduct, that conduct will satisfy the proximate cause requirement.

The harm element is often the most straightforward. If the plaintiff suffered a physical harm as a result of the defendant's breach of his or her duty, then this element is satisfied. More difficult questions—not directly relevant here—arise when the plaintiff suffers only emotional or psychological harm. In the example of Abby and Bill, all four of these elements would be satisfied if Abby was acting unreasonably when she collided with Bill, e.g., if she were sending a text message moments before the crash.

Claims of medical malpractice take place against this backdrop of the tort of negligence. As with car crashes and other injuries, a patient asserting a medical malpractice claim must establish the elements of duty, breach, causation, and harm by a preponderance of the evidence. In medical malpractice claims, however, these elements look somewhat different from other negligence claims. In most cases, the element of duty is not in contention since a clinician clearly owes a duty to his or her patient. An important exception to this general rule that clinicians owe their patients a duty is the curbside consult, which occur when a treating clinician asks another clinician for informal advice. In such a case, the physician offering the informal advice "does not owe a duty to the patient because no physician-patient relationship is created" [2].

Conditional on a duty and a breach of that duty being established, causation is generally not in question. The types of harm associated with medical malpractice (e.g., a misdiagnosis that led to delayed care or a wrong-site surgery), often do not involve difficult questions of but-for or proximate causation. Sometimes, however, trickier questions of causation do arise. For example, if a drug is known to cause a specific side effect and the patient suffers that side effect after a clinician prescribes too much of the drug, a thorny question of causation may arise. If the patient would have suffered the side effect had the patient been given the correct dose, then the clinician did not cause the relevant harm. If the patient suffered the side effect only because of the overdose, then the clinician did cause the relevant harm. Separating the former instance from the latter often requires substantial expert testimony from physicians and scientists with backgrounds in pharmacology.

While courts may find it relatively easier to address some elements of a medical malpractice claim than in other types of negligence claims, one aspect of medical malpractice is often quite difficult: evaluating whether the defendant clinician's actions were unreasonable because they fell below the standard of care. Instead of applying the typical reasonable person standard to determine whether the clinician acted reasonably, courts apply the reasonable physician, reasonable nurse practitioner, or other reasonable clinician standard. This hypothetical reasonable clinician possesses all the skills of such a clinician and behaves reasonably at all times. And determining whether the defendant clinician acted as a reasonable clinician turns on whether the clinician acted as a reasonable clinician, i.e., met the requisite standard of care.

Unlike other negligence claims, where the jury has a general idea of what constitutes reasonable behavior, such as reasonably driving, reasonably maintaining property, etc., the jury has little idea what constitutes reasonable behavior in the delivery of healthcare. Clinicians complete years of training and must routinely address complex and technical issues when caring for patients. Lay jurors lack this training and often cannot appreciate the complexities of modern healthcare delivery. Accordingly, courts rely on experts to testify as to the relevant standard of care. These experts are drawn from the ranks of the relevant profession and aid the jury by explaining what a reasonable clinician would do under the pertinent circumstances. With some exceptions, the plaintiff is generally required to introduce expert testimony, and defendants may introduce competing testimony from different experts.

Because of the critical importance of expert testimony, many disputes within medical malpractice litigation turn on the identity of the experts, the extent of their knowledge, and the content of their testimony. States have developed rules, either within their common law tradition or through legislative action, to govern which experts may testify. In addition to competence and scientific requirements that all experts must satisfy, experts in medical malpractice claims can be subject to at least two other sets of rules. The first concerns the expert's professional specialty in relation to that of the defendant's, and the second concerns the expert's knowledge of local and national standards of care.

Beginning with the first requirement, states differ on whether a testifying expert must practice in the same specialty (or subspecialty) as the defendant. For example, interpreting a state statue, the Supreme Court of New Jersey held that "[a] medical expert must be a specialist in the same field in which the defendant physician specializes" [3]. The Supreme Court of Michigan has similarly held that "if a defendant physician is a specialist, the plaintiff's expert witness must have specialized in the same specialty as the defendant physician at the time of the alleged malpractice" [4]. The Michigan court went further, holding that "[i]f a defendant physician specializes in a subspecialty, the plaintiff's expert witness must have specialized in the same subspecialty as the defendant physician at the time of the occurrence that is the basis for the action" [4]. And if the defendant was board certified, testifying experts must also be board certified in the relevant specialty [4]. States with these types of rules thus impose additional burdens on experts by requiring that they not only have relevant knowledge about the defendant clinician's acts or omissions that gave rise to the malpractice claim, those experts must specialize in the same field.

In contrast to this specialty match requirement, other states specifically reject the idea that a testifying expert must specialize in the same area as the defendant. In New York, for example, "[a] medical expert does not have to be a specialist in the same field as a defendant doctor" [5]. As long as the testifying expert can establish knowledge and expertise in the relevant area of medicine, that expert may testify even though he or she does not specialize in the same area as the defendant. In *Frank v. Smith*, a New York appellate court permitted an anesthesiologist to testify in a case against an orthopedic surgeon because the expert "set forth sufficient knowledge of the nature and location of the pertinent nerves as well as relevant aspects of the surgery" that was the focus of the malpractice claim [5]. Similarly, the Supreme Court of Georgia held that "[a]n expert testifying about the standard of care in a medical malpractice case need not actively practice in the same specialty or practice area as the defendant doctor" [6]. Of course, experts must still demonstrate knowledge of the relevant procedure or otherwise demonstrate that their testimony will elucidate the standard of care and whether the defendant satisfied it. However, states like New York and Georgia reject the idea that only clinicians who practice in the same specialty as the defendant may testify in a medical malpractice case.

Turning to the second requirement, states generally agree that clinicians must satisfy a standard of care set by their profession and that only experts may opine on this standard of care. However, states differ on how to define this standard of care. These differences generally turn on how broadly to look when determining what a reasonable clinician would do. In theory, healthcare is based on scientific evidence and is delivered virtually identically across the country. In practice, multiple studies have demonstrated that the provision of healthcare varies substantially from place to place [7]. Some states embrace to the idea that, as a science-based enterprise, the delivery of healthcare does not vary from place to place, as the science does not vary from place to place. Other states focus on the regional variation in care and recognize that different circumstances (such as the availability of medical technologies or access to insurance) may impact the delivery of care. In general, there are three rules for the standard of care: a strict local rule, a local rule, and a national rule.

The strict locality rule, which is no longer popular among states, requires that any testifying expert be familiar with the delivery of care within the community where the alleged malpractice occurred. "Historically, the strict locality rule [was] based on the rationale that there exists gross inequality between physicians practicing in large urban areas and those practicing in more remote rural communities" [8]. The rule existed "to prevent the small town practitioner from being held to the standard of practice of the more sophisticated urban areas" [8]. As healthcare evolved and became more standardized, the strict locality rule gave way to the local rule. The Tennessee Supreme Court explained in 2011 that "[a]s our society became more interconnected with improved transportation and communications, the strict locality rule gave way to a more relaxed modified locality rule in many states" [9].

Under the locality rule, an expert may demonstrate that he or she is familiar with the standard of care in the defendant's community as with the strict locality rule, or he or she may demonstrate familiarity with a similar community [9]. To comply with this rule a "plaintiff must demonstrate that an opinion witness has knowledge of the standard of care applicable in the defendant's community or in a community that the party demonstrates to be similar to that where the defendant practices medicine" [10]. Such a rule does not completely abandon the idea that variations in the availability of resources and other factors may impact the standard of care, but it relaxes the requirement that an expert demonstrate familiarity with the defendant's actual community.

Finally, the national rule abandons the requirement that a testifying expert demonstrate familiarity with a local standard of care and permits that expert to testify as to the national standard of care, i.e., what reasonable clinicians across the country would do in the defendant's situation. Where courts do not uniformly apply the national standard, board certified physicians are more likely to be subject to it. For example, the Supreme Court of Rhode Island has explained that "except in extreme cases, a witness who has obtained board certification in a particular specialty related to the procedure in question, especially when that board certification reflects a national standard of training and qualification, should be presumptively qualified to render an opinion" [11]. Similarly, the Supreme Court of Idaho explained that "for an out-of-area expert to testify that a national standard of care applied to the defendant specialist by virtue of his certification, two elements must be met" [12]. These elements include: First, the testifying witness has obtained the same board certification as the defendant, and second, "there are no local deviations from the national standard under which the defendant-physician and witness-physician were trained" [12]. States that have moved to a national rule from a local rule generally justify this move by acknowledging changing patterns of healthcare delivery. The Supreme Court of Rhode Island, for example, was "of the opinion that whatever geographical impediments may previously have justified the need for a 'similar locality' analysis are no longer applicable in view of the present-day realities of the medical profession" [11].

While specialty-match requirements and geographical limits on testifying experts may seem esoteric and irrelevant, the reason that these detailed rules have become necessary is the importance of expert testimony. No two medical malpractice cases are exactly the same, and tort law has developed to remain flexible in allowing each court to evaluate each case on its facts. Thus, there are few, if any, bright line rules that obviously delineate malpractice from other instances of healthcare delivery. A post-surgical infection may represent an unfortunate, but unavoidable, complication in one instance and an egregious departure from the standard of care in the next. The fact tort law does not identify which instances constitute malpractice in advance, many cases turn on the ability of experts to explain why a patient's harm stems from malpractice [13].

Research has demonstrated the relevance of standards governing expert witnesses in medical malpractice cases. For example, Michael Frakes found "a 30–50 percent reduction in the gap between state and national utilization rates of various treatments and diagnostic procedures following the adoption of a rule requiring physicians to follow national, as opposed to local, standards" [14]. He further found evidence that "that physicians may indeed adhere to specific liability standards," suggesting that, far from representing merely arcane legal doctrines, the rules governing expert testimony meaningfully impact the delivery of care. The evidence uncovered by Frakes makes sense because the experts will effectively determine what was reasonable under the circumstances. If a family physician manages a normal labor and delivery but an unexpected complication arises, a nationally renowned expert in obstetrics may testify differently as to the reasonability of the family physician's actions than would a local family physician who practices several miles away.

Because of the critical importance of expert testimony and the fact that the experts will closely evaluate what the defendant clinician actually did or did not do, it is important for clinicians to routinely think about the reasonability of their behavior. This internal reasonability analysis should not, however, turn on common tropes about malpractice liability. For example, the idea that always doing more or avoiding taking any risks with patients is simply wrong. All clinicians take risks at different times, and the legal reasonability analysis will simply ask if those risks were justified. For example, a family physician evaluating whether to prescribe a medication that includes a small risk of death to a patient must determine whether that risk outweighs the risk of harm that the patient will endure without the medication. Oncologists routinely prescribe medications with severe side effects, but their actions are nonetheless reasonable based on the harms these medications help patients avoid. Just as prescribing the wrong medication may subject a clinician to liability, so too may refusing to prescribe a medication when the benefits of that medication outweigh the risks. The same is true for procedures. Watchful waiting may be justified in some instances because a surgery is risky, but watchful waiting can equally subject a clinician to liability if the surgery, despite its risks, offers the patient a better path forward.

2.2   Medical Battery

In contrast to the focus on the standard of care and what a reasonable clinician would have done in the medical malpractice context, cases of medical battery focus on whether the clinician received consent to touch the patient. "The elements of a medical battery claim are: (1) an intentional act on the part of the defendant; (2) a resulting offensive contact with the plaintiff's person; and (3) a lack of consent to the defendant's conduct" [15]. And "[a]s the name suggests, medical battery is an intentional tort" which means that the defendants intent in taking the action matters and not whether the defendant acted reasonably [16]. As an intentional tort, the crux of this claim is whether there was "an unpermitted touching of the plaintiff by the defendant or by some object set in motion by the defendant" [16]. Whether the defendant clinician deviated from the standard of care does not matter, and expert testimony is generally not required. "An operation undertaken without consent (battery), even if perfectly performed with good medical results, may entitle a plaintiff to at least nominal and even punitive damages" [17]. A separate claim—informed consent, which is discussed later in this chapter—concerns itself with whether the clinician provided the patient with enough information to make an informed choice about his or her care. "[T]he question in a medical battery case is much simpler: Did the patient consent at all?" [16].

Outside of the healthcare context, the intentional tort of battery occurs when, for example, one person punches another. It similarly occurs when someone strikes someone else with a thrown object and when one person touches another in an offensive, if not necessarily physically harmful, way. Within the healthcare context, the harmfulness or offensiveness of a clinician's touch is often not in question. Surgeries, while they may ultimately improve health, undoubtedly cause physical harm. A scalpel, no matter how skillfully wielded, cuts through human tissue, causing physical harm. Many routine physical examinations, while not causing physical harm, would nonetheless be considered offensive if the patient did not consent. For example, a breast examination and prostate examination may lead to better health in the long term as they allow clinicians to discover potentially deadly conditions. Performed without consent however, these examinations are highly offensive and would therefore constitute battery.

While the intent behind an action or the harmfulness or offensiveness of that action regularly dominate non-medical battery claims, the existence and nature of the consent given tend to dominate medical battery claims. "A medical battery occurs when a physician performs a procedure without the patient's consent" [18] because the lack of consent "render[s] the [procedure] an unauthorized touching" [19]. Leaving aside for a moment the situation where a patient presents to a healthcare facility unconscious, a clinician batters a patient when that clinician lacks any consent to touch the patient. "Examples of a battery in a medical setting [where the clinician lacks consent to touch the patient] include cases … where a surgeon uses an anesthetic specifically prohibited by the plaintiff, and where a different surgeon than the one authorized by the plaintiff performs the procedure" [20]. In these examples, the clinician has no permission to touch the patient and therefore batters the patient when the touching occurs.

In evaluating whether consent has been given, courts take the circumstances into account. Perhaps most obvious is the situation in which an unconscious patient arrives at an emergency department. Clearly, this patient cannot provide affirmative consent to treatment, but clinicians provide care in this situation every day. In this situation, clinicians rely on exception to the consent requirement which allows treatment when a patient cannot consent but requires emergency treatment. More specifically, "a medical professional is not required to obtain consent to medically treat a patient if 'an emergency arises and treatment is required in order to protect the patient's health, and it is impossible or impractical to obtain consent either from the patient or from someone authorized to consent for him'" [21]. This exception does not provide clinicians with a blank check to treat patients once an emergency arises, however. "The mere existence of an emergency that places a patient at risk of future harm does not give a physician 'a license to force medical treatment and ignore a patient's exercise of the right to refuse medical treatment'" [22]. If a patient has previously refused a particular treatment, the fact that an emergency arises does not allow clinicians to provide that treatment.

Whether in an emergency situation or not, medical battery claims need not be based solely on the complete absence of consent. A plaintiff can succeed on a medical battery claim by establishing "substantial variance of the procedure from the consent granted" [23]. Thus, battery "occurs when a physician obtains the patient's consent to perform one type of treatment, but performs a substantially different treatment for which the plaintiff gave no consent" [24]. In a recent California case, for example, a patient "consented to have a small mass removed from his scrotum" through surgery [24]. During surgery, however, the surgeon realized that the mass was much larger and believed it was potentially malignant. Without consulting either the patient or his designated proxy, who was in a nearby waiting room, the surgeon performed a much more extensive surgery, removing a large part of the patient's scrotum and penis. This left the patient impotent and caused several other severe and long-term harms to the patient. The California court evaluating the medical battery claim concluded that, because the surgeon exceeded the scope of consent, he had battered the patient [24].

Overall, medical battery claims tend to be legally simpler than medical malpractice claims because they turn not on complicated investigations into the standard of care but on inquiries into whether consent was given for the treatment provided. Medical battery claims also tend to be rarer than their malpractice cousins for several reasons. For one, clinicians do not actually batter their patients that often. For separate ethical reasons, clinicians are typically careful to obtain consent before treating a patient. And healthcare facilities often require patients to sign a variety of forms that provide consent for the treatment under consideration and for related treatment that may become necessary. Additionally, even among claims that are best classified as battery, both plaintiffs and defendants have an incentive to bring these claims under a negligence theory. Medical malpractice insurance may not provide coverage for battery claims, while such insurance almost invariably covers malpractice claims. For plaintiffs seeking the largest recovery, reaching a clinician's insurance policy via a malpractice claim generally offers the best avenue of recovery. For defendants seeking to avoid satisfying a judgment with their own assets, encouraging plaintiffs to assert malpractice, and not battery, claims is in their best interest. Accordingly, even among the battery claims that eventually reach a courtroom, many of them may be concealed in a malpractice claim.

2.3   Informed Consent

Where battery claims concern themselves with the existence of consent, informed consent claims focus on the nature of that consent and whether the patient was able to make an informed choice about his or her treatment. These types of claims may be more familiar to many clinicians since they involve considerations of what information a patient must be given before he or she can provide informed consent. In some states, informed consent claims originally arose as a species of battery, and in other states, they arose as a species of negligence (i.e., malpractice). Most states now agree that informed consent claims are a type of negligence claim. And medical malpractice claims and informed consent claims are often considered distinct claims, though each is, at its core, a form of negligence [25]. The Supreme Court of Oklahoma, for example, has held that "[i]f the physician obtains a patient's consent but has breached this duty to inform, 'the patient has a cause of action sounding in negligence for failure to inform the patient of his options, regardless of due care exercised at treatment'" [26].

Informed claims generally hinge on whether the clinician has provided the patient with a sufficient explanation of the risks associated with a particular course of treatment and an explanation of the reasonable alternatives to that treatment to allow the patient to make an informed choice as to his or her care. Though states differ somewhat in the particulars of their informed consent claims, the Supreme Court of Iowa has provided a general statement of what a plaintiff must prove in an informed consent claim. The elements include:

1. The existence of a material risk [or material information] unknown to the patient;
2. A failure to disclose that risk [or information] on the part of the [clinician];
3. Disclosure of the risk [or information] would have led a reasonable patient in plaintiff's position to reject the medical procedure or choose a different course of treatment;
4. Injury [i.e., harm] [27].

The first element generally requires that the clinician disclose material information to a patient, and this element turns on what qualifies as material. In evaluating materiality, courts in different states take slightly different approaches. One approach asks what the reasonable clinician would have disclosed to the patient. Another approach asks what the actuals patient would have considered material. But the most popular approach is the "objective reasonable-patient standard" [27]. Under this standard, information is material if an objective, reasonable patient in the plaintiff's position would have found the information material. Importantly, "[a]lthough the scope of disclosure is measured by the information a reasonable patient would need to know in order to make an informed and intelligent decision, the physician need not disclose every conceivable risk" [28]. Though not universally required, expert testimony is often needed to establish information or risks as material [28, 29].

While the objective, reasonable-patient standard may provide little *ex ante* information to clinicians that they can use in determining what information they must disclose to patients, it provides courts with sufficient flexibility to determine on a case-by-case basis whether a clinician has disclosed all material information. Indeed, the Maryland Court of Appeals (the highest court in Maryland) determined that the question of whether a reasonable person would have deemed information to be material is a factual question that must be answered by a jury. That court also refused to adopt a systematic list of what information must be disclosed in general [30]. Similarly, the Supreme Court of Kentucky "has expressly recognized [that] informed consent 'is a process, not a document'" [29]. Documents are, however, relevant in evaluating whether the patient has consented.

Having determined that some information was material, the next element requires that the clinician failed to disclose that information. The third element then requires an evaluation of whether a reasonable patient would have chosen a different course of treatment had that patient had access to the relevant, material information. To establish this element, "a plaintiff must come forward with evidence that a patient could reasonably be expected to decline the medical procedure had the patient been informed of the kind of risk or danger which resulted in the harm." This evidence is essential because "[e]ven though a patient's subjective testimony regarding whether he or she would have consented to the medical procedure may be considered, it is not controlling because it is tied to hindsight and may or may not be reasonable" [31].

Finally, the element of harm is less straightforward in informed consent claims than the two types of claims discussed above. In general, the harm "analysis in an informed consent case involves a comparison between the condition a plaintiff would have been in had he or she been properly informed and not consented to the risk, with the plaintiff's impaired condition as a result of the risk's occurrence" [19]. Some states, however, will recognize the existence of an harm purely from the loss of the patient's autonomy. In failing to provide all the material information, a clinician can injure the patient even if the underlying procedure goes exactly as planned. The harm suffered by the patient is the inability to make informed choices about his or her treatment.

Overall, though informed consent claims are often considered a type of negligence, they proceed differently than medical malpractice claims. These claims exist to ensure that clinicians provide patients with sufficient information to enable them to make informed choices. The coexistence of informed consent and medical malpractice claims, however, does not mean that a plaintiff must limit himself or herself to one claim. In general, depending on the type(s) of harm(s) that occurred, a plaintiff may assert both medical malpractice, informed consent, and medical battery claims simultaneously. State rules of civil procedure generally authorize plaintiffs to assert multiple claims at the same time, even if some of those claims are inconsistent with one another. A plaintiff may not prevail on multiple inconsistent claims, but that plaintiff may need to assert them in the first instance and resolve any evidentiary issues during litigation.

2.4   Vicarious Liability

As healthcare delivery has evolved from the solo practitioner visiting patients and the local physician treating all his or her patients in the local hospital to team-based medicine with new types of licensed clinicians serving in new roles and dedicated hospitalists treating hospitalized patients, vicarious liability has become more important. Theories of vicarious liability allow a plaintiff to hold the principal—the person or organization in charge—when the principal's agent—the clinician—injures the patient in some way. In general, patients may hold any type of clinician liable for any of the above torts as long as the elements are satisfied. When the clinician who commits one of those torts, however, is supervised or employed by another clinician or by a healthcare organization, theories of vicarious liability allow the patient to hold the supervising clinician or organization liable under certain conditions.

Two separate vicarious liability doctrines are particularly relevant. While these doctrines may bear slightly different names in different states and may be treated slightly differently by those states, they form the core of the legal basis for holding principals liable for the torts of their agents. First, the doctrine of *respondeat superior* provides that "[a]n employer is subject to liability for torts committed by employees while acting within the scope of their employment" [32]. In this context, "employee" and "employer" do not necessarily coincide with their typical workplace definitions or even their definitions for the purposes of tax law. Instead, these terms replace the older terms "master" and "servant." Though the newer terms replaced the older, they retain their older, more general meanings.

The scope of *respondeat superior* "is limited to the employment relationship and to conduct falling within the scope of that relationship because an employer has the right to control how work is done" [32]. Though all principals may possess some right of control over their agents, the right necessary for the application of *respondeat superior* "is more detailed than the right of control possessed by all principals" [32]. When attempting to apply *respondeat superior*, courts must answer two broad questions. First, the court must ascertain whether the individual who committed the tort functions as an employee. In general, "an employee is an agent whose principal controls or has the right to *control the manner and means* of the agent's performance of work"(emphasis added) [32]. Second, the employee must be acting within the scope of employment at the time of the tort. "An employee acts within the scope of employment when performing work assigned by the employer or engaging in a course of conduct subject to the employer's control" [32]. However, "[a]n employee's act is not within the scope of employment when it occurs within an independent course of conduct not intended by the employee to serve any purpose of the employer" [32].

Without answering each of these broad questions in the affirmative, courts may worry about the fairness of imposing vicarious liability on employers. Additionally, a primary justification for vicarious liability is that, by imposing liability on employers, those employers will take steps to prevent their employees from committing torts in the first place. And employers will often be in a better position to take these steps than courts acting *ex post*. Thus, only the actions of employees, whom employers can control, will subject their employers to liability under the theory of *respondeat superior*. And only when employees act within the scope of their employment will employers be subjected to liability.

The theory of *respondeat superior* often becomes relevant when nurse practitioners or physician assistants deliver care. Many states require that physicians supervise these clinicians when they deliver care, often imposing by statute the level of control that *respondeat superior* requires [33]. For example, the Supreme Court of Tennessee addressed the liability of both a physician assistant's supervising physician and the clinic where the physician assistant worked. After the physician assistant allegedly negligently injured the patient, the court relied on a state statute providing that the "[s]ervices rendered by the physician assistant must be provided under the supervision, direction, and ultimate responsibility of a licensed physician" [32]. The court determined that this was sufficient to subject the supervising physician to vicarious liability. Additionally, the court explained that a "health care facility which employs a physician assistant may face liability for the physician assistant's negligence under the agency theory of *respondeat superior*" [32].

While *respondeat superior* is the appropriate theory of liability for clinicians functioning as employees, not all clinicians work as employees. Physicians, in particular, may or may not function as employees of the clinics and hospitals where they deliver care. If not classified as employees, physicians generally work as independent contractors, which means that hospitals have less control over how they deliver care than if they functioned as employees. *Respondeat superior* is not appropriate for independent contractors, but the related doctrine of apparent agency may allow plaintiffs to hold an agent's purported principal liable. For example, in a Florida case, a patient suffered a harm because a radiologist allegedly failed to diagnose an abdominal abscess prior to surgery [34]. This failure led to post-surgical complications, and the patient sought to hold the hospital where the surgery was performed liable for the radiologist's negligence. In response, the hospital asserted that the radiologist "was an independent contractor and was not an agent, servant, or employee of" the hospital [34].

The Florida court recognized that hospitals use different staffing models, with some employing physicians and others entering into contractual arrangements with separate legal entities that provide medical services through their physicians. Acknowledging that allowing hospitals to escape liability under these circumstances may prove unfair to plaintiffs who rely on the hospital to supply competent physicians, the court explained that the doctrine of apparent agency may offer plaintiffs an avenue of relief. The rationale for the doctrine of apparent agency is that the principal—here, the hospital—should be prohibited from denying the authority of the agent—here, the radiologist—"when the principal permitted an appearance of authority in the agent and, in so doing, justified a third party's reliance upon that appearance of authority as if it were actually conferred upon the agent" [34]. If the doctrine of apparent agency applies, then the hospital will be liable for the acts and omissions of the physician even though the physician is not an employee of the hospital and the hospital lacks authority to control the physician.

As with *respondeat superior*, states approach the requirements for apparent agency slightly differently, but its application generally requires "(a) a representation by the purported principal; (b) a reliance on that representation by a third party; and (c) a change in position by the third party in reliance on the representation" [34]. In other words, the purported principal must say or otherwise communicate that the purported agent has the authority to act on behalf of the principal. The third party—usually the plaintiff—must rely on that communication and change his or her course of action based on that reliance. If all these elements are satisfied, then the plaintiff may hold the purported principal liable for the actions of the agent even though no formal employment relationship exists. Thus, "[u]nder certain circumstances … a hospital may be held vicariously liable for the acts of physicians, even if they are independent contractors, if these physicians act with the apparent authority of the hospital" [34].

With respect to both *respondeat superior* and apparent agency, states differ on whether plaintiffs may hold principals liable for the non-negligent acts of their agents. Accordingly, plaintiffs can hold individual or institutional principals liable for the medical malpractice or violations of informed consent committed by their agents (where these claims are a species of negligence), but not all states permit plaintiffs to hold principals vicariously liable for intentional torts like battery.

2.5   Federal Tort Claims Act

As mentioned at the beginning of the chapter, a somewhat different set of rules apply to clinicians practicing in federal settings. The United States enjoys sovereign immunity from suit, meaning that it cannot be sued without its consent. This sovereign immunity extends to federal employees. At various points in its history, the United States has given its consent to suit, and most relevant to tort law is the Federal Tort Claims Act (FTCA). Under this act, individuals harmed by the torts of federal employees may assert their claims in federal court in certain circumstances. The FTCA provides that the "[t]he United States shall be … liable in the same manner and to the same extent as a private individual under like circumstances" [35]. This waiver of sovereign immunity comes with several caveats, however. In addition to requiring that claims be asserted in federal court, the FTCA forbids the imposition of punitive damages and pre-judgment interest, requires that all trials be conducted without a jury, and exempts certain actions from suit (including a number of intentional tort claims and claims based on the (non-)performance of a discretionary function). The FTCA also provides that suits against federal employees will proceed against the United States and that those employees may not separately face a suit.

Sometimes, the defendant's federal employee status is obvious. Suing staff at a national park, for example, would obviously involve the FTCA. In the healthcare context, however, it is not always immediately obvious which clinicians are federal employees. Clinicians at Veterans Affairs hospitals are federal employees, and less obviously, so are clinicians at federally qualified health centers (FQHC). FQHCs receive federal funding and grant money from the United States Public Health Service [36]. Because of this, FQHC employees are deemed employees of the Public Health Service, and tort claims against them must proceed under the FTCA [36]. Even if these employees appear to be private employees of individual hospitals, suits against them are treated as suits against the United States under the FTCA.

One final wrinkle of the FTCA that often presents a problem is its time limitation. Aggrieved patients have two years from the accrual of their claim to present it to the appropriate federal agency [37]. A claim accrues when the injured party knows both existence and cause of the harm [37]. Many plaintiffs have had their suits time barred under this limitation provision because they filed outside the two-year timeframe.

2.6   Tort Reform

Over the last fifty years, tort law has come to play a relatively more important role in the delivery of healthcare. And as clinicians have increasingly faced liability for their actions, policymakers have become concerned about tort law playing too big of a role in healthcare delivery. One important concern is the threat of defensive medicine. Defensive medicine is "a deviation from sound medical practice that is induced primarily by a threat of liability" [38]. While researchers debate both the existence and pervasiveness of defensive medicine, it has the potential to drastically increase costs while offering no benefit to (and potentially harming) patients. Estimates of the total cost of defensive medicine vary widely, but one estimate placed it around $55 billion [39].

To combat defensive medicine and to reduce malpractice liability pressure on physicians generally, many states have adopted various reforms to their system of tort law. Many (though, not all) of these reforms work by reducing the amount of damages plaintiffs may recover for their injuries. Two types of monetary damages may be awarded to plaintiffs following a trial victory. First, courts award compensatory damages to compensate plaintiffs for the harms they have suffered, and compensatory damages are by far the most important in medical malpractice cases generally. Compensatory damages come in two varieties. Economic damages compensate plaintiffs for economic harms like lost wages, medical bills, rehabilitation costs, etc. Noneconomic damages compensate plaintiffs for more difficult to quantify harms like pain and suffering or humiliation. Second, courts award punitive damages to punish and deter reprehensible behavior. Courts rarely impose punitive damages and reserve them for the most egregious cases. Tort reforms target these different types of damages. By reducing the monetary damages that are awarded, tort reform reduces the severity (i.e., cost) of individual malpractice claims from the clinician's perspective. Tort reform can also reduce the frequency of medical malpractice claims by reducing damages. Attorneys may be unwilling to represent injured patients when the monetary damages available may not be sufficient to cover the costs of asserting the claim. Without attorneys, patients will be unable to recover appropriate compensation for their injuries. And even if patients are able to find willing attorneys, they still may not receive full compensation for their injuries under damages caps.

California, for example, adopted one of the most famous tort reforms in its Medical Injury Compensation Reform Act (MICRA). This 1975 reform altered the ability of courts to impose noneconomic damages. Proponents of MICRA and other caps on noneconomic damages generally argue that noneconomic damages are inherently difficult to quantify. Difficulty in quantifying harms like pain and suffering leads to highly variable, and often very large, noneconomic damages awards. By capping these awards, proponents contend, clinicians need not worry as much about liability and can therefore avoid practicing defensive medicine. Many other states have similarly implemented caps on noneconomic damages [40]. Evaluations of these caps have yielded mixed results, with some evidence suggesting that they clearly reduce malpractice liability pressure [41] and defensive medicine and other evidence finding that they have no (or even a perverse) effect [42].

While researchers have focused heavily on noneconomic damages caps, states have enacted a variety of other types of reform. For example, punitive damages caps restrict the amount of damages courts may impose to punish or deter defendants. Collateral source rule reform allows defendant clinicians to introduce evidence that plaintiff patients received compensation for their injuries from other sources (such as their health insurer). Joint and several liability reform restricts the ability of plaintiffs to extract a full damages award from a single defendant when multiple defendants were at fault. Contingency fee reform caps the share of a plaintiff's award his or her attorney may retain as payment for legal services. In addition to these reforms, several states have implemented procedural hurdles that make it more difficult for plaintiffs to assert a medical malpractice (or informed consent) claim in the first place. Some states require that plaintiffs include with their initial filing an expert witness report from another physician that states the nature and cause of the plaintiff's harm.

Among these, and other, reforms that states have enacted, some are more effective than others. One tort reform that became quite popular over the last two decades is the passage of apology laws, and these laws offer an important cautionary tale. In the early 2000s, apology laws became quite popular as an alternative form of tort reform. Apology laws simply facilitate apologies and do not directly deny recovery to plaintiffs. Based on research demonstrating that patients are less likely to assert a medical malpractice claim and are more likely to settle those claims that they do assert for lower amounts when they receive an apology, states attempted to facilitate apologies from clinicians to patients.

To facilitate apologies, states passed apology laws which made statements of sympathy or apology inadmissible in medical malpractice trials. States did so under the theory that, by preventing plaintiffs from using apologies as evidence of liability, clinicians would be more willing to deliver apologies. Based on the evidence available at the time, many of these laws were passed. With their passage, more apologies should have led to a decrease in medical malpractice litigation and lower settlement amounts. However, once passed, apology laws had the opposite effect. Instead of reducing medical malpractice litigation, they increased litigation, with the attendant effects on malpractice insurance premiums and defensive medicine one would expect in connection with increased litigation [43–45]. Apology laws had this perverse effect on malpractice litigation because the existing research had not accounted for the signaling function of apologies. When a clinician apologizes effectively to a patient, the patient may feel less anger towards the clinician (and therefore be less likely to pursue legal options). However, if the patient did not know whether his or her harm stemmed from a normal, if unfortunate, adverse event or from underlying malpractice, an apology may signal that malpractice ultimately caused the harm. Having received this signal from the clinician that malpractice occurred, patients may seek other admissible evidence of malpractice and pursue a claim they otherwise would not have. In this way, apology laws can have the perverse effect of increasing malpractice litigation.

The debate over tort reform, with staunch proponents on both sides, is far from over. And the example of apology laws illustrates both the willingness of state governments to pursue novel tort reform and the problems associated with trying new types of reform. The coming years will likely see a new wave of tort reform, and clinicians will want to carefully evaluate the evidence they are given before deciding whether to support such reform.

3   Application

Case #1: Sheeley v. Mem'l Hosp., 710 A.2d 161(R.I. 1998) Dr. Ryder attended the birth of a healthy baby delivered by Ms. Sheeley. During labor, Dr. Ryder, a second-year family practice resident physician, performed an episiotomy on Ms. Sheeley. Dr. Ryder performed a repair of the episiotomy after the birth. Ms. Sheeley developed complications in the area in which Dr. Ryder had performed the episiotomy and required surgery to correct a rectovaginal fistula that developed. Ms. Sheeley asserted a medical malpractice claim against Dr. Ryder and sought to introduce the testimony of Dr. Leslie, a board-certified obstetrician/gynecologist (OB/GYN). Dr. Leslie, who had practiced as an OB/GYN for several decades and served as a clinical professor of obstetrics and gynecology, planned to testify about the standard of care as it related to the performance of an episiotomy.

Dr. Ryder sought to prevent Dr. Leslie from testifying because Dr. Leslie was not an expert in the same field as Dr. Ryder. Dr. Ryder essentially argued that Dr. Leslie was overqualified based on his training and extensive experience in the field of obstetrics and gynecology. The court rejected these arguments and held that "[t]he appropriate standard of care to be utilized in any given procedure should not be compartmentalized by a physician's area of professional specialization or certification." Instead, "the focus in any medical malpractice case should be the procedure performed and the question of whether it was executed in conformity with the recognized standard of care, the primary concern being whether the treatment was administered in a reasonable manner." The court explained that any physician with knowledge or familiarity with the relevant procedure—here an episiotomy—could testify as to the standard of care.

In reaching this conclusion, the court rejected the traditional justifications for the local rule and the specialty-specific rule. While juries may consider the resources available to a defendant physician (such as the presence of advanced diagnostic equipment, the defendant physician's training, and the length of time the defendant physician has practiced), none of these issues are determinative as to whether a given expert may testify about the standard of care. Accordingly, the court embraced a national rule and rejected a specialty-specific rule for testifying experts. The court remanded the case back to the trial court for additional proceedings based on its conclusion that Dr. Leslie should have been allowed to testify.

Case #2: Burchell v. Fac. Physicians & Surgeons of Loma Linda Univ. Sch. of Med., 54 Cal. App. 5th 515 (2020) Mr. Burchell sought care when he discovered a small mass in his scrotum that caused him some pain. After preliminary tests, Mr. Burchell agreed to undergo surgery to remove the mass and send it away for testing. During surgery, the surgeon, Dr. Barker, discovered that the mass was much larger and more extensive than anticipated. Believing the mass was malignant, Dr. Barker removed the mass, which involved excising tissue from both the scrotum and penis, without consulting either Mr. Burchell, who was anesthetized, or his medical proxy, who was in the facility's waiting room. Later testing revealed the mass was benign, but Mr. Burchell suffered extensive harm as a result of its surgical removal. This harm included impotency, infections, deformation, and pain.

Burchell asserted claims of professional medical negligence (medical malpractice) and medical battery against Dr. Barker and the Faculty Physicians & Surgeons of the Loma Linda University School of Medicine (FPS) as his employer. A trial court entered judgment of over $10 million in noneconomic damages in addition to over $20,000 in economic damages (which the parties had stipulated to). A pre-existing agreement provided that only FPS was liable for these damages. FPS appealed, arguing that California's MICRA applied to the noneconomic damages award and that, accordingly, the award should be reduced to the $250,000 limit imposed by MICRA.

The appellate court rejected FPS's arguments, explaining that MICRA applies only to claims based on negligence. The court concluded that Mr. Burchell's medical battery claim was not based on negligence. Recognizing that some failures to obtain consent are properly characterized as a deviation from the standard of care, the court explained that the failure to obtain consent in this case fell into the category of the intentional tort of battery. Unlike a traditional informed consent claim—a type of negligence in California—where a clinician fails to properly inform the patient of all material risks and reasonable alternatives to a proposed course of treatment, the surgeon in this case failed to obtain any consent to perform the more extensive surgery. Mr. Burchell originally consented to the removal of a small mass, but the surgeon performed a substantially different procedure by excising a much larger mass and tissue from the surrounding scrotum and penis. The surgeon did so without consulting either Mr. Burchell or his medical proxy. This, according to the court, constituted medical battery. And because the surgeon committed the intentional tort of battery, the limitations in MCIRA, which apply only to negligence, were not applicable.

4   Summary

In summary, tort law plays an important role in the delivery of healthcare. The most common tort in the healthcare context is medical malpractice, which is a form of negligence. This tort ensures that patients receive reasonable care. If a clinician deviates from the standard of reasonable care, the patient may recover for any harms suffered as a result. Medical malpractice claims often turn on the testimony of experts, as jurors cannot determine the standard of care for themselves. The importance of experts makes the rules governing their testimony crucial to medical malpractice claims. Some states require that any testifying expert not only have familiarity with the relevant procedure or treatment but also practice in the same specialty as the defendant clinician. Other states require only familiarity with the relevant procedure or treatment. Similarly, states differ on whether the testifying expert must be familiar with the defendant clinician's community (or similar community) or must be familiar with the national standard of care. While seemingly small, these legal details are often the biggest points of contention in medical malpractice cases.

While medical malpractice may be the most common tort in the healthcare context, other torts also have roles to play. Instead of ensuring that patients receive at least reasonable care, informed consent claims protect patients' autonomy by requiring clinicians to provide all material information that is relevant to a patient's treatment decision. If a clinician fails to so inform a patient, the patient can hold that clinician liable for any harms that result from the failure to provide material information. As with medical malpractice, the details of an informed consent claim vary from state to state, but these claims generally serve to ensure that patients remain in charge of their care. In contrast to informed consent claims, which focus on whether a patient has made an informed choice to consent to a course of treatment, battery claims focus on whether a patient has consented to be touched at all. The fact that a clinician possesses a medical, nursing, or some other healthcare-related license, does not imbue that clinician with blanket permission to touch others. Instead, clinicians must receive consent before touching their patients. Usually this consent is freely given, but in circumstances where clinicians touch someone in a substantially different way than the original consent covered, these clinicians may be held liable for battering their patient.

Because of the team-based nature of modern healthcare, clinicians are not the only ones at risk of liability for the harms they cause. Their employers, including institutional providers and other clinicians, may also be liable for acts and omissions that lead to harm. With so many clinicians delivering care in the modern healthcare system, patients can often rely on the relationships among clinicians and various institutional providers to assert vicarious liability claims. These claims may provide patients with an important avenue of obtaining monetary damages, but they expose institutional providers and clinicians to liability, even if those providers and clinicians did not cause the relevant harm themselves.

The increasing importance of tort law within the healthcare system has not gone unnoticed. Over the past several decades, states have passed a variety of reforms designed to tamp down the role of tort law in the delivery of healthcare. While the evidence supporting the need for these reforms and the efficacy of these reforms is mixed, tort reform continues to dominate many healthcare policy debates. Understanding these debates and the tort law that underlies them will only become more important for clinicians going forward.

References

1. Tort. In: Black's Law Dictionary. 11th ed. 2019.
2. Irvin v. Smith. Vol. 272, Kan. 2001. p. 112.
3. Nicholas v. Mynster. Vol. 213, N.J. 2013. p. 463.
4. Woodard v. Custer. Vol. 719, N.W.2d. 2006. p. 842.
5. Frank v. Smith. Vol. 127, A.D.3d. 2015. p. 1301.
6. Spacht v. Troyer. Vol. 288, Ga. App. 2007. p. 898.
7. Finkelstein A, Gentzkow M, Williams H. Sources of Geographic Variation in Health Care: Evidence from Patient Migration. Q J Econ. 2016 Nov;131(4):1681–726.
8. Orcutt v. Miller. Vol. 95, Nev. 1979. p. 408.
9. Shipley v. Williams. Vol. 350, S.W.3d. 2011. p. 527.
10. Brown v. United States. Vol. 355, F. App'x. 2009. p. 901.
11. Sheeley v. Mem'l Hosp. Vol. 710, A.2d. 1998. p. 161.
12. Dlouhy v. Kootenai Hosp. Dist. Vol. 167, Idaho. 2020. p. 639.
13. Blumstein JF, McMichael BJ, Storrow AB. Constraints on Medical Liability Through Malpractice Safe Harbors. JAMA Health Forum. 2020;1(8):e200961.
14. Frakes M. The Impact of Medical Liability Standards on Regional Variations in Physician Behavior: Evidence from the Adoption of National-Standard Rules. American Economic Review. 2013 Feb;103(1):257–76.
15. Sekerez v. Rush Univ. Med. Ctr. Vol. 2011, IL App (1st). 2011. p. 090889.
16. Shuler v. Garrett. Vol. 743, F.3d. 2014. p. 170.
17. Whitley-Woodford v. Jones. Vol. 600, A.2d. 1992. p. 946.
18. Carter v. Pain Ctr. of Arizona. Vol. 367, P.3d. 2016. p. 68.
19. Howard v. Univ. of Med. & Dentistry of New Jersey. Vol. 800, A.2d. 2002. p. 73.
20. Schwaller v. Maguire. Vol. 2003, WL. 2003. p. 22976339.
21. In re Est. of Allen. Vol. 365, Ill. App. 3d. 2006. p. 378.
22. Curtis v. Jaskey. Vol. 326, Ill. App. 3d. 2001. p. 90.
23. Holzrichter v. Yorath. Vol. 2013, IL App (1st). 2013. p. 110287.
24. Burchell v. Fac. Physicians & Surgeons of Loma Linda Univ. Sch. of Med. Vol. 54, Cal. App. 5th. 2020. p. 515.
25. Salvador v. Main St. Fam. Pharmacy. Vol. 251, So.3d. 2018. p. 1107.
26. Parris v. Limes. Vol. 277, P.3d. 2012. p. 1259.
27. Andersen v. Khanna. Vol. 913, N.W.2d. 2018. p. 526.
28. White v. Leimbach. Vol. 959, N.E.2d. 2011. p. 1033.
29. Univ. Med. Ctr., Inc. v. Shwab. Vol. 628, S.W.3d. 2021. p. 112.
30. Goldberg v. Boone. Vol. 912, A.2d. 2006. p. 698.
31. Acord v. Porter. Vol. 58, P.3d. 2020. p. 747.
32. Restatement (Third) Of Agency. 2006.
33. McMichael BJ. Healthcare Licensing and Liability. Indiana Law Journal. 2020;95:821–81.
34. Roessler v. Novak. Vol. 858, So.2d. 2003. p. 1158.
35. FTCA. 28 U.S.C. § 2674.
36. P.W. by Woodson v. United States. Vol. 990, F.3d. 2021.
37. Skwira v. United States. Vol. 344, F.3d. 2003. p. 64.
38. Studdert DM, Mello MM, Sage WM, DesRoches CM, Peugh J, Zapert K, et al. Defensive medicine among high-risk specialist physicians in a volatile malpractice environment. JAMA. 2005 Jun 1;293(21):2609–17.
39. Mello MM, Chandra A, Gawande AA, Studdert DM. National costs of the medical liability system. Health Aff (Millwood). 2010 Sep;29(9):1569–77.
40. Avraham R. Database of State Tort Law Reforms (7.1) [Internet]. 2021. Available from: https://papers.ssrn.com/sol3/papers.cfm?abstract_id=902711
41. Cotet AM. The Impact of Noneconomic Damages Cap on Health Care Delivery in Hospitals. American Law and Economics Review. 2012 Mar 1;14(1):192–234.

42. Paik M, Black B, Hyman DA. Damage caps and defensive medicine, revisited. J Health Econ. 2017 Jan;51:84–97.
43. McMichael BJ. The Failure of 'Sorry': An Empirical Evaluation of Apology Laws, Health Care, and Medical Malpractice. Lewis & Clark Law Review. 2018;
44. McMichael BJ, Van Horn RL, Viscusi WK. Sorry Is Never Enough: How State Apology Laws Fail to Reduce Medical Malpractice Liability Risk. Stan L Rev. 2019;
45. McMichael BJ. Insuring Apologies. 2021;48:651–718.

Liability in Reproduction and Birth

Jamie Abrams

Contents

Key Points

- Clinicians may owe dual duties of care when treating pregnant persons and the fetuses they carry. A duty to the fetus, however, does not undermine the strong principles preserving the pregnant person's autonomy in medical decision-making.
- Informed consent and robust culturally competent patient communication are critical to managing liability in caring for pregnant persons given the extensive documented history of gendered, classed, and racialized oppression in reproductive medicine.
- Liability can arise at each stage of the clinical relationship, including conception, pregnancy, labor, and post-delivery care. Claims can be styled as informed consent, negligence, wrongful death, wrongful birth, wrongful life, and wrongful pregnancy. Generally, physicians in each community will set the professional standards of care which govern liability.
- Standards of care governing assisted reproductive technologies remain highly fluid and rapidly changing from state to state along with evolving norms in law, society, politics, and culture.

J. Abrams (✉)
Washington College of Law, American University, Washington, DC, USA
e-mail: jamieabrams@wcl.american.edu

© The Author(s), under exclusive license to Springer Nature Switzerland AG 2022   151
A. S. Pasha (ed.), *Laws of Medicine*,
https://doi.org/10.1007/978-3-031-08162-0_9

Legal Concepts

- Assisted Reproductive Technologies—conception involving medical interventions.
- Wrongful Birth/Wrongful Life/Wrongful Pregnancy—claims alleging that the clinician's negligence prevented the avoidance of or termination of a pregnancy, thus causing pregnancy and/or birth.
- Informed Consent—the failure to disclose material risks to a patient and the subsequent manifestation of those risks to the patient.
- Negligence—the failure to conform with professional standards of care causing harm to the patient.
- Professional Standards of Care—the specific duties that clinicians owe to their patients.
- Causation—a legal element limiting plaintiff's recovery to only the damages that are attributable to the clinician's breach.

1   Introduction

The complex law of reproduction and birth requires clinicians to be savvy with medicine, law, society, and culture. As to the law, reproduction and birth are both like other aspects of law and medicine and extraordinarily unique. Clinicians face many avenues of potential liability following reproduction and birth caregiving. There are three genres of possible claims: negligence, informed consent, and wrongful birth/wrongful life/wrongful pregnancy. Intentional tort liability can also arise from inappropriate touching rising to a civil battery, but this liability is outside this chapter's scope.

Negligence and informed consent cases, introduced throughout other chapters, are rooted squarely in standards of care governing obstetric and gynecological work. Standards of care govern nearly every aspect of childbirth and the law, such as the use of electronic monitoring, training personnel, drug administration, and diagnosing conditions like hypertension or preeclampsia. Standards of care are shaped within the medical community by peers, within institutions through hospital protocols, and by governing bodies such as the American College of Obstetricians and Gynecologists (ACOG).

Only issues of common knowledge not requiring medical understanding (like a sponge left in a surgical site) would be left to a lay jury without expert testimony as to the governing standard of care. These issues of common knowledge can sometimes arise in pregnancy, such as when pregnant persons suffer burns, falls, or other injuries that a lay person can identify as falling short of a standard of care. Usually, expert testimony from medical peers is required to support liability, thus embedding a high degree of legal liability management within the profession itself.

Negligence claims allege that specific standards of care were breached and, had the standard of care been complied with, the resulting harm to the patient would not have occurred. Informed consent claims allege that a specific material risk was unknown to the plaintiff and, had that risk been known, the patient would not have consented. In either claim, an *actual* breach in the standard of care or the failure to inform of material risks must cause the harm. Breaches of the standard of care that do not cause harm are not actionable. For example, if a clinician performs a cesarean section while under the influence of alcohol, to use an extreme example, there is no tort liability for this breach on its own. The liability arises only if harms flow from the breach.

Causation can be analyzed through an "if → then" counterfactual: if the clinician did not breach a standard of care, then would the plaintiff's harm have occurred? This can be particularly dicey in childbirth because other factors, such as the patient's conduct, underlying health conditions, and caregiving of other actors and entities can complicate liability. In complex cases, the plaintiff need only prove that the clinician's breach in the standard of care was *at least a substantial factor* in the patient's harm. It is best to understand multiple possible defendants as an "and" rather than an "or" when assessing who might face liability. Most plaintiffs will cast the widest net possible by including each clinician and institution in the caregiving chain [1]. Hospital liability, for example, will not immunize clinician liability, and vice versa. Think of liability as a pie that can be sliced up among various wrongdoers.

Reproduction and birth can also yield liability under the umbrellas of wrongful death/wrongful birth/wrongful pregnancy. These claims are complex and unpacked more thoroughly below. Wrongful death is rooted in the basic negligence claim described above, but the plaintiffs are the *loved ones* of the deceased (e.g., the surviving children and spouse). Wrongful birth and wrongful life involve claims in which a child would not have been born had the birthing person been able to abort the pregnancy. Here, the clinician's negligence involves failing to provide material information to the pregnant person, not in the medical procedures themselves. These claims are uniquely intertwined with debates about public policy and morality because they attempt to place a value on the cost of life.

The law of reproductive medicine is uniquely complex at birth because there are multiple possible plaintiffs, the fetus(es) and the pregnant person. Accordingly, clinicians may have to navigate multiple duties [2]. Notably, it is a distortion of law and medicine to position potential fetal harms dominantly at the subordination of potential maternal harms. Simply stated, healthy babies do not negate maternal harms and fetal risks do not *per se* justify forced or compelled maternal interventions. As ACOG advises, "[p]regnancy is not an exception to the principle that a decisionally capable patient has the right to refuse treatment, even treatment needed to maintain life" [3].

Reproductive medicine is also complex because it reflects socially and politically constructed hierarchies and norms that are contextually relevant to minimizing liability risks and aligning with prevailing practice standards. Clinicians thrive when they understand their patient's lived experiences before arriving for clinical care, which often includes understanding gendered, classed, and racialized hierarchies in birthing practices [4]. Elizabeth Kukura's article, *Better Birth*, provides a robust critical account of how traditional institutional maternity care is fraught with high costs (especially for the uninsured), reports the highest rates of maternal death in the developed world, sometimes lacks evidence-based standards, and leaves many patients reporting trauma, disempowerment, and alienation [5].

Clinicians practicing within a medicalized model are savvy to understand these broader perspectives. Kukura summarizes how many pregnant persons believe that the physiological processes of birth have been over-pathologized, leaving their experiences devalued and silenced as patients. For pregnant people of color, these marginalizing experiences can layer on to bias, stereotyping, and deep historic legacies of medical gynecological abuse on women of color, deepening harms and fostering mistrust [6]. These legacies of medical bias and racism still cause real harm today, as evidenced by the disproportionate rates of Black women's maternal mortality and cesarean section rates [7]. These concerns also disproportionately impact pregnant persons living in rural or low-income communities and pregnant persons of native and indigenous descent.

Alternative models are available to bypass the traditional medicalized system, but those paths are often outside of insurance, and they have harder entry barriers due to the "legal infrastructure of childbirth." [8] Clinicians will benefit from the added peripheral vision of understanding these alternative models that use less technology, more egalitarian approaches (e.g., midwifes), and more centering of the pregnant person's autonomy. According to its critics, the medical model distinctly centers the doctor and technology on fetal wellbeing. Often, this model myopically displaces larger and more complex values, such as the pregnant person's autonomy and wellbeing.

Reproduction and birth can yield substantial liability. Notably, though, *more* interventions or more fetal focus may not lessen liability. Rather, likelihood of suit is influenced more by factors such as poor communication, lack of physician-patient trust, patient frustration, and patronizing behavior [9]. This Chapter provides an overview of liability and risk management at each stage of reproduction and birth. The first sections analyze claims for informed consent, wrongful death, wrongful birth, and wrongful life. The next sections examine liability at various temporal moments in reproduction and birth.

2 Discussion

2.1 Informed Consent

Informed consent is a thorny area of obstetric care but a vital tool of liability management. This topic frames all others. A patient holds a fundamental constitutional right to control their body [10, 11]. Informed consent requirements can cover a wide range of patient counseling including pre-conception genetic counseling, assisted reproduction, prenatal fetal testing, inducing/inhibiting labor, using anesthesia, and childbirth interventions. Effective informed consent must also build on strong cultural competencies understanding the varied experiences that different communities, religions, and cultures might bring to their relationship with the medical system [6].

Informed consent is its own cause of action. It is a genre of negligence with a historic underpinning of battery (a non-consensual touching). Historically, battery (the intentional touching of another in a harmful or offensive manner) was the vehicle to bring what we now think of as an informed consent claim. Battery was not quite the right approach though because the patient typically did not object to the touching. Rather, they objected to the lack of information that led them to consent to the touching. Negligence is also not the right approach because the patient's adverse outcome stems from the foreseeable effects of the treatment when done non-negligently, not from a breach in the clinician's standard of care. Informed consent claims are thus a unique genre of medical liability.

To bring an informed consent claim, the causal line must hold between the risk that the clinician *should* have warned of and the harm that manifests. The patient must show that, had she known of the undisclosed risk, they *would* have objected to the procedure. For example, a patient might argue that they would not have consented to a cesarean section had they known of uterine rupture risks. The patient is not alleging that the doctor *caused* the uterine rupture, only that they would not have consented to the procedure had they known of the risk. The patient then must be suing for the harm of the uterine rupture, not something else.

Absent an emergency or therapeutic justification, a clinician should obtain consent from competent patients before treatment. The clinician should advise the patient of the material risks that they might face *and* the alternatives available. It is not feasible to advise of all risks in patient care. If the risk is not material, then the doctor has no duty to inform. Accordingly, the clinician must be aware of the lens through which materiality is viewed. Informed consent in a clinician's geography of practice could take one of three different lenses. Some jurisdictions look at what customary disclosures are from the perspective of the reasonable physician under the same or similar circumstances. While this standard is the easiest for practitioners to follow, it has been criticized for not achieving the goal of patient autonomy. Other jurisdictions look at reasonable disclosures through the lens of the hypothetical "reasonable patient." This is a more patient-centered lens, but it too is criticized for being impractical in practice settings subject to retroactive manipulation (everything looks material in hindsight!). Just an extreme minority of jurisdictions look through a subjective lens of what *each* individual patient considers material, rather than the hypothetical "reasonable patient" [12].

The clinician owes a duty to advise of material risks *and* available alternatives. This would include every drug prescribed, intervention engaged, and procedure followed *unless* an exception applies. There are several relevant exceptions under which the clinician will not be required to obtain informed consent, but these exceptions should be scrutinized to apply under only the narrowest of circumstances. Emergency circumstances can alleviate the clinician's informed consent duty and create a presumptive consent. Particularly thorny for childbirth, another exception exists for therapeutic reasons if the clinician believes that informed consent will exacerbate an underlying condition. ACOG explains that there is a historic misinterpretation of the ethics of this therapeutic exception and that invoking it risks the clinician presuming to know what is best for a patient and risks "coercive misuse"

[13] ACOG explains that "although it is never ethically acceptable to withhold information without the patient's knowledge and consent, it is acceptable to communicate information over time based on the patient's stated preferences and ability to understand the information [13]. Both informed consent exceptions should be applied only very selectively and narrowly in childbirth [13].

Many clinical settings will have standard consent forms for standard components of obstetric care. To be effective, these informed consent forms need to be specific enough to procedures and risk. General waivers are harder to defend in litigation. The goal in informed consent conversations should be actualizing patient autonomy, deploying strong cultural competencies and communication skills, and documenting discussion rigorously and accurately.

2.2   Wrongful Death

Wrongful death claims vary in each state. These claims arise when a negligent act causes the death of a person, such as a birthing person or a child during delivery. These claims are usually brought in tandem with a negligence claim by the underlying decedent's estate. They allow the beneficiaries (often children, parents, and spouses) to recover for their *own* losses as loved ones. Qualifying beneficiaries can recover independently for what *they* lost as parents, spouses, and children of the decedent, such as comfort, society, companionship, and earnings.

If wrongful death cases arise out of the death of the birthing person, the spouse and/or children would have a claim for their losses under the state statute. When these claims arise in the case of fetal death, the issues are more complicated state to state. If the child is born and then dies thereafter from injuries caused *in utero*, all states will allow the qualifying beneficiaries (usually the parents) to sue clinicians for negligence causing death. This is known as the "born alive" rule or the "one breath" rule. By virtue of having been born, however brief life was, the child becomes a person within the meaning of state wrongful death statutes.

States differ as to liability when the death occurs *in utero*. Historically, these claims were disallowed for both policy and causation reasons. Today, many states allow liability for negligence causing death *in utero* after the fetus has reached the point of viability, meaning capable of independent existence outside the womb. The question is: when precisely does viability occur? Some courts find the line too imprecise. After viability, clinicians can be liable for the injuries caused to fetuses *in utero*. For deaths occurring after viability but before delivery, the plaintiff would still have to overcome causation issues and prove that the negligence caused the death, not other causes. This will require the plaintiff to address other possible contributory causes. In cases of stillbirth, determining whether the stillbirth was caused by malpractice or natural causes can be uniquely complex [14].

Modern courts are trending toward moving the threshold to sue for *in utero* harms earlier than the viability line. The question is whether a pre-viable fetus is a "person" or "decedent" within the meaning of the state wrongful death law. Some states are beginning to interpret these statutes to apply to even pre-viable tortious deaths through either judicial statutory interpretation or legislative reform. Clinicians would be prudent to assume the potential for liability even in pre-viable fetal injuries, as some courts will alternatively allow these types of claims to be folded into the lawsuit brought by pregnant persons.

When liability is allowed for the death of a fetus *in utero*, what is the appropriate measure of damages? In a typical wrongful death case, the measure of damages looks to lost relationships and lost earnings. *In utero* death damage calculations are even more speculative than infant deaths. Courts do allow for these recoveries, though, and wrongful death claims can also include the funeral and burial costs.

Wrongful death claims arise out of the negligent death of a decedent and compensate the loved ones for the emotional and economic losses. Clinicians do not hold a patient-physician relationship with these loved ones suing as plaintiffs, but they may face derivative liability for their losses.

2.3   Assisted Reproductive Technologies

Liability for harms in reproductive technologies is a rapidly changing legal landscape with few clear rules shaping it. Assisted Reproductive Technologies (ART) encompasses "all treatments or procedures that include the handling of human oocytes or embryos," including increasingly complex interventions such as "therapeutic donor insemination, ovarian stimulation, ova and sperm retrieval, in vitro fertilization, [and] gamete intrafallopian transfer" [15]. The American Medical Association directs ART clinicians to, among other guidances:

- "value the well-being of the patient and potential offspring as paramount;"
- support patient informed decision-making including "investigational techniques to be used (if any); risks, benefits, and limitations of treatment options and alternatives, for the patient and potential offspring; accurate, clinic-specific success rates; and costs;"
- "[p]rovide patients with psychological assessment, support and counseling or a referral to such services;"
- not discriminate by "race, socioeconomic status, or sexual orientation or gender identity;"
- and help develop "peer-established guidelines and self-regulation" [15]

ART legal issues can arise in many scenarios, such as negligent destruction of genetic material, mishandled reproductive material, switched donors, and contaminated samples. Dov Fox categorizes these claims as procreation imposed, procreation deprived, and procreation confounded [16]. He describes this genre of cases as "reproductive negligence" [17]. These cases often befuddle courts because of their uniqueness, policy implications, and emotiveness. ART is generally a "buyer-beware" legal arena with plaintiffs rarely prevailing [16]. Reproductive negligence cases also particularly struggle to identify breach remedies.

Procreation Imposed: Wrongful Birth and Wrongful Life Claims

Wrongful birth and wrongful life claims are unique civil suits brought against clinicians arising out of childbirth. These claims involve pregnant persons who sought a medical procedure to avoid pregnancy but became pregnant due to a medical error; pregnant persons who would have terminated their pregnancies had they known of a fetal condition; and pregnant persons whose child alleges they never would have been born if the pregnant person would have terminated the pregnancy. This area of law is rapidly evolving, heavily state specific, exceedingly emotive, and filled with complex policy considerations.

Wrongful birth and wrongful life claims are closely related. They differ in *who* is bringing the claim. Generally, a wrongful birth claim is brought by the parents alleging that they would not have continued with a pregnancy but for the clinician's negligence. There are also other related subcategories of this claim as explored below. Wrongful life claims, in contrast, are brought by the children who are born, alleging they never would have been born but for medical negligence depriving their pregnant parent of the chance to terminate the pregnancy. While wrongful life and wrongful birth cases arise from the same underlying breach in the standard of care, state courts favor parents' wrongful birth cases over children's wrongful life cases.

Claims brought by parents generally fall into two different categories. First, wrongful conception and wrongful pregnancy claims seek compensation more narrowly for the cost of carrying a pregnancy as distinct from the cost of raising a child born of that pregnancy. Wrongful conception and wrongful pregnancy claims are generally framed temporally as *pre*-conception torts arising when a clinician negligently completes a sterilization, an IVF procedure, a contraception dispensation, an IUD insertion, or similar missteps before conception. Here, the plaintiff can generally recover with ease the cost of the negligent underlying procedure or prescription, the pain and suffering of the unintended pregnancy, complications from the pregnancy, the costs of delivery, lost wages, and lost consortium.

Second, parents can also bring wrongful birth claims premised on the plaintiff's harm being the *birth* of a child and the costs associated with raising the child. These claims are notably distinct from wrongful pregnancy / conception because they involve an intermediary argument alleging that the parent would have terminated the pregnancy but for the negligence. The pregnant person is alleging that, had they known of the fetal anomaly, they would have aborted the fetus [18]. The provider did nothing to *cause* the impairment but failed to allow the pregnant person to exercise autonomy over that information. For example, possible breaches include the failure to read a genetic test result accurately, failure to perform a necessary genetic test, or failure to diagnose a pregnancy in time for termination. In *B.F. Reproductive Medicine Assoc. of N.Y., LLP*, 92 N.E.3d 766, 768 (N.Y. 2017), for example, a fertility clinic negligently screened donor eggs, missing a chromosomal abnormality, and failed to advise the recipient before she birthed a child with a rare genetic condition. Discussed more fully below, these claims raise complex damage considerations with differing state resolutions.

Before *Roe v. Wade,* 410 U.S. 113 (1973), wrongful birth claims were not recognized. After *Roe,* some courts concluded that the right to terminate a pregnancy includes the right to adequate guidance regarding the fetus based on testing and diagnostics. The causal line is that, had the clinician informed the pregnant woman of an impairment, the child would not have been born because the pregnant person would have exercised her constitutional right to terminate the pregnancy. Some courts and scholars, such as disability rights advocates, criticize these claims because of the stigmatizing implications of the testimony to the born child. Others criticize these claims for the consequentialist lens through which the decision-making is viewed without the pregnant woman actually confronting the decision. Changing political and social acceptance of abortion at the Supreme Court level or in state legislatures may cause states to revisit these rules[1].

Today, many states do recognize wrongful birth claims but differ in measuring recoverable damages. Some courts disallow damage claims for the cost of raising a child because of the policy implications, refusing to compensate for the "intangible, unmeasurable, and complex benefits of motherhood and fatherhood" [19]. Courts denying this fuller damage recovery tend to use language emphasizing the blessing, joy, and affection of childrearing when disallowing childrearing costs as damages. Courts that disallow child-rearing costs will still allow medical expenses for delivery, pain and suffering, lost wages, and even emotional distress, which can be substantial.

On the other extreme, some courts adopt a "full recovery" rule concluding that any moral concerns should not get in the way of a plaintiff's malpractice recovery. Views on abortion, sterilization, or contraception should not alter the basic tort principles compelling a "full recovery" to restore the plaintiff to the position before malpractice. This approach seeks to deter negligent conduct.

Two compromise positions exist. When a child is born with a costly and complex disabling condition, some courts strike a balance allowing recovery for the "special damages" of living with the particular condition (e.g., medical care, occupational therapy), but not the "general damages" of living (e.g., housing, clothing). These courts avoid putting a value on the cost of being born.

Other courts split the difference by awarding some general damages covering the costs of childrearing with a "benefits offset" to account for the intangible emotional rewards of raising the child. These can be challenging jury valuations. These courts acknowledge that the joy of raising a child does not alter the financial costs in doing the same. One does not displace the other. The parents seek economic compensation, courts remind litigants, not the cessation of the relationship.

The corollary wrongful life claim is brought by the child from the same underlying breach, alleging that they would not have been born but for the defendant's

---

[1] As this book went to print, *Roe v. Wade* was overturned, abolishing a federal constitutional protection for the right to terminate a pregnancy, thus previewing a widely diverging pathwork of state laws poised to emerge in the coming months and years.

negligence. The claim's premise is the availability of abortion to terminate the pregnancy and proof that the pregnant person would have elected this option. Only a very small minority of courts accept wrongful life claims because of perceived moral differences in the child alleging its own harm in being born. These claims are prohibited in most states. Courts are generally more comfortable compensating the loss in parental autonomy than the metaphysical question of being born.

These claims also carry policy complexities for those living with disabilities. Just the mere existence of these claims, disability rights advocates argue, "may exact a heavy price not only on the psychological well-being of individuals with disabilities, but also on the public image and acceptance of disability in society" [20]. Critics of these cases argue that allowing recovery from fetal diagnoses stigmatizes disability in that the lawsuit's focus is necessarily on the existence of the impaired child as compared to the loss of decision-making autonomy. In response, competing perspectives assert that existing government systems offer minimal healthcare and social support for the parents of children with disabilities. Supporters of these claims argue that concerns are easily remedied by focusing on *how* the claims are plead, argued, and covered by the media [21].

For clinicians, these state nuances are interesting and insightful, but ultimately the standards of patient care are not altered by these policy complexities.

Procreation Confounded

Other cases arise from the mix-up of genetic material. Genetic mix-ups can occur in *who's* material is used. For example, the pre-embryos of couple X or the sperm of male X is used instead of the pre-embryos of couple Y or the sperm of male Y. One particularly high-profile case, *Cramblett v. Midwest Sperm Bank, LLC.*, 2017 Ill. App. 2d 160694 (2017), involved a donor mix-up in which parents seeking a White donor with blond hair and blue eyes instead received the sperm of an African American donor [22]. This case was first framed as a wrongful birth suit. This might catch readers' attention as unusual, if not offensive, as wrongful birth cases typically involve a disability or defect. The plaintiff was alleging that raising a healthy, mixed-race child was their injury. The plaintiff struggled to plead injury while simultaneously celebrating that her daughter was a "dream come true" [22]. This claim yielded public debates about appropriate remedies and the moral / ethical complexities of reproductive technologies.

These genetic material cases can involve messy intersections with the family law system too if the person holding the inadvertent genetic connection pursues parentage, custody, or visitation rights. States do not do genetic testing at birth in the ordinary course, so these challenges often do not arise until bonding and relationships have already been established.

Procreation Denied

Denying parents the right to procreate can also yield liability. Again, historical and cultural context can strengthen modern clinical practices. A troubling history of forced sterilizations performed on persons of color, immigrants, indigenous persons, poor persons, and persons with disabilities haunts many communities in their interactions with the medical system today.

While most "procreation denied" cases are not as systemic or troublesome today, this historical context is still vital to caregiving. More common today, clinicians can be sued for the negligent destruction of embryos in ART, although the proper theory and remedy still confounds courts. For example, in one of the earliest such cases, plaintiffs froze harvested and fertilized eggs subject to informed consent agreements and written contracts. After most embryos were damaged during a facility relocation, plaintiffs sued for negligent infliction of emotional distress (NIED), alleging emotional anguish and pain and suffering [23]. The court wrestled with whether pre-embryos were persons or property and what type of claim was appropriate, if any, among property, contract, and tort theories. The defendants argued that the plaintiffs really sought an emotional damages recovery via a breach of contract claim, which is not allowed under well-established rules. The court ultimately allowed the plaintiffs' NIED claims to prevail because of their physical loss of the pre-embryos, which were recoverable in tort as "irreplaceable property." This case offered future plaintiffs a path to sue, but its legal rationale was not well-defined.

The negligent destruction of genetic material, such as pre-embryos, has not historically been successful as wrongful death claims. *Jeter v. Mayo Clinic Ariz.,* 121 P.3d 1256 (Ariz. Ct. App. 2005), for example, held that "person" could not possibly include a cryopreserved, three-day-old pre-embryo. As fights over reproductive politics continue, clinicians should monitor changing personhood definitions under federal and state law.

Many ART clinical relationships will be governed heavily by contract. Code of Medical Ethics Opinion 4.2.5 directs clinicians to discuss the handling of stored embryos proactively, including when and how the embryos can be used by a surviving party upon the death of a partner or donor, given to other patients for reproduction, donated to research, thawed, and disposed of [24].

No government agency currently regulates ART, federal and state law are sparse, and rules are generally lacking to "prevent or mitigate reproductive negligence" [16]. Michele Goodwin concludes that "[ART] provides an open canvas on which to sketch the very complicated social and legal identities that spring forth from this mix of technology, biology, law, and commerce" [25]. Clinicians should monitor developments in regulation, legislation, and standards of care vigilantly.

2.4 Abortion

Abortion is a heavily regulated area of law and medicine. States have generally refrained from imposing criminal consequences on women who procure an abortion illegally, instead focusing on the criminal liability *of doctors* who perform abortions outside legal boundaries.

Improper abortions can yield traditional medical malpractice liability as well as other administrative and potentially criminal liabilities [26]. Traditional malpractice cases can arise from surgical abortions. These claims are generally like any other malpractice claim. The most common claims are for occurrences like a perforated uterus or insufficient postoperative care. Complications in childbirth are dramatically higher than in abortion care. With the emergence of "medical abortion," the surgical interventions are substantially less frequent and abortion is even safer. Nonetheless, rights of "conscientious objectors" can complicate caregiving and create access barriers. Given that many states are down to just a small number of abortion providers, many women must travel long distances to obtain an abortion and may be in an entirely different geography for any post-abortion care or complications. Should complications arise, timely and professional care is owed, no different than any other procedure. As this book went to print and *Roe v. Wade* was overturned, state legislatures were battling out how to regulate potential liability as patients were compelled to cross states lines for abortion care. Some states sought to add protections for providers while others sought to ratchet up liability for providers, thus leaving foundational questions unanswered about the applicability of the constitutional right to travel and the privileges and immunities clause of the U.S. Constitution.

Roe v. Wade protected the right to terminate a pregnancy, but it did so squarely as a decision between a woman and her doctor [27]. *Roe v. Wade* did not entrust women to make this decision alone, which is both controversial and unconventional from a patient autonomy perspective. Informed consent for abortion is often heavily regulated by state law. Many states, particularly throughout the Midwest and South, have legislatively regulated abortion informed consent language with great specificity and arguably great paternalism. These states override the clinician's judgment and the patient's circumstances by scripting the doctor and often the timing. These often read more like political scripts, in stark contrast to typical doctor-patient counseling. Clinicians performing abortions will become familiar with these requirements. Although not technically "informed consent," many states also impose legal requirements for minors seeking to terminate a pregnancy, either by requiring parental notification or parental consent (with opportunities for judicial override).

2.5 Labor and Delivery

Labor and delivery can yield many lawsuits in caregiver transitions, data monitoring, and birthing interventions. Clinicians have the same duty to attend to patients in labor and delivery as in other care settings. Managing the unpredictable timing and length of labor and delivery necessarily requires teamwork monitoring the patient. Many lawsuits allege that clinicians were not adequately monitoring maternal or fetal health. This could include a failure to attend to the patient promptly or responsively.

The standard of care does not require constant monitoring. It requires a process for transitioning caregiving effectively. Sign-in records are important to understand by whom and when the patient was seen. Communication and practice management are key. Liability can include vicarious liability for those to whom care was delegated.

Labor and delivery are unique care sequences involving the collection of more diagnostic data than gestation to this point. Many cases arise from how those data are analyzed, addressed, and communicated to others. Electronic Fetal Monitoring (EFM) is often used to display and record data indicative of fetal distress. EFM allows for faster interventions and baseline monitoring. EFM also creates a litigation record in a trial. Many cases arise from clinicians failing to monitor vital signs. Reliance on EFM can also tip the scale problematically toward monitoring for *fetal* distress and overriding the vitals and sensory perceptions of the birthing person.

EFM needs to be accompanied by also monitoring of the birthing person's vital signs frequently for distress.

Clinicians exercise best judgment throughout labor and delivery in consultation with the birthing person's medical autonomy. Cases are heavily fact specific. There are lawsuits for birth progressing too slow, too fast, using forceps, not using forceps, using anesthesia, not using anesthesia, rushing delivery, delaying delivery, intervening unnecessarily, and failing to intervene [28–30]. Consider the use of Pitocin, as an example. Pitocin can increase risks to the fetus and the birthing person because of the intensity of contractions. Its overadministration can cause harms and liability [31]. These risks are managed with strong informed consent practices, combined with careful monitoring and communication [32]. Of course, the plaintiff must again prove causation. If there were an issue with the administration of Pitocin, for example, a plaintiff would have to prove that *this intervention* was at least a substantial factor in causing the harms that occurred. Often Pitocin is being administered already because of identified delivery concerns and risks. The plaintiff must show that *this* medical intervention was at least a substantial factor in the harm. Other causes, such as existing health complications, can co-exist allowing multiple concurrent factual causes.

Whether an intervention was necessary is not a "one-size-fits-all" answer. They are heavily informed by general medical risks, patient autonomy, and individualized patient concerns. For example, the birthing person's economics, family obligations, personal beliefs, mental health, health complications, and family support may factor in. A decision to perform a cesarean delivery, for example, may face unique complexities for someone with no paid family leave or existing caregiving responsibilities because of the recovery time and limitations.

Public awareness and activism have recently exposed how costly and interventionist the United States birthing system is. Many advocates contest whether medicalized interventions are necessary and observe how misaligned the United States is with other Western developed countries. One example arises from vaginal births after cesarian deliveries (VBAC). Liability can arise from both sides of the decision matrix [33]. A notable historic pendulum swing shapes liability here [34]. In the 1990s, VBACs were considered more low risk because they were performed more regularly. Institutions and providers throughout the 2000s came to disfavor VBACs and some even banned them outright. In larger geographies, a patient might have been able to find a provider to support this birth plan, but more rural and under-resourced areas could not. This led to critiques that women's reproductive choices were being denied unconstitutionally and that women were forced to undergo a compulsory surgery that would not occur in other contexts. Wholesale bans ignored the individual needs of pregnant persons, such as the history of prior pregnancies; the weight, age, health of the mother; and fetal monitoring, size, positioning, and gestational age [34].

Today, the pendulum has centered with more clinicians supportive of this birthing approach. The history of VBAC policies and their controversies are a powerful insight. Clinicians could crudely, yet accurately, calculate that the risk of liability from a fetal birthing injury dramatically outweighs the risk of liability for a violation of a birthing woman's autonomy and thus override the patient's wishes. Birthing injuries to the child can be catastrophic for plaintiff and clinician alike. Birth-related neurological injuries are some of the most emotive cases that come before civil courts. The damages are often extraordinarily high because the injuries last a lifetime. These cases often involve extraordinarily high legal fees. Plaintiff's counsel often work on a contingent fee basis allowing all plaintiffs to pursue these cases to the greatest degree of the law.

Liability risks do not dictate medical standards of care though. Practicing medicine with a hierarchy positioning the fetus over the birthing person is deeply problematic in both law and medicine.

> [B]ecause the fetus has become the dominant patient in childbirth and the far riskier putative plaintiff in modern obstetric malpractice cases, this reality diminishes and subordinates the rights and remedies of birthing women as patients and plaintiffs in problematic ways. … It distorts the standard of care that doctors distinctly owe to both the woman and the fetus. It tilts the dualities of childbirth toward the fetus. It valorizes medical judgments in response to uncertainty in childbirth and villainizes maternal responses that do not conform to an essentialized, self-sacrificial, and historically myopic view of childbirth [35].

Rather, standards of care are grounded in the judgments of reasonably prudent practitioners in the same or similar circumstances. That determination can change over time and across communities (e.g., rural v. urban). In extreme cases, courts have intervened in emergency circumstances to help navigate complex disputes. Even just the threat of these judicial interventions are problematic to patient autonomy. These can include forced medical interventions like blood transfusions, cesarean sections, and hospital births.

Virginia and Florida adopted no-fault regimes for birth-related neurological harms. These statutes function like workers' compensation programs whereby plaintiffs trade off the ease of recovery (fast and prompt) for the quantity of recovery (foregoing damage categories like punitive damages) and the breadth of recovery (foregoing any other claims the parents might have, for example). This allows obstetricians to pool liability risks. In these states, compensation is more administrative. While these reflect a no-fault regime, they still require proof that an injury did occur. There is no indication that these compensation fund approaches are likely to expand.

Understanding the law of reproduction includes understanding its social and political context, including understanding perceived harms that might not even yield a lawsuit. Birth rights advocates have coined the term "obstetric violence" to govern a range of harms women experience in childbirth, which reflect a pattern of disrespect, abuse, and neglect of pregnant women in childbirth [9]. Obstetric violence can yield claims arising from medical abuse, medical coercion, or complaints of disrespectful treatment. "Obstetric violence" often involves multiple violations that cumulatively rise to this level.

These issues create relatively small risks of legal liability, but they undermine patient trust. Birth rights advocates would classify forced surgeries like cesareans and episiotomies as the worst extreme of obstetric violence followed by unconsented medical procedures (e.g., labor induction, membrane braking, forcep assists, and placenta removal), sexual violations (e.g., frequent unnecessary vaginal exams) , the sexualization of interventions (e.g., a vaginal tear repair, referred to as a 'husband stitch"), physical restraint, and other forms of abuse [9]. These behaviors can not only be physically harmful to birthing persons and their babies, but emotionally and psychologically harmful, as well [9].

Clinicians should keep the birthing person squarely centered as a patient with their own rights and decision-making abilities amidst this context.

2.6   Postpartum Care

Standards of postpartum care are also important in managing birth. Clinician liability has arisen from failure to monitor for signs of infection, hemorrhaging, and other complications. Black and Native American women are 2–3 times more likely to die in childbirth than White women. State and federal efforts are working actively to better monitor data and improve outcomes. Reflecting the importance of ongoing patient monitoring, Section 9812 of the American Rescue Plan Act of 2021 (H.R. 1319) expanded postpartum public benefits coverage. Section 9812 authorized states to extend Medicaid coverage to include the twelve-month postpartum period after delivery.

After birth, fit parents retain the right to make medical decisions regarding their children, a right protected by the Fourteenth Amendment of the United States Constitution [36, 37]. The state retains a *parens patriae* power to intervene as necessary if the child's mental or physical health is in jeopardy.

2.7   Application (Cases and Examples)

Application 1: Wrongful Birth and Wrongful Life

After Claudia Mae Graves gave birth, her physician attempted to sterilize her at the hospital, but performed the procedure negligently and she became pregnant again. This child was born with a birth defect, and Graves sued for negligence. The hospital moved for summary judgment arguing that Georgia did not recognize a cause of action for wrongful pregnancy, and if such cause of action did exist, the damages Graves sought were a direct result of the child's birth. The trial court denied the hospital's motion. The Georgia Supreme Court reversed in part, concluding that a cause of action for wrongful pregnancy or wrongful conception should be recognized in Georgia, as the underlying claims are "no more than a species of malpractice." With respect to damages, the court held that, while expenses for unsuccessful medical procedures, pain and suffering, medical complications, cost of delivery, lost wages, and loss of consortium could be recovered, the cost of raising a child could *not* be recovered. To permit recovery for the cost of child rearing and education would be to recognize that human life and parenthood are compensable losses.

In 2020, Georgia's Supreme Court had an opportunity to clarify the scope of recoverable damages. Wendy and Janet Norman purchased the sperm of Donor #9623 after Xytex represented to them that Donor #9623 was one of its "best" sperm donors with a high IQ and no history of mental health issues or criminal activity. Wendy was inseminated with Donor #9623's sperm and gave birth to a son, A.A, who was subsequently diagnosed with a variety of health issues, including ADHD, an inheritable blood disorder (for which Wendy was not a carrier), depression, and suicidal ideations. The Normans later discovered that Donor #9623 made false representations to Xytex during screening procedures. The Normans sued Xytex for several claims, all of which the trial court dismissed on the basis that the Normans were asserting "wrongful birth claims" not recognized under Georgia law. On appeal, the Supreme Court of Georgia reaffirmed the rule that the law does not recognize claims for damages that depend on life as an injury but concluded that not all claims based on prenatal injuries require a characterization of the child's life as an injury. The delay in treating or mitigating A.A.'s condition had the Norman's known about Donor #9623's medical history and the difference in price between the cost of the sperm received and the fair market value of the sperm Xytex sold to them are examples of such recoverable awards.

Source: *Fulton-DeKalb Hosp. Auth. v. Graves*, 314 S.E.2d 653 (Ga. 1984); *Norman v. Xytex Corp.*, 348 S.E.2d 835 (2020); *Atlanta Obstetrics & Gynecology Group v. Abelson*, 398 S.E.2d 557 (Ga. 1990) (declining to recognize wrongful birth actions).

Application 2: Informed Consent in Childbirth

Janice Schreiber was pregnant with her third child, Kimberly. Her first two children were delivered by cesarean sections. By the time of her third pregnancy, medical research suggested that a vaginal birth after a cesarean ("VBAC") did not pose any additional risks to the mother or the child. Schreiber's doctor recommended this course, and she consented to a VBAC. Upon hospital admission, she told her doctor she changed her mind and wanted the cesarean, but her doctor encouraged her to continue with the VBAC. She had signed standard informed consent forms for *both* a VBAC and a cesarean.

When her labor did not progress, the doctor manually broke her amniotic fluid sac to speed up labor. Schreiber experienced sharp stomach pains and she reiterated her request for a cesarean on several other occasions. The doctor wanted to give it more time believing the VBAC was the better treatment. When the fetal heart rate decreased one hour and forty minutes after Schreiber began asking to proceed with the cesarean section, the doctor performed an emergency cesarean section twenty minutes after the decline in fetal heart rate. Janice's uterus had ruptured and Kimberly was deprived of oxygen, causing her to be born paralyzed.

The Schreibers sued the doctor and his insurer for violation of Janice's right to informed consent. They dropped their general medical malpractice case. The trial court concluded that the doctor did not violate Schreiber's right to informed consent because there had not been a substantial medical change in circumstances which required the doctor to advise of new risks unconsidered by the original consent. The

Court of Appeals reversed, holding that where there is more than one medically acceptable option, "the competent patient has the absolute right to select from among [those] treatment options after being informed of the relative risks and benefits of each approach."

The Supreme Court of Wisconsin found that Schreiber had successfully revoked her VBAC consent. As such, this presented alternative viable modes of medical treatment, constituting a substantial change in circumstances requiring the doctor to revisit informed consent. The revocation of consent could indicate a legal change in circumstances. The court applied an objective test considering what a reasonable patient would want to know and whether that information would have changed a reasonable patient's course of action. The court concluded that a reasonable patient would have chosen the cesarean. The doctor's failure to abide by Schreiber's request deprived her of her treatment choice and resulted in the injuries to the child.

Source: *Schreiber v. Physicians Ins. Co.*, 588 N.W.2d 26 (Wis. 1999); *McQuitty v. Spangler*, 976 A.2d 1020 (Md. 2009) (reaffirming that informed consent claims are separate from medical malpractice claims with the former requiring communication to allow a patient to make an informed choice with full disclosure of material risks and not requiring proof of a physical harm).

Application 3: Post-Natal Care and Maternal Death

Kyira Adele Dixon presented to the hospital for a repeat elective cesarean delivery. Once in the operative suite, a foley catheter was inserted. Dixon delivered her child approximately thirty minutes later. Two hours after being taken to the Post Anesthesia Care Unit, Dixon began exhibiting symptoms of uterine atony (failure of the uterus to contract) and blood-tinged urine was observed. Further tests were ordered revealing acute blood loss and an enlarged hematoma. Hours later, her condition had worsened to reveal blood in the catheter, no urine output, abnormal labs, symptoms suggesting acute blood loss, and an enlarged hematoma. Her doctors discussed surgical intervention to identify the bleeding source. An ultrasound revealed abdominal fluid. One doctor wished to continue the current treatment as opposed to surgery. Dixon was given a blood transfusion to support circulation and coagulation during a hemorrhage. About forty-five minutes after her doctors consulted, she was taken to surgery where doctors discovered three liters of blood in her abdomen. She died almost two hours later. The autopsy concluded that the cause of "death was due to hemorrhagic shock, due to acute hemoperitoneum."

Dixon's husband and two sons sued the hospital, physicians, and nurses alleging NIED and wrongful death. The negligence underlying the wrongful death claim asserted that the defendants had failed "to appreciate and properly manage Kyira's post-partum hemorrhage in a timely manner," and "to return Kyira to surgery in a timely manner." The husband and sons sought damages for Dixon's lost "love, companionship, comfort, affection, society, solace, moral support, care, counsel, physical assistances, services, financial support, and protection" as well as funeral expenses.

Source: Complaint for Damages, Johnson v. Cedars-Sinai Med. Ctr., No. BC655107 (Cal. Super. Mar. 22, 2017).

3   Summary

Liability for harms in reproduction and birth can arise in informed consent claims, negligence claims, wrongful death claims, wrongful birth claims, and wrongful life claims. They can arise at any point in the reproductive process from assisted reproduction to delivery to postpartum caregiving. While some areas are heavily regulated, such as abortion care, others are surprisingly unregulated, such as assisted reproductive technologies. Successful practice management involves a strong cultural context to the history of medicalized childbirth, particularly within certain communities, combined with strong communication and informed consent with patients.

The Supreme Court's overturning of *Roe v. Wade* just as this book went to press, set into motion a dramatic schism in state legislative approaches. In the coming months and years, we can expect to see some states moving toward stronger fetal personhood framings (even overriding entirely the rights of pregnant persons) and further regulation of ART, while other states will move to liberalize the autonomy of pregnant persons to engage in private medical decision-making with their providers. This schism, in turn, will strain liability frameworks that are community-centered and governed by national bodies. All providers delivering care relating to childbirth and reproduction should follow these abortion-specific legal developments closely as they hold more sweeping collateral implications on the liability of reproduction and birth.

Acknowledgements   The author extends her heartfelt thanks to University of Baltimore School of Law students Neha Khan (Class of 2022) and Alexa Mellis (Class of 2022) for their research and editing support and University of Louisville School of Law student Mary Elizabeth Howard (Class of 2022).

References

1. Kirkland v. Blaine Cty. Med. Ctr., 4 P.3d 1115 (Idaho 2000) (Medical Malpractice Verdicts, Settlements, & Experts, Aug., 1999, p.27).
2. Johnson v. Thompson, 650 S.E.2d 322 (Ga. App. 2007).
3. *Comm. Op., No. 664 – Refusal of Medically Recommended Treatment During Pregnancy*, AM. COLL. OF OBSTETRICIANS & GYNECOLOGISTS (June 2016), https://www.acog.org/clinical/clinical-guidance/committee-opinion/articles/2016/06/refusal-of-medically-recommended-treatment-during-pregnancy.
4. Shirish S. Sheth, *Reproductive Health and Obstetricians & Gynecologists,* ANNALS N.Y. ACAD. SCI. 1 (2003).
5. Elizabeth Kukura, *Better Birth,* 93 TEMP. L. REV. 243 (2021).
6. Colleen Campbell, *Medical Violence, Obstetric Racism, and the Limits of Informed Consent for Black Women,* 26 MICH. J. RACE & L. 47 (2021).
7. FARAH TANIS ET AL., INST. FOR GENDER & CULTURE AT BLACK WOMEN'S BLUEPRINT, THE SEXUAL ABUSE TO MATERNAL MORTALITY PIPELINE (2019).
8. *Development in the Law – Intersections in Healthcare and Legal Rights: Chapter Three: The Legal Infrastructure of Childbirth,* 134 HARV. L. REV. 2209, 2220 (2021).
9. Elizabeth Kukura, *Obstetric Violence,* 106 GEO. L.J. 721, 771-72 (2018).

10. Union P.R. Co. v. Botsford, 141 U.S. 250, 251 (1891).

11. Planned Parenthood v. Casey, 505 U.S. 833, 851 (1992).

12. Scott v. Bradford, 606 P.2d 554 (Okla. 1979).

13. *Comm. Op., No. 918 – Informed Consent and Shared Decision Making in Obstetrics and Gynecology*, AM. COLL. OF OBSTETRICIANS & GYNECOLOGISTS (Feb. 2021), https://www. acog.org/clinical/clinical-guidance/committeeopinion/articles/2021/02/informed-consent-and-shared-decision-making-in-obstetrics-and-gynecology.

14. Jill Wieber Lens, *Tort Law's Devaluation of Stillbirth,* 19 NEV. L. J. 955 (2019).

15. *Opinion 4.2.1 – Assisted Reproductive Technology*, AM. MED. ASS'N, https://www.ama-assn. org/delivering-care/ethics/assisted-reproductive-technology.

16. DOV FOX, BIRTH RIGHTS AND WRONGS: HOW MEDICINE AND TECHNOLOGY ARE REMAKING REPRODUCTION AND THE LAW 6-7 (2019).

17. Dov Fox, *Reproductive Negligence*, 117 COLUM. L. REV. (2017).

18. Thornhill v. Midwest Physician Ctr., 787 N.E.2d 247 (Ill. App. Ct. 2003).

19. Gleitman v. Cosgrove, 227 A.2d 689, 693 (N.J. 1967).

20. Wendy F. Hensel, *The Disabling Impact of Wrongful Birth and Wrongful Life Actions*, 40 HARV. C.R.-C.L. L. REV. 141, 144, 174 (2005).

21. Sofia Yakren, *"Wrongful Birth" Claims and the Paradox of Parenting a Child with a Disability*, 87 FORDHAM L. REV. 583 (2018).

22. Alberto Bernabe, *Do Black Lives Matter? Race as a Measure of Injury in Tort Law*, 18 SCHOLAR: ST. MARY'S L. REV. & SOC. JUST. 41 (2016).

23. Frisina v. Women & Infants Hosp. of R.I., Nos. 95-4037, 95-4469, 95-5827, 2002 WL 1288784 (R.I. Super. Ct. May 30, 2002).

24. *Opinion 4.2.5 - Storage & Use of Human Embryos*, AM. MED. ASS'N, https://www.ama-assn. org/delivering-care/ethics/storage-use-human-embryos.

25. Michele Goodwin, *A View from the Cradle: Tort Law and the Private Regulation of Assisted Reproduction*, 59 EMORY L. J. 1039, 1048 (2010).

26. Karlin v. Foust, 188 F.3d 446 (7th Cir. 1999).

27. Roe v. Wade, 410 U.S. 113 (1973).

28. Anonymous Male Infant v. Anonymous 120, Super. Ct., Wash. D.C. (Medical Malpractice Verdicts, Settlements & Experts, Oct., 2000, p.42).

29. Anonymous v. Anonymous Hosp., Super. Ct., Wash. D.C. (Medical Malpractice Verdicts, Settlements & Experts, Aug., 2000, p.36-37).

30. Eagleman v. Korzeniowski, 924 So. 2d 855 (Fla. Dist. Ct. App. 2006).

31. Jones v. Chi. Osteopathic Hosp., 738 N.E.2d 542 (Ill. App. Ct. 2000).

32. Marlow v. Moore, Cir. Ct., Shelby Cty., Tenn. (Medical Malpractice Verdicts, Settlements & Experts, April, 1999, p.30-31).

33. *Compare* Baker v. McNamee, No. CV970141857S, 2005 Conn. Super. LEXIS 1167 (Apr. 27, 2005) *with* Bond v. Kalla, No. 543295, 1998 Conn. Super. LEXIS 1061 (Apr. 13, 1998).

34. Elizabeth Kukura, *Choice in Birth: Preserving Access to VBAC*, 114 PENN. ST. L. REV. 955 (2010).

35. Jamie R. Abrams, *Distorted and Diminished Tort Claims for Women*, 34 CARDOZO L. REV. 1955 (2013).

36. Troxel v. Granville, 530 U.S. 57, 68-69 (2000).

37. Meyer v. Nebraska, 262 U.S. 390, 399 (1923)

Product Liability

Michael Vinluan

Contents

Key Points

- Product liability is the legal doctrine in civil law that gives consumers a right to file suit against a manufacturer who sold a defective product for injuries suffered as a result of the foreseeable use of that product.
- A product is considered defective when its design, manufacturing process or warning instruction renders it unreasonably dangerous to consumers.
- Any product sold in the marketplace or placed in the "stream of commerce" can be a subject of a product liability lawsuit.
- Any entity responsible for putting the product at issue in the marketplace can be named as defendants in a product liability suit, including the manufacturer, distributors, retailers, trade associations, and in medical product cases, prescribers.
- Product liability is based on several theories of liability such as negligence, strict liability, breach of implied warranty, breach of express warranty, and misrepresentation and fraud.
- Other causes of actions include violation of consumer protection and deceptive trade practices laws.

M. Vinluan (✉)
University of Maryland, Baltimore, Baltimore, MD, USA

Medical-Legal Consultant, Washington, DC, USA
e-mail: mvinluan@umaryland.edu

- Some of the defenses used in product liability cases include: Assumption of risk, product misuse, state of the art, statute of limitation and statue of repose, federal preemption, learned intermediary doctrine, and alternative causation.
- A mass tort is a civil action involving numerous plaintiffs versus a manufacturer; examples of which are multidistrict litigations and class actions.

Legal Concepts
- Product Defect: In product liability law, a product is considered defective when its design, manufacturing process or warning instructions renders it unreasonably dangerous to consumers.
- Design Defect: A product has a design defect if the foreseeable risk of harm posed by the product could have been reduced or avoided had the manufacturer adopted a reasonable alternative design.
- Manufacturing Defect: A product has a manufacturing defect when the product departs from its intended design.
- Failure to Warn: Liability based on failure to warn is based on the manufacturer's inadequate instructions or warnings when the foreseeable risk of harm of using the product could have been reduced or avoided by such instruction or warning.
- Negligence: Negligence is one of the fundamental theories of liability in tort law. Negligence occurs when a person or entity fails to exercise reasonable prudence. Negligence has four elements that must be proven by the plaintiff: (1) a legal duty owed by the defendant to the plaintiff, (2) breach of that duty, (3) causation, and (4) the resulting damages.
- Strict Liability: Under a strict liability cause of action, the plaintiff must prove that (1) the defendant manufactured (or sold) the product; (2) the product was defective when it left the defendant's possession, and (3) the defect caused harm to the plaintiff. Unlike in a negligence cause of action, in strict liability, the plaintiff does *not* need to prove a wrongdoing on the part of manufacturer so long as the product at issue is defective.
- Breach of Implied Warranty of Merchantability: Arises when the manufacturer sells a product that is not generally fit for the purpose for which similar products are normally used.
- Breach of Implied Warranty of Fitness for a Particular Purpose: Arises when the manufacturer sells a product that is not fit for the product's specific purpose.
- Breach of Express Warranty: Based on a guarantee made by the seller that is relied on by the buyer, in the form of advertisement, product labels, or statements given during the sale or after the sale of the product
- Misrepresentation: Occurs when a defendant negligently presented relevant facts that are not true about the product and that these facts were relied upon by the plaintiff.
- Fraud: A type of misrepresentation where the defendant knowingly and intentionally or recklessly provided a material statement that is false to induce the plaintiff to use the product.
- Other causes of actions: Plaintiffs may also file actions based on violations of certain federal and state regulations, such as violation of consumer protection laws and unfair trade practices laws.

1   Introduction

Product liability is the legal doctrine in civil law that gives a consumer a right to file suit against a manufacturer who sold a defective product for injuries suffered as a result of the foreseeable use of that product. Any manufacturer who sells a defective product that is unreasonably dangerous to the consumer is subject to liability [1]. A product is considered defective when its design, manufacturing process or warning instructions renders it unreasonably dangerous to consumers. Any product sold in the marketplace or placed in the "stream of commerce" can be subject to a product liability lawsuit. As such, product liability cases have involved a wide range of products from cigarettes, drugs, medical devices, vehicles, baby powder, weed killers to many other product types.

Product liability can involve thousands of consumers of a single product or just one consumer, such as the highly publicized McDonald's hot coffee case, where a plaintiff was awarded more than half a million dollars for burn injuries she suffered from the coffee that spilled out on her lap [2].

In addition to the manufacturer, any entity responsible for putting the product at issue in the marketplace can be named as a defendant in a product liability suit. As such, along with the manufacturer, the product's distributors and retailers may also be named as defendants. In few cases, trade associations who promoted the product or set standards related to the product at issue have also been named as defendants. For example, Personal Care Products Council (PCPC), a national trade association representing global cosmetics and personal care product companies was sued along with the manufacturer of talcum powder for alleged risks of ovarian cancer with talcum powder use. PCPC was later granted summary judgment and taken out in the lawsuits [3]. Likewise, medical doctors too have been included as defendants in product liability cases involving pharmaceutical and medical device products.

Product liability is based on several theories of liability including negligence, strict liability, breach of implied warranty, breach of express warranty, and misrepresentation and fraud. Other causes of actions include violation of consumer protection and deceptive trade practices laws.

2   Discussion

Product Defect. In product liability law, a product is considered defective when its design, manufacturing process or warning instructions renders it unreasonably dangerous to consumers. The Restatement (Third) of Torts defines states, "A product is defective when, at the time of the sale or distribution, it contains a manufacturing defect, is defective in design, or is defective because of inadequate instructions or warnings" [4].

Design Defect. A product has a design defect if the foreseeable risk of harm posed by the product could have been reduced or avoided had the manufacturer adopted a reasonable alternative design [5]. To prove a design defect, some states or jurisdictions ask, under the consumer expectation test, whether the danger of the design cannot be contemplated by an ordinary consumer. Under the consumer expectation test, the plaintiff must prove that a foreseeable user like him or her would not have expected the danger of using that product. Some jurisdictions adopt the risk-utility test and ask whether the risk of using the product outweigh its benefits. Under the risk-utility test, the plaintiff must prove that the defendant had an alternative design that is reasonable (in cost, for example) and safer but failed to use that alternative design.

Not all inherently dangerous products, however, are subjected to product liability suits. For example, manufacturers of knives, are not sued for design defects even if knives are dangerous by design because (1) the danger of this product is a common knowledge and (2) there is generally no safer way to design knives.

Manufacturing Defect. A product has a manufacturing defect when the product departs from its intended design [6]. Thus, the defect occurred during the manufacturing process instead of during the design phase. The manufacturer must have deviated from design specifications so that the specific item at issue is no longer compliant with the manufacturer's intended finished product. As a result, products that are affected generally are single items or a batch of items but not the entire product line. For example, many pharmaceutical companies have been sued for products that are allegedly contaminated with harmful impurities during the manufacturing process.

Failure to Warn. Liability based on failure to warn is based on the manufacturer's inadequate instructions or warnings when the foreseeable risk of harm of using the product could have been reduced or avoided by such instruction or warning [7]. Also known as marketing defect or warning defect, some of the biggest product liability cases in history are based on failure to warn defects. For example, patients sued the manufacturer of Risperdal alleging that the manufacturer failed to warn them that this drug can cause male breast enlargement.

Legal Bases for Product Liability Law. Product liability lawsuits are based on several legal grounds or "causes of action." A single product liability lawsuit typically includes multiple "counts" of causes of actions. A plaintiff need not win each cause of action and the trier of fact (typically the jury) decides whether the plaintiff has proven his or her case in each of the counts.

1. **Negligence.** Negligence is a fundamental theory of liability in tort law. Negligence occurs when a person or entity fails to exercise reasonable prudence. Plaintiffs must prove the following four elements to prove a negligence claim: (1) the defendant owed a legal duty to the plaintiff, (2) the defendant breached that duty, (3) this breach caused the plaintiff to be harmed, and (4) the plaintiff had resulting damages.

   Under the first element, a duty of reasonable care is owed to a foreseeable plaintiff. Of note, the plaintiff does not need to be the actual buyer of the product at issue, he or she only has to be a foreseeable user of the product. Under the breach element, the plaintiff must prove that the defendant breached that duty, creating a product defect. The plaintiff must then establish that each defendant in the chain of distribution breached their duty.

After the breach element is established, the plaintiff must prove that the defendant's conduct was (1) an actual cause (also known as cause in fact) and (2) a proximate (also known as legal cause) of the alleged harm. Within the actual cause requirement, the plaintiff must prove that "but for" the defendant's breach of duty, his or her injury would not have occurred (commonly referred to as the "but for test"). Some jurisdictions use the "substantial factor" test, asking if the defendant's conduct was a substantial factor in causing the alleged injury. Certain jurisdictions apply both tests. The substantial factor test is useful when there are several defendants involved in the product defect. Under the proximate cause requirement, the plaintiff must prove that the injury he or she suffered is an injury that is not too remote, meaning it was foreseeable from the reasonable use of the product. There must be a natural and continuous sequence of events between the plaintiff's use of the defective product and his or her alleged injury. A new and independent cause of the alleged injury may "break the chain of causation" and the product at issue might not be proven as the proximate cause of the patient's injury.

Under damages, plaintiffs can claim for both personal injury and property damages. These damages are called compensatory damages, which are meant to make the injured plaintiff "whole" again to an extent possible via monetary award. Personal injury damages often include economic damages, such as medical expenses and loss of income, and non-economic damages, such as pain and suffering, mental anguish, and disfigurement. A plaintiff's spouse may also file for loss of consortium claims due to loss of affection, intimacy, and care.

In some cases, a plaintiff may ask for punitive damages. Unlike compensatory damages, punitive damages are meant to punish the defendant for certain willful wrongdoing, or for reckless or wanton disregard for the safety of consumers. Punitive damages are meant to deter the defendants from continuing unlawful conduct [8].

These damages can also be claimed for the following causes of actions.

2. **Strict Liability.** Another theory under tort law is strict liability. In a strict liability cause of action, the plaintiff must prove that (1) the defendant manufactured (or sold) the product; (2) the product was defective when it left the defendant's possession, and (3) the defect caused harm to the plaintiff. Unlike a negligence cause of action, in strict liability, the plaintiff does *not* need to prove a wrongdoing on the part of manufacturer so long as the product at issue is defective. The plaintiff must, however, prove that the defective product was the actual and proximate cause of his or her injury.

Strict liability may be applied to a failure to warn cause of action when the manufacturer failed to warn the plaintiff about dangers of the product. This theory may also be applied to a design defect cause of action when the product's design (or formula) to begin with is inherently dangerous. Likewise, strict liability may be applied to a manufacturing defect cause of action when, despite a safe design (or formula), the product at issue deviated from its intended design through the manufacturing process.

3. **Breach of Warranties.** Causes of action based on breach of warranties are based upon a breach of contract. Plaintiffs must prove that a warranty existed and the product he or she used does not conform to that warranty.

    Breach of Implied Warranty. Implied warranties arise from an understanding that the product sold is generally acceptable and reasonably fit for ordinary purposes for which it is being sold. A breach of implied warranty of merchantability arises when a manufacturer sells a product that is not generally fit for the purpose for which similar products are normally used. Likewise, a breach of implied warranty of fitness for a particular purpose arises when the manufacturer sells a product that is not fit for the product's specific purpose.

    Breach of Express Warranty. An express warranty is based on a guarantee made by the seller that is relied on by the buyer, in the form of advertisement, product labels, or statements given during the sale or after the sale of the product. An express warranty, therefore, involves an affirmation of fact or promise made by the seller to the buyer that becomes the "basis of the bargain" [9]. Failure to perform that fact or promise constitutes a breach.

4. **Misrepresentation and Fraud.** Under misrepresentation, the plaintiff must show that the defendant negligently presented relevant facts that are not true about the product, and that these facts were relied upon by the plaintiff. Relevant facts are material facts such as the nature and safety of the product at issue. These material facts, often found in product labels and advertisements, must have been provided by the defendant to induce substantial reliance of the plaintiff.

    Fraud is a type of misrepresentation. However, under a cause of action based on fraud the plaintiff must show that the defendant knowingly and intentionally or recklessly provided a material statement that is false to induce the plaintiff to use the product. Concealing or omitting material facts also constitutes fraud. In essence, fraud is based on bad faith.

5. **Violation of Statutes.** In addition to the causes of actions discussed above, plaintiffs may also file actions based on violations of certain federal and state regulations, also known as statutory violations. Such causes of action may include violation of consumer protection laws and unfair trade practices laws. These actions typically allege deceptive acts or practices and unfair methods of competition premised on the actual language of the statute cited in the cause of action.

Defenses in Product Liability. Defendants in product liability suits can assert one or more defenses to a plaintiff's claim. Typical defenses are defined below.

1. **Assumption of risk.** An assumption of risk defense is where a plaintiff voluntarily and unreasonably used a product despite knowing that it is defective.
2. **Product misuse.** Product misuse is another defense where the plaintiff used a product in a manner different from its intended or foreseeable use.
3. **State of the Art.** Another defense commonly seen is based on the "state of the art," which refers to all the available knowledge and technology available at a

certain time. "State of the art" is a defense where the manufacturer must prove that there is no other alternative design or method of manufacturing the product that is both safer and feasible at the time it was designed or manufactured. In some cases, defendants will also argue that during the time the product was designed or manufactured, no mechanical or scientific knowledge or technology existed that could have predicted of the "danger" of using the product. In other words, the defendant could not have known or predicted the alleged danger of the product at issue at the time it was designed or manufactured because the means to detect the defect was still unavailable.

4. **Statute of Limitation and Statue of Repose.** Statute of limitations and statue of repose are deadlines to file suit as set forth by each state's regulation. A statute of limitations is the deadline to file a lawsuit after the injured person discovered, or reasonably should have discovered, that the injury was possibly caused by the product at issue. Related to the statute of limitations defense is the statute of repose. In product liability, statute of repose limits the total number of years from the date of purchase or manufacture after which a consumer can no longer file a product liability suit related to that product. Both the statute of limitations and statute of repose depend on a specific number of years set by each state's regulation.

5. **Federal Preemption.** Typically encountered in medical product liability cases, federal preemption simply means that even though the plaintiff can prove that the product at issue violated certain state laws, the U.S. Constitution's Supremacy Clause provides that certain federal laws, such as the Food, Drug and Cosmetic Act (FDCA), supersede the state law at issue and therefore "preempt" and bar the lawsuit from being further litigated. For example, drug manufacturers have claimed that in certain cases it is impossible to comply with both state and federal requirements [10]. Thus, when in compliance with FDA drug labeling requirements, drug manufacturers bring forward arguments of dismissal predicated on preemption [11].

    Note, however, that compliance with federal and state product safety regulations is evidence that a product is not defective. Nonetheless, compliance with government safety regulations is not an absolute proof that the product is not defective and does not absolve the manufacturer from liability. As such, FDA approval does not bar patients from filing product liability suits against drug and medical device manufacturers. Compliance with FDCA and FDA regulations may not protect a manufacturer from state law product liability claims [12]. Similarly, a prior FDA recall involving the product at issue does not guarantee that the plaintiff will win his or her case.

6. **Learned Intermediary Doctrine.** The learned intermediary doctrine is typically seen in medical product liability cases involving failure to warn actions. Here, a manufacturer may argue that its duty to warn of the risk of its product is discharged after the manufacturer has already warned the prescribing clinician of the product risk and that the manufacturer has no duty to warn each patient [13]. This defense further asserts that the prescribing clinician is the better party to know- about the risk and benefits of the drug or medical device at issue to a

specific patient. Thus, if the prescribing clinician who is aware of the risks of the drug, prescribes the drug at issue, the manufacturer cannot be a producing cause of the injury.

7. **Alternative Causation.** In product liability cases involving an adverse medical outcome, defendants may also attack the causation element of the plaintiff's case. Such arguments include providing evidence that the alleged injury, or medical condition, of the plaintiff was caused by something else unrelated to the product at issue, also known as alternative causation. As such, pre-existing conditions, risk factors, genetic predispositions, environmental exposures, and family history become relevant in analyzing whether the product at issue caused the alleged injury or something else.

Product Liability Litigation in General. A consumer becomes a plaintiff once he or she files a product liability suit, typically in a state court in his or her home state where they have a "home court" advantage. Defendant manufacturers, however, prefer to defend their cases in a federal court in the state where they are either incorporated or headquartered. There are several reasons for their preference to a federal court, including alleviating any perceived prejudice against an out-of-state defendant and familiarity with Federal Rules of Civil Procedure. Many defendants thus "remove" the case from state courts to a federal court. Transferring to a federal court is possible if the plaintiff and defendant are from different states, or if the case arise under or involve a federal law or the U.S. Constitution [14].

In most states, the plaintiff has the burden of proof to prove that the product at issue was defective. For example, in causes of action based on negligence, to meet the burden of proof, the plaintiff must prove by "a preponderance of evidence" that the product at issue was defective and the defect caused him or her harm. Preponderance of evidence simply means "more likely than not." In terms of probability, it means more than 50% true than not true. This evidentiary standard is a lower threshold in comparison to the "beyond reasonable doubt" standard in criminal cases.

Because the burden of proof rests on the plaintiff, the defendant does not need to prove that their product was not defective. Instead, the defendant only needs to show the jury that the plaintiff failed to prove the necessary elements of the case. This is done by exposing weaknesses or poking holes in the plaintiff's case, such as questioning the credibility of the plaintiff's experts, or proving that the alleged injury was caused by something else and not the product at issue. Defendants may, also, file various substantive and procedural-based defenses throughout the litigation.

If a plaintiff is successful in proving that the product at issue was defective based on one or more of the theories of liabilities discussed above, and the product defect caused him or her harm, the plaintiff is awarded money as compensatory damages to pay for past, present, and future medical expenses, loss of income, and pain and suffering. The jury decides how much money to award the plaintiff. Because product liability is within the civil law system, as opposed to criminal law system, a civil penalty does not carry jail time. Imprisonment, while uncommon, may be imposed if there was a criminal intent that led to the product defect. If with criminal intent, prosecution as a criminal case typically involve company executives who knowingly or willfully commit fraud and misrepresentation to sell a product [15].

Many plaintiffs, however, do not argue their cases before a jury. In many cases, the defendant, without admitting any wrongdoing, may agree to settle the case out of court in exchange for the plaintiff's withdrawal of the lawsuit or agreement not to file suit. Settlements can occur anytime before a case is filed up to the trial days. Settlements can also occur even after a plaintiff wins a trial, in exchange for the defendant not to file any post-verdict appeal or request for a new trial.

Mass Torts. Product defects typically involve numerous consumers and potential plaintiffs. A mass tort is a civil action involving more than one plaintiff against a manufacturer, typical of product liability cases.

One type of mass tort is called multi-district litigation or MDL. Plaintiffs typically file product liability complaints in their home state courts for convenience and "home-court advantage," among many reasons. However, when faced with hundreds or thousands of plaintiffs, defendants prefer to transfer these numerous lawsuits to one federal court. This federal legal procedure creates a multi-district litigation or MDL. However, this federal court does not hear all those hundreds or thousands of cases. Instead, the court chooses a few cases under a select group of plaintiff attorneys. These cases become what is known as "bellwether cases." Depending on the outcome of these bellwether trials, defendants may choose to either settle the remaining cases or litigate them.

Another type of mass tort is the class action. Similar to MDLs, class actions typically involve numerous consumers in a common set of facts. However, unlike MDLs, a class action involves one or several plaintiffs who litigate their case on behalf of the whole "class." A few representative plaintiffs take an active role in the litigation on behalf of the entire class. Any court ruling binds all persons in the same class sometimes even if a class member did not know they were part of the class action. Thus, unless they opted out of the class action, a person may receive compensation even if they did not file a complaint. A person may also choose not to be a part of the class action and file his or her case separately from the class.

3   Applications

3.1   Federal Preemption

Thousands of patients filed suit against Merck claiming that its blockbuster osteoporosis drug Fosamax (alendronate sodium) caused them atypical femur fractures. These lawsuits allege that Fosamax is defective because Merck failed to warn prescribers and patients about this adverse effect. Plaintiffs filed in various state courts, and eventually many were transferred to a Multi District Litigation in New Jersey. (Of note, Merck has also been sued by thousands of plaintiffs in a separate MDL concerning Fosamax and osteonecrosis of the jaw, alleging that bisphosphonates makes it difficult for bones to heal following tooth extractions.)

When the FDA approved Fosamax in 1995, its label did not warn of the then-speculative risk of atypical femoral fractures associated with this drug. FDA subsequently ordered Merck to add a warning about atypical femoral fractures to the Fosamax label in 2011. Patients who took Fosamax prior to 2011 claimed that Merck failed in its state law duty to warn them and their doctors about these fractures. In defense, Merck, argued that the Plaintiffs' state-law-failure-to-warn claims should be dismissed due to preemption, arguing that the FDA would reject Merck's attempt to warn patients about atypical femoral fractures. Previously, FDA rejected Merck's 2008 attempt to warn about the risk of "stress fractures" with Fosamax. The trial court agreed with Merck. On appeal to the Court of Appeals for the Third Circuit, the appellate court vacated and remanded. Ultimately, the case reached the U.S. Supreme Court which held in 2019 that there must be a "'clear evidence' that the drug manufacturer fully informed the FDA of the justifications for the warning required by state law and that the FDA, in turn, informed the drug manufacturer that the FDA would not approve a change to the drug's label to include that warning" [16]. The case was remanded back to the Court of Appeals for further consideration based on this opinion.

3.2   Jurisdictions

As briefly discussed above, most plaintiffs prefer to sue manufacturers in the plaintiff's home state. Defendant manufacturers however, want these suits in the state where they are either incorporated or headquartered. The products liability suits at issue in this matter stemmed from car accidents involving Ford. Based in Michigan and incorporated in Delaware, Ford argued that their contacts with the two states at issue, Montana and Minnesota, were insufficient to confer jurisdictions on their courts. Ford contended that the cars involved were not designed or manufactured in those states.

The U.S. Supreme Court has long held that corporations may be sued for alleged product defects in states where they are incorporated or where their headquarters are located. However, on March 25, 2021, the Court in a unanimous 8-0 decision, held that state courts may entertain suits where plaintiff product liability claims arose from injuries occurring in their home states where a company advertises and markets its products, and where the company fosters ongoing consumer relations [17].

Justice Elena Kagan explained, "By every means imaginable—among them, billboards, TV and radio spots, print ads and direct email—Ford urges Montanans and Minnesotans to buy its vehicles, including (at all relevant times) Explorers and Crown Victorias [...] they encourage Montanans and Minnesotans to become lifelong Ford drivers."

Clinical Vignette—Learned Intermediary Doctrine
Case: Jim, an 18-year-old recent high school graduate, consulted his primary care physician for acne and was prescribed Solodyn (minocycline HCl), a tetracycline-class drug indicated to treat moderate to severe acne vulgaris. When he got his prescription from a pharmacy, he received an informational insert that state "[t]he

safety of using [Solodyn] longer than 12 weeks has not been studied and is not known." Jim did not receive the full prescribing informational material for Solodyn where it warns: "The long-term use of minocycline in the treatment of acne has been associated with drug-induced lupus-like syndrome, autoimmune hepatitis and vasculitis." The prescribing material further explain that: "Autoimmune syndromes, including drug-induced lupus-like syndrome, autoimmune hepatitis, vasculitis and serum sickness have been observed with tetracycline-class drugs, including minocycline. Symptoms may be manifested by arthralgia, fever, rash and malaise. Patients who experience such symptoms should be cautioned to stop the drug immediately and seek medical help" [18].

About a year later, Jim received another prescription for Solodyn and took it for twenty weeks as directed. Soon after, Jim was hospitalized and diagnosed with hepatitis and drug-induced lupus. While the hepatitis resolved, he learned he would have lupus for the rest of his life. Jim filed a product liability suit against the manufacturer of Solodyn. Can the manufacturer use the learned intermediary doctrine defense to successfully dismiss Jim's lawsuit?

Discussion: Yes, the manufacturer can raise the learned intermediary doctrine and get the case dismissed. State courts have held that if the manufacturer provided adequate warnings for the drug to the prescribing physician, who serves as the learned intermediary, as a matter of law, the manufacturer satisfied its duty to warn and would be entitled to a summary judgment [19, 20]. The learned intermediary doctrine, however, does not prevent Jim from filing suit against his doctor for failing to inform him about the risk for drug-induced lupus and hepatitis.

4   Summary

In summary, product liability is the legal doctrine in civil law that gives consumers a right to file suit against a manufacturer who sold a defective product for injuries suffered as a result of the foreseeable use of that product. A product is considered defective when its design, manufacturing process or warning instruction renders it unreasonably dangerous to consumers. Any product sold in the marketplace or placed in the "stream of commerce" can be a subject to a product liability lawsuit. Any entity responsible for putting the product at issue in the marketplace can be named as defendants in a product liability suit, including the manufacturer, distributors, retailers, trade associations, and in medical product cases, medical doctors. Product liability is based on several theories of liability including negligence, strict liability, breach of implied warranty, breach of express warranty, and misrepresentation and fraud. Other causes of actions include violation of consumer protection and deceptive trade practices laws.

References

1. Restatement (Third) of Torts: Products Liability.
2. Legal Myths: The McDonals's "Hot Coffee" Case, available at https://www.citizen.org/article/legal-myths-the-mcdonalds-hot-coffee-case/

3. In re Johnson & Johnson Talcum Powder Products Marketing, Sales Practices, and Products Liability Litigation, MDL No. 16-2738 (2021).
4. Restatement (Third) of Torts: Product Liability §2.
5. Restatement (Third) of Torts: Product Liability §2 (b)
6. Restatement (Third) of Torts: Product Liability §2 (a).
7. Restatement (Third) of Torts: Product Liability §2 (c).
8. Philip Morris USA v. Williams, 127 S. Ct. 1057 (2007).
9. Uniform Commercial Code § 2-313.
10. Mutual Pharmaceutical Co. V, Bartlett, 133 S. Ct. 2466 (2013).
11. Merck Sharp & Doheme Corp v. Albrecht, 587 U.S. __ 2019.
12. Wyeth v. Levine, 129 S. Ct. 1187 (2009).
13. Urbaniak v. Am. Drug Stores, LLC, 2019 IL App (1st).
14. How to Remove a Case to Federal Court. American Bar Association, April 01, 2021. https://www.americanbar.org/groups/litigation/committees/mass-torts/practice/2021/how-to-remove-a-case-to-federal-court/
15. Bynum R. Former Georgia Peanut Managers Get Prison Time for Salmonella Outbreak. Insurance Journal [Internet]. 2015 Oct 7 [cited 2021 Sep 10]. Available from: https://www.insurancejournal.com/news/southeast/2015/10/07/384091.htm.
16. Merck Sharp & Doheme Corp v. Albrecht, 139 S. Ct. 1668 (2019).
17. Ford Motor v. Montana Eight Judicial District, 141 S. Ct. 1017 (2021).
18. Solodyn, Highlights of prescribing Information. https://www.accessdata.fda.gov/drugsatfda_docs/label/2011/050808s014lbl.pdf
19. Watts v. Medicis Pharm. Corp., 239 Ariz. 19, 365 P.3d 944 (2016).
20. 50-State Survey: The Learned Intermediary Doctrine, https://www.jdsupra.com/legalnews/50-state-survey-the-learned-70847/

Part V

Administrative Law

Administrative Law—And Its Application in Health Law

Jeffrey S. Lubbers

Contents

Key Points

- Federal administrative agencies play a crucial role in the promulgation and implementation of national health policy.
- U.S. federal administrative law is not a single, compact body of law, but rather is a collection of a number of principles that are contained in the Constitution, the Administrative Procedure Act, the substantive statutes that agencies administer, opinions of the courts, and duties imposed by the President and others in the Executive Branch.
- Key agencies are the Department of Health and Human Services and its component agencies, the Social Security Administration, and other agencies that provide disability benefits.
- Each of these agencies have important regulatory and adjudicative responsibilities and also operate under statutes that often contain agency-specific or program-specific procedures that differ from the APA

J. S. Lubbers (✉)
Washington College of Law, American University, Washington, DC, USA
e-mail: jlubbers@wcl.american.edu

© The Author(s), under exclusive license to Springer Nature Switzerland AG 2022   185
A. S. Pasha (ed.), *Laws of Medicine*,
https://doi.org/10.1007/978-3-031-08162-0_11

Legal Concepts
- Rule Definition: an "agency statement of general applicability and future effect designed to implement, interpret, or prescribe law or policy."
- Adjudicative order Definition: "a final disposition of an agency in a matter other than rulemaking but including licensing."

1   Introduction

Administrative law is the group of requirements and restrictions on how administrative agencies take actions that affect members of the public. In the United States, administrative law is not a single, compact body of law, but rather is a collection of a number of principles that are contained in the Constitution, the Administrative Procedure Act (APA) [1], the substantive statutes that agencies administer, opinions of the courts, and duties imposed by the President and others in the Executive Branch. It governs both what agencies must do and what they may not do when taking actions, as well as the judicial review of their actions. It is concerned with the *process* by which agencies take actions, not the substance of those actions, although the substantive issues may influence what procedure is appropriate and review by the courts. It applies to government agencies that exercise power delegated to them from Congress; thus it is not directly concerned with the decisionmaking of the President, Congress, or the courts, although each of these play large roles in administrative law. Finally, administrative law addresses the relationship between the government and the private sector; it is not concerned with the internal management of government except as it affects the public.

Obviously the administrative agencies in the Executive Branch are the focal point of U.S. administrative law. Many of them have an important role to play in health law and policy. The fifteen Departments are part of the President's Cabinet and are closely aligned with the President and his policies. One, of course, is the Department of Health and Human Services (HHS). It contains many agencies as subparts, such as the Food and Drug Administration, which regulates the safety and efficacy of food, drugs, and medical devices; the Centers for Disease Control and Prevention, which protects our nation against health threats, and responds when they arise; and the Centers for Medicare and Medicaid Services, which oversees the financial aspects of Medicare, Medicaid, and the Affordable Care Act. The major medical research institution, the National Institutes of Health, is also part of HHS. Thus, this one single department contains a number of sub-agencies that together span a variety of activities—from regulation, to enforcement, to payments of benefits, to research. The Social Security Administration (SSA) was also part of HHS until 1984, when Congress made it an independent agency [2]. Each of these, agencies, regardless of whether they are called departments, commissions, or administrations, or whether they are part of another agency, are "agencies" within the terms of the APA and are governed by administrative law.[1]

---

[1]This chapter will primarily focus on administrative law at the federal level, and hence will be concerned with federal agencies. The same overall concepts generally apply at both the state and local levels, although the actual implementation and emphasis differ jurisdiction by jurisdiction.

2   Discussion

2.1   Importance of State Regulation of Medicine

Although Congress is authorized to regulate interstate commerce under Article I, section 8 of the Constitution, and therefore can assign the regulation of drugs, medical devices, and the health and safety of the workplace to administrative agencies, it cannot regulate the practice of medicine per se. The Court of Appeals for the District of Columbia Circuit stated this plainly in a recent case: "States, not the federal government, traditionally have regulated the practice of medicine" [3]. This means that state and local governments, acting through their own administrative agencies are the primary regulators of medicine in the United States. Each of the 50 states and Washington, DC (and U.S. territories) have their own departments of health [4]. In effect, these are the state equivalents of the federal HHS. In addition, each state also has a state medical board, which administers or accredits medical licensing examinations, conducts fitness evaluations, organizes continuing medical education, and investigates and adjudicates complaints that can lead to disciplinary measures against physicians. As explained by the Federation of State Medical Boards (FASB), whose website contains a wealth of information about Board structures, memberships, and authorities around the country:

> The structure and authority of medical boards vary from state to state. Some boards are independent and maintain all licensing and disciplinary powers, while others are part of a larger umbrella agency, such as a state department of health ….

> State medical boards are typically made up of volunteer physicians and members of the public who are, in most cases, appointed by the governor. In recent years, non-physician board members—often referred to as "public members"—have become common. [5]

Each state has its own administrative procedure act, which tend broadly to follow the federal APA, and state courts are often informed by federal court decisions on administrative law. Thus, this chapter focuses on the nexus of federal administrative law and federal health laws. But it should also be noted that because the Due Process Clause of the Fourteenth Amendment applies to the states, both state and federal courts can and do opine on the fairness of the procedures of state medical boards. For example, in one case [6], the Supreme Court allowed members of a state board of medical examiners to exercise both investigative and adjudication functions.

Parts 2.2 & 2.3 of this chapter will provide an overview of federal administrative law, with an emphasis on rulemaking, adjudication, and judicial review.[2] Part 2.4

The allocation of responsibilities between the federal government and the states and between the states and local governments within the states is usually analyzed in "constitutional law" and in "state and local government law".

[2] There are, of course many other aspects of administrative law that cannot be detailed here, such as constitutional separation-of-powers restrictions; agency investigative, information collection, and enforcement powers; openness laws, including the Freedom of Information Act; and suits for money damages against the government or officials. These matters are described more fully in [7].

will discuss how these requirements and doctrines apply (often in very distinctive ways) to the major health law agencies in the U.S., primarily those in HHS, the Departments of Labor and Veterans Affairs, and SSA.

2.2   The Administrative Process—the Dichotomy Between Adjudication and Rulemaking

A pair of Supreme Court decisions early in the 20th century defined the great divide of American administrative law between rulemaking and adjudication [8]. The distinction drawn by the Court in those two cases is that a hearing is required when the government is making a decision that applies existing law or policy based on someone's individualized circumstances. These are adjudicatory decisions. Individual hearings are not required, however, if the action is to make policy that applies generally, as opposed to a specific individual, and has future effect. In administrative law, this is rulemaking. This distinction is embodied in the APA, which forms the basis for modern federal administrative procedure, and prescribes differing procedures for each.

Rulemaking

A "rule" is defined by the APA as an "agency statement of general … applicability and future effect designed to implement, interpret, or prescribe law or policy" [9]. The procedures set out in the APA for developing rules are quite streamlined: an agency must publish a general notice of proposed rulemaking in the *Federal Register* that sets out the legal authority for the agency's action and the terms or substance of the proposed rule [10]. The agency must then give the public the opportunity to submit written comments on the proposed rule. The agency is directed to consider all relevant matter that was submitted by the public. When it issues the final rule, the agency is normally required to publish it in the *Federal Register* at least 30 days before its effective date [11], and it must include in the publication a "concise general statement of [the rule's] basis and purpose" [12]. Given the nature of the process, this is called "informal" or "notice-and-comment" rulemaking. In part, these labels are to distinguish this process from the other rulemaking procedure that is set out in the APA, which is called "formal" or "trial type" rulemaking. If the statute requires the agency to develop the rule "on the record" after an opportunity for an agency hearing, then the APA requires a procedure similar to that used for adjudication. Very few statutes do so, however, and the Supreme Court has held that unless the statute is quite explicit about the use of formal procedures, the agency may use the informal process [13]. As a result, formal rulemaking is very rarely used.

Although in the first few decades after enactment of the APA agencies tended to rely more on case-by-case adjudication to make policy, that began to change in the 1960s, when Congress enacted numerous new regulatory statutes. Moreover, rulemaking was seen as a more open, participatory, and precise way to develop policy. In addition, the minimal requirements of the APA were expanded by the courts [14], and as a result, new procedures emerged that agencies must follow when issuing rules. Also, the courts expanded judicial review to ensure that the agencies adhered to the procedures and that the rules were well-grounded in the facts presented by the agency. In effect, there was a great push to make the rulemaking process more *rational*. The courts required that agencies explain the basis for their rules far more than had previously been the case. As a result, the "concise general statement" of the APA became a "preamble" that may run for hundreds of pages in the small type of the *Federal Register*. Moreover, to the extent the factual basis of a rule can actually be developed, the agency is required to do so; it must generate the scientific and technical information that supports the agency's position and explain any contrary evidence. Of course, in some instances some facts are not determinable and the agency must extrapolate from what is known to support the rule. In any event, the agency is required to place its information in the rulemaking file that may be examined by the public. Also, the agency needs to explain how it progressed from the facts to the rule. As a result of these requirements, if someone submits comments that cast significant doubt on the factual predicate of an agency's proposal, the agency is required to make appropriate changes before issuing the final rule.

It should be noted that, as with many aspects of modern life, rulemaking has become "e-rulemaking", and the great majority of comments are filed (and stored) electronically on the federal government's centralized portal, *www.regulations.gov*. This has not only made it easier for interested persons to file comments, it has created some new challenges for agencies in terms of privacy issues, the volume of comments, and handling comments with copyrighted material or confidential business information [14].

Presidential Involvement in Rulemaking

As rulemaking became the more significant means of developing government policy, not surprisingly it also became seen as a political undertaking, and hence the White House became more involved than it had been under the original structure of the APA. Through a series of presidential executive orders, an agency must submit a proposed rule to the Office of Information and Regulatory Affairs (OIRA), which is part of the Office of Management and Budget, which in turn is part of the Executive Office of the President, generally known collectively as "the White House". Presidents since Nixon have issued such orders. Under the most important existing order, President Clinton's Executive Order 12866 [15], OIRA reviews the proposed rule to make sure it is consistent with the President's regulatory policy, that it is the most cost effective means to achieve the regulatory objective, that alternatives (including economic incentives and other non-regulatory approaches) have been considered, and that the rule would impose the least burden on society that is consistent with achieving the goal of the action [15]. If the rule would have an annual effect on the economy of more than $100 million or adversely affects a sector of the economy, it is deemed an "economically significant regulatory action" [15], and the agency must also prepare a regulatory impact analysis that analyses the costs and benefits of the proposed rule. The agency must again submit the final rule for OMB clearance before publishing it in the *Federal Register*. Thus, the President and his immediate office have become far more involved in regulatory decisionmaking than was the case 50 years ago [16].

These developments converted a flexible, abbreviated process into one that is far more complex. The APA also provides that "each agency shall give an interested person the right to petition for the issuance, amendment, or repeal of a rule" [17]. An agency is required to respond to such a petition "promptly" and provide a brief explanation for its actions. But the courts are highly deferential to the agencies in allowing them to control their own agenda and allocate their inherently scarce resources. However, where an agency denies a petition on the grounds that it lacks jurisdiction or where the petition asserts that an agency rule needs to be modified due to changed conditions, courts are more likely to review with more intensity.

Interpretive Rules, Policy Statements, and Guidelines

Substantive rules (often informally referred to as "regulations") are an exercise of authority delegated by Congress to make law. Assuming they have been properly issued and have withstood any judicial review, they have the same force and effect as a statute. Thus, they are binding on those to whom they apply. But in addition, agencies also issue a variety of guidance documents that have varying practical effects. One type is an "interpretive rule" in which the agency provides the public with its views as to the meaning of a statutory or regulatory term. Another is a statement of general policy in which the agency describes how it plans to exercise its discretion in the future or provides guidance on how to comply with a requirement. To encourage agencies to provide the public with this useful information, these guidance documents are not required to be published for notice and comment [10] nor be published at least 30 days before they become effective [12]; they are, however, required to be published in the *Federal Register* or made available on the agency's website [18].

While there may be a clear conceptual difference between regulations and guidance documents, distinguishing between them is not an easy task for reviewing courts [19]. Agencies frequently argue that an issuance is a non-binding guidance document in order to avoid complying with the rigors of the rulemaking process, but then turn around and ask a court to defer to it in an enforcement action, which would, as a practical matter, give binding effect to something that purportedly was not binding. Courts are then pressed to decide whether the statement was in fact a regulation, which should have complied with the notice-and-comment procedures.

Adjudication Under the APA

Prior to 1946, and even through the mid-1960s, the focus of U.S. administrative law was primarily on agency adjudication since that was largely how agencies functioned. As a result, the APA is quite prescriptive as to how agencies are to undertake the adjudicative hearings specified by the Act. Agency adjudications most often occur when statutes or regulations are being enforced, and the respondent seeks a hearing, or when a private party applies for a benefit (such as social security) or a license, and it is denied. Basically, the APA procedures apply to "every case of adjudication required by statute to be determined on the record after opportunity for any agency hearing" [20]. But the APA is silent as to what procedures are to be followed if some sort of hearing is required, but the governing statute does not contain this "on the record" language. The question as to whether a hearing must be held in adjudications under such statutes and, if so, what type of hearing, is determined by the Due Process Clause of the Constitution. In administrative law literature, hearings "on the record" are called "formal adjudication" and the rest are considered "informal" [21]. Even where statutes do not explicitly trigger the adjudicatory sections of the APA, agencies often, nevertheless, voluntarily comply with most of those provisions.

Formal Adjudication

The APA provides many details as to how formal adjudication proceedings are conducted. The process begins by the agency providing a notice of (1) the time, place, and nature of the hearing; (2) the legal authority and jurisdiction under which the hearing is to be held; and (3) the matters of fact and law that are asserted [22]. The agency must then give all parties the opportunity to present facts, arguments, and offers of settlement [23]. An Administrative Law Judge (ALJ) then presides over any hearing [24]. ALJs are judicial officers within agencies who are independent of the policymaking and enforcement parts of the agency. The ALJ has the normal judicial authority to regulate the hearing [25] and make the initial or recommended decision in the case [25]. Typically, the agency will also provide at least one level of appeal or review within the agency itself [26]. In addition, the APA provides for "separation of functions", meaning that staff engaged in investigation or prosecution may not communicate with decisionmakers about a case while it is in progress [27]. The formal rules of evidence do not apply in agency proceedings, and hearsay evidence is normally allowed [28]. *Ex parte* communications—those between the decisionmaker and fewer than all of the parties or outsiders—are prohibited [29]. Agencies and agency officials are allowed to have a particular point of view or general bias. Thus, for example, it is fully appropriate for a Federal Trade Commission that is avowedly pro-competition to hear a case in which it is charging a company with anti-competitive activity [30]. And, it is similarly acceptable for a state medical board to investigate whether or not there is probable cause to bring an action and then to sit in judgment as to whether or not the offence occurred despite the commingling of investigative and adjudicative functions [6]. There are limits, however, and courts will disqualify an adjudicator if the person has a specific, personal bias, such as having already decided major issues of fact [31], animus against a litigant, or a personal interest in the outcome [32].

Due Process and Informal Adjudication

If the governing program statute does not require a hearing "on the record", the question arises as to the scope of the hearing the agency must hold before taking action that someone views as adverse to their interests. The issue then becomes when does the Due Process Clause apply and when it does, what does it require?

In a series of decisions in the 1960s and 1970s the Supreme Court expanded the notion of "property" interests to include government entitlements such as welfare payments or a tenured government job [33], and also expanded the notion of "liberty" interests to include stigmatic actions by the government that also impaired a person's legal status [34]. Once it is established that someone has a property or liberty interest at stake, the next issue that must be decided is "what process is due" under the Due Process Clause. The Court has determined that it depends on the circumstances and requires the balancing of three factors, known as the "*Mathews* balancing test": [35] (1) the nature of the private interest that will be affected by the official action; (2) the risk of erroneous deprivation of such interest though the procedures used by the government, and the probable value if any of additional procedural safeguards; and (3) the government's interest, including the function involved and the fiscal and administrative burden that the additional procedures would cause. Thus, unlike formal adjudication where the procedures are clearly specified in the APA, the process of informal adjudication under the Due Process Clause is rather variable and ad hoc. Hearings in these instances are sometimes extremely brief: an opportunity for the adversely affected person to tell her side of the controversy and present whatever evidence she may have to support it.

In sum, the process for "formal" adjudication is quite detailed in the APA, but applies only when the statute in question has the requisite triggering language. If the statute lacks such language, the procedures of any resulting adjudications are in effect governed by the *Mathews* balancing test and thus can range from cursory "paper" proceedings to close-to-formal trial-like proceedings, depending on the circumstances. One other element of due process, the right to be represented by counsel in all federal administrative proceedings, is provided for in the APA [36]. Congress has also authorized agencies to use alternative means of dispute resolution to resolve issues if all the parties agree to do so. Thus, instead of relying solely on adjudication, either formal or informal, the parties may use mediation and arbitration to resolve their dispute [37].

Choice of Procedures

Agencies have broad discretion in choosing the procedural means of achieving their goals and in the use of their resources. If an agency has the statutory authority to develop its policy either through rules or through a series of case-by-case decisions, the Supreme Court has made it clear that the choice is for the agency alone [38]. This choice is limited only by unusual circumstances in which the decision would be an abuse of discretion, such as imposing liability in an adjudicatory proceeding for past actions that were taken in good faith reliance on the then-current state of the law as developed by the agency. Moreover, the Supreme Court has made clear that agencies cannot engage in retroactive rulemaking without clear authorization from Congress [39].

2.3   Judicial Review of Administrative Action

In general, judicial review is based on the administrative record assembled by the agency in the proceeding that is being challenged. So, normally, reviewing courts do not allow new evidence to be admitted. The judicial review of actions taken by administrative agencies has several components that are provided by the APA, agency-specific statutes, the Constitution, and decisions by the courts. They are: (1) whether the action is judicially reviewable at all; (2) who may bring the action; (3) when can they bring it; and (4) what is the scope of judicial review for the factual, legal, and policy issues raised by challenging parties? [49]

Availability of Judicial Review

The judicial review sections of the APA begin by saying that they apply "except to the extent that—(1) statutes preclude judicial review; or (2) agency action is committed to agency discretion by law" [41]. Implicit in the first provision is the recognition that Congress may prohibit a court from reviewing the actions taken under a particular program. However, courts do not favor such limitations and have developed a "strong presumption that Congress intends judicial review of administrative action" [42].

The second clause—"committed to agency discretion by law"—means that Congress has not provided any standards by which the action is to be gauged by a court; in other words, there is no law to apply. In a case involving the FDA, the Supreme Court ruled that an agency determination whether or not to take an enforcement action is usually deemed exempt from review [43]. But, in general, with relatively rare exceptions, some sort of judicial review is available, provided the other, following, conditions are met.

Standing to Sue

If judicial review is available, a key question is: who can invoke it? The Constitution [Article III, section 2] limits the jurisdiction of the federal courts to "cases" and "controversies". Thus, courts are not authorized to provide advisory opinions, but rather they may only resolve actual disputes. This limitation undergirds the law of standing, which has frequently been interpreted and applied by the courts in various factual situations. To have standing, the plaintiff must allege an "injury in fact" [44] that is "actual or imminent" and fairly traceable to the defendant's allegedly unlawful conduct that is also distinct and palpable, and hence not abstract, conjectural, or hypothetical [45]. Further, the court must be in a position that it can redress the wrong if the plaintiff is successful [45].

Timing of Judicial Review

When an action may be brought is determined by three interrelated, and often overlapping, doctrines: ripeness, exhaustion of remedies, and finality.

Ripeness is a determination as to whether the agency's action is suitable for judicial resolution at the present time; whether it is "ready" for the court or whether more is needed. In deciding whether an issue is ripe for decision, the court balances whether the legal issue is currently fit for judicial resolution and the hardship to the parties for withholding judicial review until further developments. This test was developed in a case challenging an FDA drug labeling rule [46].

The APA provides that if the agency has a multi-step internal appeals process and specifies by rule that a party must complete all the steps before seeking judicial review, courts should honor that [47]. This doctrine of *exhaustion of administrative remedies* is based in part on the view that the agency should be given a full opportunity to address the disputed issue and perhaps remedy it so that a court challenge would be unnecessary. Exhaustion is not jurisdictional, however, so that in appropriate situations courts may waive its application where the agency's views are not relevant, where the agency cannot grant the relief sought, or where the petitioner would be prejudiced by the delay [48]. Note that one important consequence of the exhaustion doctrine is that an issue may not be raised for the first time in front of a court when challenging an action taken following an agency hearing. Even in challenges to agency rules, courts may prevent a challenger from raising an issue that had not been presented to the agency first in the comment period [49].

For similar reasons, and because the APA so stipulates, courts will only review *final agency actions* [47]. The courts do not want to become ensnared in an issue that is still evolving in the agency since the legal issues may be sharpened and the facts better defined if the agency completes its work. The Supreme Court has prescribed a rather strict test for finality: "As a general matter, two conditions must be satisfied for agency action to be 'final.' First, the action must mark the 'consummation' of the agency's decisionmaking process)—it must not be of a merely tentative or interlocutory nature. And second, the action must be one by which 'rights or obligations have been determined,' or from which 'legal consequences will flow'" [50].

The Forum for Judicial Review

There is no single rule that determines which court will hear a particular category of appeals. The APA leaves the matter to be dictated in particular statutes; but it also provides that if the relevant statute fails to specify the reviewing court, then challengers must file for declaratory or injunctive relief in the appropriate federal district court [51]. Thus, appeals of rules that are primarily based on the rulemaking record and of orders issued after formal adjudications are generally directed by statute to be filed in the courts of appeals. Appeals that tend to need factual development or reviews of high volume agency adjudications are channeled to the district courts. But challengers need to consult the specific statutory requirements for review of any particular action.

Scope of Review

The Administrative Procedure Act defines the scope of review as directing the reviewing court to overturn agency actions if they are (A) arbitrary and capricious or an abuse of discretion, (B) unconstitutional, (C) in violation of a statute, (D) in violation of required procedure (if such violation is prejudicial), or (E) unsupported by substantial evidence (in cases of formal adjudication) [40].

Review of the Agency's Fact Finding

The APA provides that an agency's decision made in a formal adjudication must be supported by "substantial evidence". The courts have interpreted this term as meaning the kind of evidence that a reasonable mind might accept as adequate to support a conclusion. In deciding whether a record contains substantial evidence, the court looks at the evidence that supports the proposition discounted by the evidence that fairly detracts from it [52]. In doing so, the court looks to the totality of the record (including both the ALJ's decision and the final agency decision).

For informal agency actions, including informal adjudications and most rule-makings, the substantial evidence test normally does not apply. Instead, the factual basis for rules and informal decisions is reviewed under the "arbitrary-and-capricious" test. In an early case involving an informal adjudication [53], the Supreme Court determined that it should apply that test by conducting a "searching and careful" review of the facts, but it cautioned that its role was limited to ensuring that the decision was supported in the materials before the agency and that the court is not authorized to substitute its judgment for that of the agency. This mode of review has also been applied to the review of the factual basis for, and the choice of policy made, in rulemakings as well [54]. This review has several components: (1) whether the decision was based on the consideration of the relevant factors set out in the legislation; (2) whether any impermissible factors might have entered into the calculus of the decision, or (3) whether there was a clear error of judgment [54]. In particular, the agency must consider the relevant factors, inclusively and exclusively, and must explain the basis for its decision. That is, it must describe the facts, and the evidence supporting the facts, and the relationship between the facts found and the choice made [54]. The intensity of review increases if the agency is changing direction.

Review of Legal Interpretations

Many challenges to agency action are based on a claim that the agency action was *ultra vires* (violated the governing statute). This raises the question as to how much weight a reviewing court should give the agency's interpretation of its own statute. Traditionally, courts looked at the text and legislative history of the statute and would then impose their views on the agencies. The Supreme Court redefined that relationship in a case that has become one of the most cited cases in American law, the *Chevron* case [55]. In that case, the Court established a two-step test for analyzing such questions: First, has Congress "directly spoken to the precise question at issue"? If so, "that is the end of the matter; for the court, as well as the agency, must give effect to the unambiguously expressed intent of Congress." If not, if Congress was silent or its intent was ambiguous, then "the question for the court is whether the agency's answer is based on a permissible construction of the statute" [55].

The Court indicated that in answering the "step one" question, the courts are to employ traditional tools of statutory construction—so no deference is given to the agency at this step. But step two requires the court to defer to a reasonable agency interpretation. The resulting acceptance of the agency's filling of the gaps is called "*Chevron* deference". While experts still argue about the case's jurisprudential significance, it clearly has had an important effect on the relationship between agencies and courts, with the courts sustaining agency interpretations of their statutes more than previously.

The Supreme Court subsequently decided several cases clarifying that application of the *Chevron* doctrine is not appropriate for *all* agency pronouncements that fill gaps in their statutes. In general, the application of *Chevron* appears to be now primarily limited to interpretations made in the exercise of notice-and-comment rulemaking or formal adjudication by agencies. It should be noted that, among the conservative majority of justices on the Supreme Court, there is an increasing skepticism about judicial deference to agency legal interpretations [56].

Non-Judicial Means of Accountability

The courts play the primary role in ensuring that agencies are held accountable for their actions. But there are other accountability mechanisms as well that bear mentioning, such as congressional oversight hearings, agency ombudsmen, the Government Accountability Office (an arm of Congress), and most significantly for this chapter, inspectors general.

Congress has established offices of inspector general in all departments and most large agencies. "IGs" as they are called, are independent officials, meaning there are some limitations on their removal and they file their reports simultaneously to the President and the Congress [57]. Their main role is to protect against "waste, fraud, and abuse". But in some instances and in some agencies their role is broader and includes ensuring that the agency is administering its duties legitimately. Over time, they have become important and independent anti-corruption and pro-accountability power centers within the executive branch. The IG office in HHS is one of the largest and plays an important role especially in investigating and enforcing fraud statutes relating to Medicare and Medicaid.

2.4   Application of Administrative Law to Health Law

As mentioned above, although regulation of the practice of medicine is a matter for each state, Congress, through the Interstate Commerce Clause, has assigned many aspects of the regulation of national health law to federal agencies. These agencies are primarily those in the Departments of HHS, Labor, and Veterans Affairs, as well as SSA.[3]

The Department of Health and Human Services

Like most departments, HHS has an Office of the Secretary that itself contains some operational offices such as the Office of General Counsel, the Office of Inspector General, and two important adjudication offices—the Departmental Appeals Board (DAB) and the Office of Medicare Hearings and Appeals (OMHA). The Department also contains eleven sub-agencies that engage in a variety of regulatory, research,

---

[3] Space does not permit a discussion of other government health care programs, such as the TRICARE program for active and retired military personnel and their dependents, the Indian Health Service (part of HHS), which provides direct medical and public health services to members of federally recognized tribes, or the various subsidized health care insurance plans offered to government employees. Of course, other regulatory programs administered by agencies such as the Environmental Protection Agency, National Highway Traffic Safety Administration, and the Consumer Product Safety Commission are also intended to protect our health.

policy development, and benefits activities [58]. The main ones for our purposes are the Food and Drug Administration (FDA), Centers for Medicare & Medicaid Services (CMS), Centers for Disease Control and Prevention (CDC), and National Institutes of Health (NIH). HHS also is the home of the U.S. Surgeon General, appointed by the President and confirmed by the Senate for a four-year term, who issues reports and advisories and also oversees the U.S. Public Health Service Commissioned Corps, a group of over 6,000 uniformed officers working throughout the federal government to promote public health [59].

Food and Drug Administration

The FDA regulates the safety and efficacy of human and veterinary drugs and medical devices, as well as vaccines, biologics, radiation emitting products, blood products, cosmetics, and tobacco products. It (along with the Food Safety and Inspection Service of the Department of Agriculture) also regulates food standards and safety [60]. In doing so, the FDA has issued many regulations, which can be found in Title 21 of the Code of Federal Regulations (C.F.R.), Parts 1 to 1499. But the FDA also relies extensively on non-binding guidance documents that are published on the FDA's website and arranged by topic [61]. As the FDA's website makes clear: "Guidance documents represent FDA's current thinking on a topic. They do not create or confer any rights for or on any person and do not operate to bind FDA or the public. You can use an alternative approach if the approach satisfies the requirements of the applicable statutes and regulations" [61].

Although the FDA engages in extensive enforcement of its regulations and statutes, it relies heavily on industry compliance with warning letters, and notices of potential violations [62]. If such violations are not rectified, the company or individual involved can suffer various sanctions, including loss of federal contracts, disapproval of related premarket approval applications, denial of Certificates to Foreign Governments, and civil money penalties, or in some cases referral for criminal prosecution to the Department of Justice. One example of regulatory emphasis is proceedings to potentially disqualify clinical investigators [63]. It is relatively rare for alleged violators to seek formal hearings before the FDA, although the agency does have rules of practice for such hearings. These rules are written in a rather general way to allow the use of ALJs where necessary and other presiding officers where allowed [64, 65]. One program that has resulted in a fair number of hearings by ALJs (lodged in HHS's DAB) is the enforcement of the Tobacco Control Act [66]. FDA brings numerous actions seeking civil money penalty and no-tobacco-sale orders for retailers who have sold tobacco products to underage buyers [67]. Its rules for civil money proceedings provide for ALJ hearings [68]. In the important area of drug regulation, FDA engages in the same type of post-market oversight and use of various compliance techniques, but it has avoided the need for numerous hearings in the new drug application phase by emphasizing the need for clinical trials showing safety and effectiveness in the application and using the technique of summary judgments to deny such applications if they lack evidence of a successful clinical trial. The Supreme Court approved of this procedural stratagem [69].

Centers for Medicare & Medicaid Services

The CMS has significant rulemaking responsibilities under numerous statutes, including Medicare and the Affordable Care Act (ACA). An example of its regulations under the ACA is its extensive set of certification requirements for Qualified Health Plans that may be offered through the Act's marketplaces. These regulations are codified in 45 C.F.R. subpart K. Its other many program regulations are codified in 42 C.F.R. Chapter IV, subchapters A–I. Subchapter B covers the Medicare program. As the Supreme Court recently noted:

> Today, Medicare stands as the largest federal program after Social Security. It spends about $700 billion annually to provide health insurance for nearly 60 million aged or disabled Americans, nearly one-fifth of the Nation's population. Needless to say, even seemingly modest modifications to the program can affect the lives of millions [70].

Due to the importance of these regulations, Congress included a Medicare-specific set of rulemaking procedures that deviate somewhat from the APA's, such as requiring a 60-day comment period and providing for the use of "interim-final" rules, with some conditions such as time limits on their effectiveness [71]. The main advantage for the agency in using interim-final rulemaking in place of regular notice-and-comment rulemaking is that the agency can issue a rule that is final on an interim basis, while seeking public comment after the fact. The agency is still obliged to consider the comments in deciding whether or not to modify the final rule, but this is especially handy for CMS because of the highly technical nature of many of its regulations and the need to promulgate provider reimbursement rules expeditiously. Other CMS-administered statutes also provide authorization to use interim final rules [72].

But, like the FDA, CMS also relies heavily on sub-regulatory guidance, primarily "manuals", that offer "day-to-day operating instructions, policies, and procedures that are based on statutes, regulations, guidelines, models, and directives" [73].

CMS also engages in administrative adjudication. It has its own Offices of Hearings and Inquiries, which performs several functions in HHS's complicated web of adjudications [74]. Among other things, it:

- Provides professional staff support to the Provider Reimbursement Review Board (PRRB) and the Medicare Geographic Classified Review Board (MGCRB).
- Conducts Medicare and Medicaid Hearings on behalf of the Secretary or the Administrator that are not within the jurisdiction of the DAB, the SSA's Office of Hearings and Appeals, the PRRB, the MGCRB, or the States [74].

These CMS adjudicative actions are a mix of informal evidentiary hearings and formal APA adjudications. The two "review boards" provide informal adjudication to providers in the Medicare program. The most significant of these boards is the PRRB, "an independent panel to which a certified Medicare provider of services may appeal if it is dissatisfied with a final determination by its Medicare contractor or by [a reviewing official in CMS]" [75, 76]. It is a five-member board appointed by the HHS Secretary for a renewable three-year term. According to the regulations, all members "shall be knowledgeable in the field of cost reimbursement. At least one shall be a certified public accountant. Two Board members shall be representative of providers of services." [77]. However, the first hearing in such cases is by a Medicare Administrator Contractor (MAC) (sometimes called a fiscal intermediary). MACs are multi-state, regional contractors responsible for administering both Medicare Part A and Medicare Part B claims,[4] and serve as the first level decision-maker when health care providers enrolled in the program dispute a reimbursement amount [78]. These MAC hearings are evidentiary hearings conducted by MAC-chosen hearing officers. A dissatisfied provider can then obtain a review by a CMS reviewing official. If such review does not satisfy the provider and the claim exceeds $10,000, the PRRB can hear the case if it determines first that it has jurisdiction to grant a hearing on each of the specific matters at issue in the hearing request. A list of its jurisdictional determinations is on its website [79]. PRRB decisions may be appealed to or reviewed by the CMS Administrator [80]. Judicial review for final agency actions is specified by statute to be in the appropriate federal district court [81].

The other CMS review board, the MGCRB, is more specialized. It makes determinations on geographic reclassification requests of hospitals who are receiving payment under the inpatient prospective payment system but wish to reclassify to a higher wage area for purposes of receiving a higher payment rate [82, 83]. As indicated above, CMS also has its own hearing officers who hear cases not heard by other adjudicative entities within CMS or HHS. One such type of case arises when CMS denies an application to qualify as a Part C Medicare Advantage or Part D Prescription Drug Plan or takes adverse action against an existing Plan. A list of such decisions is available on the CMS website [84].

OMHA

There is one other large-scale adjudication program that stems from the Medicare program, although it is not located in CMS—it involves formal adjudications conducted by ALJs in the HHS Office of Medicare Hearings and Appeals (OMHA). As the Department's organizational chart states:

> OMHA reports directly to the Secretary of HHS. OMHA administers the nationwide [ALJ] hearing program for appeals arising from individual claims for Medicare coverage and payment under Medicare Parts A, B, C and D, as well as appeals arising from claims for entitlement to Medicare benefits and disputes of premium surcharges [58].

The individual appeals of Medicare coverage, payment, and premiums heard by OMHA ALJs go through a five step process that is similar to the provider appeals heard by the PRRB. The ALJ hearings occur at the third stage of a five-step process.

---

[4]These are the "traditional Medicare" programs. Part A is Hospital Insurance and Part B is Supplemental Medical Insurance. Later, Part C (an alternative private "managed care" program that beneficiaries may choose instead of Parts A and B) and Part D (optional subsidized additional prescription drug benefits) were added.

The two preceding steps are first, a redetermination by the MAC and second, a reconsideration by a Qualified Independent Contractor (QIC). Appeals from QIC decisions (over $100, indexed to inflation from 2003) can be taken next to an OMHA ALJ for a formal hearing and a de novo decision. OMHA has over 100 ALJs, and has recently hired more to successfully pare down a large backlog [85]. Adverse ALJ decisions can be appealed to the Medicare Appeals Council, with a right to judicial review in federal district court (if worth over $1000 as indexed above). The Medicare Appeals Council is not part of OMHA; rather it is an arm of the HHS DAB, which houses the Administrative Appeals Judges who sit on the Council [86]. As this brief summary illustrates, the Medicare Appeals Process is complicated, and beneficiary groups have critiqued its complexity [87].

The previous section mentioned the Departmental Appeals Board, which is the centralized adjudication office for the Department. It is headed by a Board Chair and four other members appointed by the Secretary. In addition to housing the Medicare Appeals Council, it also houses a cadre of over dozen ALJs who conduct formal hearings and decide a wide variety of cases, including (among others) appeals from (1) civil money penalties (CMPs) and other enforcement actions taken by CMS against nursing home providers, clinical laboratories, home health care agencies, and other health care providers; (2) CMPs imposed by the HHS Office of the Inspector General against health care practitioners; and (3) CMPs and No Sale Tobacco Orders imposed by the FDA [88]. Most of these ALJ decisions are appealable to the DAB's Board.

In addition to hearing appeals from ALJs, the Board also provides review (which may include an evidentiary hearing) of certain types of final decisions of HHS operating components, such as determinations made concerning discretionary and mandatory grant programs [89]. The Board also plays an important role in implementing alternative dispute resolution (ADR) across the Department under the Administrative Dispute Resolution Act [37]. DAB staff provide support for ADR activities such as mediation, facilitation of dispute settlements, and negotiated rulemaking committees [86].

Centers for Disease Control and Prevention

The CDC suddenly became a nationally-known agency due to its role in providing non-binding but influential guidance to health departments and the general public on how to deal with various aspects of Covid-19—vaccine administration, testing, contract tracing, etc. [90]. But what is less known is that the CDC Director also has extensive powers to order quarantines and other measures to deal with the spread of communicable diseases, all backed up by civil and criminal penalties for violations [91]. In conjunction with this authority, the CDC Director hears written appeals from individuals subject to a federal order restricting travel [92]. In one recent case the CDC's "conditional sailing orders" governing the reopening of the cruise line industry were successfully challenged by the State of Florida [93].

The CDC also houses the National Institute for Occupational Safety and Health (NIOSH), which was created by the Occupational Safety and Health Act of 1970, and serves as the nation's research agency on worker safety and health issues [94].

Thus, it works closely with the Occupational Safety and Health Administration, described below.

National Institutes of Health

NIH is the nation's leading funder of primary and secondary medical research. It consists of 21 Institutes and six Centers [95]. Given its mission of conducting research and support of research through grants, contracts, and cooperative agreements, the agency has issued extensive guidelines and regulations on various compliance matters—protection of human subjects and animals involved in research, scientific integrity, and compliance with the terms of grants and contracts [96]. In many of these matters NIH relies on departmental regulations and policies, including HHS's rules on research misconduct, which provide for adjudications by the DAB and/or ALJs [97].

Social Security Administration

The Social Security Administration administers one of the largest adjudication systems in the world. Each year, over 1500 ALJs and an Appeals Council staffed by about 60 "Administrative Appeals Judges" hear and review half a million decisions on appeal from initial decisions to deny (or revoke) disability benefits for not meeting the Act's definition of disability: "inability to engage in any substantial gainful activity by reason of any medically determinable physical or mental impairment which can be expected to result in death or which has lasted or can be expected to last for a continuous period of not less than 12 months." [98].

There are four levels of administrative decisionmaking for Social Security claims—and most claims must pass through all four before a decision is subject to judicial review [99]. The process begins at local SSA offices, where the first level initial disability determinations are contracted out to state-run Disability Determination Service (DDS) offices. SSA, together with the DDSs, makes the initial decision on an applications and the initial decision on whether to terminate benefits in "continuing disability review" cases. In case of appeal, SSA and DDS also handle the first level of review, known as "reconsideration". Further administrative appeals are handled by SSA, through its Office of Hearings Operations, which houses the Office of the Chief ALJ and nine Regional Chief ALJs (along with rank-and-file ALJs), and the Office of Analytics, Review, and Oversight, home of the Appeals Council [100]. The Appeals Council reviews administrative hearing decisions on appeal by a claimant or, in a few cases, on its own initiative. Like the Medicare Appeals Council members, these judges are not ALJs and lack the statutory independence and APA protection enjoyed by ALJs. In the SSA hearings themselves, unlike most ALJ hearings in the federal government, the agency is not represented by counsel; rather it is the responsibility of the ALJ to make sure the administrative record is complete.

SSA also issues regulations, "rulings", a Handbook [101], and a public version of its Program Operations Manual System [102]. Most of its regulations are procedural in nature [103], but Social Security Rulings are a distinctive feature of SSA's activities. They are issued under the Commissioner's authority and are precedential in nature. They "may be based on case decisions made at all administrative levels of adjudication, Federal court decisions, Commissioner's decisions, opinions of the Office of the General Counsel, and other policy interpretations of the law and regulations." [104]. They are published in the *Federal Register* without public comment. The SSA's website takes pains to explain that these rulings "do not have the force and effect of the law or regulations, [but] they are binding on all [SSA] components ... and are to be relied upon as precedents in adjudicating other cases." [104].

It should be noted that there is a separate agency that administers retirement/survivor and unemployment/sickness insurance benefits programs for railroad workers and their families. This is administered by the federal Railroad Retirement Board, located in Chicago [105].

Department of Labor

The U.S. Department of Labor houses two important health related programs: the Black Lung Benefits Program and the Occupational Safety and Health Administration.[5]

Black Lung Benefits Program

In 1969, Congress passed and President Nixon signed the Coal Mine Health and Safety Act [106], a bill to provide benefits to coal miners (and their survivors) who could show they were disabled by pneumoconiosis—defined as "a chronic dust disease of the lung arising out of employment is an underground coal mine." [107, 108]. Under this program, the miner's last employer is responsible for paying benefits thorough an adjudicative program administered by the Department of Labor [109]. DOL administers the program through its Division of Coal Mine Workers' Compensation, consisting of a national office in Washington and eight district offices around the country [110].

Claims are filed with the district office which can authorize or conduct medical tests and X-rays, and accept the claimant's medical evidence. The district director can make a determination on the claimant's eligibility and whether there is a responsible employer (if there is not, payment from a federal trust fund, funded by coal companies, is available). The Department's rules [111] specify that a claimant has a right to an ALJ hearing [112] concerning any contested fact. Parties to such cases may also include the mine operator (who, if identified, must be notified) and its insurance carrier.

Appeals from ALJ decisions may be taken to the Department's Benefits Review Board (BRB), which makes final orders for the Department on these matters. The BRB, established in the 1972 amendments to this Act [113], consists of a Chair, Vice Chair, and three other members [114]. They serve at the pleasure of the Secretary of Labor [115]. It also hears appeals from ALJs in cases involving the Longshore and Harbor Workers' Compensation Act [116]. BRB orders may be appealed to the U.S. Court of Appeals for the circuit in which the injury occurred.

---

[5] The Department also administers the ERISA program (and the Affordable Care Act provisions) that govern employer-funded health insurance coverage that most Americans still rely on.

OSHA

A year after the Coal Act was enacted, Congress also enacted the Occupational Safety and Health Act [117], which created OSHA to regulate the health and safety of most workplaces across the country with ten or more workers. This herculean task obviously would require the issuance of health and safety standards (through rulemaking), investigations, and the bringing of civil or criminal enforcement actions against violators. But, given the strong opposition by employers to this legislation [118], Congress made the issuance of OSHA health and safety standards more difficult than it would ordinarily be under the APA, and it also assigned the adjudication of challenges to enforcement citations to a separate independent commission, the Occupational Safety and Health Review Commission (OSHRC) with its own three-member commission and its own ALJs. When OSHA loses a case brought to OSHRC by a respondent, OSHA may itself seek judicial review against OSHRC in the U.S. Court of Appeals. This "split-enforcement model" is rare—it has been used in the Mine Safety and Health context as well, with the DOL Mine Safety and Health Administration's citations subject to challenge in the independent Federal Mine Safety and Health Review Commission [119].

OSHA's rulemaking provisions [120], among other things, require the agency to convene a public hearing upon request by an interested person, and explain why any standard will be better than an existing voluntary consensus standard. The Act also subjects such standards to substantial evidence review, rather than the usual arbitrary-and-capricious test. These provisions have contributed to making the OSHA rulemaking process very time-consuming. The agency itself estimates a time frame of 52–135 months from start to finish [121].

Although OSHA has a seemingly fairly expansive list of health and safety standards [122]—many of them are repackaged industry standards, and it has left blank many "reserved" sections in the C.F.R. for needed rules. The agency was criticized for failing to issue any standards concerning workers' exposure to Covid 19, and only recently began to do so. On June 17, 2021, it issued an emergency temporary standard (ETS) concerning occupational exposure to COVID-19 in health care workplaces [123], and on September 9, President Biden announced that OSHA was developing a new ETS that will require all employers with 100 or more employees to ensure their workforce is fully vaccinated or require any unvaccinated workers to produce a negative test result on at least a weekly basis [124].

The OSH Act also was designed as a program of "regulatory federalism" in that it allowed OSHA to approve workplace safety and health programs operated by individual states or U.S. territories [125]. OSHA is supposed to monitor these plans to make sure they are effective. According to OSHA, there are currently 22 state plans covering both private sector and state and local government workers, and six state plans covering only state and local government workers [126].

Department of Veterans Affairs

The final health related agency is the Department of Veterans Affairs (VA), known as the Veterans Administration until 1988 when it was elevated to Cabinet status. Most people are unaware that the VA is the second largest department in the government, behind only the Department of Defense in number of employees. That is largely because of the VA's extensive network of hospitals. In fact, the Department's Veterans Health Administration "is America's largest integrated health care system, providing care at 1,293 health care facilities, including 171 medical centers and 1112 outpatient sites of care of varying complexity (VHA outpatient clinics), serving 9 million enrolled Veterans each year." [127].

But the VA also administers a large disability benefits program. To be eligible for such benefits an applicant must have served on active duty, or active or inactive duty for training, and be rated as at least partially disabled due to a service-connected condition. VA employs salaried or contracted physicians to examine the severity of a claimant's condition, after which VA rating personnel make determinations on whether the condition is service connected, and if so what the disability (percentage) rating is. Eligibility and disability ratings may be appealed through the VA adjudication process. Appeals are taken to the Board of Veterans' Appeals where a Veterans Law Judge (VLJ) reviews the case. The Board consists of a Chair and Vice Chair and about 100 VLJs [128], all of whom are appointed by the President upon the recommendation of the Secretary of the VA Department. The Board can grant or deny benefits or remand the case for more information.

Appeals of BVA decision can be filed with the U.S. Court of Appeals for Veterans Claims (COVA)—a specialized court in Washington, DC [129]. Claimants (but not the VA) may appeal COVA decisions to the U.S. Court of Appeals for the Federal Circuit. COVA was established in 1988. Prior to that, VA benefits decisions were not judicially reviewable at all.

3   Application

Case # 1: Bowen v. Georgetown University Hospital, 488 U.S. 204 (1988) Under the Medicare program, the federal government reimburses health care providers for expenses incurred in providing medical services to Medicare beneficiaries. The Medicare Act, 42 U.S.C. § 1395x(v)(1)(A), authorizes the Secretary of Health and Human Services (Secretary) to promulgate cost-reimbursement regulations and also provides that "[s]uch regulations shall … (ii) provide for the making of suitable retroactive corrective adjustments where, for a provider of services for any fiscal period, the aggregate reimbursement produced by the methods of determining costs proves to be either inadequate or excessive." In 1981, the Secretary issued a rule that revised the formula for calculating the reimbursement to hospitals, but this rule was challenged by seven hospitals in Washington DC, including Georgetown Hospital, which were especially affected by the change. The federal district court invalidated the rule (thus reinstating the existing formula that was beneficial to the hospitals) because it had been adopted without the notice-and-comment procedures required by the Administrative Procedure Act (APA).

The Department accepted that result, but in 1984 Secretary Bowen reissued the 1981 rule (this time after notice and comment). The 1984 rule was retroactive in that it purported to allow HHS to recoup the sums previously paid to the hospitals, as a result of the district court's ruling. After the hospitals exhausted their administrative appeals, they challenged the 1984 rule as impermissibly retroactive. The district court granted summary judgment for the hospitals, and the court of appeals affirmed, on the ground that under the APA, "rules" had to be of "future effect" (5 U.S.C. § 551(4)).

The Supreme Court unanimously affirmed, although only one concurring justice relied on the APA's language. Instead the opinion for the Court's other eight members created a blackletter rule: "Retroactivity is not favored in the law. Thus, congressional enactments and administrative rules will not be construed to have retroactive effect unless their language requires this result." Although the Secretary pointed to the language in the Medicare Act authorizing him to promulgate regulations to "provide for the making of suitable retroactive corrective adjustments," the Court concluded that that retroactivity provision "directs the Secretary to establish a procedure for making case-by-case adjustments to reimbursement payments where the regulations prescribing computation methods do not reach the correct result in individual cases." And that therefore, "the structure and language of the statute require the conclusion that the retroactivity provision applies only to case-by-case adjudication, not to rulemaking."

Thus, apart from the beneficial result for the hospitals who brought the suit, the case set forth an important principle of administrative law—namely that federal agencies cannot engage in retroactive rulemaking unless Congress has clearly provided them the authority to do so.

Case # 2: <u>Food and Drug Admin. v. Brown & Williamson Tobacco Corp., 529 U.S. 120 (2000)</u> As mentioned in this chapter, the Food and Drug Administration (FDA) has jurisdiction over a wide range of medicinal products, including drugs and medical devices. The FDA's statute defines "drug" to include "articles (other than food) intended to affect the structure or any function of the body." 21 U.S.C. § 321(g)(1)(C). It defines "device," in part, as "an instrument, apparatus, implement, machine, contrivance, ... or other similar or related article, ... which is ... intended to affect the structure or any function of the body." § 321(h). The Act also grants the FDA the authority to regulate so-called "combination products," which "constitute a combination of a drug, device, or biological product." § 353(g)(1).

In August 1995, the FDA for the first time proposed a rule to regulate the sale of cigarettes and smokeless tobacco to children and adolescents. 60 Fed. Reg. 41,314. The rule, which included several restrictions on the sale, distribution, and advertisement of tobacco products, was designed to reduce. the availability and attractiveness of tobacco products to minors. A public comment period followed, during which the FDA received over 700,000 submissions—the highest number it had ever received on a proposed rule.

One year later, in August 1996, the FDA issued its final rule entitled "Regulations Restricting the Sale and Distribution of Cigarettes and Smokeless Tobacco to Protect Children and Adolescents." *Id.,* at 44,396. The FDA determined that nicotine is a "drug" and that cigarettes and smokeless tobacco are "drug delivery devices," and therefore it had jurisdiction under the FDCA to regulate tobacco products as customarily marketed.

A group of tobacco companies, advertisers, and retailers challenged the rule. The district court upheld the FDA's jurisdiction but enjoined the part of the rule limiting advertising and promotion of tobacco. products. The court of appeals reversed and found that the FDA lacked jurisdiction entirely. At the Supreme Court, the issue was whether the FDA had jurisdiction over tobacco products.

This case was widely viewed as a test of the Supreme Court's "*Chevron* doctrine," adopted in the 1984 decision in *Chevron U.S.A. Inc. v. Natural Res. Def. Council, Inc.* [55]. In that case, the Court established a two-step test for analyzing whether to uphold agency interpretations of their own statutes: First, has Congress "directly spoken to the precise question at issue"? If so, "that is the end of the matter; for the court, as well as the agency, must give effect to the unambiguously expressed intent of Congress." If not, if Congress was silent or its intent was ambiguous, then "the question for the court is whether the agency's answer is based on a permissible construction of the statute." The government naturally argued that (1) the statutory text clearly authorized the agency's action, but (2) if not, the text was at least ambiguous and the agency's interpretation was a reasonable one, given the health problems caused by the products. The challengers pointed out that the under the FDA's interpretation, its conclusion that these products were "unsafe" would have required a ban—a step the FDA had declined to take. They also argued that that the FDA had never in its long history claimed this jurisdiction., and that Congress knew that and had considered and rejected specific bills to give it that jurisdiction.

The Supreme Court ultimately decided 5-4 to uphold the challenge and affirm the court of appeals. The Court did use the *Chevron* framework, but the majority decided under Step One that it was clear that Congress had not intended to give the FDA the power to regulate tobacco products. In the course of its opinion, in addition to agreeing with the points raised by the challengers, the majority suggested some new limiting principles for its *Chevron* doctrine: (1) courts should not review statutory definitions in isolation of the context of the entire statute, (2) the agency had previously taken the opposite position when it denied petitions to regulate cigarettes, (3) "we are confident that Congress could not have intended to delegate a decision of such economic and political significance to an agency in so cryptic a fashion." This later point became known as the "major question" exception to *Chevron* and was illustrated in a later. case by the colorful phrase, "Congress does not hide elephants in mouseholes."

Although the four dissenters argued that the FDA should prevail under either Step One or Step Two of *Chevron*, the FDA's rule was vacated. Nine years later, however, there was a happy ending for the FDA: Congress passed the Family Smoking Prevention and Tobacco Control Act of 2009 (Pub. L. No. 111-31), which authorized the FDA to regulate tobacco products.

4   Summary and Conclusion

The above description illustrates how the standard administrative law models of adjudication, rulemaking, and judicial review only partly describe the wide variety of actions taken by U.S health-related agencies. Perhaps the most apparent dichotomy is between enforcement oriented agencies and those primarily devoted to providing benefits to eligible beneficiaries. In general, most of these agencies do act on their delegated authority to issue binding regulations, but also rely heavily on non-binding guidance documents to make their policies known. With respect to dispute resolution assigned to them by Congress, the agencies have an extensive mix of formal and informal adjudications, with ALJs playing a central role in some adjudicative programs, but also a wide variety of non-ALJ adjudicators and boards making both first-level and review-level decisions. Finally, although most of these agencies' actions are judicially reviewable, some are precluded from review, and others have to be reviewed in a particular court—a federal district court, court of appeals, or specialized court.

Because Administrative Law is a procedural course in law school, the cases discussed in class can originate in any agency, but a surprisingly high percentage of them come from the health-related agencies discussed above. For example in the casebook I use in my class [130], there are 64 "principal" cases that are reproduced in abridged fashion followed by many more "note cases" that amplify them. Of those 64, 13 of them (20%) were health-law related cases. It is thus an understatement to conclude that health lawyers must have a good understanding of Administrative Law—on both the federal and state levels.

Acknowledgement Thanks to my colleague, Professor Lindsay Wiley for her helpful comments on an earlier draft.

References

1. 5 U.S.C. §§ 551–559, 701-706, and other scattered sections in title 5.
2. Pub. L. No. 103–296, title I, § 101, 108 Stat. 1465 (1994), codified at 42 U.S.C. § 901.
3. Judge Rotenberg Educ. Center, Inc., v. U.S. Food & Drug Admin., 3 F.4th 390, 400 (D.C. Cir. 2021).
4. Ctrs. for Disease Control: State & Territorial Health Department Websites. https://www.cdc.gov/publichealthgateway/healthdirectories/healthdepartments.html. Accessed 19 Sept 2021.
5. Fed'n of State Med. Bds.: Guide to Medical Regulation in the United States. https://www.fsmb.org/u.s.-medical-regulatory-trends-and-actions/guide-to-medical-regulation-in-the-united-states/. Accessed 19 Sept 2021.
6. Withrow v. Larkin, 421 U.S. 35 (1975).
7. Lubbers JS, Administrative Law in the United States. In: Seerden R, editor. Administrative Law of the European Union, its Member States, and the United States: A Comparative Analysis. 4th ed. Cambridge: Intersentia; 2018. p. 357–414.
8. Londoner v. Denver, 210 U.S. 373 (1908) and Bi-Metallic Inv. Co. v. State Bd. of Equalization, 239 U.S. 441 (1915).
9. 5 U.S.C. § 551(4).
10. 5 U.S.C. § 553(b).
11. 5 U.S.C. § 553(d).
12. 5 U.S.C. § 553(c).
13. Fla. East Coast Ry. Co., v. United States, 410 U.S. 224 (1973).
14. Lubbers JS. A Guide to Federal Agency Rulemaking. 6th ed. Chicago: American Bar Association; 2018.
15. William Jefferson Clinton. Ex. Ord. 12,866, Regulatory Planning and Review (Sept. 30, 1993), 58 Fed. Reg. 51,735 (Oct. 4, 1993).

16. Strauss PL. Overseer, or "The Decider"? The President in Administrative Law. *Geo. Wash. L. Rev.* 2007; 75: 696.
17. 5 U.S.C. § 553(e).
18. 5 U.S.C. § 552(a)(1).
19. Funk WF. A Primer on Nonlegislative Rules. Admin. L. Rev. 2001;53:1321.
20. 5 U.S.C. § 554(a).
21. Asimow M. Best Practices for Evidentiary Hearings Outside the Administrative Procedure Act. Geo. Mason L. Rev. 2019;26:923.
22. 5 U.S.C. § 554(b).
23. 5 U.S.C. § 554(c).
24. 5 U.S.C. § 556(b).
25. 5 U.S.C. § 556(c).
26. 5 U.S.C. § 557(b).
27. 5 U.S.C. § 554(d).
28. 5 U.S.C. § 556(d).
29. 5 U.S.C. § 557(d)(1).
30. FTC v. Cement Inst., 333 U.S. 683 (1948).
31. Texaco, Inc. v. FTC, 336 F.2d 754 (D.C. Cir. 1964).
32. Gibson v. Berryhill, 411 U.S. 564 (1973).
33. Goldberg v. Kelly, 395 U.S. 254 (1970); Bd. of Regents v. Roth, 408 U.S. 564 (1972).
34. Paul v. Davis, 424 U.S. 693 (1976).
35. Mathews v. Eldridge, 424 U.S. 319 (1976).
36. 5 U.S.C. § 555(b).
37. Administrative Dispute Resolution Act, Pub. L. No. 101-552, 104 Stat. 2736, codified as amended at 5 U.S.C. §§ 571–584.
38. NLRB v. Bell Aerospace Co., 416 U.S. 267 (1974); *SEC v. Chenery Corp.*, 332 U.S. 194 (1947).
39. Bowen v. Georgetown Hosp., 488 U.S 204 (1988).
40. 5 U.S.C. § 706.
41. 5 U.S.C. § 701(a).
42. Bowen v. Mich. Acad. of Family Physicians, 476 U.S. 667 (1986).
43. Heckler v. Chaney, 470 U.S. 821 (1985).
44. Ass'n of Data Processing Serv. Orgs. v. Camp, 397 U.S. 150 (1970).
45. Lujan v. Defenders of Wildlife, 504 U.S 555 (1992).
46. Abbott Labs. v. Gardner, 387 U.S. 136 (1967).
47. 5 U.S.C. § 704, as interpreted by Darby v. Cisneros, 509 U.S. 137 (1993).
48. McCarthy v. Madigan, 503 U.S. 140 (1992).
49. Lubbers JS. Fail to Comment at Your Own Risk: Does Issue Exhaustion Have a Place in Judicial Review of Rules? Admin. L. Rev. 2018;70:109.
50. Bennett v. Spear, 520 U.S. 154, 177–78 (1997).
51. 5 U.S.C. § 703.
52. Universal Camera Corp. v. NLRB, 340 U.S. 474 (1951).
53. Citizens to Preserve Overton Park v. Volpe, 401 U.S. 402 (1971).
54. Motor Vehicle Mfrs. Ass'n v. State Farm Mut. Auto. Ins. Co., 463 U.S. 29 (1983).
55. Chevron U.S.A. Inc. v. Natural Res. Def. Council, Inc., 467 U.S. 837 (1984).
56. City of Arlington, Tex. v. FCC, 569 U.S. 290, 312 (2013) (Roberts, C.J., dissenting).
57. Inspector General Act of 1978, Pub. L. No. 95-452, 92 Stat. 1101, codified as amended at 5 U.S.C. App. §§ 1–12.
58. Dept. of Health & Human Servs.: Organizational chart. https://www.hhs.gov/about/agencies/orgchart/index.html. Accessed 19 Sept 2021.
59. Dept. of Health & Human Servs.: About the Surgeon General. https://www.hhs.gov/surgeon-general/about/index.html. Accessed 19 Sept 2021.
60. Food and Drug Admin.: What does FDA regulate? https://www.fda.gov/about-fda/fda-basics/what-does-fda-regulate. Accessed 19 Sept 2021.

61. Food and Drug Admin.: Search for FDA Guidance Documents. https://www.fda.gov/regulatory-information/search-fda-guidance-documents. Accessed 19 Sept 2021.
62. Food and Drug Admin.: About Warning and Close-Out Letters. https://www.fda.gov/inspections-compliance-enforcement-and-criminal-investigations/warning-letters/about-warning-and-close-out-letters. Accessed 19 Sept 2021.
63. Food and Drug Admin.: Clinical Investigators—Disqualification Proceedings. https://www.accessdata.fda.gov/scripts/SDA/sdNavigation.cfm?sd=clinicalinvestigatorsdisqualificationproceedings&previewMode=true&displayAll=true. Accessed 19 Sept 2021.
64. 1 C.F.R. § 16.42 (2021).
65. Food and Drug Admin.: FDA Staff Manual Guides, Volume II—Delegations of Authority Regulatory—General Redelegations of Authority Hearings, SMG 1410.29 (Sept 29, 2011). https://www.fda.gov/media/80027/download. Accessed 19 Sept 2021.
66. Family Smoking Prevention and Tobacco Control Act, Pub. L. No. 111-31, 123 Stat. 1776 (2009).
67. Food and Drug Admin.: CTP Compliance & Enforcement. https://www.fda.gov/tobacco-products/compliance-enforcement-training/ctp-compliance-enforcement#civilmoneypenalty. Accessed 19 Sept 2021.
68. 21 C.F.R. § 17.3(c) (2021).
69. Weinberger v. Hynson, Westcott & Dunning, 412 U.S. 609 (1973).
70. Azar v. Allina Health Servs., 139 S. Ct. 1804, 1808 (2019).
71. 42 U.S.C. § 1395hh(a)(2).
72. 42 U.S.C. § 300gg–92:
73. Ctrs. for Medicare and Medicaid Servs.: Internet-Only Manuals. https://www.cms.gov/Regulations-and-Guidance/Guidance/Manuals/Internet-Only-Manuals-IOMs. Accessed 19 Sept 2021.
74. Ctrs. for Medicare and Medicaid Servs.: Offices of Hearings and Inquiries. https://www.cms.gov/About-CMS/Agency-Information/CMSLeadership/Office_OHI. Accessed 19 Sept 2021.
75. Dept. of Health and Human Servs.: CMS Regulations & Guidance- Review of Provider Reimbursement Review. https://www.hhs.gov/guidance/document/review-provider-reimbursement-review-board-prrb-0. Accessed 19 Sept 2021.
76. 42 C.F.R. § 405, Subpart R (2021).
77. 42 C.F.R. § 405.1845(a) (2021).
78. Ctrs. for Medicare and Medicaid Servs.: What is a MAC? https://www.cms.gov/Medicare/Medicare-Contracting/Medicare-Administrative-Contractors/What-is-a-MAC. Accessed 19 Sept 2021.
79. Ctrs. for Medicare and Medicaid Servs.: List of PRRB Jurisdictional Decisions. https://www.cms.gov/Regulations-and-Guidance/Review-Boards/PRRBReview/List-of-PRRB-Jurisdictional-Decisions. Accessed 19 Sept 2021.
80. 42 C.F.R. § 405.1875.
81. 42 U.S.C. § 1395oo(f)(1).
82. Ctrs. for Medicare and Medicaid Servs.: Medicare Geographic Classification Review Board. https://www.cms.gov/regulations-and-guidance/review-boards/mgcrb. Accessed 19 Sept 2021.
83. 42 U.S.C. § 1395ww(d)(10) and 42 C.F.R. § 412.230 (2021).
84. Ctrs. for Medicare and Medicaid Servs.: List of Medicare Advantage/Prescription Drug Plan Decisions. Https://www.cms.gov/Regulations-and-Guidance/Review-Boards/CMS-Hearing-Officer/List-of-MA-PD-Decisions. Accessed 19 Sept 2021.
85. RAC Monitor, The ALJ Backlog Dissolves—and SMRC Audits Escalate. https://racmonitor.com/the-alj-backlog-dissolves-and-smrc-audits-escalate/. (14 Apr 2021). Accessed 19 Sept 2021.
86. Dept. of Health and Human Servs.: Who are the Board Members & Judges? https://www.hhs.gov/about/agencies/dab/about-dab/who-are-the-board-members-and-judges/index.html. Accessed 19 Sept 2021.

87. AARP Public Policy Institute: Improving the Medicare Appeals Process. (Dec 2009), https://www.masslegalservices.org/system/files/library/AARP%20Fact%20Sheet%20-%20 Medicare%20Appeals.pdf. Accessed 19 Sept 2021.
88. Dept. of Health and Human Servs.: Appeals to DAB Administrative Law Judges. https:// www.hhs.gov/about/agencies/dab/different-appeals-at-dab/appeals-to-alj/index.html. Accessed 19 Sept 2021.
89. Dept. of Health and Human Servs.: Appeals to Board. https://www.hhs.gov/about/agencies/ dab/different-appeals-at-dab/appeals-to-board/index.html. Accessed 19 Sept 2021.
90. Ctrs. for Disease Control and Prevention: Covid-19. https://www.cdc.gov/coronavirus/2019-nCoV/index.html. Accessed 19 Sept 2021.
91. 42 U.S.C. §§ 264, 265; 42 C.F.R. Part 70 (2021).
92. 42 C.F.R. § 70.5 (2021).
93. State v. Becerra, No. 8:21-cv-839-SDM-AAS, 2021 WL 2514138 (M.D. Fla. June 18, 2021).
94. Ctrs. for Disease Control and Prevention: About NIOSH. https://www.cdc.gov/niosh/about/ default.html. Accessed 19 Sept 2021.
95. Nat'l Insts. of Health: List of NIH Institutes, Centers, and Offices. https://www.nih.gov/ institutes-nih/list-nih-institutes-centers-offices. Accessed 19 Sept 2021.
96. Nat'l Insts. of Health: NIH Policy Manual, Chapter 1111, Laws and Regulations, https:// policymanual.nih.gov/1111. Accessed 19 Sept 2021.
97. 42 C.F.R. Part 93 (2021).
98. 42 U.S.C. § 423(d)(1).
99. Johnson v. Shalala, 2 F.3d 918, 920 (9th Cir. 1993).
100. Soc. Sec. Admin.: Hearings and Appeals. https://www.ssa.gov/appeals/about_ac.html. Accessed 19 Sept 2021.
101. Soc. Sec. Admin.: Online Social Security Handbook. https://www.ssa.gov/OP_Home/hand-book/handbook.html. Accessed 19 Sept 2021.
102. Public Policy Home. https://secure.ssa.gov/apps10/. Accessed 19 Sept 2021.
103. 20 C.F.R. §§ 400-499 (2021).
104. Soc. Sec. Admin.: Social Security and Acquiescence Rulings. https://www.ssa.gov/OP_ Home/rulings/rulings-pref.html. Accessed 19 Sept 2021.
105. U.S. Railroad Retirement Bd: Railroad Retirement Handbook 2021. https://rrb.gov/sites/ default/files/2021-07/2021%20Railroad%20Retirement%20Handbook.pdf. Accessed 19 Sept 2021.
106. Pub. L. No. 91-173, 83 Stat. 742 (1969).
107. Pub. L. No. 91-173, § 491(b), 83 Stat. 793 (1969).
108. Hamby C. Soul Full of Coal Dust—A Fight for Breath and Justice in Appalachia. New York, Little, Brown and Co.; 2020. p. 60-67.
109. Dep't of Labor: Black Lung Program. https://www.dol.gov/agencies/owcp/dcmwc. Accessed 19 Sept 2021.
110. Dep't of Labor: DCMWC Offices and Leadership. https://www.dol.gov/agencies/owcp/ dcmwc/districtoffices. Accessed 19 Sept 2021.
111. 20 C.F.R. Part 725 (2021).
112. Dep't of Labor: About the Office of Administrative Law Judges. https://www.dol.gov/agen-cies/oalj/about/ALJMISSN. Accessed 19 Sept 2021.
113. Pub. L. 92-576, 86 Stat. 1251 (1972).
114. Dep't of Labor: BRB Board Members. https://www.dol.gov/agencies/brb/members. Accessed 19 Sept 2021.
115. Kalaris v. Donovan, 697 F.2d 376 (D.C. Cir. 1983).
116. 33 U.S.C. §§ 901-950.
117. Pub. L. No. 91-596, 84 Stat. 1590 (1970), codified at 29 U.S.C. §§ 651-678.
118. MacLaury J. The Job Safety Law of 1970: Its Passage Was Perilous. https://www.dol.gov/ general/aboutdol/history/osha. Accessed 19 Sept 2021.
119. Johnson GR, Jr. The Split-Enforcement Model: Some Conclusions from the OSHA and MSHA Experiences. Admin. L. Rev. 1997;39:315.

120. 29 U.S.C. § 655.
121. Dep't of Labor: OSHA Rulemaking Process Flow Chart. https://www.osha.gov/sites/default/files/OSHA_FlowChart.pdf. Accessed 19 Sept 2021.
122. 29 C.F.R. § Part 1910 (2021).
123. 85 Fed. Reg. 32,376 (June 21, 2021), to be codified at 19 C.F.R. 1910, subpart U.
124. The White House: Path out of the Pandemic—President Biden's Covid-19 Action Plan. https://www.whitehouse.gov/covidplan/#vaccinate. Accessed 19 Sept 2021. However, the Supreme Court invalidated this rule, Nat'l Fed'n of Indep. Bus. v. Dep't of Labor, 142 S. Ct 661 (2022).
125. 29 U.S.C § 667.
126. Dep't of Labor: State Plans. https://www.osha.gov/stateplans. Accessed 19 Sept 2021.
127. Dep't of Veterans Affairs: Labor Veterans Health Administration. https://www.va.gov/health/. Accessed 19 Sept 2021.
128. Dep't of Veterans Affairs: Board of Veterans' Appeals, Annual Report Fiscal Year (FY) 2020 p. 6. https://www.bva.va.gov/docs/Chairmans_Annual_Rpts/BVA2020AR.pdf. Accessed 19 Sept 2021.
129. US Court of Appeals for Veterans Claims: About the Court. http://m.uscourts.cavc.gov/About.php. Accessed 19 Sept 2021.
130. Asimow M, Levin RM. State and Federal Administrative Law. 5th ed. St. Paul: West Academic: 2020.

Licensing Boards

Andrew M. Knoll

Contents

Key Points
- Medical Licensing Boards regulate quality by investigating and prosecuting professional misconduct, which is very broadly defined.
- Common categories are negligence, improper prescribing, impairment, fraud and moral unfitness, and being disciplined by another jurisdiction.
- Complaints are fully investigated, and then the Licensing Board will decide whether to close the investigation or press charges for professional misconduct.
- If charged, a clinician will have to choose between taking the case to a hearing or entering into a Consent Order—the administrative equivalent of a plea bargain.
- If disciplined, the clinician may face fallout from credentialing organizations, such as third-party payers, hospitals, and specialty Boards.
- State Physician Health Programs offer a process to facilitate treatment, monitoring, and continued practice of medicine for clinicians suffering from a substance use disorder or other medical/psychiatric impairment.

A. M. Knoll (✉)
Cohen Compagni Beckman Appler & Knoll, PLLC, Syracuse, NY, USA
e-mail: aknoll@ccblaw.com

© The Author(s), under exclusive license to Springer Nature Switzerland AG 2022
A. S. Pasha (ed.), *Laws of Medicine*,
https://doi.org/10.1007/978-3-031-08162-0_12

Legal Concepts
- Common categories of professional misconduct are: (1) negligence/incompetence, (2) inappropriate prescribing, (3) clinician impairment, (4) fraud and moral unfitness (e.g., patient boundary issues), and (5) referral or reciprocal discipline.
- Judicial review of Licensing Board action is limited and deferential to the agency.

1   Introduction

Quality and professionalism in medicine are regulated at the State level and through medical Licensing Boards (LBs) under State law. Accordingly, each State has its definitions of professional misconduct and its own investigating and prosecuting procedures. There is substantial overlap and consistency between different States; and this chapter will outline the process in generic terms. Your State will have its definitions of professional misconduct and procedures. The reader is recommended to contact an experienced attorney in his/her State should he/she have to interact adversely with the LB.

2   Discussion

Professional misconduct falls into different categories, with the following being the most common: (1) negligence/incompetence, (2) inappropriate prescribing, (3) clinician impairment, (4) fraud and moral unfitness (e.g., patient boundary issues), and (5) referral or reciprocal discipline.

Negligence is the same as with medical malpractice, except that causation and damages are not required. In other words, a clinician can be guilty of professional misconduct based on negligence for deviating from the standard of care even if there was no patient harm.

Inappropriate prescribing predominantly refers to controlled substances. In this opioid crisis era, substantial scrutiny is facilitated by an LB's ability to search prescription drug monitoring programs' databases.

Impairment refers to a clinician's impairment by alcohol, drugs, or medical or psychiatric conditions. It is discussed in more detail below.

Fraud and moral unfitness can encompass what is commonly thought of as fraud, such as billing and insurance fraud and fraudulent entries in a medical record. Boundary violations with patients constitute a major area of concern in the moral unfitness misconduct category.

Lastly, all States consider it professional misconduct to have been disciplined by another State. In many States, it is also considered professional misconduct to be convicted of or plead guilty to a crime, such as DWI, which can be either a misdemeanor or a felony. With reciprocal discipline, the issue is not guilt or innocence (guilt having been established by the conviction or guilty plea) but rather the penalty that will be imposed.

An LB proceeding begins with a complaint. A complaint may be filed by the patient or family, a hospital, or even a third-party payer ("Payer"). From there, the case will be assigned to an investigator. The investigator will typically start with a background investigation of both the clinician and the facts and circumstances surrounding the complaint. This involves speaking with the complainant, talking to any witnesses, reviewing hospital and licensing files, and possibly conducting a criminal background check. If the matter involves patient care, the patient's chart will be requested. This is permissible without patient authorization under the Health Insurance Portability and Accountability Act of 1996 (HIPAA) as a "health care oversight" activity [1]. This also may be the first notice to the clinician that he or she is under investigation. The investigation may ultimately find no evidence of misconduct, but early involvement of counsel, even if behind the scenes, is recommended. Professional liability insurance may provide coverage of legal fees incurred by an administrative investigation, so this is also an appropriate time to consult the carrier.[1]

Once the background investigation has been sufficiently completed, the clinician will be interviewed. This is the clinician's opportunity to tell his/her side of the story. Typically, it is to the clinician's advantage to participate in an interview, but there are occasions when it is prudent to decline. For example, in a patient boundary violation case in which the clinician is guilty, the clinician's admissions may not be helpful. In cases such as this, it is sometimes more beneficial to provide a written submission in lieu of an interview, which may consist of evidence of health evaluations, CME courses taken, or participation in a clinician health program (see discussion below). Whether to participate in or decline the interview is a decision made jointly by the clinician and his/her attorney.

Before, during, or after the interview, the clinician is typically allowed to submit evidence. Medical literature can be very persuasive. For example, in a case involving knee surgery, the physician from the LB who participated in the interview believed that the thigh tourniquet time was excessive because it exceeded two hours. Medical literature was submitted that confirmed that if the tourniquet time was two hours, it was within the standard of care to make an assessment. If the total tourniquet time was expected to be less than two and a half hours, it was appropriate to not deflate the tourniquet [2]. The investigation closed shortly thereafter in light of the evidence that the standard of care had been met.

After the initial investigation is completed, the LB will review the case. It will decide whether to close the investigation or press charges of professional misconduct.[2] If the LB decides to press charges, the clinician now must decide to defend him/herself at a hearing or enter into a consent order (CO).

Hearings are trials involving the attendant costs, stressors, and uncertainties. However, there are very significant differences between civil and criminal trials because an LB hearing is conducted under administrative law. Typical due process

---

[1] Different carriers have different procedures for licensing defense. Some permit the clinician to choose counsel, other assign counsel as it would do in a medical malpractice case.

[2] Some LBs have the option of issuing an administrative warning, which is non-disciplinary in nature.

protections, especially those provided in a criminal case, do not apply; there are essentially no technicalities that could result in a dismissal, and hearsay is admissible [3]. The trier of fact (i.e., the jury) may be a single hearing officer or, as in New York, a panel of three comprising two physicians and a community member presided over by an Administrative Law Judge [4]. These juries are not randomly selected from the community and are usually very knowledgeable and sophisticated. Furthermore analogy can be made to a military court martial, where the judge, jury, and prosecutor all are employed by the prosecution.

Just as with other administrative proceedings, there are limited appeal rights following an LB hearing. One must first exhaust all administrative remedies, which means there is no right to an interlocutory appeal[3] if the clinician believes the LB is being unreasonable in its investigation and prosecution of the case, and the clinician must take the case all the way to the final decision of the LB [5]. The burden of proof is on the clinician to prove that the LB was wrong and that the clinician has suffered substantial prejudice. The standard of review is whether there is substantial evidence to support the decision and whether the decision was arbitrary or capricious. These are highly deferential standards, with "substantial evidence" meaning a reasonable person could agree with the finding, and "arbitrary or capricious" essentially meaning irrational [6]. In effect, the court could disagree with the findings of the LB or the imposed penalty but still not overturn the decision unless it is completely unsupported by the evidence or is irrational.

Thus, if an acceptable CO is offered by the LB, accepting it rather than trying the case at a hearing is almost always preferable.

A CO is essentially a plea bargain. Like all plea bargains, the advantages are certainty in the outcome, substantially less expense, and often comprise the ability to negotiate the language, which is important as these are public documents that will be reviewed by credentialing organizations such as hospitals, Payers, specialty Boards, and other State LBs.

A discipline, whether imposed at a hearing or through entering into a CO, will, unfortunately, have collateral consequences. It will be reported to the National Practitioner Data Bank (NPDB) and the Federation of State Medical Boards (FSMB). From there, the credentialing organizations mentioned above will receive notice of the discipline.

As mentioned, all States consider it professional misconduct to be disciplined by a sister State's LB, so the clinician with multiple State licenses can expect reciprocal discipline from those other States. Such can be particularly expensive and problematic for certain clinicians, such as locum tenens or those who practice telemedicine and may be licensed in dozens of States.

Specialty Boards generally require an unrestricted license to remain Board certified. Some Boards will consider each case on a case-by-case basis, while others will automatically suspend the Board Certification.

---

[3] An "interlocutory appeal" is one done during the pendency of an action and before a final judgment is given.

Payers also, unfortunately, have become increasingly intolerant of maintaining a disciplined clinician on its panels. From a legal perspective, insurance participation is like being a member of a private club, and there is little to no recourse if the Payer decides to disenroll a clinician. Typically, the Payer will allow the clinician to appeal the decision, but the Payer rarely overturns its previous decision.

Medicare exclusion is, fortunately, a relatively rare event because its effect can be career-ending. Federal regulations list criteria for mandatory exclusions (e.g., a criminal conviction for Medicare fraud, patient abuse, or a felony conviction involving controlled substances) and permissive exclusions (e.g., a misdemeanor conviction involving controlled substances or a medical license suspension) [7]. The latter is discretionary, and the Office of the Inspector General may choose not to exercise its power to exclude.

No discussion of LBs would be complete without discussing impairment and physician health programs (PHPs). Almost all states have a PHP, with one of the exceptions being California. As their name implies, PHPs assist impaired clinicians. The majority of such clinicians have diagnosed substance use disorders (SUDs), although impairment from other psychiatric and medical conditions (e.g., neurocognitive dysfunction) also comes within the purview of a PHP.

PHPs do not treat impaired clinicians but rather facilitate treatment, monitoring, and advocacy for the continued practice of medicine. Upon referral, a clinician is typically sent for evaluation, and if an SUD is diagnosed, extended inpatient or outpatient treatment is recommended. The clinician is then monitored through frequent and random forensic urine drug screens (UDS). The frequency of UDS may begin with one or more screens per week and then be reduced over time as the clinician demonstrates sustained remission. PHPs often cooperate with LBs, with the leverage for continued sobriety because a relapse could result in a license suspension [8].

Contracts with PHPs are typically five years in length because the medical literature shows better outcomes when recovery is monitored for that period of time [8]. The advantage of enrolling in a PHP is that it can offer a safe harbor whereby a clinician with an SUD diagnosis can demonstrate ongoing sobriety and fitness for duty [8].

3   Application

Clinical Vignette #1 Dr. Jones is the Chief of Emergency Medicine at a medium-sized urban hospital. A few cases, all seen by a certain physician assistant (PA), have come to the attention of the LB. As part of its investigation, the LB notes that Dr. Jones is listed as the supervising physician. He is made an additional target of the investigation and invited to an interview.

At the interview, Dr. Jones tells the investigator that for regulatory purposes, one physician has to be listed as the supervising physician and that it's common practice for the service chief to be named. However, on a day-to-day basis, the attending physician on duty supervises the PA on duty. Dr. Jones was not on duty on the days in question when these patients were seen.

Dr. Jones further explains that there are general supervision requirements, such as periodic chart reviews for quality and meeting with the PA, which he does perform. Furthermore, with experienced PAs, as in this case, the standard of care regarding the determination of whether attending physician consultation is necessary is delegated to the discretion of the PA.

Now LB, aware of these additional facts, closes the investigation soon after the interview was completed.

Clinical Vignette #2 Dr. Smith is a primary care physician in private practice in a small group practice. The LB has requested the chart of Mary Brown. Around that time, Dr. Smith had a brief sexual relationship with Ms. Brown. The relationship did not end well.

When the LB requested Ms. Brown's chart, Dr. Smith consulted an attorney. He admitted to the boundary violation, stating that he was going through a divorce and that it was a bad time in his life. The attorney recommends that Dr. Smith undergo a psychiatric independent medical examination (IME) and enroll in a three-day CME on understanding and maintaining proper boundaries.[4] The IME diagnosed prior adjustment disorder (at the time of the affair) and ongoing depression but no evidence of a personality disorder or a chemical use disorder. Dr. Smith was referred to his State's PHP, which further referred him for medication management and therapy.

By the time the LB offered an interview, Dr. Smith had significant insight into his situation, in no small part because of the CME course, and the decision was made to participate in the interview. At the interview, Dr. Smith acknowledged the transgression, took responsibility for his actions, exhibited remorse, and discussed the steps he took to remediate the problem, which, as well as the steps described above, included regularly using chaperones.

Because of the affirmative steps he took, the LB stated it would not seek revocation of Dr. Smith's license, but the offered CO was harsh: 36 months' suspension with 33 months stayed (i.e., three months of actual suspension), three years' probation, and a requirement for a permanent chaperone when seeing female patients. The discipline was reported to the NPDB and FSMB.

Because of the period of actual suspension, Dr. Smith was disenrolled from all his third-party payer panels. He had to surrender his Drug Enforcement Administration (DEA) registration because having an active license to practice is a requirement for holding the registration. His board certification was suspended. Ms. Brown sued for malpractice, and the carrier settled the case, reporting the settlement to the NPDB.

---

[4]A resource for such courses may be found on the FSMB's website, *Directory of Physician Assessment and Remedial Education Programs,* available at https://www.fsmb.org/siteassets/spex/pdfs/remedprog.pdf (last accessed December 23, 2021).

His partners at work had been supportive, and he was permitted to return to work after the period of actual suspension expired, albeit with a substantial reduction in compensation. It is now two years since entering into the CO, in which Dr. Smith has slowly rebuilt his practice. Through compliance letters from the LB and PHP, he has been readmitted to most, but not all, of the payer panels. His DEA registration has been reinstated. He hopes to become board certified again after his probation is completed but has concerns because the permanent requirement of a chaperone is considered a restricted license. While Dr. Smith is in neither the professional nor financial position that he was before the discipline, he has sufficiently recovered from the process to maintain a viable medical practice.

4   Summary

LBs regulate quality by investigating and enforcing professional misconduct, which is broadly categorized. Under the principles of administrative law, LBs have considerable power, and there is limited judicial review of their actions. Since a discipline's effects can be substantial, early consultation with knowledgeable and experienced counsel is recommended if a clinician finds him or herself the subject of an LB investigation. For clinicians who suffer from SUDs, their State PHP may provide them with a safe haven in which they can demonstrate continued sobriety and safely continue or return to the practice of medicine.

References

1. 45 C.F.R. § 164.512(d).
2. Fitzgibbons, P., DiGiovanni, C., Hares, S., Akelman, E. Safe Tourniquet Use: A Review of the Evidence, J Am Acad Orthop Surg, 20 (2012): 310–319, doi: https://doi.org/10.5435/JAAOS-20-05-310.
3. Model State Administrative Procedure Act (MSAPA), § 4-212. Each State has its own Administrative Procedure Act (SAPA), but for the purposes of this chapter, the Model Act is being used. Most States have patterned their SAPA on the Model Act.
4. New York Public Health Law §§ 230(6), (10)(e).
5. MSAPA, § 5-107.
6. MSAPA, § 5-116.
7. *See* 42 C.F.R. §§ 1001.101, 1001.401, 1001.501.
8. Editorial, Five-Year Recovery: A New Standard for Assessing Effectiveness of Substance Use Disorder Treatment, J Substance Abuse Treatment, 58 (2015): 1–5, doi: https://doi.org/10.1016/j.jstat.2015.06.024.

Part VI

Federal Regulations

Health Privacy, Security, and Information Management

Stacey A. Tovino

Contents

Key Points
- The United States does not have one federal law that protects all health data in all situations. Instead, health privacy, data security, and information management are governed by a confusing patchwork of federal and state laws that have limited (but sometimes overlapping) application.
- Depending on the situation, an individual or institution that inappropriately uses, discloses, or manages health data in the United States may risk professional discipline, loss of professional or business licensure, loss of accreditation, loss of third-party reimbursement, civil and criminal penalties, liability under tort law, and other negative consequences.
- Lawmakers continue to introduce new consumer data bills in response to the privacy and security risks raised by advances in mobile and wearable healthcare technologies as well as public health crises, including COVID-19. Many of these bills do not become law.
- Due to the limited application of many health privacy, data security, and information management laws, some uses, disclosures, and sales of health data remain unregulated.

S. A. Tovino (✉)
College of Law, The University of Oklahoma, Norman, OK, USA
e-mail: stacey.tovino@ou.edu

© The Author(s), under exclusive license to Springer Nature Switzerland AG 2022
A. S. Pasha (ed.), *Laws of Medicine*,
https://doi.org/10.1007/978-3-031-08162-0_13

Legal Concepts

- Covered Entity definition: "Covered entity means a health plan, a health care clearinghouse, or a health care provider who transmits any health information in electronic form in connection with a [standard] transaction."
- Individual Rights: The right to receive a notice of privacy practices, the right to request additional privacy protections, the right to access PHI, the right to request amendment of incorrect or incomplete PHI, and the right to receive an accounting of PHI disclosures.
- PBA Activities: Public benefit activities.
- TPO Activities: Treatment, payment, and health care operations activities.

1   Introduction

In the United States, health privacy, data security, and information management are governed by a patchwork of federal and state laws. Some of these laws are sourced in state professional practice acts that apply to healthcare professionals. Other laws are sourced in state healthcare facility licensing laws that apply to healthcare institutions. Still others are sourced in federal regulations that apply to certain, but not all, healthcare industry participants. Additional sources of health privacy, data security, and information management requirements include state consumer laws, deceptive trade practice laws, public health and safety laws, and state and local government laws. Illustrative, but not exhaustive, examples of the laws governing health privacy, data security, and information management in the United States are presented below. Case studies are included to illustrate the application of these laws in particular situations.

2   Discussion

2.1   State Professional Practice Acts

State professional practice acts, including those that apply to physicians, psychologists, nurses, social workers, marriage and family therapists, professional counselors, and other licensed healthcare professionals, are one source of health privacy and information management requirements. For example, the Oklahoma Allopathic Medical and Surgical Licensure and Supervision Act (Oklahoma Medical Practice Act) authorizes the discipline of physicians who "[w]illfully betray[] a professional secret to the detriment of the patient" and who "[f]ail to maintain an office record for each patient which accurately reflects the evaluation, treatment, and medical necessity of treatment of the patient" [1]. By further example, the Maryland Medical Practice Act authorizes the discipline of physicians who "[f]ail to keep adequate medical records as determined by appropriate peer review" [2]. Similarly, the Illinois Nurse Practice Act authorizes the discipline of nurses who "[w]illfully or negligently violat[e] the confidentiality between nurse and patient" and "fail to establish and maintain records of patient care and treatment" [3]. By final illustrative, but not exhaustive, example, the North Dakota Integrative Health Care Act authorizes the discipline of naturopaths, medical imaging technicians, and radiation therapists for the "willful or negligent violation of the confidentiality between [practitioner] and patient" as well as "[t]he lack of appropriate documentation in medical records for diagnosis, testing, and treatment of patients" [4].

In addition to provisions that specifically reference privacy and information management obligations, many healthcare professional practice acts also contain general, or catch-all, provisions that authorize discipline when a healthcare professional "brings disrepute to the profession," "undermines confidence in the profession," or "violates ethical principles." For example, the Nevada Medical Practice Act allows the discipline of physicians who "engag[e] in conduct that brings the medical profession into disrepute" [5]. By further example, the Iowa Medical Practice Act allows the discipline of physicians who engage in "unethical or unprofessional conduct," including "a violation of the standards and principles of medical ethics" [6]. As discussed in more detail in the first case application, below, these catch-all provisions have been implicated in cases involving allegations of privacy and information management wrongdoing.

Licensed healthcare professionals who fail to comply with their professional practice acts risk disciplinary action. Disciplinary action can include private reprimand, public censure, probation, community service, completion of additional clinical or ethics education, license suspension, license revocation, and/or fines payable to the licensing board. The precise type and amount of discipline depends on the state in which the healthcare professional practices as well as the class of healthcare professional involved. Illinois nurses, for example, can be subject to fines as high as $10,000 per violation of the Illinois Nurse Practice Act [3]. Oklahoma physicians can be subject to fines as high as $5000 per violation of the Oklahoma Medical Practice Act [7]. North Dakota naturopaths, imaging technicians, and radiation therapists, also can be subject to fines as high as $5000 per violation of the North Dakota Integrative Health Care Act [4].

Case Application: Sugarman v. Board of Registration, 422 Mass. 338 (Sup. Jud. Ct. Mass., Suffolk, 1996).

Dr. Sugarman was a Massachusetts-licensed psychiatrist who was retained as an expert witness in a highly publicized and acrimonious custody dispute involving allegations of sexual abuse of a minor child by the child's father. The child's mother strongly opposed any unsupervised visitation between the father and the child while the father denied all allegations of sexual abuse. Because the allegations of sexual abuse were so highly publicized, the judge in the case issued orders prohibiting the parties, their attorneys, and "all known and potential expert witnesses" from making any public statements about the case. The judge's orders specifically prohibited the parties, their attorneys, and their expert witnesses from publicly speaking or writing about the family's medical history, records, examinations, evaluations, and treatments (collectively "medical reports").

When Dr. Sugarman received the medical reports that would form the basis of her expert testimony, she was informed of the judge's orders and told that she would be subject to contempt charges if she discussed the contents of the medical reports with anyone other than the attorney who hired her. Despite her knowledge of the judge's orders, Dr. Sugarman showed selected portions of the medical reports to a *Boston Globe* reporter. Apparently, Dr. Sugarman believed that she had a moral duty to protect the child from the risk of further sexual abuse and that publicizing the medical reports would be beneficial to the child. After Dr. Sugarman showed selected portions of the medical reports to the *Boston Globe* reporter, the newspaper ran a detailed story that included many quotations and details taken directly from the medical reports. On the same day that the *Boston Globe* ran its story, Dr. Sugarman also staged a press conference in which she identified herself as a "child advocate" and referenced evidence contained in the medical reports that she believed indicated that the child was sexually abused by the child's father on at least three occasions. During the press conference, Dr. Sugarman also stated her opinion that the treatment decisions being made for the child placed the child at grave risk for further sexual abuse and serious psychological damage. At no time did the child or the child's parents authorize Dr. Sugarman to disclose any information in the medical reports to the *Boston Globe* reporter or to hold a press conference about the case.

Under the Massachusetts Medical Practice Act (Practice Act), the Massachusetts Board of Registration in Medicine (Medical Board) had the authority to discipline a physician for unprofessional conduct, including "conduct [that undermined] public confidence in the integrity of the medical profession." The Medical Board relied on this Practice Act language to impose disciplinary action on Dr. Sugarman. Dr. Sugarman appealed the disciplinary action, arguing that it was inappropriate and unauthorized by the Practice Act. On review, the Supreme Judicial Court of Massachusetts (Court) disagreed with Dr. Sugarman. The Court explained that the Medical Board had broad authority to protect the image of the medical profession under the Practice Act and that Dr. Sugarman's conversation with the *Boston Globe* reporter and press conference statements violated the family's right to health privacy as well as Dr. Sugarman's ethical obligations as a psychiatrist. The Court ultimately affirmed the Medical Board's order temporarily suspending Dr. Sugarman's license, imposing a $10,000 fine, and requiring the completion of one hundred hours of community service.

2.2   State Facility Licensing Laws

State facility licensing laws, including those that apply to hospitals, nursing homes, intermediate care facilities, ambulatory surgery centers, freestanding emergency departments, and other healthcare facilities, are a second source of health privacy and information management obligations. Some state healthcare facility licensing laws establish relatively general privacy and information management obligations. Michigan's medical facility licensing law, for example, requires hospitals, nursing homes, county medical facilities, freestanding surgical facilities, and hospices to maintain a medical record for each patient, including records of "tests and examinations performed, observations made, treatments provided" [8]. The Michigan law also requires the retention of records for seven years "in such a manner as to protect their integrity, to ensure their confidentiality and proper use, and to ensure their accessibility and availability to each patient" [8].

Nevada's medical facility licensing law also gives medical facility patients a general right to "privacy concerning the patient's program of medical care" [9]. The Nevada law further specifies that, "discussions of the care of a patient, consultation with other persons concerning the patient, examinations or treatments, and all communications and records concerning the patient" are "confidential" and that patients must consent to the presence of "any person who is not directly involved with the patient's care during any examination, consultation or treatment" [9].

Other state facility licensing laws establish slightly more specific privacy and information management requirements. Minnesota's hospital licensing law, for example, requires each hospital medical record to contain the patient's name, place of residence, admitting diagnosis, history and physical examination, progress notes, signed physician orders, operative notes, pathology and radiology reports, consultation records, anesthesia records, nurses' notes, discharge diagnosis, and autopsy report, if applicable [10]. The Minnesota law also requires hospital medical records to be completed within a "reasonable time following the discharge of the patient" and to be maintained for at least seven years following the age of majority, although a Minnesota hospital can choose to maintain the records for a longer period of time [11].

New York's mental health clinic licensing law, by further example, specifies that mental health clinic records and the information contained therein (including patient names, diagnoses, prognoses, treatments, and care coordination) are confidential and shall not be disclosed unless an exception applies [12]. The New York law lists more than a dozen exceptions to confidentiality; that is, situations in which a mental health clinic record can be disclosed without the patient's prior written authorization. These situations include, but certainly are not limited to, disclosures to judges (pursuant to court orders), coroners and medical examiners, the mental hygiene legal services office, attorneys representing clients in proceedings involving the client's involuntary hospitalization or assisted outpatient treatment, qualified researchers, and an endangered individual [12]. If an exception to confidentiality applies, the New York law specifies that the disclosure must be "limited to that information necessary and required in light of the reason for disclosure," the party receiving the information must keep the information confidential, and the mental health clinic must place a notation of the disclosure in the patient's clinical record and inform the patient of such disclosure upon the patient's request [12].

A healthcare facility that violates the privacy and information management requirements set forth in its state licensing law can lose its license to do business in the state. Under the Texas Hospital Licensing Law, for example, a hospital's business license can be probated, suspended, or revoked if the hospital fails to comply with a provision in the Texas law or a regulation adopted pursuant to the Texas law [13]. Under the Florida Hospital Licensing Law, by further example, a hospital's license can be modified, suspended, or revoked whenever there has been a "substantial failure" by the hospital to comply with the Florida law [14].

Case Application: Commonwealth v. Brownsville Golden Age Nursing Home, Inc., 103 Pa. Cmwlth. 449 (1987).

Brownsville Golden Age Nursing Home, located in Brownsville, Pennsylvania (Nursing Home), was surveyed by state authorities to determine the Nursing Home's compliance with the Pennsylvania Health Care Facilities Act (Licensing Law). The survey team found the Nursing Home to be out of compliance with Licensing Law provisions requiring the Nursing Home to "treat patients with consideration, respect, and full recognition of his dignity and individuality, including privacy in treatment and in care for personal needs." In particular, the survey team found that the Nursing Home allowed patients to be assisted with toileting in full view of persons passing by in the hallway. The survey team also found that patients were insufficiently dressed or were dressed in nightclothes in the middle of the day and that the Nursing Home did not have sufficient privacy curtains between patients who resided in shared rooms. In addition to these privacy-related requirements, the survey team also found other Licensing Law violations that posed a significant threat to the health and safety of Nursing Home patients. On judicial review, the Commonwealth Court of Pennsylvania directed the Pennsylvania Health Facility Hearing Board to revoke the license of the Nursing Home due to its multiple Licensing Law violations which included, but certainly were not limited to, the patient privacy violations.

2.3   Federal Medicare Conditions of Participation

In addition to state professional practice acts and state facility licensing laws, federal regulations are an additional source of health privacy, data security, and information management requirements. Federal regulations governing Medicare-participating hospitals, for example, provide that hospital patients have a "right to personal privacy," a "right to the confidentiality of [their] clinical records," and a "right to access their medical records ... in the form and format requested by the individual" [15]. These regulations, called the "Medicare Conditions of Participation," also contain detailed information management requirements, such as the length of time the hospital must retain medical records, the required contents of medical records, as well as additional obligations relating to the date, time, and authentication of written and verbal orders referenced in medical records [16].

Federal regulations that govern Medicare-participating long-term care (LTC) facilities are similar to those governing hospitals. For example, LTC residents have a "right to personal privacy and confidentiality of [their] personal and medical records" [17]. This privacy right applies to "accommodations, medical treatment, written and telephone communications, personal care, visits, and meetings of family and resident groups but this does not require the facility to provide a private room for each resident." This right also includes the "right to privacy in [the resident's oral,] written, and electronic communications, including the right to send and promptly receive unopened mail and other letters, packages and other materials delivered to the facility for the resident, including those delivered through a means other than a postal service" [17]. Federal regulations governing other types of Medicare-participating facilities also contain privacy and information management obligations. For example, federal regulations governing Medicare-participating home health agencies (HHAs) specify that HHA patients have a right to a "confidential clinical record" [18], that clinical records must include a variety of reports and entries, and that clinical records must be kept for a minimum of five years after discharge of the patient from the service of the HHA [19]. Medicare-participating facilities that fail to comply with their Medicare Conditions of Participation risk civil penalties as well as termination (*i.e.*, inability to receive reimbursement) from the Medicare and Medicaid Programs.

Case Application: Ivy Woods Healthcare & Rehab. Ctr. v. Thompson, 156 Fed. Appx. 775 (6th Cir. 2006).

Ivy Woods Healthcare and Rehabilitation Center (Ivy Woods), a Medicare-participating LTC facility located in North Lima, Ohio, was surveyed to determine its compliance with the Medicare Conditions of Participation (COPs) for LTC facilities. The survey team discovered numerous deficiencies relating to twelve different COPs. One such COP gave residents the "right to personal privacy and confidentiality of [their] personal and clinical records. Personal privacy includes … personal care." A surveyor found that Ivy Woods violated this COP by failing to cover a resident receiving personal care when the door to the resident's room was open, allowing other residents to see the first resident. As a result of this violation as well as other COP violations, HHS imposed a civil money penalty in the amount of $6600 against Ivy Woods.

Ivy Woods appealed the civil money penalty, first to an administrative law judge (ALJ) and then to HHS's Departmental Appeals Board (DAB). Both the ALJ and the DAB affirmed the civil money penalty. On review, the United States Court of Appeals for the Sixth Circuit agreed with the ALJ and the DAB, concluding that Ivy Woods had failed to maintain patient privacy and that substantial evidence supported the $6600 civil money penalty.

2.4   The Federal HIPAA Privacy Rule

In addition to regulations governing Medicare-participating facilities, additional federal regulations establish health privacy requirements. The privacy rule that implements section 264 the Health Insurance Portability and Accountability Act of 1996 (HIPAA Privacy Rule) is one example [20]. Like the state and federal authorities previously discussed, the HIPAA Privacy Rule has limited application. The HIPAA Privacy Rule only regulates covered entities and their business associates when they are using or disclosing certain protected health information (PHI) [21]. Covered entities are defined to include health plans, health care clearinghouses, and only those health care providers that transmit health information in electronic form in connection with certain standard transactions including the health claim transaction [22]. With four, rarely-implicated exceptions, PHI is individually identifiable health information [22].

The HIPAA Privacy Rule contains three groups of sub-regulations, including the use and disclosure requirements [23], the individual rights [24], and the administrative requirements [25], that are designed to protect patient privacy and health information confidentiality. The first use and disclosure requirement within the HIPAA Privacy Rule allows covered entities and business associates to use and disclose PHI with no prior permission from the individual who is the subject of the PHI—but only in certain situations. That is, covered entities may freely use and disclose PHI without any form of prior permission in order to carry out certain treatment (T), payment (P), and health care operations (O) activities (collectively TPO activities) [26], as well as certain public benefit activities (PBAs) [27]. For example, a covered physician is not required to obtain the patient's prior permission before disclosing the patient's PHI as part of a patient referral because treatment (T) is defined to include patient referrals. By further example, the covered physician also would not be required to obtain the patient's prior permission before disclosing the patient's PHI to the patient's insurer for reimbursement purposes because payment (P) is defined to include billing third-party insurers. By still further example, the covered physician would not be required to obtain the patient's prior permission before disclosing the patient's PHI to the physician's attorney because the definition of health care operations (O) includes the procurement of legal services. By final example, the covered physician would not be required to obtain the patient's prior permission before disclosing the patient's COVID-19 diagnosis to a public health authority, such as a state department of health or the U.S. Centers for Disease Control and Prevention, because the PBAs include infectious disease surveillance.

The second use or disclosure requirement permits a covered entity or business associate to use and disclose an individual's PHI for certain activities, but only if the individual is informed (orally or in writing) in advance of the use or disclosure and is given the (oral or written) opportunity to agree to, prohibit, or restrict the use or disclosure [28]. The certain activities captured by this second provision include, but are not limited to, disclosures of PHI: (1) from a health care provider's facility directory; (2) to a person who is involved in an individual's care or payment for care; and (3) for certain notification purposes, such as when an attending physician or a hospital social worker notifies a partner or spouse of a patient's death [28]. The third use and disclosure requirement—a default rule—requires covered entities and business associates to obtain the prior written authorization of the individual who is the subject of the PHI before using or disclosing the individual's PHI in any situation that does not fit within the first two rules [29]. The HIPAA Privacy Rule requires authorizations to contain a number of specific elements and statements [29].

In addition to its use and disclosure requirements, the HIPAA Privacy Rule also establishes five rights for individuals who are the subject of PHI (hereinafter individual rights). These individual rights include the right to receive a notice of privacy practices [30], the right to request additional privacy protections [31], the right to access PHI [32], the right to request amendment of incorrect or incomplete PHI [33], and the right to receive an accounting of PHI disclosures [34]. In addition to the use and disclosure requirements and the individual rights, the HIPAA Privacy Rule also contains a number of administrative requirements. For example, covered entities are required to designate a privacy officer who will oversee compliance with the HIPAA Privacy Rule, train workforce members regarding how to comply with the HIPAA Privacy Rule, sanction workforce members who violate the HIPAA Privacy Rule, establish complaint processes for individuals who believe their privacy rights have been violated, and develop privacy-related policies and procedures [25].

Case Application: Resolution Agreement between U.S. Department of Health and Human Services and New York Presbyterian Hospital (Apr. 21, 2016).

New York Presbyterian Hospital (Hospital), a HIPAA covered entity located in New York City, New York, allowed American Broadcasting Company (ABC) film crews to enter its emergency room and other patient care areas to film patients receiving care as part of the production of "NY Med," an ABC reality-based television series. Because the filming of a television series does not constitute a TPO activity or otherwise fall into the first or second use or disclosure requirements within the HIPAA Privacy Rule, the third (default) requirement was implicated. That is, the HIPAA Privacy Rule would have required the Hospital to obtain the prior written authorization of each patient (or the patient's legally authorized representative) before the patient could be filmed by the ABC film crews in the Hospital. The Hospital did not satisfy this requirement; that is, the Hospital did not obtain the authorization of any patients before allowing ABC to film the patients. Indeed, ABC filmed one patient who was dying and a second patient who was in significant distress without obtaining either patient's (or the patient's legally authorized representative's) prior written authorization.

The federal Department of Health and Human services (HHS), which provides civil enforcement for the HIPAA Privacy Rule, announced on April 21, 2016, that it had entered into a $2.2 million resolution agreement with the Hospital for the "egregious disclosure of two patients' [PHI] to film crews and staff" during the filming of "NY Med." Jocelyn Samuels, the Director of the Office for Civil Rights (OCR) within HHS, stated in the press release announcing the settlement that, "'This case sends an important message that OCR will not permit covered entities to compromise their patients' privacy by allowing news or television crews to film the patients without their authorization. We take seriously all complaints filed by individuals and will seek the necessary remedies to ensure that patients' privacy is fully protected.'" Samuels also stated, "By allowing individuals receiving urgent medical care to be filmed without their authorization by members of the media, [the Hospital's] actions blatantly violate the HIPAA [Privacy Rule, which was] specifically designed to prohibit the disclosure of individual's PHI, including images, in circumstances such as these" [35].

2.5   The Federal HIPAA Security Rule

In addition to the HIPAA Privacy Rule, HHS also has promulgated a HIPAA Security Rule [36]. The HIPAA Security Rule requires covered entities and business associates to implement three classes of safeguards—administrative, physical, and technical—to protect the confidentiality, integrity, and availability of electronic protected health information (ePHI). The first class of (administrative) safeguards require covered entities and business associates to designate a security official responsible for the development and implementation of the covered entity's or business associate's security policies and procedures [37]. These policies and procedures shall: (1) prevent, detect, contain, and correct security violations; (2) ensure that workforce members have appropriate access to ePHI; (3) prevent workforce members who should not have access to ePHI from obtaining such access; (4) create a security awareness and training program for all workforce members; and (5) address and respond to security incidents, emergencies, environmental problems, and other occurrences such as fire, vandalism, system failure, and natural disaster that affect systems containing ePHI and the security of ePHI, among other requirements [37].

The second class of (physical) safeguards require covered entities and business associates to implement policies and procedures that: (1) limit physical access to electronic information systems and the facilities in which they are located; (2) address the safeguarding, functioning, and physical attributes of workstations through which ePHI is accessed; and (3) govern the receipt and removal of hardware and electronic media that contain ePHI [38]. The third class of (technical) safeguards require covered entities and business associates to implement: (1) technical policies and procedures for electronic information systems that maintain ePHI to allow access only to those persons or software programs that have been granted access rights; (2) hardware, software, and/or procedural mechanisms that record and examine activity in information systems that contain or use ePHI; (3) policies and procedures to protect ePHI from improper alteration or destruction; (4) procedures to verify that a person or entity seeking access to ePHI is the one claimed; and (5) technical security measures to guard against unauthorized access to ePHI that is being transmitted over an electronic communications network [39].

Case Application: Notice of Final Determination by the U.S. Depart ment of Health and Human Services re: Children's Medical Center of Dallas (Jan. 18, 2017).

In November 2009, an employee of Children's Medical Center of Dallas, located in Dallas, Texas (Children's Medical Center) lost an unencrypted, non-password protected BlackBerry device at the Dallas/Fort Worth (DFW) International Airport. The device contained the ePHI of approximately 3800 individuals. A second security breach occurred in April 2013, when an unencrypted laptop was stolen from the premises of Children's Medical Center. The laptop contained the ePHI of approximately 2462 individuals. OCR investigated and found that Children's Medical Center had failed to implement a number of HIPAA Security Rule requirements. For example, OCR found that the medical center had failed to implement risk management plans (contrary to prior external recommendations to do so) and failed to deploy encryption or equivalent alternative measures on all of its laptops, mobile devices, and removable storage media. OCR further found that the medical center continued to issue unencrypted BlackBerry devices to nurses and allowed workforce members to continue to use unencrypted laptops and other mobile devices through 2013, four years after the DFW Airport breach.

On February 1, 2017, HHS issued a press release announcing that it had imposed, and that Children's Medical Center had paid, a $3.2 million civil money penalty to resolve violations of the HIPAA Security Rule. Robinsue Frohboese, the Acting Director of OCR, stated in the press release that, "'Ensuring adequate security precautions to protect health information, including identifying any security risks and immediately correcting them, is essential'" [40].

2.6   The Federal HIPAA Breach Notification Rule

In addition to its Privacy and Security Rules, HHS also has promulgated a Breach Notification Rule [41]. The HIPAA Breach Notification Rule requires covered entities, following the discovery of a breach of unsecured protected health information (uPHI), to notify each individual whose uPHI has been, or is reasonably believed by the covered entity to have been, accessed, acquired, used, or disclosed as a result of such breach [42]. The notification, which shall be provided without undue delay and within sixty calendar days after the discovery of the breach, shall include: (1) a brief description of the nature of the breach, including the date of the breach and the date of its discovery; (2) a description of the types of uPHI involved in the breach; (3) any steps the individual should take to protect herself from potential harm resulting from the breach; (4) a brief description of the steps taken by the covered entity to investigate the breach, to mitigate harm to individuals whose uPHI was part of the breach, and to protect against future breaches; and (5) contact information sufficient to allow individuals to ask questions or learn additional information about the breach [42].

When a breach involves the uPHI of more than 500 residents of a state or jurisdiction, the HIPAA Breach Notification Rule also requires the covered entity to notify prominent media outlets serving the state or jurisdiction [43]. When a breach involves the uPHI of 500 or more individuals, regardless of their state of residency, the covered entity is also required to notify the Secretary of HHS within sixty calendar days after the discovery of the breach [44]. Finally, when the breach involves the uPHI of less than 500 individuals, the covered entity is required to notify the Secretary of HHS no later than sixty calendar days after the end of the calendar year [44].

Case Application: Resolution Agreement between U.S. Department of Health and Human Services and Presence Health Network (Jan. 3, 2017).

In January 2014, OCR received a breach notification report from Presence Health, an Illinois-based healthcare network that includes eleven hospitals, twenty-seven LTC and senior living facilities, and multiple physicians' offices. The breach notification report indicated that in October 2013, Presence Health discovered that paper-based operating room schedules, which contained the PHI of 836 individuals, were missing from the Presence Surgery Center at the Presence St. Joseph Medical Center in Joliet, Illinois. The missing information included patient names, dates of birth, medical record numbers, dates of procedures, types of procedures, surgeon names, and types of anesthesia. OCR investigated and found that Presence Health had failed to notify each of the 836 individuals affected by the breach and prominent media outlets (as required for breaches affecting 500 or more individuals) as required by the HIPAA Breach Notification Rule.

On January 3, 2017, HHS entered into a resolution agreement with Presence Health to resolve allegations of violations of the HIPAA Breach Notification Rule. Pursuant to the resolution, Presence Health agreed to pay $475,000 to HHS and to implement a corrective action plan. The agreement between HHS and Presence Health was the first to resolve allegations of violations of the HIPAA Breach Notification Rule (versus the HIPAA Privacy or Security Rules). In the press release announcing the first-of-its-kind breach notification resolution, OCR Director Jocelyn Samuels explained, "Covered entities need to have a clear policy and procedures in place to respond to the Breach Notification Rule's timeliness requirement." Samuels further explained, "Individuals need prompt notice of a breach of their unsecured PHI so they can take action that could help mitigate any potential harm caused by the breach" [45].

2.7   State Consumer Data Protection Laws

Newer consumer data protection laws, which also protect health data, apply to certain persons that conduct business in the state or that produce products or services that are targeted to residents of the state. However, these newer consumer laws usually apply only when the data controlled or processed exceeds certain thresholds or when the data controller or processor derives a certain percentage of gross revenue from the sale of health and other consumer data. For example, the California Consumer Privacy Act (CCPA) [46], which went into effect in 2020, regulates certain businesses that collect consumer personal information (including health information) and that do business in the State of California if the business meets one of three thresholds. The thresholds include: (1) having annual gross revenues in excess of $25 million; (2) annually buying, receiving for the business's commercial purposes, selling, or sharing for commercial purposes, alone or in combination, the personal information of 50,000 or more consumers, households, or devices; and (3) deriving 50 percent or more of its annual revenues from selling consumers' personal information [47].

A second example is the Virginia Consumer Data Protection Act (VCDPA), signed into law on March 2, 2021 [48]. Effective January 1, 2023, the VCDPA will regulate certain businesses that conduct business in Virginia or that produce products or services that are targeted to residents of Virginia if, during a calendar year, the business controls or processes the personal data (including health data) of: (1) at least 100,000 consumers or (ii) at least 25,000 consumers and derives over 50 percent of gross revenue from the sale of personal data [49].

A third example is the Colorado Privacy Act (CPA), signed into law on July 7, 2021 [50]. Effective July 1, 2023, the CPA will regulate certain data controllers (including certain health data controllers) that conduct business in Colorado or that produce or deliver commercial products or services that are intentionally targeted to residents of Colorado if the controller processes the personal data of 100,000 consumers or more during a calendar year or derives revenue or receives a discount on the price of goods or services from the sale of personal data and processes or controls the personal data of 25,000 consumers or more [51].

2.8   Federal Consumer Data Protection Bills

In addition to the recently enacted state consumer data protection laws discussed above, a number of federal lawmakers have recently introduced consumer data (including health data) protection bills in Congress. These federal bills are designed to respond to the privacy, security, and information management risks associated with new healthcare technologies, including mobile applications and wearable devices, as well as the electronic infectious disease exposure notification services developed during the COVID-19 pandemic. These bills also are designed to respond to the limited application of state professional practice acts, state facility licensing laws, the Medicare Conditions of Participation, and the HIPAA Privacy, Security, and Breach Notification Rules. These bills include, but are not limited to, the COVID-19 Consumer Data Protection Act of 2020 (CCDPA) [52], the Data Care Act of 2018 (DCA) [53], the Exposure Notification Privacy Act of 2020 (ENPA) [54], the Information Transparency & Personal Data Control Act of 2021 (ITPDCA) [55], the Mind Your Own Business Act of 2019 (MYOBA) [56], the Public Health Emergency Privacy Act of 2020 (PHEPA) [57], the Protecting Personal Health Data Act of 2019 (PPHDA) [58], and the Stop Marketing and Revealing the Wearables and Trackers Consumer Health (SMARTWATCH) Data Act of 2019 [59]. As of this writing, not one of these federal bills has been signed into law.

3   Summary

In the United States, health privacy, data security, and information management are governed by a patchwork of federal and state laws. These laws include, but are certainly not limited to, state professional practice acts, state facility licensing laws, the Medicare Conditions of Participation, and the HIPAA Privacy, Security, and Breach Notification Rules, as well as newer state consumer data protection laws. These laws build on common law rights to privacy and confidentiality recognized in most states [60] as well as constitutional rights to privacy recognized in some states [61]. Depending on the situation, an individual or institution that inappropriately uses, discloses, or manages health data in the United States may risk professional discipline, loss of professional or business licensure, loss of third-party reimbursement, civil and criminal penalties, liability under tort law, and/or other negative consequences. Due to the limited application and scope of current federal and state laws, federal lawmakers continue to introduce new privacy, security, and information management bills to Congress. As of this writing, however, the United States does not have one federal law that protects all health data. As a result, many uses, disclosures, and sales of health data remain unregulated [62].

References

1. OKLA. STAT. ANN. tit. 59, § 509.
2. MD. CODE ANN., Health Occ. § 14-404.
3. 225 ILL. COMP. STAT. ANN. 65/70-5.
4. N.D. CENT. CODE ANN. § 43-57-08.
5. NEV. REV. STAT. ANN. 630.301.
6. IOWA ADMIN. CODE § 653-23.1.
7. OKLA. STAT. ANN. tit. 59, § 509.1.
8. MICH. COMP. LAWS ANN. § 333.20175.
9. NEV. REV. STAT. § 449A.112.
10. MINN. ADMIN. CODE § 4640.1000.
11. MINN. STAT. ANN. § 145.32.
12. N.Y. MENTAL HYG. LAW § 33.13.
13. TEX. HEALTH & SAFETY CODE § 241.053.
14. FLA. STAT. ANN. § 393.003.
15. 42 C.F.R. § 482.13.
16. 42 C.F.R. § 482.24.
17. 42 C.F.R. § 483.10.
18. 42 C.F.R. § 484.50.
19. 42 C.F.R. § 484.110.
20. Health Insurance Portability and Accountability Act of 1996, Pub. L. No. 104-191, 110 Stat. 1936 (Aug. 21, 1996) (codified as amended in scattered sections of 42 U.S.C.), amended in part by Health Information Technology for Economic and Clinical Health Act, Pub. L. No. 111-5, 123 Stat. 115, 226 (Feb. 17, 2009) (codified as amended in scattered sections of 42 U.S.C.). The HIPAA Privacy Rule, which implements section 264(c) of HIPAA, is codified at 45 C.F.R. Part 164, Subpart E (45 C.F.R. §§ 164.500-.534).
21. 45 C.F.R. § 160.102.
22. 45 C.F.R. § 160.103.
23. 45 C.F.R. §§ 164.502-.514.
24. 45 C.F.R. §§ 164.520-.528.
25. 45 C.F.R. § 164.530.
26. 45 C.F.R. § 164.501.
27. 45 C.F.R. § 164.512.
28. 45 C.F.R. § 164.510.
29. 45 C.F.R. § 164.508.
30. 45 C.F.R. § 164.520.
31. 45 C.F.R. § 164.522.
32. 45 C.F.R. § 164.524.
33. 45 C.F.R. § 164.526.
34. 45 C.F.R. § 164.528.
35. U.S. Dep't Health & Human Servs. Unauthorized Filming for "NY Med" Results in $2.2 Million Settlement with New York Presbyterian Hospital. 2016. Accessed 22 December 2021.
36. 45 C.F.R. §§ 164.302-.318.
37. 45 C.F.R. § 164.308.
38. 45 C.F.R. § 164.310.
39. 45 C.F.R. § 164.312.
40. U.S. Dep't Health & Human Servs. Judge Rules in Favor of OCR and Requires a Texas Cancer Center to Pay $4.3 Million in Penalties for HIPAA Violations. 2018. Accessed 22 December 2021.
41. 45 C.F.R. §§ 164.400-.414.
42. 45 C.F.R. § 164.404.
43. 45 C.F.R. § 164.406.

44. 45 C.F.R. § 164.408.

45. U.S. Dep't Health & Human Servs. First HIPAA Enforcement Action for Lack of Timely Breach Notification Settles for $475,000. 2017. Accessed 22 December 2021.

46. California Consumer Privacy Act, *codified at* CAL. CIV. CODE §§ 1798.100 - 1798.199.35 et seq. (latest revisions eff. January 1, 2023).

47. CAL. CIV. CODE § 1798.140 (West 2021).

48. Virginia Consumer Data Protection Act, S.B.1392 (Mar. 2, 2021), *to be codified at* VA. CODE ANN. §§ 59.1-571 -581 (eff. Jan. 1, 2023).

49. VA. CODE ANN. § 59.1-572(A) (eff. Jan. 1, 2023).

50. Colorado Privacy Act, S.B. 21-190, signed into law on July 7, 2021, *to be codified at* COLO. REV. STAT. § 6-1-1301 (eff. July 1, 2023).

51. COLO. REV. STAT. ANN. § 6-1-1304(1) (eff. July 1, 2023).

52. The COVID-19 Consumer Data Protection Act of 2020, S.3663, 116th Cong., 2nd Sess. (May 7, 2020).

53. Data Care Act of 2018, S.3744, 115th Cong., 2nd Sess., § 3 (Dec. 12, 2018).

54. The Exposure Notification Privacy Act, S.3861, 116th Cong., 2nd Sess., § 2(6) (June 1, 2020).

55. Information Transparency & Personal Data Control Act, H.R. 1816, 17th Cong., 1st Sess. (Mar. 11, 2021).

56. Mind Your Own Business Act of 2019, S.2637, 116th Cong., 1st Sess., § 2(12) (Oct. 17, 2019).

57. Public Health Emergency Privacy Act, S.3749, 116th Cong., 2nd Sess., § 2(8) (May 14, 2020).

58. Protecting Personal Health Data Act, S.1842, 116th Cong., 1st Sess. (June 13, 2019).

59. Stop Marketing and Revealing the Wearables and Trackers Consumer Health (SMARTWATCH) Data Act, S.2885, 116th Cong., 1st Sess. (Nov. 18, 2019).

60. *See, e.g.,* McCormick v. England, 328 S.C. 627, 494 S.E.2d 431 (Ct. App. 1997) (holding that a common law tort claim for breach of duty to maintain patient confidences would be recognized in a case involving a physician who disclosed without authorization information about a patient's emotional health during a divorce proceeding); Doe v. Roe, 400 N.Y.S.2d 668 (Sup. Ct. 1977) (holding that a physician who enters into an agreement with a patient to provide medical attention impliedly covenants to keep in confidence all disclosures made by the patient concerning the patient's physical or mental health condition as well as all matters discovered by the physician in the course of examination or treatment); Horne v. Patton, 291 Ala. 701, 287 So. 2d 824 (1973) (holding that there is a confidential relationship between a doctor and a patient that imposes a duty upon the doctor not to disclose information concerning the patient obtained during the course of treatment; further holding that the defendant physician's release of information to a patient's employer can constitute an invasion of a patient's privacy).

61. *See generally* Catherine Louisa Glenn, *Protecting Health Information Privacy: The Case for Self-Regulation of Electronically Held Medical Records*, 53 VAND. L. REV. 1605, 1609 n.25 (2000) (identifying the constitutions of Alaska, Arizona, California, Florida, Hawaii, Illinois, Louisiana, Montana, South Carolina, and Washington as establishing rights to privacy).

62. The Author has written about patient privacy, health information confidentiality, data security, and information management for more than seventeen years. *See, e.g.*, Stacey A. Tovino, *Not so Private*, 71 DUKE L.J. ___ (forthcoming 2022);; Stacey A. Tovino, *The United States*, in PRIVACY AND HEALTH: A COMPARATIVE ANALYSIS ___ (Nicola Glover-Thomas & Thierry Vansweevelt eds., Edward Elgar Publishing forthcoming 2021); Stacey A. Tovino, *At a COVID Crossroads: Public Health, Patient Privacy, and Health Information Confidentiality*, 65 ST. LOUIS U. L.J. 849 (2021); Stacey A. Tovino, *HIPAA Compliance*, in CAMBRIDGE HANDBOOK OF COMPLIANCE 895-908 (Daniel Sokol & Benjamin van Rooij eds., Cambridge University Press 2021); Stacey A. Tovino, *COVID-19 and the HIPAA Privacy Rule: Asked and Answered*, 50 STETSON L. REV. 365 (2021); Stacey A. Tovino, *Assumed Compliance*, 72 ALA. L. REV. 279 (2020); Stacey A. Tovino, *Mobile Research Applications and State Research Laws*, 48(Supp.1) J. L. MED. & ETHICS 82 (2020); Stacey A. Tovino, *Mobile Research Applications and State Data Protection Statutes*, 48(Supp.1) J. L, MED. & ETHICS 86 (2020); Stacey A. Tovino, *Going Rogue: Mobile Research Applications and the Right to Privacy*, 95 NOTRE DAME L. REV. 155 (2019); Stacey A. Tovino, *A Timely Right to Privacy*, 104 IOWA L. REV.

1361 (2019); Stacey A. Tovino, *Florida Law, Mobile Research Applications, and the Right to Privacy*, 43 Nova L. Rev. 353 (2019); Mark A. Rothstein & Stacey A. Tovino, *Privacy Risks of Interoperable Health Records Require Segmentation*, 47 J. L. Med. & Ethics 771 (2019); Stacey A. Tovino, *Privacy and Security Issues in mHealth Research*, 47(Supp.2) J. L. Med. & Ethics 154 (2019); Mark A. Rothstein & Stacey A. Tovino, *California Takes the Lead on Data Privacy*, Hastings Center Report, Sept.-Oct. 2019, at 4; Stacey A. Tovino, *Patient Privacy: Problems, Perspectives, and Opportunities*, 27 Annals Health L. 243 (2018); Stacey A. Tovino, *The EU GDPR and the HIPAA Privacy Rule: Illustrative Comparisons*, 47 Seton Hall L. Rev. 973 (2017); Stacey A. Tovino, *Teaching the HIPAA Privacy Rule*, 61 St. Louis U. L. J. 469 (2017); Stacey A. Tovino, *Complying with the HIPAA Privacy Rule: Problems and Perspectives*, 1 Loy. U. Chi. J. Reg. Compliance 23 (2016); Stacey A. Tovino, *Silence Is Golden ... Except in Health Care Philanthropy*, 48 U. Rich. L. Rev. 1157 (2014); Stacey A. Tovino, *Gone Too Far: Federal Regulation of Health Care Attorneys*, 91 Or. L. Rev. 813 (2013); Stacey A. Tovino, *HIPAA Privacy for Physicians*, 17 Pathology Case Rev. 160 (2012); Stacey A. Tovino, *Medical Privacy, in* Governing America: Major Decisions of Federal, State, and Local Government (Paul Quirk & William Cunion eds., 2011); Stacey A. Tovino, *Functional Neuroimaging Information: A Case for Neuro Exceptionalism?* 34 Fla. St. U. L. Rev. 415 (2007); Stacey A. Tovino, *Imaging Body Structure and Mapping Brain Function: A Historical Approach*, 33 Am. J. L. & Med. 193 (2007); Stacey A. Tovino, The Visible Brain: Confidentiality and Privacy Implications of Functional Magnetic Resonance Imaging (Ph.D. Dissertation, University of Texas Medical Branch, 2006); Stacey A. Tovino, *Hospital Chaplaincy under the HIPAA Privacy Rule: Health Care or "Just Visiting the Sick?"* 2 Ind. Health L. Rev. 51 (2005); Stacey A. Tovino, *Confidentiality and Privacy Implications of Functional Magnetic Resonance Imaging*, 33 J.L. Med. & Ethics 844 (2005); and Stacey A. Tovino, *The Use and Disclosure of Protected Health Information for Research under the HIPAA Privacy Rule: Burdensome Government Regulation and Unrealized Patient Autonomy*, 49 S.D. L. Rev. 447 (2004). Although the case studies presented in this chapter as well as the chapter sections pertaining to state professional practice acts, state healthcare facility licensure laws, the federal Medicare Conditions of Participation, state consumer data protection laws, and federal consumer data bills were newly written for this chapter, portions of the paragraphs summarizing the HIPAA Privacy, Security, and Breach Notification Rules are similar to language summarizing the HIPAA Rules set forth in the Author's many prior works.

Corporate Practice of Medicine

Michael F. Schaff and Jason J. Krisza

Contents

Key Points

- The corporate practice of medicine doctrine prohibits unlicensed individuals or entities from practicing medicine and places limits on physicians "partnering" with, or being employed by, such individuals or entities.
- The corporate practice of medicine doctrine is state-specific and varies from state to state, ranging from not being applicable in some states, to imposing significant restrictions in others.
- Currently, healthcare is going through a transition whereby smaller practices are consolidating into larger groups. This consolidation has been led by large medical systems and private equity investors and often implicates corporate practice of medicine prohibitions.
- In states that have strong corporate practice of medicine doctrines, formation of management services organizations (MSOs) may enable physicians to "partner" with non-licensed entities, like private equity investors, while remaining in compliance with CPOM restrictions.

M. F. Schaff (✉) · J. J. Krisza
Wilentz, Goldman & Spitzer, P.A., Woodbridge, NJ, USA
e-mail: mschaff@wilentz.com; jkrisza@wilentz.com

© The Author(s), under exclusive license to Springer Nature Switzerland AG 2022
A. S. Pasha (ed.), *Laws of Medicine*,
https://doi.org/10.1007/978-3-031-08162-0_14

Legal Concepts

- *Corporate Practice of Medicine Doctrine*: The prohibition of any unlicensed person or entity from practicing medicine, interfering with a medical professional's clinical judgment, profiting from the practice of medicine, or directly employing physicians. CPOM laws are designed to avoid corporate influence on physicians and any reduction in patient care quality that may accompany such influence.
- *Management Services Organization (MSO)*: An entity that performs non-clinical tasks associated with a medical practice, such as billing, collections, employment of non-clinical office staff, acquisition of inventory, supplies, and space. MSOs are sometimes referred to as management companies or administrative services companies.
- *Professional Entity*: Professional entities are entities solely comprised of licensed individuals. In some states, they may take the form of professional corporations or professional limited liability companies. In other states, they may simply be organized as corporations or limited liability companies with solely professional ownership.

1   Introduction

The CPOM doctrine prohibits any unlicensed person or entity from practicing medicine, directly employing physicians, or interfering with a medical professional's clinical judgment. CPOM laws vary significantly from state to state and take the form of case law, administrative decisions, statutes, and regulations. Certain states do not prohibit unlicensed individuals from employing physicians or owning and/or controlling a medical practice, while others impose significant restrictions on these activities. Healthcare professionals should be aware of the restrictions imposed by the CPOM doctrine and whether it is applicable in the state in which they practice medicine, especially if they intend to structure a healthcare arrangement with a non-licensee or non-medical entity. Because the CPOM doctrine is state-specific, readers are encouraged to seek local counsel when addressing CPOM issues and should not rely on the broad explanations of the doctrine set forth in this chapter.

In states that have strong CPOM restrictions, non-physicians cannot employ or partner with physicians to provide clinical care, even when the unlicensed individual does not perform any clinical duties. Certain states even prevent physicians from entering into arrangements with vendors (e.g., billing companies) where the vendor's compensation is based upon a percentage of the practice's collections.

Proponents of the CPOM doctrine believe it protects the integrity of the medical profession and ensures that patients are provided with the highest level of care without interference by unlicensed individuals or entities who may focus on maximizing profits at the expense of patient well-being. Opponents of the doctrine consider it an archaic rule that unnecessarily restricts relationships in modern medicine.

In recent decades, healthcare has experienced significant consolidation. Independent solo physician practitioners and small practices became overwhelmed by the healthcare regulatory environment, the documentation needed to properly carry on the business of medicine, and the need to make significant investments in technologies like electronic medical records and billing software. If solo practitioners and practices are not able to manage these issues, they typically consider either forming or joining larger practice groups or affiliating with large health and hospital systems which are vying to acquire these once independent practices. More recently, private equity groups and other unlicensed individuals and entities began to view the healthcare industry as a significant investment opportunity, comprising a large percentage of the U.S. gross national product and increasing each year. In states with active CPOM restrictions, these relationships are scrutinized and must be structured carefully.

2 Discussion

2.1 Background

The origins of the CPOM doctrine can be traced back to the American Medical Association's issuance of its Principles of Medical Ethics and its efforts to distinguish physicians from healers and other non-physicians offering remedies or cures for various ailments and afflictions [1]. While the AMA's Principles of Medical Ethics applied to AMA's members, the real threat was from non-members. During the late 1800s, AMA members began to lobby for strict licensing standards [2]. By the early twentieth century, a vast majority of states required physicians to abide by formal license requirements [2].

The CPOM doctrine is based on the idea that a physician's obligations to care for patients and exercise independent clinical judgment conflict with the interests of a corporation in maximizing its profits and reducing its costs. In a 1935 opinion, the Illinois Supreme Court explained the rationale behind the doctrine and the state's prohibition on the unlicensed practice of medicine by corporations:

> To practice a profession requires something more than the financial ability to hire and competent persons to do the actual work. It can be done only by a duly qualified human being, and to qualify something more than mere knowledge or skill is essential. The qualifications include personal characteristics, such as honesty, guided by an upright conscience and a sense of loyalty to clients or patients, even to the extent sacrificing pecuniary profit, if necessary. These requirements are spoken of generically as the good moral character which is a pre-requisite to the licensing of any professional man. No corporation can qualify [3].

2.2 Applicability of the CPOM Doctrine

As indicated above, the CPOM doctrine varies from state to state. Some take the form of case law, e.g., the Dr. Allison case from Illinois cited above, while others take the form of statute, e.g., Ark. Code Ann. 4-29-307, and still others the form of advisory opinions, e.g., New York Department of Health Opinion Letter dated

March 21, 2006 (available at https://archive.nysba.org/Sections/Health/Department_of_Health_Opinion_Letters/06-01BillingforServicesPerformedinPhysiciansOffi ces_pdf.html).

There are general principles that are prevalent in many states with active CPOM doctrines. In such states, physicians are typically prohibited from entering into professional relationships with more limited licensed professionals, e.g., chiropractors, nurse practitioners, or dentists, or non-physicians where the physician's practice of medicine is in any way controlled or directed by a non-physician. For example, in California:

> … any person who practices or attempts to practice, or who advertises or holds himself or herself out as practicing, any system or mode of treating the sick or afflicted in this state, or who diagnoses, treats, operates for, or prescribes for any ailment, blemish, deformity, disease, disfigurement, disorder, injury, or other physical or mental condition of any person, without having at the time of so doing a valid, unrevoked, or unsuspended certificate as provided in this chapter or without being authorized to perform the act pursuant to a certificate obtained in accordance with some other provision of law is guilty of a public offense, punishable by a fine not exceeding ten thousand dollars ($10,000), by imprisonment pursuant to subdivision (h) of Section 1170 of the Penal Code, by imprisonment in a county jail not exceeding one year, or by both the fine and either imprisonment [4].

Another state with a CPOM doctrine is New Jersey, where state medical board regulations provide a limited list of acceptable practice structures under which a licensed physician may practice [5]. These entities are sometimes referred to as professional entities. They can only be owned by licensed individuals. Certain states, such as New York, prohibit professionals from incorporating as general business corporations or forming limited liability companies; rather, they must incorporate or form, as the case may be, as a professional corporation or professional limited liability company. States with these types of laws prohibit anyone other than a physician, an entity owned solely by physicians, or a licensed healthcare entity (e.g., a hospital) from employing physicians [5]. In these jurisdictions, a licensing authority, commonly a state's Department of Health, issues licenses to individuals which own healthcare facilities such as hospitals, inpatient rehabilitation facilities, ambulatory surgery centers, and imaging centers. The rationale for these exceptions is that the licensing authority meticulously examines applicants, their backgrounds, and management policies to, among other things, ensure the clinicians maintain authority over the clinical aspects of the facility.

Some states, such as New York, further require that the physician owners of a practice actually be engaged in the practice of medicine within the practice entity [6]. New York also prohibits physicians from engaging in fee-splitting arrangements [7]. Fee-splitting arrangements are those whereby a physician and an unlicensed individual divide the fees received for the physician's services. While in this context it may seem logical, many billing companies charge medical practices a percentage of the practice's collections. In states with broad fee-splitting prohibitions, such as New York, this practice would be prohibited [7]. Other common examples of fee splitting include marketing arrangements and revenue splits with landlords.

2.3 Healthcare Consolidation and the CPOM Doctrine

In the early 1990s, the healthcare profession experienced its first major wave of consolidation. This trend continued as private equity investors began injecting capital into the healthcare industry in the decades that followed. These private equity firms are typically composed of unlicensed individuals seeking to "partner" with physicians or physician-owned entities. If the physicians or physician-owned entities are located or providing medical services in states that have a CPOM doctrine, they must carefully review the state's CPOM laws to ensure they structure the relationship to comply with such laws.

In many cases, in an effort to avoid violating corporate practice restrictions, the relationship is structured by bifurcating the medical practice into two distinct entities: (1) the professional practice, which employs the clinical staff, owns the clinical assets, and provides the clinical services; and (2) a non-clinical entity or management services organization, which employs the non-clinical staff, owns the non-clinical assets, and provides the non-clinical services that support the practice. The MSO is typically owned by the private equity firm or other non-physician entity, while ownership of the medical professional corporation remains with the physician(s).

Under these types of management arrangements, the medical practice sells all of its non-clinical assets (e.g., inventory and equipment) to the MSO but remains the owner, employer, and manager of all the clinical employees and clinical assets. The practice then enters into an administrative services agreement with the MSO, whereby the MSO provides the practice with administrative services or practice management tools that may include billing, purchasing, accounting, office space, supplies, human resources services, non-clinical assets and/or leased non-clinical personnel. In turn, the practice pays the MSO a fair market value fee to perform these operational functions to the extent the services do not interfere with the professional medical judgment of the physician(s) or otherwise result in MSO control over the medical practice in violation of corporate practice prohibitions. These arrangements require careful consideration of the applicable state laws to ensure that the arrangements and the terms of the management services contracts are consistent with applicable state CPOM laws.

3 Application

Case # 1 Andrew Carothers, M.D., P.C. v. Insurance Companies Represented by Bruno, Gerbino & Soriano, LLP, 26 Misc. 3d 448, 888 N.Y.S.2d 372 (N.Y. Civ. Ct. 2009)

In a dispute between a radiology practice, Andrew Carothers M.D., P.C. (the PC), and multiple no-fault insurance companies, the insurance companies refused to pay over $23 million in claims for reimbursement submitted by the PC, alleging that the PC was fraudulently incorporated in violation of New York law. Specifically, the insurance companies alleged that Dr. Andrew Carothers was not the true owner of the PC, but rather two other individuals were the actual owners, neither of whom held a medical degree or license in New York.

The jury found in favor of the insurance companies, upholding the state's CPOM doctrine and not requiring the insurance companies to pay for the claims at issue, after considering whether and the extent to which: (1) agreements between the PC and the non-licensees and their entities were the products of arm's length transactions or whether the financial and non-financial terms were designed to give the non-licensees and their entities substantial control over the PC and to channel to them the PC's profits; (2) the non-licensees exercised dominion and control over the assets of the PC, including the practice's bank accounts; (3) the non-licensees made capital investments in the PC; (4) the funds of the PC were used by the non-licensees for personal rather than corporate purposes; (5) the non-licensees had the ability to bind the PC to legal obligations with third parties; (6) the non-licensees were responsible for the hiring, firing and/or payment of salaries of the PC's employees and whether and the extent to which they dictated policy decisions; (7) the day-to-day corporate formalities were followed by the PC, including the issuance of stock, election of directors, holding of corporate meetings, keeping of contemporaneous corporate books and records and the filing of corporate income tax returns; (8) the PC and non-licensed entities had common office space, address, and telephone numbers; (9) Dr. Carothers played a substantial role in the day-to-day and overall operation and management of the PC; (10) the non-licensees assumed the financial obligations of the PC as if they were their own; (11) the funds of the PC and those of the MSO were commingled; (12) Dr. Carothers and the non-licensees shared the risks, expenses, and interest in the profits and losses of the PC; and (13) the non-licensees played a role in the professional decision-making of the PC.

Case # 2 Allstate Insurance Company v. Northfield Medical Center, P.C., 228 N.J. 596, 159 A.3d 412 (N.J. 2017)

The *Allstate* case stemmed from an insurance company seeking to avoid paying claims to a New Jersey medical practice, Northfield Medical Center, P.C., alleging the practice failed to comply with regulatory requirements regarding the corporate practice of medicine. In this case, a New Jersey chiropractor formed an MSO and a medical practice. The medical practice was owned by a physician, and the MSO was owned by a chiropractor. In reviewing the case, the trial court focused on the "harsh facts" and the subjective intention of the parties. Based upon the facts and the testimony of expert witnesses, the New Jersey Supreme Court affirmed the trial court's decision and determined that the arrangement was structured to make it appear the physician was "in charge" of the medical practice; however, the court found an abundance of proof in the contracts and penalties imposed upon the physician owner that, in reality, the true control of the practice was held by the chiropractor-owned MSO. Specifically, the trial court stated that the model "promoted and assisted in the creation of a practice structure that was designed to circumvent regulatory requirements with respect to control, ownership, and direction of a medical practice."

In reaching its holding, the court identified the following factors when making this decision: (1) the physician owner was not present or involved in the practice; (2)

the MSO used terminator stock option agreements enabling the management company to terminate the physician-owner's stock ownership in the practice; (3) the lease between the MSO and the medical practice did not allow termination by the medical practice (and provided automatic renewal each year unless the management company decided not to renew); (4) the compensation was calculated entirely by the management company; (5) only the MSO could cancel the management services agreement between the MSO and the medical practice; (6) the physician owner of the medical practice had no signature authority over the bank account of the medical practice; (7) the owners of the professional corporation which owned the medical practice did not invest any money or make any capital contributions to the practice; (8) the MSO held a security interest in the practice assets; (9) there were multiple interlocking agreements restricting the medical practice's ability to terminate the relationship; (10) there was a lease breakage fee (not based on fair market value) intended to penalize the physician owner for breaking the lease; and (11) the physician designated as owner of the medical practice was asked to sign an undated resignation letter.

4   Summary

The corporate practice of medicine doctrine is state-specific and each state with a CPOM prohibition has its own laws surrounding ownership and control of medical practices and employment of physicians. Generally, states do not permit unlicensed individuals to exert control over physicians or impact their clinical decision-making. As physicians seek to creatively find ways to address the changing environment of medicine and to strengthen their practice, they must be aware of the applicable laws, including, but not limited to CPOM laws, that may affect how they may structure their practices.

References

1. Am. Med. Ass'n, Principles of Medical Ethics. Available at https://www.ama-assn.org/about/publications-newsletters/ama-principles-medical-ethics.
2. Markenson, A. and Humphreys, A., n.d. *What is … the corporate practice of medicine and fee-splitting?*.
3. Dr. Allison, Dentist, Inc. v. Allison, 360 Ill. 638, 641-642 (Ill. 1935).
4. Cal. Bus. & Prof. Code § 2052(a).
5. N.J. Admin. Code § 13:35-6.16(f).
6. N.Y. Bus. Corp. Law § 1507.
7. N.Y. Educ. Law § 6509-a.

Transparency and the Federal Physician Payments Sunshine Act

Catherine London

Contents

Key Points
- The Physician Payments Sunshine Act (PPSA) requires applicable drug and device manufacturers to disclose any payments and other transfers of value made to clinician and teaching hospital "covered recipients" unless an exception applies.
- The PPSA requires applicable manufacturers and group purchasing organizations (GPOs) to report any physician ownership or investment interests held by physicians and their immediate family members.
- Reported information is made publicly available annually on the Open Payments website for use by patients, institutions, media, and other stakeholders.
- The PPSA's reporting requirements apply only to applicable manufacturers and GPOs, and not to clinicians or teaching hospitals. Providers, however, can review reported payments and transfers of value made in their name and dispute data if they believe it is inaccurate.
- The PPSA preempts state or local laws requiring reporting of the same categories of information. States can, however, impose additional disclosure mandates for information that is not covered by the PPSA, and may also require reporting of

C. London (✉)
London Legal Consulting, LLC, Minneapolis, MN, USA
e-mail: catherine@londonlegalmn.com

© The Author(s), under exclusive license to Springer Nature Switzerland AG 2022   247
A. S. Pasha (ed.), *Laws of Medicine*,
https://doi.org/10.1007/978-3-031-08162-0_15

the same information if the purpose is for public health surveillance, investigation, or public health or health oversight.
- Applicable manufacturers and GPOs can incur civil monetary penalties for violating PPSA reporting requirements.

Legal Concepts
- Applicable Manufacturers: Include (1) entities that operate in the U.S. and are engaged in the production, preparation, propagation, compounding, or conversion of a covered drug, device, biological, or medical supply that is not solely for use by or within the entity itself or by the entity's patients, and (2) entities under common ownership or control with such an entity which provide assistance or support regarding the production, preparation, propagation, compounding, conversion, marketing, promotion, sale or distribution of a covered drug, device, biological, or medical supply.
- Covered Drugs, Devices, Biologicals, or Medical Supplies: Products for which payment is available under Medicare, Medicaid, or the Children's Health Insurance Program, either separately or as part of a bundled payment, and that require a prescription to be dispensed or premarket approval by or notification to the FDA.
- Covered Recipients: Include teaching hospitals, physicians (doctors of medicine and osteopathy, dentists, podiatrists, optometrists, and chiropractors), physician assistants, nurse practitioners, clinical nurse specialists, certified registered nurse anesthetists, and certified nurse midwives who are licensed by the state in which they practice.
- Teaching Hospitals: Institutions that received payment for Medicare direct graduate medical education, indirect medical education (IME), or psychiatric hospital IME programs in the most recent year for which information is available.

1   Introduction

Collaborations between healthcare providers and the pharmaceutical and medical device industries are crucial to the development of new products and technologies that benefit both individual and public health. These collaborations are prevalent and take many forms. In a 2007 national survey of U.S. physicians, 94% of physicians reported having some type of relationship with industry [1]. These relationships included receiving payments for consulting, lecturing, or enrolling patients in clinical trials, accepting food or drug samples from industry representatives, and obtaining reimbursement for continuing medical education costs. Though common, financial relationships between clinicians and pharmaceutical and medical technology companies have come under increasing scrutiny in recent decades, prompting criticism that such relationships undermine clinical objectivity and give rise to real and apparent conflicts of interest [2].

In 2009, the Institute of Medicine published an influential report on conflicts of interest in medical research, education, and practice [3]. The report critically examined the risks posed by relationships between industry and the medical community, explaining that "wide-ranging financial ties to industry may unduly influence professional judgment involving the primary interests and goals of medicine." The report recommended legislative action to promote disclosure of these financial relationships and called on Congress to create a national program to publicly report payments made by industry to health care professionals and healthcare institutions. The Medicare Payment Advisory Commission issued similar recommendations calling for regulatory efforts to address these issues [4].

The federal Physician Payments Sunshine Act (PPSA) emerged out of these and other efforts to increase transparency around the financial relationships between industry and the medical community. Enacted as part of the Patient Protection and Affordable Care Act of 2010, the PPSA requires certain "applicable manufacturers" of drugs, devices, or biological or medical supplies to report on an annual basis to the Centers for Medicare and Medicaid Services (CMS) any payments or other transfers of value made to certain clinicians and teaching hospitals [5]. The PPSA also requires certain manufacturers and group purchasing organizations (GPOs) to report any physician ownership or investment interests held by physicians and their immediate family members. CMS collects and publishes the reported information on the publicly available Open Payments website each year.

The PPSA does not impose any obligations directly on clinicians or teaching hospitals, but it does permit providers to dispute reported information if they believe it is inaccurate. Thus, it is critical for healthcare providers to understand the scope and applicability of the PPSA's requirements. This Chapter provides an overview of the PPSA, describes its applicability to manufacturers and healthcare providers, discusses its impact on providers, identifies key terms and concepts, and applies such concepts to common real-world scenarios.

2 Discussion

Before the federal PPSA was enacted, efforts to regulate interactions between industry and healthcare providers included voluntary industry measures, state "sunshine laws" requiring transparency and public disclosure of industry information, and federal disclosure requirements relating to sponsored research [6].

2.1 Voluntary Industry Efforts

Over the decades, several organizations have issued codes of conduct establishing ethical standards governing provider-industry collaborations. Such organizations include the American Medical Association, the American Association of Medical Colleges, and trade organizations for the pharmaceutical and medical device industries [7, 8].

Notably, in 1993, the Advanced Medical Technology Association (AdvaMed), a trade association representing the medical technology industry, introduced its "Code of Ethics in Interactions with Health Care Professionals." These voluntary guidelines were intended to help medical technology companies structure their interactions with healthcare professionals to minimize risks under federal and state fraud and abuse laws [9].

The Pharmaceutical Research and Manufacturers of America (PhRMA) also published its "Code on Interactions with Health Care Professionals" in 2002, which is a voluntary code of ethics governing pharmaceutical company relationships with healthcare professionals. These industry guidelines articulate standards of conduct intended to prevent "interfere[nce] with the independence of a healthcare professional's prescribing practices" [10].

Although compliance with these industry codes of conduct is not mandatory, many manufacturers submit annual certifications to AdvaMed or PhRMA confirming they have policies and procedures in place to facilitate compliance with these codes.

2.2 State Laws

In the 1990s and early 2000s, several states enacted laws aimed at banning, limiting, or requiring disclosure of certain provider-industry relationships. These laws varied widely in terms of scope and applicability to manufacturers and healthcare providers. California, Connecticut, Colorado, Maine, Massachusetts, Minnesota, Nevada Vermont, West Virginia, and the District of Columbia passed laws requiring disclosure of gifts or payments to healthcare providers and/or imposing restrictions on pharmaceutical and medical device manufacturers making such gifts or payments or engaging in other marketing activities.

States laws in Massachusetts, Minnesota, and Vermont directly prohibited certain types of gifts and payments to healthcare practitioners [11–13]. Other states such as California and Connecticut did not impose direct gift bans, but instead required manufacturers to adopt comprehensive compliance programs in accordance with the U.S. Department of Health and Human Services Office of the Inspector General's compliance program guidance for manufacturers and policies consistent with the PhRMA or AdvaMed Codes (as applicable) [14, 15].

2.3 Federal Transparency Efforts

Federal regulations have long required reporting of medical industry relationships in the clinical research context. The National Institutes of Health (NIH) requires grantees and institutions to manage and disclose significant financial relationships with industry [16, 17]. The Food and Drug Administration (FDA) regulations also require companies seeking product approval by the FDA to disclose certain financial relationships with clinical investigators testing the product in FDA-regulated clinical trials [18]. These federal efforts, however, are limited in scope and extend only to interactions that involve federal funding.

2.4   Federal Physician Payments Sunshine Act (PPSA)

The PPSA built upon previous transparency efforts and established the first comprehensive reporting system with uniform requirements for public reporting of relationships between healthcare providers and industry. The PPSA is broader than previous federal transparency efforts and evolved out of efforts to create accountability and transparency regarding payments made by pharmaceutical and medical device companies to physicians. First introduced in 2007 by Senators Chuck Grassley and Herb Kohl, the PPSA initially failed to pass and was later amended and enacted as section 6002 of the Patient Protection and Affordable Care Act [5]. CMS published the PPSA's implementing regulations on February 8, 2013, following an extensive public comment period [19].

The PPSA was intended to shine a light on payments and other benefits provided by industry to clinicians and health care institutions that could influence prescribing practices and potentially compromise independent medical judgment. Compliance with PPSA reporting requirements does not shield entities or covered recipients from potential liability associated with the reported transactions, including liability under healthcare fraud and abuse laws. However, inclusion on the Open Payments website does not necessarily suggest wrongdoing or illegal conduct. The intent, rather, is to reveal the nature and extent of relationships and discourage inappropriate relationships that could compromise patient care and increase healthcare costs.

In general, the PPSA requires "applicable manufacturers" of drugs, devices, biologicals, and medical supplies to annually report "payments and other transfers of value" made during the reporting year to clinician and teaching hospital "covered recipients." In addition to general payments or transfers of value, the PPSA requires separate reporting of payments or other transfers of value made in connection with research and subject to a protocol or written agreement. It also requires applicable manufacturers and GPOs to report ownership and investment interests held by physicians and their immediate family members.

Applicable Manufacturers

"Applicable manufacturers" are defined as (1) entities that operate in the United States and are engaged in the production, preparation, propagation, compounding, or conversion of a covered drug, device, biological, or medical supply that is not solely for use by or within the entity itself or by the entity's own patients, and (2) entities under common ownership or control with such an entity which provide assistance or support with respect to the production, preparation, propagation, compounding, conversion, marketing, promotion, sale or distribution of a covered drug, device, biological, or medical supply. "Covered drugs, devices, biologicals, or medical supplies" are products for which payment is available under Medicare, Medicaid, or the Children's Health Insurance Program, either separately (such as through a fee schedule or formulary) or as part of a bundled payment, and that require a prescription to be dispensed or premarket approval by or notification to the FDA [20].

Covered Recipients

The PPSA initially defined "covered recipients" to include only physicians and teaching hospitals. Physicians include doctors of medicine and osteopathy, dentists, podiatrists, optometrists, and chiropractors who are licensed by the state in which they practice [21]. In 2018, the SUPPORT for Patients and Communities Act expanded the definition of covered recipients to include certain advanced practice providers (APPs), including physician assistants, nurse practitioners, clinical nurse specialists, certified registered nurse anesthetists, and certified nurse midwives, effective January 1, 2021 [22].

Physicians and APPs must be licensed in the U.S. to be considered covered recipients. A physician or APP holding an active license will be considered a covered recipient regardless of whether the healthcare provider actively treats patients in the state where they are licensed. Accordingly, payments made by an applicable manufacturer to non-practicing physicians or APPs, such as researchers and administrators, are reportable. In addition, payments made by an applicable manufacture to a U.S.-licensed healthcare provider may be reportable even if the payments are for services rendered outside of the U.S. (e.g., speaking at an international seminar). Medical residents and healthcare provider employees of a manufacturer are not considered covered recipients for purposes of the PPSA [23].

"Teaching hospitals" are defined as institutions that received payment for Medicare direct graduate medical education (GME), indirect medical education (IME), or psychiatric hospital IME programs in the most recent year for which information is available [20].CMS maintains a list of teaching hospitals that is available on the CMS website and is updated on an annual basis. Applicable manufacturers and GPOs rely on this list to correctly identify teaching hospitals and determine whether payments made to a particular institution are reportable.

Reportable Payments and Transfers of Value

As a general rule, applicable manufacturers are required to report the following types of payments or transfers of value made to covered recipients, unless an exception applies: (1) direct payments or other transfers of other value by a manufacturer to a covered recipient; (2) indirect payments or transfers of value by a manufacturer to a covered recipient; and (3) payments or other transfers of value provided by a manufacturer to a third party at the request of or designated by the manufacturer on behalf of a covered recipient [24]. The PPSA's implementing regulations broadly define a "payment or other transfer of value" as a "transfer of anything of value" and do not specify how manufacturers should calculate the value of reportable payments or transfers [20].

Most payments or transfers of value reported by applicable manufacturers fall into the category of direct payments or transfers of value, such as royalties, consulting fees, meals, and travel reimbursement. Indirect payments and transfers of value are more difficult to identify and correctly report and are defined as "payments or other transfers of value made by an applicable manufacturer … to a covered recipient … through a third party, where the applicable manufacturer … requires, instructs, directs, or otherwise causes the third party to provide the payment or transfer of value, in whole or in part, to a covered recipient(s)" [20]. For example, if a manufacturer provides a payment to a clinic for consulting services and requests that the services be performed by a particular physician, this would be considered an indirect payment to the physician because the manufacturer intends that the payment ultimately be transmitted, at least in part, to the specific physician.

The third category of reportable payments and transfers of value is intended to capture payments or transfers that are not ultimately made to covered recipients but are either (i) requested by covered recipients, or (ii) designated on their behalf. According to CMS, if a covered recipient directs that a manufacturer provides a payment or other transfer of value to a specific entity or individual, rather than receiving it personally, then the payment is being made "at the request" of such covered recipient and must be reported [25]. For example, if a clinician waives his or her right to receive a payment, and instead requests to donate the payment to a charity, this would be reportable in the clinician's name.

Food and Beverage

Providing meals and refreshments to healthcare professionals is a common industry practice. The PPSA regulations set forth special rules for reporting food and beverage items provided to covered recipients and clarify that manufacturers are not required to report buffet meals, snacks, soft drinks, or coffee made generally available to all participants of large-scale conferences or events. In addition, when determining the cost of food and beverage among covered recipients in a group setting where the cost of each covered recipient's meal is not separately identifiable (such as a platter provided to a group of clinicians), manufacturers calculate the value per person by dividing the total cost of the food or beverage by the total number of individuals who took part in the meal [26]. See Application for examples.

Excluded Payments and Transfers of Value

Although the PPSA reporting requirements cover a wide range of financial relationships between industry and covered recipients, several types of payments and transfers of value are exempt from reporting. Notably, manufacturers are not required to report payments or transfers of value less than $10 unless the aggregate amount of such payments provided to a covered recipient in a calendar year exceeds $100.[1] Other exemptions include the following:

- Product samples intended for patient use and not intended to be sold (including coupons and vouchers);
- Educational materials and items that directly benefit patients or are intended to be used by or with patients, including the manufacturer's education of patients regarding the manufacturer's products;
- The loan of a covered device or device under development, or the provision of a limited quantity of medical supplies, for a short-term trial period not to exceed

---

[1] These threshold dollar amounts are increased annually. The minimum reportable payment amounts for 2022 are $11.64 and $116.35 in the aggregate.

90 cumulative days per calendar year (or a quantity of 90 cumulative days of average daily use per calendar year), to permit evaluation of the product by the covered recipient;

- Items or services provided under a contractual warranty (e.g., service or mainte-nance agreements), including replacement of a product, if the terms of the war-ranty are set forth in the purchase or lease agreement for the product;
- A transfer of value to a clinician when the clinician is a patient, research subject, or participant in data collection for research and is not acting in the clinician's professional capacity;
- Discounts, including rebates;
- In-kind items used for the provision of charity care;
- Dividends or other profit distributions from, or ownership of an investment inter-est in publicly traded security or mutual fund;
- Payments for the provision of healthcare to employees and their families, if the manufacturer is self-insured or directly reimburses for healthcare expenses;
- Transfers to a clinician who is also a non-medical professional (e.g., lawyer), so long as the transfer is solely for the non-medical professional services of the clinician (e.g., legal services);
- Transfer of value to a clinician if the transfer is payment solely for the services of the clinician with respect to an administrative proceeding, legal defense, pros-ecution, or settlement or judgment of a civil or criminal action and arbitra-tion; and
- Transfers made solely in the context of a personal, non-business-related relation-ship [27].

Reported Information

Manufacturers are required to categorize reported payments or transfers of value as one of the following: consulting fees; compensation for services other than consulting, including serving as faculty or as a speaker at an event other than a continuing education program; honoraria; gifts; entertainment; food and bever-age; travel and lodging; education; charitable contributions; debt forgiveness; roy-alties or licenses; current or prospective ownership or investment interest; compensation for serving as faculty or as a speaker for a medical education pro-gram; long term medical supply or device loan; grants; space rental or facility fees; or acquisitions.

For each payment or transfer reported, manufacturers must identify the covered recipient's name and business address and, for clinician recipients, the clinician's specialty, National Provider Identifier (if applicable), and state professional license number(s) and state(s) in which the license is held. Reports must also identify the amount and date of the payment or other transfer of value as well as the form it took (e.g., cash or cash equivalent, in-kind items or services, stock, stock option, or other ownership interest, or dividend, profit, or other return investment). In some cases, manufacturers must also report the name of the drug, device, biological, or medical supply related to the transaction [28].

Reportable Ownership or Investment Interests

Applicable manufacturers and GPOs are also required to report any known ownership or investment interest in the manufacturer or GPO held by a physician (including physician employees) or the physician's immediate family members unless an exception applies. Immediate family members include the following: spouses; natural or adoptive parents, children or siblings; stepparents, stepchildren, stepbrothers or stepsisters; fathers-, mothers-, daughters-, sons-, brothers- or sisters-in-law; grandparents or grandchildren; or spouses of grandparents or grandchildren [20].

"Ownership or investment interests" include the following: (1) stock, stock options (other than those received as compensation, until they are exercised); (2) partnership shares; (3) limited liability company memberships; and (4) loans, bonds, or other financial instruments secured with an entity's property or revenue or a portion thereof. Ownership or investment interests do not include ownership or investment interests in a publicly traded security or mutual fund, an interest that arises from a retirement plan offered by the manufacturer or GPO to a physician (or a member of his/her immediate family) through employment, stock options and convertible securities received as compensation (until exercised or converted to equity, at which time they must be reported), an unsecured loan subordinated to a credit facility, interest in an entity that arises from an employee stock ownership plan, or a titular ownership or investment interest that excludes the ability or right to receive financial benefits thereof (e.g., distribution of profits, dividends, proceeds of sale, or similar returns on investment).

Manufacturers and GPOs reporting ownership and investment interests must indicate whether the interest was held by the physician or an immediate family member, and must identify the dollar amount invested by each physician or immediate family member and the value and terms of each interest.

Research Payments

Recognizing the unique and complex nature of research relationships among clinicians, teaching hospitals, and industry, the PPSA sets forth special reporting requirements for research-related payments, which must be reported separately from other general payments or transfers of value. "Research" is defined as a systematic investigation to develop or contribute to generalized knowledge about public health, including behavioral and social sciences research [20, 29]. To be reportable as a research payment, the payment must be subject to a written agreement, a research protocol, or both between the manufacturer and the entity conducting the research [30].

Covered Recipient Review and Dispute of Reported Data

The PPSA does not impose any reporting obligations directly on clinicians or teaching hospitals, nor does it subject them to penalties if information reported in their name is inaccurate. Nevertheless, clinicians and teaching hospitals have a vested interest in ensuring that information reported about them is indeed accurate. Erroneous reporting can lead to reputational harm and unnecessary scrutiny by the public, media, and even regulators. At the very least, inaccurate reporting can paint a misleading portrait of a clinician's legitimate professional activities.

To help safeguard against inaccurate reporting, the PPSA allows covered recipients to review and dispute information reported in their name before it is made publicly available on the Open Payments website. This process begins after an applicable manufacturer or GPO submits its annual reports to CMS and CMS notifies covered recipients when reported information is ready for review. Covered recipients then have a 45-day period during which they can review data attributed to them for the previous year and initiate a dispute if they believe any reports are inaccurate or incomplete. Covered recipients must be registered in the CMS Identity Management system to review and verify the accuracy of reported information within the Open Payments system.

If a covered recipient disagrees with information reported in their name, they can initiate a dispute, which is sent to the manufacturer or GPO to be resolved directly between the parties. Manufacturers and GPOs must notify CMS of resolved disputes and changes to reported information within 15 days of the end of the 45-day review period to ensure posted information reflects such changes. If the dispute is not resolved and reported during this 60-day review and notification period, CMS will publicly report the manufacturer's or GPO's version of the data, but the data will be marked as disputed. Corrected data for disputes resolved outside of the pre-publication review and dispute period will be published in subsequent years [31].

Some manufacturers and GPOs voluntarily notify each covered recipient before reporting payments or transfers of value in the covered recipient's name. This may take the form of a letter notifying the covered recipient that the manufacturer or GPO is required to report the specific payment or transfer of value identified. This type of notification is not required by the PPSA but can often prevent the need for formal disputes after the information has been reported to CMS.

Penalties for Noncompliance

Under the PPSA, applicable manufacturers and GPOs may be subject to civil money penalties (CMPs) of $1000 to $10,000 for each payment or other transfer of value or ownership or investment interest not reported, up to an annual maximum penalty of $150,000. Knowingly failing to report can result in CMPs of $10,000 to $100,000 for each violation, up to an annual maximum penalty of $1,000,000 [32]. These penalties apply only to entities required to report under the PPSA and not to healthcare providers or teaching hospitals. It is important, however, for healthcare providers to understand that manufacturers have an incentive to over-report if there is any question about whether a specific payment or transfer of value must be reported.

Relationship to State Sunshine Laws

The PPSA preempts any state or local laws requiring reporting of the same type of information concerning payments or other transfers of value made by applicable manufacturers to covered recipients unless the information is being collected by a federal, state, or local government agency for public health surveillance, investigation or other public health purposes or health oversight [33]. However, some states (including Vermont, Massachusetts, and Connecticut) require reporting of information for payments or other transfers of value that is not required to be reported under federal law. Accordingly, healthcare providers need to understand that state laws may impose additional disclosure obligations regarding payments or transfers of value made to providers in these states.

Use of PPSA Information

The primary goal of the PPSA was to increase access to and awareness of industry-provider relationships and allow the public to make better-informed decisions about their care [34]. Patients, however, are not the only audience for data reported through the Open Payments Program. Information in the database has been used by media, researchers, regulators, counsel, and healthcare institutions, among other stakeholders [35]. For example, institutions can use reported data to monitor clinicians' prescribing trends and validate information reported by clinicians in conflict of interest disclosures. The database has also served as a tool for researchers and the media to analyze industry payments and prescribing patterns [36, 37]. In addition, regulators have leveraged the information to assist in identifying and investigating potential violations of healthcare fraud and abuse laws. [38]. Given the PPSA's far-reaching impact, it is important for healthcare providers to understand how reported data are being used not only by patients but also by these secondary audiences.

3   Application

Case Example 1: Food and Beverage for a Physician Group Practice A pharmaceutical sales representative brings a catered lunch costing $165 to a ten-physician group practice located in Massachusetts. Six of the ten physicians and five support staff take part in the meal. Because the cost per participant is $15 ($165 divided among 11 participants) and this exceeds the $10 threshold for reporting under the PPSA, the meal would need to be reported for the six physicians who took part in the meal, but not for the four other physicians who did not participate.

In addition, Massachusetts state law prohibits pharmaceutical and medical device companies and their agents from paying for meals for healthcare practitioners if the meals are offered without an informational presentation made by an agent. Modest meals provided in a healthcare practitioner's office or hospital setting are permitted if they occur in connection with such presentations. Accordingly, the catered lunch would be permissible under state law only if the pharmaceutical sales representative offers the lunch as part of such an informational or educational presentation, and the meal would not need to be reported in Massachusetts.

Case Example 2: Grant Funds Paid to a Specialty Society A Colorado-licensed physician receives a grant award from a professional specialty society that is funded in part by a medical device company. Whether this payment is reportable under the PPSA depends on several factors. First, recall that the PPSA requires reporting of *indirect* payments or other transfers of value provided by an applicable manufacturer to a covered recipient. An indirect payment is a payment or transfer of value made by an applicable manufacturer to a covered recipient through a third party where the manufacturer requires, instructs, directs, or otherwise causes the third party to provide the payment or transfer of value, in whole or in part, to a covered recipient. In this case, if the medical device company gave money to the specialty society earmarked for funding a physician grant, the grant would likely constitute a reportable indirect payment to the physician [39].

On the other hand, perhaps the manufacturer provided an unrestricted donation to the specialty society to use at its discretion, and the specialty society chose to use part of the donation to make grants to physicians. In that case, the manufacturer did not require, instruct, or direct the specialty society to use the donation for grants to physicians and the payment might not constitute a reportable indirect payment.

In addition, an applicable manufacturer is not required to report indirect payments or other transfers of value when the manufacturer does not know the covered recipient's identity. If the medical device company in this scenario did not have actual knowledge of the physician's identity or act in deliberate ignorance or reckless disregard of the information, it would not be required to report the payment. But if the company allowed the specialty society to use its name in a grant award, the company would likely be able to determine the grant recipient's identity and could be deemed to be acting with deliberate ignorance if it did not request this information from the specialty society.

Case Example 3: Research Payments to Academic Medical Center A medical device manufacturer contracts with an academic medical center to fund a clinical trial conducted at the academic medical center to evaluate the manufacturer's FDA-approved cardiac implant device. Dr. Barbara Evans is a cardiologist at the institution who will serve as the trial's principal investigator. Her colleagues, Dr. Chuang and Dr. Roberts, will serve as co-investigators.

The clinical trial funding paid to the academic medical center by the manufacturer would likely be reportable as a research-related payment. To be reportable as a research payment, the payment must be subject to a written agreement, a research protocol, or both between the manufacturer and the institution conducting the research. In this case, there is a "contract" between the academic medical center and the manufacturer and there is likely a study protocol as well. Because this relationship involves payment for a research-related activity, the manufacturer would be required to report the payment separately from other general payments. The report would need to include the name of the institution and information about each principal investigator covered recipient. Sub-investigators such as Dr. Chuang and Dr. Roberts are not required to be identified in the report. The manufacturer would need to disclose the total amount of the research payment, including all research-related costs for activities outlined in the agreement or protocol, the name of the research study, and any covered drugs, devices, biological or medical supplies related to the study.

4   Summary

Transparency has become a preferred regulatory approach in healthcare to promote accountability, facilitate patient decision-making, and discourage inappropriate behavior. The PPSA was the first large-scale federal transparency initiative to cast a light on provider-industry financial relationships. Although the PPSA does not impose obligations directly on healthcare providers, it is critical for providers to understand the scope of the law and how it applies to their professional relationships. Healthcare providers should be aware of publicly reported industry payments and ensure the accuracy of any information reported in their name.

Acknowledgments   None.

References

1. Campbell EG, Gruen RL, Mountford J, Miller LG, Cleary PD, Blumenthal D. A national survey of physician-industry relationships. N Engl J Med. 2007 Apr 26;356(19):1742-50.
2. Brennan TA, Rothman DJ, Blank L, et al. Health industry practices that create conflicts of interest: a policy proposal for academic medical centers. JAMA 2006;295:429-433.
3. Institute of Medicine (US) Committee on Conflict of Interest in Medical Research, Education, and Practice; Lo B, Field MJ, editors. Conflict of interest in medical research, education, and practice. Washington (DC): National Academies Press (US); 2009.
4. Medicare Payment Advisory Commission. Report to the Congress: Medicare payment policy. Washington (DC): MedPAC; 2009. Chapter 5, Public reporting of physicians' financial relationships; p. 313-43.
5. Patient Protection and Affordable Care Act, Pub. L. No. 111-148, § 6002, 124 Stat. 119, 689 (2010) (codified as amended at 42 U.S.C. § 1320(a)).
6. Gorlach I, Pham-Kanter G. Brightening up: the effect of the Physician Payment Sunshine Act on existing regulation of pharmaceutical marketing. J Law Med Ethics. 2013;41(1):315-22.
7. American Medical Association, Council on Ethical and Judicial Affairs. Opinion 8.061 - Gifts to physicians from industry. 1992.
8. Association of American Medical Colleges. Industry funding of medical education, report of the AAMC Task Force. Association of American Medical Colleges; Washington (DC); 2008.
9. Advanced Medical Technology Association. Code of ethics on interactions with health care professionals [Internet]. 2021. Available from: https://www.advamed.org/wp-content/uploads/2021/05/AdvaMed-Code-of-Ethics-2021.pdf.
10. Pharmaceutical Research and Manufacturers of America. Code on Interactions with healthcare professionals. 2021. Available from: https://phrma.org/-/media/Project/PhRMA/PhRMA-Org/PhRMA-Org/PDF/A-C/Code-of-Interaction_FINAL21.pdf.
11. 105 Mass. Code Regs. §§ 970.000, et seq.
12. Minn. Stat. Ann. § 151.461.
13. Vt. Stat. Ann. tit. 18, § 4631a.

14. Cal. Health & Safety Code § 119402.
15. Conn. Gen. Stat. § 21a-70e.
16. 42 C.F.R. §§ 50.601-605.
17. 45 C.F.R. § 94.1-5.
18. 21 C.F.R. pt. 54.
19. 78 Fed. Reg. 9458-9528 (Feb. 8, 2013).
20. 42 C.F.R. § 403.902.
21. 42 U.S.C. § 1320a–7h(e)(11).
22. Substance Use-Disorder Prevention that Promotes Opioid Recovery and Treatment for Patients and Communities Act, Pub. L. No. 115-271.
23. 42 U.S.C. § 1320a–7h(e)(6)(B).
24. 42 C.F.R. § 403.904(a).
25. 78 Fed. Reg. 9472 (Feb. 8, 2013).
26. 42 C.F.R. § 403.904(g).
27. 42 C.F.R. § 403.904(h).
28. 42 C.F.R. § 403.904(c).
29. 42 C.F.R. §§ 403.904(f).
30. 42 C.F.R. § 403.904(f)(1).
31. 42 C.F.R. § 403.908(g).
32. 42 C.F.R. § 403.912.
33. 42 C.F.R. § 403.914.
34. Pham-Kanter G, Mello MM, Lehmann LS, Campbell EG, Carpenter D. Public awareness of and contact with physicians who receive industry payments: a national survey. J Gen Intern Med. 2017;32(7):767-74.
35. Saver RS. Deciphering the Sunshine Act: transparency regulation and financial conflicts in health care. Am J Law Med. 2017;43(4):303-343.
36. Ornstein C, Tigas M, Grochowski Jones R. Now there's proof: docs who get company cash tend to prescribe more brand-name meds. ProPublica [Internet] 2016 Mar 17. Available from: https://www.propublica.org/article/doctors-who-take-company-cash-tend-to-prescribe-more-brand-name-drugs.
37. Rathi VK, Samuel AM, Mehra S. Industry ties in otolaryngology: initial insights from the physician payment sunshine act. Otolaryngol Head Neck Surg. 2015;152(6):993-999.
38. In 2020, the DOJ announced the first enforcement action involving alleged non-compliance with the PPSA. Medical device manufacturer Medtronic USA Inc. agreed to pay $1.11 million to resolve these allegations in addition to $8.1 million to resolve claims related to alleged Anti-Kickback Statute and False Claims Act violations. Department of Justice. Medtronic to pay over $9.2 million to settle allegations of improper payments to South Dakota neurosurgeon. Justice News [Internet] 2020 Oct 29. Available from: https://www.justice.gov/opa/pr/medtronic-pay-over-92-million-settle-allegations-improper-payments-south-dakota-neurosurgeon.
39. 78 Fed. Reg. 9490 (Feb. 8, 2013).

Labor and Employment Issues for Clinicians

Anthony Rizzotti and Lauren DiGiovine

Contents

Key Points
- Non-compete agreements must be reviewed under applicable state law.
- State law with respect to non-compete agreements can vary greatly from state to state.
- The NLRA applies almost exclusively to issues like an employee's right to engage in protected concerted activity, such as the right to join or not join a union.
- There are numerous federal laws prohibiting discrimination in the workplace, including Title VII of the Civil Rights Act, the Age Discrimination in Employment Act, and the Americans with Disabilities Act.
- Almost every state has its own laws and regulations prohibiting discrimination in the workplace.
- Federal and state laws prohibit discrimination based upon a protected classification which includes, among others, age, gender, disability, sexual orientation, race and national origin.

A. Rizzotti (✉) · L. DiGiovine
Littler, Mendelson P.C., Boston, MA, USA
e-mail: arizzotti@littler.com; ldigiovine@littler.com

© The Author(s), under exclusive license to Springer Nature Switzerland AG 2022   261
A. S. Pasha (ed.), *Laws of Medicine*,
https://doi.org/10.1007/978-3-031-08162-0_16

- Misclassifying an employee as an independent contractor is common, and states are increasingly scrutinizing employers' treatment of workers as independent contractors.

Legal Concepts
- In many states, restrictive covenants are enforceable to the extent that the restrictions are reasonable in time and scope; however, some states prohibit restrictive covenants for physicians as contrary to public policy.
- It is unlawful to discriminate against an employee in the terms and conditions of employment (*e.g.*, hiring, promotion, compensation, discharge) because of a protected classification.
- While there are various tests to determine whether a worker is an employee or an independent contractor, the determination often comes down to the extent of control exercised by the employer over the worker.

1   Introduction

There are myriad federal, state, and local laws that impact the healthcare workplace. These laws touch upon virtually every aspect of the employment relationship. This chapter is intended to provide clinicians with an overview of key labor and employment issues relevant to providers in the healthcare industry. Topics include covenants against competition (non-compete agreements), traditional labor law issues (such as employees' right to join a labor union), employment discrimination, and misclassification of a worker as an independent contractor under such laws as the Fair Labor Standards Act (FLSA).

There is one important distinction between employment law and traditional labor law worth mentioning. Labor law is almost exclusively controlled by federal law, specifically the National Labor Relations Act (NLRA). In contrast, employment law—most notably for purposes of this chapter, laws prohibiting discrimination in the workplace—requires a review of federal, state, and, in some cases, local laws and regulations.

This publication is not intended to be an in-depth legal analysis of these issues, but rather a guide to some of the most common and challenging issues that may arise for providers.[1]

2   Restrictive Covenants

2.1   Discussion

The existing landscape on this topic could change significantly as President Biden has asked the Federal Trade Commission to pursue rulemaking that would limit the reach of non-competition agreements [1]. On July 9, 2021, President Biden signed

---

[1] The authors would like to thank Littler Mendelson, PC and its many talented lawyers and their resources, which assisted in the drafting of this chapter.

the Executive Order on Promoting Competition in the American Economy ("Order") [1]. The Order is directed at "the promotion of competition and innovation by firms small and large, at home and worldwide" [1]. Notably, the Order directs the Federal Trade Commission (FTC) to evaluate enacting rules to "curtail the unfair use of non-compete clauses and other agreements that may unfairly limit worker mobility" [1]. At this time, it is unclear what steps, if any, the FTC will take in response to this Order.

The laws governing non-competes varies considerably from state to state, and therefore, the initial question is which state's law controls. For example, in Texas, employer non-compete agreements are enforceable if ancillary to an otherwise enforceable agreement and if they contain "limitations as to time, geographical area, and scope of activity to be restrained that are reasonable and do not impose a greater restraint than is necessary to protect the goodwill or other business interest" of the employer [2]. In California, on the other hand, non-competition agreements are unenforceable as contrary to public policy. Specifically, California law provides that "[E]very contract by which anyone is restrained from engaging in a lawful profession, trade or business of any kind is to that extent void" [3].

Moreover, even in states that otherwise recognize the enforceability of non-competition agreements, certain professions—like physicians—cannot be subject to such agreements or have additional legal requirements to be valid. For example, in Massachusetts, a state that otherwise recognizes the enforceability of non-competition agreements consistent with a 2018 statute, physicians cannot be subject to such an agreement [4]. In Texas, physician non-competition agreements are enforceable but must meet additional requirements intended to protect the interests of patients, such as permitting the physician, despite the agreement, to continue to treat patients with acute illnesses [5].

Consequently, when considering use of a non-competition agreement, the potential provider should first determine which state's law applies. Often, agreements will contain a so-called choice-of-law provision, which identifies the controlling law. Even then, a choice-of-law provision cannot be used to skirt the public policy of a state with jurisdiction over the matter. For instance, a California employer cannot choose the law of Texas in its agreements to get around California law.

Once the applicable law is determined, the contract then must be reviewed for compliance. Many states that recognize non-competition agreements have provisions like that of Texas. The agreement is enforceable if it is narrowly drawn and necessary to protect the employer's legitimate business interests. Thus, courts will examine the length of the non-compete restriction (*e.g.*, 6 months, 1 year), the geographic reach of the restriction, and whether the agreement is necessary to protect the employer's trade secrets, confidential information or goodwill. In many states, judges are granted wide discretion to reform the contract's duration and scope. These cases can often turn on how the judge assesses those factors. Below is a discussion of two cases that illustrate this analysis.

2.2   Application

Case 1: *Valley Med. Specialists v. Farber*, **194 Ariz. 363, 982 P.2d 1277 (1999).** In this case, the Supreme Court of Arizona rejected a professional corporation's restrictive covenant in its shareholder/employment agreement. In 1985, Dr. Farber started working for Valley Medical Specialists (VMS). In 1991, Dr. Farber, an internist and pulmonologist, treated AIDS and HIV-positive patients and performed brachytherapy—a procedure that irradiates the inside of the lung in lung cancer patients. Dr. Farber eventually became a shareholder, minority officer, and director. As a result, he entered into new stock and employment agreements that contained restrictive covenants. Dr. Farber left VMS in 1994 and started working in an area covered by the restrictive covenants.

The plaintiff, VMS, sued Dr. Farber seeking, among other things, damages and to enforce the restrictive covenant. On appeal, the Arizona Supreme Court held: (1) covenants not to compete between physicians will be strictly construed for reasonableness; and (2) the restrictive covenant at issue was unreasonable and unenforceable. In arriving at this conclusion, the court noted, "the doctor-patient relationship is special and entitled to unique protection. It cannot be easily or accurately compared to relationships in the commercial context. Considering the great public policy interest involved in covenants not to compete between physicians, each agreement will be strictly construed for reasonableness" [6].

Case 2: *Total Health Physicians, S.C. v. Barrientos*, **151 Ill. App. 3d 726, 502 N.E.2d 1240 (1986).** In this case a medical corporation sued two physicians, former employees of the medical corporation, whose employment with the medical corporation had been terminated. In 1984, the physicians entered into agreements with their employer which, in exchange for compensation, required that they "…were not to practice medicine on their own, nor were they to practice medicine for any of plaintiff's competitors (*i.e.*, any physician or group of physicians not affiliated with plaintiff) without the express consent…" of their employer [7]. The circuit court ruled in favor of the physicians, holding that the restrictive covenants were void and unenforceable because they placed an unreasonable geographic restriction upon the physicians' practice of medicine. The medical corporation appealed.

The appellate court reversed the circuit court's decision and held the restrictive covenants *were not* void and unenforceable because at the time, Illinois law permitted the covenants to be limited to an area that was reasonable to protect the interests of the employer.

The appellate court noted that it was required to determine whether the covenants not to compete were "void and unenforceable because they placed unreasonable geographic restrictions upon each defendant's practice of medicine" [7]. The appellate court stated that it was irrelevant whether the area *defined* in or by the restrictive covenant is unreasonable or vague, because the parties or the court may limit the restrictive covenant to an area that *is* reasonable to protect the interests of the employer and comply with Illinois law [7].

In evaluating enforceability, the court relied on four factors:

(1) whether the contract is supported by valuable consideration; (2) whether the restraint imposed is limited or partial; (3) whether the restraint is greater than is necessary to protect the promisee; and (4) whether enforcement of the contract would be injurious to the public or cause undue hardship to the promisor [7].

Ultimately, as noted above, the court determined that the restrictive covenant could be valid and enforceable if the geographic restrictions were reframed.

These two cases are emblematic of the divergent views on restrictive covenant agreements in the United States, particularly as applied to clinicians. In the first case, the court held that the restrictive covenant was unreasonably restrictive. Public policy was a factor, particularly where the clinician was a specialist during the AIDS crisis. In the second case, the court was more deferential, citing the parties' ability to reframe the agreement so the geographic parameters could be reasonable.

3   Discrimination

3.1   Discussion

Title VII of the Civil Rights Act of 1964, as amended, prohibits employers from discriminating against an individual on the basis of race, color, national origin, sex, sexual orientation, gender identity, or religion with respect to hiring, discharge, compensation, promotion, or other terms, conditions and privileges of employment [8]. Since that landmark piece of federal law was enacted, numerous federal laws have been ratified to prohibit discrimination based upon other protected classifications including the Americans with Disabilities Act [9] (prohibiting discrimination based on disability), the Age Discrimination in Employment Act [10] (prohibiting discrimination based on age), and the Equal Pay Act [11] (prohibiting sex discrimination with respect to compensation). In addition, almost every state has enacted laws that similarly prohibit discrimination, and in some cases, cities have adopted laws of their own.

All these laws have the same basic premise. An employer cannot discriminate against people because they are members of a protected class. In addition, these laws have been statutorily defined and interpreted broadly by federal and state courts to prohibit retaliation and harassment based upon a protected classification. Moreover, state and federal courts have construed these laws to include both obvious and ambiguous discrimination. For example, it would be a plain violation to refuse to hire or discharge an employee because of race.

Less apparent employer policies such as dress and grooming codes have also been litigated frequently and, in some cases, found to be discriminatory [12, 13]. For example, in E.E.O.C. v. Abercrombie & Fitch Stores, Inc., 575 U.S. 768 (2015), a Muslim job applicant wore a headscarf to a job interview at an Abercrombie store. Defendant Abercrombie denied her employment after her interviewer assumed the scarf was worn for religious reasons and that hiring her would create a conflict, even though the applicant did not raise the issue in the interview [12]. The Court rejected

Abercrombie's argument that actual knowledge of the need for an accommodation was necessary, holding "an applicant need only show that his need for accommodation was a motivating factor in the employer's decision [12]. A less recent case was *Hollins v. Atlantic Co.*, 188 F.3d 652 (6th Cir. 1999), in which the court denied summary judgment and held that Defendant's grooming policy, which Plaintiff claimed was not applied to white women, would have been pretext for its treatment of racial minorities [13].

Legislation has been introduced to address this type of discrimination. For example, as recently as March 2021, S.888, the Creating a Respectful and Open World for Natural Hair (CROWN) Act of 2021, was introduced in the U.S. Senate [14]. The bill recognizes "individuals who have hair texture or wear a hairstyle that is historically and contemporarily associated with African Americans or persons of African descent systematically suffer harmful discrimination in schools, workplaces, and other contexts [14]." On that basis, the bill would prohibit this type of discrimination against those participating in federally assisted programs, housing programs, public accommodations, and employment [14].

At the federal level, these laws are enforced by the Equal Employment Opportunity Commission (EEOC), and many states have a similar administrative body. Ultimately, however, many cases end up before the court.

Of interest for healthcare providers, courts have consistently found that an employer's obligation to provide a discrimination-free workplace takes precedence over a patient's racial or gender preferences. While courts have ruled that employers cannot discriminate based on patient preference relating to race or national origin, gender preference has been open to more interpretation. A healthcare employer can honor a patient's request to not have an opposite-sex caregiver assisting with care without violating antidiscrimination employment laws, but only as to care that involves issues of intimate personal privacy, such toileting or examination of private areas [15]. There must be a specific patient request related to personal privacy, rather than a blanket policy of exclusion.

3.2   Application

Case 1: *Chaney v. Plainfield Health Care Ctr.*, **612 F.3d 908 (7th Cir. 2010).** In this case, the Seventh Circuit rejected a nursing home policy honoring the racial preferences of its residents when assigning care providers. Ms. Chaney, an African American nurse assistant, was given written instruction that a resident in her assigned unit "Prefers No Black CNAs." It was undisputed that such policy existed, and Chaney went along with the policy because she feared that she would be fired if not. In several instances, Chaney had to refrain from helping a patient, even though she was best suited to do so in a timely way and had to find a white CNA to help the patient.

This case pitted the requirements of Title VII against the racial preferences of patients in a healthcare setting. The court held that catering to the racial preferences of residents is an insufficient justification for otherwise violating Title VII protections against disparate treatment. As a result, it struck down the policy and found

Plainfield liable. In so doing, the court recognized that certain privacy interests of patients, such as preferring same-sex health providers, did not violate the law. However, it made clear that there was no basis under the law for race-based preferences to be protected.

Case 2: *Young v. United Parcel Service*, **135 S. Ct. 1338 (2015).** In this case, the United States Supreme Court addressed the question of whether employers must provide light duty and other workplace accommodations to pregnant employees in the same manner they provide accommodations to employees who are injured on the job. This issue is relevant to a healthcare workplace because clinicians may be required to lift heavy equipment and patients. The Supreme Court has yet to answer this very specific question. However, the Supreme Court *has* provided general guidelines for pregnant employees navigating challenging workplace accommodation policies and practices under Title VII, as amended by the Pregnancy Discrimination Act (PDA) [16].

In *Young v. United Postal Service*, the Supreme Court evaluated the requirements for bringing a disparate impact claim under the PDA. Here, the plaintiff worked as a part-time delivery driver. As an essential function of the job, all drivers were required to lift items weighing up to 70 pounds. After undergoing in vitro fertilization, the plaintiff became pregnant; she then requested a leave of absence. Upon her return, she informed her employer that, at the advice of her physician, she could not lift more than 20 pounds. Accordingly, Plaintiff requested an accommodation to work light duty.

The company denied the plaintiff's return to work because the ability to lift more than 20 pounds was an essential job function. UPS argued that light duty was not available to plaintiff because she did not fall into any of the categories prescribed by the company that would entitle her to light duty (i.e., on-the-job, ADA, DOT). As a result, she remained on an unpaid leave of absence during the term of her pregnancy.

The plaintiff sued her employer. She argued that employers were required under the PDA to provide pregnant employees with light duty work if the employer offers, or had previously offered, light duty work to other employees as an accommodation. The Court sided with the plaintiff. Justice Breyer, in his majority opinion, held that the central inquiry was whether the disparate treatment imposed upon pregnant employees gave rise to an inference of intentional discrimination [17]. In so holding, the Court described the test for determining if an employer's policy gave rise to an inference of intentional discrimination:

> Thus, a plaintiff alleging that the denial of an accommodation constituted disparate treatment under the Pregnancy Discrimination Act … may make out a prima facie case by showing, as in *McDonnell Douglas*, that she belongs to the protected class, that she sought accommodation, that the employer did not accommodate her, and that the employer did accommodate others 'similar in their ability or inability to work.'
>
> The employer may then seek to justify its refusal to accommodate the plaintiff by relying on 'legitimate, nondiscriminatory' reasons for denying her accommodation. …But, consistent with the Act's basic objective, that reason normally cannot consist simply of a claim that it is more expensive or less convenient to add pregnant women to the category of those … whom the employer accommodates. [17]

In remanding to the lower court for further review, Justice Breyer noted that Plaintiff's employer maintained three separate and distinct policies for accommodating disabilities [17]. When these policies were viewed together, it created a genuine dispute as to whether some employees were treated less favorably than others under similar circumstances, and whether there is an undue hardship on pregnant employees [17].

In the wake of *Young* the EEOC sued a medical services provider. The EEOC's lawsuit alleged the employer violated Title VII of the Civil Rights Act of 1964, as amended by the PDA, by failing to accommodate a pregnant employee's medical restrictions and instead placing her on leave [18]. The EEOC's lawsuit alleged the pregnant employee applied for two vacant desk positions that would have permitted her to work, notwithstanding her pregnancy-related medical restrictions; however, the hospital denied her requests and eventually terminated her employment [18]. The EEOC's complaint asserted that the pregnant employee's termination was inconsistent with the hospital's policies and practice toward similarly situated non-pregnant employees with medical restrictions who were frequently accommodated with light-duty work.

The lawsuit settled on or about August 29, 2019, when the EEOC advised employers "woman should not have to choose between her pregnancy and her job. Employers should not refuse to accommodate pregnant workers based on considerations of cost or convenience when they accommodate other workers who are similar in their ability to work" [19].

4 Labor Law

4.1 Discussion

In most industries, organized labor has been on the decline for decades. In the healthcare industry, that is not the case. Clinicians remain very much a focus of union organizing efforts.

Section 7 of the NLRA protects the right of employees to choose to join or not join a union. The National Labor Relations Board (NLRB) is the federal agency charged with enforcing the NLRA. While a full discussion of union organizing and the NLRA is beyond the scope of this chapter, there are some issues of interest worth highlighting for the healthcare industry.

For example, as they do for other employers, unions have restricted access to the employees of hospitals and healthcare facilities. Although the U.S. Supreme Court has held "[n]o restriction may be placed on the employees' right to discuss self-organization among themselves, unless the employer can demonstrate that a restriction is necessary to maintain production or discipline[,]" it observed that "no such obligation is owed non-employee organizers" [20].

Put more generally, the Supreme Court has held that an employer may bar non-employee union organizers if the employer's notice or order does not discriminate against the union (e.g., allowing other outside third parties access to the company's premises) [20]. This is because the NLRA only requires employers to refrain from interfering with or discriminating against employees in the exercise of their rights [20].

The rules related to employee solicitation are far less stringent. The cases discussed below illustrate some of the relevant issues.

4.2 Application

Case 1: *Baptist Medical Systems v. NLRB*, **876 F.2d 661 (8th Cir. 1989).** Here, the hospital denied union representatives access to its cafeteria for the purpose of union organizing. The NLRB objected because the cafeteria was located on the ground floor and open to employees, patients, and the public. The hospital argued that it maintained and imposed rules that prohibited visitors, patients, and other non-employees from soliciting or distributing literature on any hospital property for any purpose at any time [21]. The hospital also maintained and imposed rules that prohibited employees from soliciting or distributing materials during work time and prohibited all solicitation and distribution in patient care areas [21].

The court held the hospital was permitted to prohibit union organizing activity in the cafeteria because "an employer does not have an affirmative duty to allow the use of its facilities by nonemployees for organizational purposes" [21]. The court further held that "inviting the public to use an area of its property… does not surrender its right to control the uses to which that area is put" because union organizing activity was not associated with the normal use of the cafeteria [21]. Finally, the court noted that such union activity "could be particularly disturbing in a hospital setting" [21].

This case is relevant to clinicians who *are* unionized or are subject to *becoming* unionized. Medical centers, hospitals, and other "like" employers have a right to restrict union access to their premises. However, those employers generally cannot exercise that right while permitting other nonemployee organizations unfettered access, especially the right to solicit on their premises. There are two exceptions to this general rule [22]:

- *Beneficent Acts Exception:* The employer may permit discrete, isolated "beneficent" acts by outside organizations [22]. These would be for a charitable purpose [22]. If the solicitations are commercial instead of charitable, and/or occur on a regular instead of sporadic basis, then the solicitations would not qualify for the "benevolent acts" exception, and the employer could not deny union access predicated on its "no solicitation" policy [22].
- *Business Function Exception* The employer may permit solicitations that relate to its business functions [22]. Examples include blood drives, pharmaceutical sales, and fundraising [22]; fringe benefits offered to employees (e.g., tax-sheltered annuity plans, health insurance plans); and medical textbooks for clinicians [23].

What solicitations have *not* been permitted? Those solicitations that are neither beneficent, nor reasonably related to hospital operations. Examples include solicitations from a credit union; distributions and referrals regarding family-care resources from a child and family services organization; and solicitations for flowers and jewelry [23].

Case 2: *UPMC, 362 NLRB No. 191 (Aug. 27, 2015).* In this case, the NLRB determined that a group of Pennsylvania hospitals maintained a nonsolicitation policy that violated the NLRA [24]. The language at issue in the solicitation policy read as follows:

> No staff member may distribute any form of literature that is not related to UPMC business or staff duties at any time in any work, patient care, or treatment areas. Additionally, staff members may not use UPMC electronic messaging systems to engage in solicitation (see also Policy HS-IO147 Electronic Mail and Messaging).
> …
> All situations of unauthorized solicitation or distribution must be immediately reported to a supervisor or department director and the Human Resources Department and may subject the staff member to corrective action up to and including discharge [24].

The Board took issue with the second provision requiring employees to immediately report instances of unauthorized solicitation, as the policy "defines 'unauthorized solicitation' to include solicitation protected by Section 7" [24]. Therefore, the rule "reasonably tends to chill employees in the exercise of their Section 7 rights" [24].

The hospital had argued that special circumstances related to patient safety justified the standard. Specifically, the hospital cited studies finding a correlation between employee distraction and patient safety and identifying computers and other electronic communication devices as sources of distraction. The Board, however, deemed these reasons insufficient.

5   Independent Contractors

5.1   Discussion

In most industries, healthcare included, the treatment of independent contractors is rapidly evolving. An employer's failure to properly classify an individual as an employee or independent contractor may have serious repercussions with both state and federal administrative bodies (*e.g.*, Internal Revenue Service). Misclassification may mean years of backpay, workers' compensation, insurance premiums, or back taxes.

This issue is critically important in healthcare because physicians are frequently treated as independent contractors [25]. Thus, in medical malpractice matters, misclassification impacts how and whether that physician is indemnified by the employer [25]. The legal landscape is also quite complicated. There is no "silver bullet" assessment. Several tests are used by government agencies and accepted by courts to classify workers. There are three primary tests to determine whether an individual qualifies as an employee or independent contractor: (1) the right-to-control/common law test, (2) the economic realities test, and (3) the ABC test.

The "right to control" or common law test has been adopted by several agencies and courts, including the Internal Revenue Service (IRS). The test is concerned with an organization's (1) behavioral control, (2) financial control, and (3) the type of relationship with the worker [26]. This formula was affirmed by the Supreme Court in *Nationwide Mutual Ins. Co. v. Darden*, 503 U.S. 318 (1992), which relied on common law principles of agency [27]. There, the Court held:

> In determining whether a hired party is an employee under the general common law of agency, we consider the hiring party's right to control the manner and means by which the product is accomplished. Among the other factors relevant to this inquiry are the skill required; the source of the instrumentalities and tools; the location of the work; the duration of the relationship between the parties; whether the hiring party has the right to assign additional projects to the hired party; the extent of the hired party's discretion over when and how long to work; the method of payment; the hired party's role in hiring and paying assistants; whether the work is part of the regular business of the hiring party; whether the hiring party is in business; the provision of employee benefits; and the tax treatment of the hired party. [27]

The economic realities test has been adopted by the U.S. Department of Labor (DOL) and the Occupational Safety and Health Administration (OSHA). The test is derived from the *United States v. Silk*, 331 U.S. 704 (1947), *abrogated in part by* 503 U.S. 318 (1992), and *Rutherford Food Corp. v. McComb*, 331 U.S. 722 (1947), to determine whether a worker is an independent contractor or employee for purposes of the FLSA. These factors have been memorialized by the DOL's Wage and Hour Division [28]:

1. The extent to which the services rendered are an integral part of the principal's business.
2. The permanency of the relationship.
3. The amount of the alleged contractor's investment in facilities and equipment.
4. The nature and degree of control by the principal.
5. The alleged contractor's opportunities for profit and loss.
6. The amount of initiative, judgment, or foresight in open market competition with others required for the success of the claimed independent contractor.
7. The degree of independent business organization and operation.

A third way of approaching this issue is the "ABC" test. The ABC test is used in and, to a lesser extent, has been memorialized into statute by more than thirty states, including healthcare hubs California and Massachusetts. In order to be considered an independent contractor, a worker must meet three criteria (some states require only two criteria to be met):

- The worker is free from the control and direction of the hiring entity in connection with the work's performance, both under the contract for the performance of the work and in fact.
- The worker performs work that is outside the usual course of the hiring entity's business.

- The worker is customarily engaged in an independently established trade, occupation, or business of the same nature as the work performed [29].

California's AB5—an example of the codified ABC test—has an explicit exemption for physicians [30]. However, this exemption is for *physicians only*—no other clinicians. Healthcare companies still must use the ABC test to determine if other clinicians (*e.g.*, physician assistants, nurse practitioners, Certified Registered Nurse Anesthetists) are properly classified. Failure to meet any of the three criteria means that the proposed independent contractor must be classified as an employee.

Thus, clinician-employers in California must be vigilant to ensure that they do not misclassify any would-be employees as independent contractors.

5.2 Application

Case 1: *Cilecek v. Inova Health Sys. Servs.*, **115 F.3d 256 (4th Cir. 1997).** In this case, the plaintiff, Dr. James W. Cilecek, a physician under contract to provide emergency medical services at hospitals, brought a Title VII lawsuit against his employers. The issue on appeal was whether Dr. Cilecek had standing to bring a Title VII claim because, the employer argued, he was an independent contractor rather than an employee, and as such, was not covered by Title VII.

In the instant case, the defendant owned and operated several health care facilities. In March 1989, it contracted with Emergency Physicians of Northern Virginia, Ltd. ("Emergency Physicians"), under which Emergency Physicians agreed to staff Fairfax Hospital and ACCESS of Reston with emergency physicians. Simultaneously, Dr. Cilecek, an existing Fairfax Hospital emergency medicine physician, contacted the CEO of Emergency Physicians to negotiate the terms and conditions of his independent contractor arrangement with the organization [31].

The court applied the control, or common law, test to the context of healthcare providers, to determine whether this clinician was an employee or an independent contractor. The court considered:

> (1) the control of when the doctor works, how many hours he works, and the administrative details incident to his work; (2) the source of instrumentalities of the doctor's work; (3) the duration of the relationship between the parties; (4) whether the hiring party has the right to assign additional work to the doctor or to preclude the doctor from working at other facilities or for competitors; (5) the method of payment; (6) the doctor's role in hiring and paying assistants; (7) whether the work is part of the regular business of the hiring party and how it is customarily discharged; (8) the provision of pension benefits and other employee benefits; (9) the tax treatment of the doctor's income; and (10) whether the parties believe they have created an employment relationship or an independent contractor relationship [31].

In applying these factors to Dr. Cilecek and Emergency Physicians, the court concluded that his relationship with the defendant and Emergency Physicians was incompatible with that of an employee and that he was not entitled to sue under Title VII.

Case 2: *Robb v. United States*, **80 F.3d 884 (4th Cir. 1996).** In this case, plaintiff-appellant John Robb, a patient who had been treated at an Air Force base hospital by physicians who had contracted with the Air Force to provide services brought an action under the Federal Tort Claims Act (FTCA) based on alleged negligence of those physicians in failing to diagnose a cancerous lesion on his lung. He then appealed the lower court's decision to dismiss the FTCA claim against the United States, arguing that the United States government was liable for the alleged negligence of the two physicians.

In reviewing the appeal, the court of appeals held that the physicians who were employed by third parties (a health care provider and a radiologist) who contracted to provide service at base were "independent contractors" and came within the independent contractor exception to FTCA's waiver of sovereign immunity. In arriving at that decision, the court relied on the control test. For example, the court noted that the physicians exercised independent medical judgment, established "stand alone clinics," and were immune from such legislation as the Gonzales Act, 10 U.S. Code § 1089 (the federal law delineating physician liability in military medical malpractice cases), to which they would be subject if they were government employees [32].

6   Summary

In summary, there are many employment and labor laws affecting clinicians in the workplace. This chapter only provides a high-level survey of restrictive covenants, labor laws, discrimination, and independent contractor relationships. Clinicians operating in an ever-evolving legal landscape must be vigilant, and should be aware that (1) state law with respect to non-competes varies greatly from state to state, (2) there are numerous federal and state laws protecting employees from discrimination in the workplace, (3) physicians are generally treated as independent contractors, but this is not the case for all clinicians, and (4) the NLRA applies almost exclusively to an employee's right to engage in protected concerted activity, such as the right to join or not join a union.

References

1. *See* Exec. Order No. 14036, 86 Fed. Reg. 36987 (Jul. 14, 2021).
2. *See* Texas Free Enterprise and Antitrust Act of 1983, Tex. Bus. & Com. Code Ann. §§ 15.50-15.52 (Aug 29, 1983).
3. *See* CAL. BUS. & PROF CODE §§ 16600 *et. seq.*
4. *See* Mass. Gen. Laws ch. 112, § 12X.
5. *See* Tex. Bus. & Com. Code Ann. §15.50(b).
6. *Valley* Med. *Specialists v. Farber*, 194 Ariz. 363, 369, 982 P.2d 1277, 1283 (1999).
7. *Total Health Physicians, S.C. v. Barrientos*, 151 Ill. App. 3d 726 (1986).
8. Title VII of the Civil Rights Act of 1964, as amended, 42 U.S.C. §§ 2000e *et. seq.*
9. Americans with Disabilities Act as Amended, 42 U.S.C. §§ 12101 *et. seq*
10. Age Discrimination in Employment Act of 1967, 29 U.S.C. §§ 621 *et. seq.*
11. Equal Pay Act (EPA) of 1963, 29 U.S.C. § 206(d) (June 10, 1963).

12. *E.E.O.C. v. Abercrombie & Fitch Stores, Inc.*, 575 U.S. 768 (2015).
13. *Hollins v. Atl. Co.*, 188 F.3d 652 (6th Cir. 1999).
14. Creating a Respectful and Open World for Natural Hair Act of 2021 or the CROWN Act of 2021, S. 888, 117th Cong. §2(b)(1) & §2(c) (2021).
15. *See Spragg v. Shore Care,* 293 N.J. Super. 33, 50–55 (1996).
16. Joseph P. Harkins, Alexis C. Knapp, Steven E. Kaplan, *The Heavy Burden of Light Duty: Young v. UPS*, Littler Insight [Internet]. 2015 Mar 31 [cited **insert date**]. Available from: https://www.littler.com/heavy-burden-light-duty-young-v-ups.
17. *Young v. United Parcel Serv., Inc.*, 135 S. Ct. 1338 (2015).
18. *EEOC v. Nix Hosp. Sys., L.L.C.*, W.D. Tex. No. 5:18-cv-01004 (Sept. 25, 2018); *see also* EEOC [Internet], Press Release, *EEOC Sues Nix Hospital System, LLC D/B/A Nix Healthcare System Services for Pregnancy Discrimination.* 2018 Sep 25 [cited **insert date**]. Available from: https://www1.eeoc.gov/eeoc/newsroom/release/9-25-18d.cfm?renderforprint=1.
19. NIX Hospital Settles EEOC Pregnancy Discrimination Suit, EEOC, Press Release [Internet], *Hospital Refused to Accommodate and Terminated Pregnant Worker, Federal Agency Said.* 2019 Aug 29 [cited **insert date**] Available from: https://www.eeoc.gov/newsroom/nix-hospital-settles-eeoc-pregnancy-discrimination-suit.
20. *NLRB v. Babcock & Wilcox Co.*, 351 U.S. 105 (1956).
21. *Baptist Med. Sys. v. N.L.R.B.*, 876 F.2d 661 (8th Cir. 1989).
22. *Lucile Salter Packard Children's Hosp. at Stanford v. NLRB*, 97 F.3d 583 (D.C. Cir. 1996).
23. *Lucile Salter Packard Children's Hosp. at Stanford*, 318 NLRB 433 (1985), *enforced*, 97 F.3d 583 (D.C. Cir. 1996).
24. *UPMC*, 362 NLRB No. 191 (Aug. 27, 2015).
25. *See* Sara Hoffman Jurand, Hospitals Doctors May Be Liable for Contractors' Negligence, Trial, OCTOBER 2002, at 78; *see also* 38 Am. Jur. Proof of Facts 2d 445 (Originally published in 1984).
26. Irs.gov. Publication 15-A: Employer's Supplemental Tax Guide [Internet]. [cited 2021 Jul 14]. Available at: https://www.irs.gov/pub/irs-pdf/p15a.pdf.
27. *Nationwide Mut. Ins. Co. v. Darden*, 503 U.S. 318 (1992).
28. FACT SHEET 13: EMPLOYMENT RELATIONSHIP UNDER THE FAIR LABOR STANDARDS ACT (FLSA) [Internet]. [cited 2021 Jul 14]. Available from: https://www.dol.gov/agencies/whd/fact-sheets/13-flsa-employment-relationship.
29. 35 STATE OF CALIFORNIA DEPT. OF INDUSTRIAL RELATIONS. "INDEPENDENT CONTRACTOR VERSUS EMPLOYEE." [Internet]. [cited 2021 Jul 13]. Available from: https://www.dir.ca.gov/dlse/faq_independentcontractor.htm; *see also* M.G.L. c. 149, s. 148B.
30. LABOR AND EMPLOYMENT—UNEMPLOYMENT INSURANCE, 2019 Cal. Legis. Serv. Ch. 296 (A.B. 5) (WEST).
31. *Cilecek v. Inova Health Sys. Servs.*, 115 F.3d 256(4th Cir. 1997).
32. *Robb v. United States*, 80 F.3d 884 (4th Cir. 1996).

Part VII

Healthcare Fraud and Abuse

In 2019, healthcare spending in the United States amount to $3.8 trillion, or 17.7% of gross domestic product. The federal government accounted for nearly a third of this spending, including $799.4 billion on Medicare and $613.5 billion on Medicaid [1]. With the enormous amounts of money involved it is no wonder that these programs are the frequent target of fraud and abuse costing taxpayers billions of dollars every year. Reports by the Government Accounting Office ("GAO") estimate the amount of money lost to improper payments by Medicare and Medicaid in 2017 alone was $88.7 billion [2].

Efforts to combat fraud and abuse in the healthcare system has been a government priority for decades. The government has potent anti-fraud and abuse statutes and a powerful enforcement apparatus to enforce these laws. An understanding of these laws by clinicians and other participants in the healthcare system is essential to remain in compliance with various healthcare regulations and to avoid the severe consequences that can result of running afoul of one of the many statutory and regulatory requirements for participants in the healthcare system.

This chapter summarizes five of the central fraud and abuse statutes the government uses to police the healthcare industry. First, we address the False Claims Act, which is the primary statute the government uses in civil investigations of healthcare fraud. Second, we address the Anti-Kickback Statute, which is a criminal statute that prohibits providing benefits in exchange for referrals or the ordering of medical services or products paid for by federal programs. Third, we address the Stark Law, which prohibits financial relationships between clinicians (and family members) and entities to which they refer certain healthcare services paid for by federal programs. Fourth, we address the administrative enforcement processes available to the Office of Inspector General for the Department of Health and Human Services including exclusion from participation from federal healthcare programs and civil monetary penalties. Finally, we address the criminal healthcare statute used to address the most egregious healthcare fraud schemes.

References

1. *National Healthcare Expenditure Data from Centers for Medicare and Medicaid Services*, https://www.cms.gov/Research-Statistics-Data-and-Systems/Statistics-Trends-and-Reports/NationalHealthExpendData/NHE-Fact-Sheet
2. Testimony Before the Subcommittee on Oversight, Committee on Ways and Means, House of Representatives, *Medicare: Actions Needed to Better Manage Fraud Risks*, July 18, 2018, GAO-18-660T; Testimony Before the Committee on Homeland Security and Governmental Affairs U.S. Senate, *Medicaid: Actions Needed to Mitigate Billions in Improper Payments and Program Integrity Risks.* June 27, 2018, GAO-18-598T

The False Claims Act

Jonathan H. Ferry and Lyndsay E. Medlin

Contents

Key Points

- The FCA is the central civil enforcement tool in healthcare fraud cases.
- The FCA entails treble damages and penalties of $11,665 to $23,331 *per false claim.*
- The government can prove a violation without showing a clinician had actual knowledge that claims are false.
- Falsity in claims must be material to the government's decision to pay.
- Whistleblowers may bring cases in the name of the government based on insider information.
- Whistleblowers get a share of any recovery, and a defendant must pay whistleblower attorneys' fees if a whistleblower prevails.

Legal Concepts

- What is a "false" claim
- Whether the falsity in a claim in "material" to the government's decision to pay the claim

J. H. Ferry (✉) · L. E. Medlin
Bradley Arant Boult Cummings LLP, Charlotte, NC, USA
e-mail: jferry@bradley.com; lmedlin@bradley.com

- What level of intent is necessary to establish liability under the FCA
- Importance of whistleblower provisions in the FCA
- Standards for dismissal of FCA suit by government when brought by whistleblower
- Treble Damages and penalties per claim can result in ruinous liability

1   Introduction

Initially passed in 1863 to combat Civil War profiteers and fraudsters, the False Claims Act ("FCA") has evolved into the government's most potent civil enforcement tool in healthcare fraud cases. With treble damages, and penalties between $11,665 to $23,331 per claim, the FCA threatens financial ruin to clinicians who run afoul of its strictures.

This is because the FCA presents a particular threat to healthcare providers. Because of the vast sums of money spent on federal healthcare programs, the Department of Justice ("DOJ") and the Office of Inspector General of the Department of Health and Human Services ("OIG") have made healthcare fraud enforcement a priority item for decades. DOJ assigns highly trained and expert attorneys to a specific division for healthcare fraud cases. Additionally, U.S. Attorney's Offices across the United States have designated healthcare prosecutors. On the investigative side, OIG and the Federal Bureau of Investigation ("FBI") employ agents specializing in healthcare fraud. With the addition of government investigators from the Department of Defense (which operates the Tricare program), the Office of Personnel Management (the Federal Employee Healthcare Program), and other federal programs, the United States has an extremely potent team of investigators and lawyers at its disposal to enforce the FCA. The monetary results of enforcement activity speak for themselves and demonstrate the focus on healthcare (see Table 1).

For a more general comparison, since the FCA was amended in 1986 to strengthen its enforcement mechanisms, the United States has recovered a total of about $64.5 billion under the FCA, and of that amount, about $43.4 billion, just over 67%, is from healthcare cases [1].

Table 1   Source: U.S. Dept. of Justice, *Fraud Statistic Overview,* October 1, 1986 to September 30, 2020, available at https://www.justice.gov/opa/press-release/file/1354316/download

Year	Total FCA Recoveries	Healthcare Recoveries	% Health Care Recoveries
2016	$4,963,562,635	$2,724,854,556	54.9
2017	$3,431,203,229	$2,147,136,189	62.6
2018	$2,904,655,162	$2,535,353,343	87.3
2019	$3,083,512,430	$2,623,132,161	85.1
2020	$2,231,454,855	$1,859,670,739	83.3

2   Discussion

2.1   The Statute

The FCA makes it illegal to submit or to cause to be submitted a false and fraudulent claim to the United States. It is also an FCA violation to avoid payment of a known obligation to the United States. Specifically, there may be an FCA violation against a person or entity who:

> knowingly presents, or causes to be presented, a false or fraudulent claim for payment or approval;

> knowingly makes, uses, or causes to be made or used, a false record or statement material to a false or fraudulent claim;

> knowingly makes, uses, or causes to be made or used, a false record or statement material to an obligation to pay or transmit money or property to the Government, or knowingly conceals or knowingly and improperly avoids or decreases an obligation to pay or transmit money or property to the Government[;] or

> conspires to commit [one of these violations] [2].

2.2   The Elements of the False Claims Act

Generally speaking, there are four elements to an FCA claim: (1) there must be a "claim," as the word is defined under the FCA; (2) the claim must be either legally or factually false; (3) the defendant must have acted with "knowledge" of the falsity; and (4) the falsity of the claim must be material to the government's decision about whether to pay the claim. Importantly, under the FCA, the government needs only to prove each element by a preponderance of the evidence [3]. Often described as a "50 percent plus one standard," this standard of proof allows the government to pursue cases for money damages and penalties that it could not otherwise pursue under the heightened burden of proof associated with a criminal healthcare statute.

The Claim
As with many terms used in the FCA, "claim" has both a statutory, textual definition, as well as an understood meaning based on legal developments in the case law. The definition of a claim is critical not only to whether liability will attach, but also to the monetary amount of potential penalties. In the context of a healthcare fraud case, thousands or even tens of thousands of claims can be at issue. With penalties as high at $23,331 for *each* false claim, healthcare companies face the possibility of enormous financial exposure.

The FCA statutorily defines a claim as "any request or demand… for money or property… whether or not the United States has title to the money or property" either (1) "presented to an officer, employee or agent of the United States" or (2) "made to a contractor, grantee, or other recipient, if the money or property is to be spent or used on the Government's behalf or to advance a Government program or interest" and the government has provided or will reimburse any portion of the money or property requested [2]. In the healthcare context, particularly with Medicare Part B, the application of this definition is straightforward, yet incredibly all-encompassing. Every request for payment submitted to a Medicare Administrative Contractor ("MAC") is a claim for FCA purposes. For example, clinicians typically submit the electronic equivalent of the CMS-1500 that includes detailed information about the services provided including Current Procedural Terminology ("CPT") codes that represent a specific service. Each time a clinician submits such an electronic claim to a government program, such as Medicare, that is a single claims for purposes of the FCA. As discussed in more detail in the *Damages* section below, the breadth of the definition of a claim under the FCA has dire implications for clinicians.

Falsity

Under the FCA, a threshold matter is whether a claim is "false." Specifically, an FCA plaintiff must show that claims are "false or fraudulent" [2]. A claim may be false under a theory of either "factual falsity" or "legal falsity." In the situation of a factually false claim, the defendant "submits information that is untrue on its face" [4]. Further, a factually false claim generally involves "an incorrect description of goods or services provided or a request for reimbursement for goods or services never provided [5]." In the healthcare context, the CMS-1500 claim form includes multiple factual representations that can be falsified. For example, if a clinician submits claims using Current Procedural Terminology ("CPT") codes for services that were never provided, often referred to as "upcoding" cases, the CPT code is factually false. The FCA plaintiff must show that the service provided was not the service described by the CPT code. Another fact which may be misrepresented in the claim form is the patient's diagnosis. In many cases Medicare will pay for certain services only if supported by the appropriate diagnosis code. A common healthcare fraud allegation is that a clinician falsified the diagnosis code to justify a service actually provided.

In contrast, a legally false claim is one that is "predicated upon a false representation of compliance with a federal statute or regulation or a prescribed contractual term" [6]. This theory of falsity is extremely common in healthcare FCA suits. Regulations governing federal healthcare programs fill thousands of pages. Legal falsity can thus potentially result from a violation of any one of these regulations.

Although legal falsity is always based on a false certification, such certification can be either "express" or "implied" [6]. The theory of express certification requires that the entity submitting the claim "falsely certifie[d] compliance with a particular statute, regulation or contractual term, where compliance is a prerequisite to payment" [6]. The theory of implied certification "is based on the notion that the act of submitting a claim for reimbursement itself implies compliance with governing federal rules that are a precondition to payment" [6]. Under this theory, the submitter of the claim may not have attested to any actual certification, but courts nonetheless interpret the act of submitting the claim for payment as implying such a certification. As discussed in more detail below in the *Materiality* section, the Supreme Court endorsed the application of the implied certification theory under certain circumstances [7].

In the context of healthcare claims, express and implied certification theories are both at play and often intertwined. CMS-1500 claims forms have multiple "express" certifications including:

> I have familiarized myself with all applicable laws, regulations, and program instructions, where available from the Medicare contractor;
>
> This claim… complies with all applicable Medicare and/or Medicaid laws, regulations, and program instructions for payment including but not limited to the Federal Anti-kickback statute, and the Physician Self-Referral law (commonly known as the Stark Act)…
>
> The Services on this form were medically necessary…

As is evident from the breadth of the certifications, which often refer to "all applicable laws," a violation may depend on the interpretation of any one of thousands of regulations governing the various federal healthcare programs.

Knowledge

Simply put, the difference between a mistake and fraud is intent. Importantly, a mistake may result in administrative repayment or recoupment by the MAC, but acting "knowingly" may result in a violation of the FCA with the attendant damages, penalties, and potential administrative sanctions [2]. It is important for clinicians to note, however, that "knowingly" as defined by the FCA may not be what a layperson means by knowing. Instead, the FCA defines knowingly as follows:

1. the terms "knowing" and "knowingly"---
   (a) mean that a person, with respect to information –
       (i) has actual knowledge of the information;
       (ii) acts in deliberate ignorance of the truth or falsity of the information; or
       (iii) acts in reckless disregard of the truth or falsity of the information; and
   (b) require no proof of specific intent to defraud… [2]

This standard for scienter gives the government extensive leeway in parsing an entity's conduct. Courts have expressed several opinions of the meaning of "deliberate ignorance" or "reckless disregard" under the FCA. In *United States v. Krizek*, the D.C. Circuit Court of Appeals stated that "[u]se of reckless disregard as a substitute for the forbidden intent prevents the defendant from "deliberately blind[ing] himself to the consequences of his tortious conduct" [8]. The court went on to note that reckless disregard could also be characterized as "a palpable failure to meet the appropriate standard of care." In *United States v. King-Vassel*, the Seventh Circuit Court of Appeals noted that reckless disregard is meant to capture those "who failed 'to make such inquiry as would be reasonable and prudent to conduct under the circumstances'" [9].

Courts find deliberate ignorance or reckless disregard in several patterns of activity prevalent in the healthcare context. Failure to follow the government's interpretation of a regulation, even if a defendant's interpretation is reasonable, may meet the requisite scienter requirements of the FCA [10]. Thus, clinicians must ensure that they familiarize themselves with government guidance regarding the plethora of rules and regulations governing the Medicare and Medicare systems. Failure to investigate warnings or other red flags foreshadowing the possibility that false claims may be submitted to the government carries significant penalties [11]. Furthermore, an entity that fails to make an adequate investigation into the veracity of the information it provides to the government can be found to have acted with reckless disregard [12]. Failure to maintain an effective compliance program may also fulfill the FCA's reckless disregard standard [13]. Courts have even found that, if "a[n] [entity's] structure prevented it from learning the fact that made its claims for payment false, then the plaintiff may establish that the company acted in deliberate ignorance or reckless disregard of the truth of its claims" [11]. Importantly, the scienter requirement of the FCA is highly dependent on the facts. Whether a person or entity has acted appropriately to "meet the standard of care" or made such inquiry "as would be reasonable and prudent to conduct under the circumstances" may be highly dependent on who that person or entity is and the level of expertise and diligence expected of them [14, 15].

Several cases demonstrating the level of care required by clinicians are enlightening. In *Krizek,* the court found reckless disregard on the part of an individual psychiatric provider who failed to properly supervise his billing [8]. The *Krizek* case is an example of the principle that lack of supervision over the billing process can result in liability for clinicians when false claims are submitted even without their actual knowledge. Dr. Krizek was held liable for submitting claims for timed therapy codes that indicated he was billing over 24 h a day in some cases. The court noted several deficiencies with the billing at the practice, including that: (1) the biller submitted claims with little or no factual basis; (2) the biller made no effort to establish how much time Dr. Krizek spent with a particular patient; and, perhaps most importantly, (3) Dr. Krizek made no effort to review bills submitted in his name. Opinions like *Krizek* show that plaintiffs and the government can demonstrate reckless disregard based on allegations that a clinician ignored red flags, failed to address known compliance risks, and failed to maintain adequate compliance protocols. In effect, the FCA puts a burden on healthcare providers to take affirmative steps to mitigate risks of non-compliance.

Materiality

In addition to being false, to give rise to liability, a claim or statement on which an FCA action is based must also be "material" [2]. Although defined in the FCA, the term "materiality" was the subject of varied interpretation until a seminal Supreme Court decision issued in 2016 [7]. The analysis of materiality essentially asks whether an established falsity in a claim or in a record or statement associated with a claim would matter to the government in making a decision to pay the claim.

As statutorily defined, materiality means: "having a natural tendency to influence, or be capable of influencing, the payment or receipt of money or property" [2]. Courts have interpreted this definition to require an objective standard of materiality. Thus, the question is not, whether the government would have, in fact, refused payment of the claims but for the falsity, but rather, whether the falsity had "an objective, natural tendency, to affect the government's decision" [16]. Under this standard, a plaintiff need not provide any proof that the government would not have paid the claim had it actually known of the falsity. Instead, the analysis focuses "on the potential effect of the false claim when it is made, not on the actual effect of the false statement when it is discovered" [17].

In 2016, the Supreme Court drastically changed materiality jurisprudence with its decision in *Escobar* [7]. At issue in *Escobar* was whether the implied certification theory could support a finding of falsity in an FCA case. A patient in a mental health facility died after she was prescribed medication by personnel at the facility. The plaintiff alleged that few facility employees were licensed to provide mental health services, and Massachusetts Medicaid regulations required such licenses. The defendant submitted claims to Medicaid, and although it did not expressly certify compliance with the specific regulations at issue, defendant also failed to disclose the deficiencies to the state. The First Circuit Court of Appeals held that "every submission of a claim implicitly represents compliance with relevant regulations, and that any undisclosed violation of a precondition of payment (whether or not expressly identified as such) renders the claim 'false or fraudulent.'" On appeal from the First Circuit's ruling, the Supreme Court accepted that, even in the absence of an express certification of compliance, the violation of a rule or regulation may be the basis for falsity, "at least in some circumstances."

After acknowledging the possibility of a viable implied certification claim, the court then turned to the question of whether the FCA required the falsity at issue to relate to a rule or regulation classified as a "condition of payment." The court rejected the defendant's argument that only violations of a "condition of payment" could be the basis for implied certification liability, but also opined that not every violation of a regulation or rule classified as a condition of payment *always* gives rise to liability.

Instead, the court held that "a misrepresentation about the compliance with a statutory, regulatory, or contractual requirement must be material to the Government's payment decision in order to be actionable under the False Claims Act." Establishing materiality as the central question in the case, the court went on to describe how the FCA's materiality requirement should be enforced. Whether the government classifies the requirement as a condition of payment is relevant, but not dispositive, to materiality. Other factors including, but not limited to, whether the government consistently refuses to make payment in the mine run of cases based on noncompliance with the requirement could indicate materiality. Conversely, the government making payment despite knowledge of the violation of the requirement may be indicative, though again, not dispositive of a lack of materiality. *Escobar* established a fluid, multi-factor analysis for materiality that has opened the door for clinicians to argue that many of the thousands of rules and regulations applicable to federal healthcare programs may not be material, and thus, may not serve as the basis for an FCA case.

Damages

The FCA establishes that a person who violates the FCA "is liable to the United States Government for a civil penalty of not less than $5,000 and not more than $10,000, as adjusted by the Federal Civil Penalties Inflation Adjustment Act of 1990…, plus 3 times the amount of damages which the Government sustains because of the act of the person" [2]. Due to inflation adjustment, penalties are now $11,665 to $23,331 per false claim.

Generally speaking, damages models are as varied as the fact patterns in FCA cases. But, a few patterns emerge frequently in healthcare cases. First, cases of incorrect coding will result in damages calculated as the difference between the code submitted in the false claim and the correct code. Other common FCA cases pursued by the government are those where liability is based on the lack of medical necessity for the services provided. In such cases, damages are often the full amount paid, as the government takes the position that *no* services were necessary. However, cases where services are performed by unqualified individuals or practitioners with lower level qualifications may result in damages calculated on the entire claim amount or only a portion. The key question in such a case is whether the government received *some* payable service even if what was received was not the service for which the government intended to pay. Finally, violations of the Stark Act and Anti-Kickback Statute, discussed further herein, may also give rise to FCA liability. In such cases, the government takes the position that violation of either statute is a complete bar to payment, and damages are the total amount of claims affected by the violation of the statute [18].

In addition to damages, the FCA requires penalties for each claim. Often in healthcare cases, hundreds or even thousands of claims are at issue. The penalties provision can thus result in enormous and ruinous liability for clinicians [19]. Further, the per claim penalties can be completely disproportional to the government's actual damages. For example, a claim upcoded from a level 3 evaluation and management to a level 4 evaluation and management may result in actual damages of less than $100, but could conceivably result in a penalty of tens of thousands of dollars for that single claim. And, often, cumulative penalties in healthcare cases reach into the many millions of dollars. Although, the government occasionally waives such penalties in the context of settlement negotiations, it maintains its right to seek them if the case goes to court. This provides the government with enormous leverage in settlement negotiations.

Whistleblower Provisions

Although the government frequently investigates healthcare providers and brings FCA suits of its own accord, the FCA's whistleblower provisions are an extremely important part of the enforcement environment. Under what is referred to as the *qui tam* provisions in the FCA, private individuals may file lawsuits in the name of the United States to enforce the FCA [20]. The whistleblower, known as the "relator" under the statute, is entitled to a significant share of any recovery against the defendant. Relator initiated suits have resulted in the lion's share of government recoveries under the FCA in recent years [21].

The relator may be anyone with knowledge of the allegations—such as a current or former employee, a competitor, a customer, or a consultant. When a *qui tam* lawsuit is filed, the complaint is kept under seal and is not provided to the defendant until the presiding federal court orders otherwise [20]. While the complaint is under seal, the government may investigate the relator's claims and must decide whether it will elect to intervene, taking responsibility for prosecuting the lawsuit, or decline to intervene, leaving the relator to litigate his or her complaint [20]. The FCA incentivizes whistleblowers to bring claims by providing them with a share of any proceeds of the action or settlement as mentioned above [20].

2.3 Qui Tam Procedure

A person may bring a civil action for violation of the FCA "for the person and for the United States Government," but "the action must be brought in the name of the government" [20] To begin, the whistleblower must serve the government with a copy of the complaint and a written disclosure of all material evidence and information the person possesses regarding the allegations [20]. The complaint is filed in camera and under seal, meaning it is not publicly available and its existence is known only to the relator, the government, and the court. The government then has 60 days to investigate the matter and decide whether it will intervene or decline to intervene in the matter [20]. FCA practitioners consider the 60-day deadline laughable, as the government routinely requests extensions of that deadline for "good cause" as allowed by the court. The FCA gives the government two options with respect to intervention: (1) proceed with the action, in which case the action shall be conducted by the government; or (2) notify the court that it declines to take over the action, in which case the person bringing the action, the relator, shall have the right to conduct the action [20].

3 Government Intervention Decision

One of the most important events in *qui tam* litigation is the government's intervention decision. The decision has important legal and practical implications for the defendant and the relator. First, if the government intervenes, it takes over the case and has the option of including relator and relator's counsel as it sees fit [3]. Typically the government, will file its own complaint to include more detailed allegations of misconduct revealed during the government's investigation. The government complaint may also add claims and defendants. Government intervention also means highly experienced lawyers from the DOJ will be in charge of the case and will have the backing of experienced healthcare fraud investigators and many other substantial litigation resources available only to the federal government. Additionally, government intervention is usually accompanied by a press release from the DOJ announcing its decision to intervene and transforming the unsubstantiated allegations of a private party relator into dangerous allegations from governmental authorities claiming that the defendant has engaged in fraud on the public fisc. The reputational damage alone resulting from the government's decision to intervene can be significant. Government intervention is generally very bad news for a defendant.

If, on the other hand, the government declines to intervene, the case is left to the relator and his or her counsel to pursue [20]. Although the DOJ insists that declination of intervention is not a comment on the merits of the case, declination of intervention is often seen as a refutation of relators allegations. Defendants certainly claim that to be the case. Nonetheless, declination does not mean the case will go away. Although, some relators will abandon a case if the government does not intervene, a substantial number of cases with very high damage awards have been won by relators taking the case forward after the government has declined to intervene [22].

It is also important to note that even if the government declines intervention, its involvement in the case is not over. The government often has extensive relevant information and is the subject to discovery requests from both relators and defendants. Additionally, the government can intervene at a later time for "good cause" and retains the power, at least theoretically, to dismiss an FCA suit over the objections of the relator at any time [20].

4 Dismissal of a Qui Tam Action

An increasingly important, but controversial, feature of the FCA is the government's statutory authority to dismiss an FCA suit brought by a relator at any point in the case. The FCA states "[t]he Government may dismiss the action notwithstanding the objections of the person initiating the action if the person has been notified by the Government of the filing of the motion and the court has provided the person with an opportunity for a hearing on the motion" [20].

Although rarely utilized, this provision has been paid more attention in recent years. In 2018, DOJ formally outlined the various factors DOJ attorneys should consider when deciding whether to dismiss a *qui tam* FCA suit [23]. Moreover, although still rare, DOJ has used the power more frequently in recent years [24]. This renewed attention to DOJ's dismissal power has even caught congressional attention leading to new bipartisan legislation designed to revise the FCA and limit DOJ's discretion, primarily proposed by Senator Chuck Grassley [25].

As is evident from the statutory language, an effort by DOJ to dismiss a *qui tam* action is likely to draw spirited objection from the relator and her counsel, and a relator is entitled to a "hearing" on the issue under the statute. However, courts are split on the amount of deference to give the government in its decision to dismiss these cases. In *Swift v. United States*, the D.C. Circuit Court of Appeals held that the government's authority to dismiss a case under the FCA was "unfettered" and relators' rights under the statute are limited to demanding a hearing so that relators have "the formal opportunity to convince the government not to end the case" [26]. To the contrary, in *Sequoia Orange Co. v. Baird-Neece Packaging Corp.*, the Ninth Circuit Court of Appeals found that courts must engage in a limited review of the government's decision [27]. The court endorsed a "rational relationship" test that required the government to identify a valid purpose for dismissal and a rational relation between dismissal and accomplishment of that purpose. Multiple courts have reviewed government dismissal decisions in recent years under both the *Swift* and *Sequoia Orange* standards [28–30].

Although the law is unsettled on this issue in many circuits, practically speaking, in any given case, it is unlikely that the government will exercise this power to curb a non-meritorious *qui tam* action. As previously mentioned, the government takes the position that declination of intervention does not reflect on the merits of an FCA case. Thus, in a case where the government has investigated and chosen not to intervene, declination, rather than dismissal, is the most likely outcome for a variety of reasons. Relator may uncover information during discovery that the government did not in its investigations which may cause the government to change its mind about intervention. Further, declination is unreviewable and cannot be challenged by the relator, whereas a motion to dismiss will likely draw a relator's challenge. And, unless the D.C. Circuit's *Swift* standard is applied, a court may investigate the basis for the government's dismissal, including the thoroughness of the government's investigation. Such an investigation is burdensome to the government and could potentially expose investigative procedures the government does not wish to reveal. Thus, declination of intervention remains the path of least resistance for the government when it desires to minimize its role in an FCA suit.

4.1 Retaliation Provisions

In addition to the *qui tam* provision of the FCA, a relator may also have a personal cause of action against an employer under the statute's retaliation provisions. The FCA states that if an employee is "discharged, demoted, suspended, threatened, harassed, or in any other manner discriminated against in the terms and conditions of employment" because of lawful acts to stop a violation of the FCA, then the employee will have a cause of action for reinstatement and two times the amount of backpay, interest on back pay, and compensation for special damages [20].

The retaliation provisions require that clinicians be very careful in their treatment of any employee, contractor, or agent of the company who raises concerns under the FCA, or even who raises other compliance issues. It is commonplace for whistleblowers to include retaliation claims as a matter of course in *qui tam* complaints. The government often settles the FCA fraud claims with a defendant, but leaves the retaliation claims to be settled separately. Even in cases where the actual FCA fraud claims have been dismissed, the retaliation claims may continue, leading to extensive and expensive litigation [31, 32]. Importantly, relators are entitled to attorney's fees and costs associated with litigating retaliation claims if they prevail [20].

4.2   Relator's Share and Attorney's Fees

The FCA provides substantial financial incentives for both relators and their counsel. First, relators are entitled to 15–25% of the government's recovery in cases where the government intervenes in the lawsuit. When the government does not intervene and the relator recovers under the FCA, the relator is entitled to 25–30% of the recovery [20]. The remainder goes to the government as damages and penalties for the FCA violations.

The relator's share is often the subject of negotiations between relator and the United States. Defendants play no role in these discussions. The United States, applies a number of factors to determine the appropriate percentage for a relator's share including whether: (1) the relator promptly reported the fraud; (2) the relator attempted to stop the fraud; (3) the *qui tam* suit caused the defendant to stop committing the fraud; (4) the complaint involves a significant safety issue; (5) the complaint exposes a significant nationwide practice; (6) the relator provides significant first-hand knowledge of the fraud; (7) the government was previously unaware of the fraud; (8) the relator provided substantial assistance during the investigation; (9) the relator is credible as a witness; (10) the relator's counsel provided substantial assistance to the government; (11) the relator and counsel supported and cooperated with the government; (12) the case goes to trial; (13) the amount of recovery was relatively small or significant; and, finally, (14) the filing of complaint has had a substantial adverse impact on the relator [33]. Occasionally the United States and relators cannot agree on the appropriate share and relators can petition the courts to award a greater share. The Court's apply the same factors listed above. The relator's share can also be reduced or eliminated if the court determines relator planned or initiated the FCA violation [20]. The aggregate payments to relators under the FCA over the last 5 years demonstrate the enormous financial incentive to bring a *qui tam* action (see Table 2).

Defendants in *qui tam* actions frequently ask whether they can recover their attorney's fees from relators if defendants prevail in defending the case. Unfortunately for defendants, the standard for such fee shifting is quite onerous, and requires the court find that the "claim of the person bringing the action was clearly frivolous, clearly vexatious, or brought primarily for the purposes of harassment" [20].

In addition to a relator's share, relators are also entitled to obtain their expenses, including their attorney's fees and costs, from the defendant. DOJ often considers this a separate dispute among defendants and relators and their counsel. Thus, many settlements between defendants and DOJ will omit settlement of this issue. Much

Table 2 Source: U.S. Dept. of Justice, *Fraud Statistic Overview,* October 1, 1986 to September 30, 2020, available at *https://www.justice.gov/opa/press-release/file/1354316/download*

Year	Aggregate Award to Relators
2016	$553,982,692
2017	$537,586,579
2018	$341,858,856
2019	$365,623,744
2020	$309,416,126

like retaliation claims, attorney's fees claims can result in contentious and expensive litigation for *qui tam* defendants even after resolution with the government of the substantive fraud allegations [34].

5 Application

The government and relators may pursue FCA investigations and cases based on several theories of liability. Cases summarized below are exemplars of the types of cases routinely pursued by enforcement authorities.

5.1 Medical Necessity

The government broadly pursues cases based on the alleged lack of medical necessity for procedures or services across the healthcare industry. Such cases are usually based on testimony from government experts that the medical records of the patients do not support the need for the services rendered. The government takes the position that, because the services were not necessary, any claims submitted to a federal healthcare program for payment on the services are false.

Three important federal appellate court decisions address the question of whether a disagreement between medical experts about medical necessity can be the basis for falsity in an FCA suit. In *United States v. AseraCare, Inc.*, the United States sought to establish hospice claims as false with expert evidence that the patients were not actually terminally ill [35]. The Eleventh Circuit Court of Appeals framed the relevant question as: "When can a physician's clinical judgment regarding a patient's prognosis be deemed 'false?'" The court went on to note that "physicians applying their clinical judgment about a patient's projected life expectancy could disagree, and neither physician … be wrong." Thus, the Eleventh Circuit concluded that a subjective but honest disagreement on medical prognosis could not be the basis for falsity under the FCA. Instead, the government must demonstrate something more "objective" to show falsity, including, for example: (1) that the certifying physician failed to familiarize himself with the medical record; (2) that the physician did not in fact subjectively believe the prognosis to be true; or (3) proof that no reasonable physician could have concluded that a patient was terminally ill.

To the contrary, in *U.S. ex rel. Druding v. Care Alternatives*, the Third Circuit Court of Appeals found that "medical opinions can be false" even if honestly held and reasonable [36]. Thus, an after-the-fact expert opinion that the prognosis was incorrect presents a triable issue of fact for the jury regarding whether the claim may be "false." The Third Circuit stated directly that it disagreed with the reasoning in *AseraCare* in the Eleventh Circuit.

Shortly after the Third Circuit's *Druding* opinion, the Ninth Circuit Court of Appeals also weighed in on this question in *U.S. ex rel. Winter v. Gardens Reg'l Hosp. & Med. Ctr., Inc.* [37]. The Ninth Circuit concluded that a medical opinion can be false or fraudulent for the same reasons that other opinions could be false or fraudulent, including that "the opinion is not honestly held, or if it implies the existence of facts that do not exist."

The Supreme Court decided not to take up this question, leaving in place this apparent legal split among circuit courts. Clinicians often believe that government cases concerning medical necessity amount to little more than the unjustified second guessing of legitimate medical judgments. The current law, however, suggest that cases based on the government's view of medical necessity, as opposed to the attending clinician's view, will remain viable for FCA enforcement.

5.2   Upcoding

Another common theory of liability pursued by DOJ in FCA cases against clinicians is the practice of "upcoding." In these cases, DOJ takes the position that, although the services provided for payment are payable, the CPT code included in the claim form falsely represented a higher-level and more expensive service. This theory of liability is often pursued when the DOJ believes a clinician is submitting higher than warranted level evaluation and management codes. DOJ may target such clinicians based on data showing that they bill more high-level codes than peer clinicians. In these cases, damages are calculated by the difference between the code billed and the code enforcement authorities assert is the correct code. Damages in such cases can be significant depending on the circumstances. A major hospital service provider paid $60 million based on allegations that its employed hospitalists up-coded their services in response to pressure from the company [38]. Hospitalists billed Medicare using certain CPT codes that corresponded to a level of service with *suggested* times spent with patients. The relator alleged in a *qui tam* complaint that the company encouraged its hospitalists to enter codes representing a higher level of service than what was actually provided.

6   Summary

The FCA remains the government's most significant civil enforcement tool for healthcare fraud and abuse. Although the FCA is generally applicable to any program in which government funds are available, healthcare related investigations, cases, and settlements have made up the vast majority of FCA cases in most years. Although the FCA poses significant risk to healthcare providers with compliance issues the government faces significant hurdles to establishing FCA liability. First, the government must show that the subject claims are false, that the falsity in the claim is material to the government's decision to pay the claims, and that the healthcare provider acted with the requisite scienter: (1) actual knowledge; (2) reckless disregard; or (3) willfull blindness as to the falsity of the claims.

In addition to government oversight and enforcement, healthcare providers must also be wary of internal whistleblowers that are empowered under the *qui tam* provisions of the FCA to bring suits on behalf of the government and claim significant shares of any recovery. Diligent compliance measures that include familiarity with the relevant laws, regulations, and guidance, policies and procedures, and monitoring of potential risk areas are essential mitigating risk of FCA liability.

Practicitions may wish to consider at least the following when assessing FCA risk:

1. What are the relevant laws, regulations, and guidance applicable to the services my practice provides;
2. Do I have a compliance program that monitors compliance with the relevant laws, regulations, and guidance;
3. Do I have programs to confirm "medical necessity"—as defined by Medicare— exists for services provided by my practice; and
4. In the event errors are found, determine how they occurred and whether the error is the kind problem that would affect the payment decision from the government.

References

1. *Fraud Statistic Overview,* October 1, 1986 to September 30, 2020, United States Department of Justice, available at *https://www.justice.gov/opa/press-release/file/1354316/download*
2. 31 U.S.C. § 3729
3. 31 U.S.C. § 3731
4. *United States v. Kellogg Brown & Root Servs., Inc.,* 800 F. Supp. 2d 143 (D.D.C. 2011)
5. *United States v. Sci. Applications Int'l Corp.,* 626 F.3d 1257 (D.C. Cir. 2010)
6. *Mikes v. Straus,* 274 F.3d 687 (2d Cir. 2001), abrogated on other grounds by *Universal Health Servs., Inc. v. U.S. ex rel. Escobar,* 136 S. Ct. 1989 (2016))
7. *Universal Health Servs., Inc. v. U.S. ex rel. Escobar,* 136 S. Ct. 1989 (2016)
8. *United States v. Krizek,*111 F.3d 934 (D.C. Cir. 1997)
9. *United States v. King-Vassel,* 728 F.3d 707 (7th Cir. 2013)
10. *Visiting Nurses Association of Brooklyn v. Thompson,* 378 F.Supp.2d 75 (E.D.N.Y. 2004)
11. *U.S. ex rel. Wuestenhoefer v. Jefferson,* 105 F.Supp. 3d 641 (N.D. Miss. 2015)
12. *United States v. Raymond & Whitcomb Co.,* 53 F.Supp. 2d 436 (S.D.N.Y 1999)
13. *U.S. ex rel. Hunt v. Merck-Medco Managed Care LLC,* 336 F.Supp. 2d 430 (E.D. Pa. 2004)
14. *United States v. Molina Healthcare of Illinois, Inc.,* 17 F.4th 732 (7th Cir. 2021)
15. *OIG Compliance Program for Individual and Small Group Physician Practices,* Fed. Reg., Vol. 65. No. 194 at 59434 (October 5, 2000)
16. *United States v. United Technologies Corp.,* 626 F.3d 313 (6th Cir. 2010)
17. *U.S. ex rel. A+ Homecare, Inc. v. Medshares Management Group, Inc.,* 400 F.3d 428 (6th Cir. 2005) *superseded by statute as stated in U.S. ex rel. Harper v. Muskingum Watershed Conservancy Dist.,* 842 F.3d 430 (6th Cir. 2016)
18. *United States v. Rogan,* 517 F.3d 449 (7th Cir. 2008)
19. U.S. Dept. of Justice, United States Resolves $237 Million False Claims Act Judgment against South Carolina Hospital that Made Illegal Payments to Referring Physicians (Oct. 16, 2015), https://www.justice.gov/opa/pr/united-states-resolves-237-million-false-claims-act-judgment-against-south-carolina-hospital
20. 31 U.S.C. § 3730
21. U.S. Dept. of Justice, *Fraud Statistic Overview,* October 1, 1986 to September 30, 2020, available at *https://www.justice.gov/opa/press-release/file/1354316/download*
22. *Ruckh v. Salus Rehab., LLC,* 963 F.3d 1089 (11th Cir. 2020)

23. U.S. Dept. of Justice, M.D. Granston, *Factors for Evaluating Dismissal Pursuant to 31 U.S.C. 3730(c)(2)(A)* (Jan. 10, 2018)
24. Reuters, A. Frankel, DOJ doubles down in brief to discredit 'Wall Street-backed' False Claims Act whistleblower (Feb. 25, 2019), https://www.reuters.com/article/us-otc-fca/doj-doubles-down-in-brief-to-discredit-wall-street-backed-false-claims-act-whistleblower-idUSKCN1QE2IX
25. Chuck Grassley News Releases, Senators Introduce Bipartisan Legislation To Fight Government Waste, Fraud (Jul. 26, 2021), https://www.grassley.senate.gov/news/news-releases/senators-introduce-of-bipartisan-legislation-to-fight-government-waste-fraud
26. *Swift v. United States*, 318 F.3d 250 (D.C. Cir. 2003)
27. *Sequoia Orange Co. v. Baird-Neece Packaging Corp.*, 151 F.3d 1139 (9th Cir. 1998)
28. *United States v. UCB, Inc.*, 970 F.3d 835 (7th Cir. 2020), cert. denied sub nom. *Cimznhca, LLC v. United States*, 2021 WL 2637991 (U.S. June 28, 2021)
29. *U.S. ex rel. Borzilleri v. AbbVie, Inc.*, 837 F. App'x 813 (2d Cir. 2020)
30. *United States v. Eli Lilly & Co., Inc.*, 4 F.4th 255 (5th Cir. 2021)
31. *Erickson v. Biogen, Inc.*, No. C18-1029-JCC, 2020 WL 885743 (W.D. Wash. Feb. 24, 2020)
32. *U.S. ex rel. Grant v. United Airlines Inc.*, 912 F.3d 190 (4th Cir. 2018)
33. *U.S. ex rel. Shea v. Verizon Communications, Inc.*, 844 F.Supp.2d 78 (D.C.D. 2012)
34. *U.S. ex rel. Doghramji v. Cmty. Health Sys., Inc.*, No. 3:11 C 442, 2019 WL 4887190 (M.D. Tenn. Oct. 2, 2019)
35. *United States v. AseraCare, Inc.*, 938 F.3d 1278 (11th Cir. 2019)
36. *U.S. ex rel. Druding v. Care Alternatives*, 952 F.3d 89 (3rd Cir. 2020)
37. *U.S. ex rel. Winter v. Gardens Reg'l Hosp. & Med. Ctr., Inc.*, 953 F.3d 1108 (9th Cir. 2020)
38. U.S. Department of Justice, *Healthcare Service Provider to Pay $60 million to Settle Medicare and Medicaid False Claims Act Allegations*, https://www.justice.gov/opa/pr/healthcare-service-provider-pay-60-million-settle-medicare-and-medicaid-false-claims-act (Feb. 6, 2017)

The Anti-Kickback Statute

Jonathan H. Ferry and Lyndsay E. Medlin

Contents

Key Points
- The AKS prohibits providing remuneration to a referral source to induce referrals.
- The AKS prohibits providing remuneration to any individual to induce them to purchase, lease, order or recommend any good, facility, or service.
- The AKS violation requires a knowing and willful act on the part of participants in the transaction.
- Remuneration is very broadly defined and can include any benefit.
- Induce means to endeavor to influence the decision-making of the individual receiving the remuneration.

Legal Concepts
- Remuneration is broadly defined to be anything of value
- AKS is an intent based statute that imposes liability if the intent of the remuneration is to induce the receiving party to purchase or refer goods for services to a certain entity
- Applicable goods and services paid for by federal healthcare programs, but many states have "all payor" AKS-likes statutes that must be considered

J. H. Ferry (✉) · L. E. Medlin
Bradley Arant Boult Cummings LLP, Charlotte, NC, USA
e-mail: jferry@bradley.com; lmedlin@bradley.com

© The Author(s), under exclusive license to Springer Nature Switzerland AG 2022   293
A. S. Pasha (ed.), *Laws of Medicine*,
https://doi.org/10.1007/978-3-031-08162-0_18

- Compliance with safe-harbor can mitigate risk of AKS liability
- "One-purpose" test significantly increases risk for any business transaction in which a referral source is paid by a healthcare provider

1   Introduction

The Anti-Kickback Statute, 42 U.S.C. § 1320a-7b ("AKS"), penalizes both recipients of, and those that pay, purported kickbacks in exchange for referrals of services and items paid for by federal healthcare programs. Violation of the AKS is a felony and carries potential penalties of fines up to $100,000, imprisonment for not more than 10 years, or both. Violations of the AKS may also serve as grounds for civil monetary penalties, and as a basis for FCA liability [1]. A claim that includes items or services resulting from a violation of the AKS constitutes a false or fraudulent claim sufficient to satisfy the first element of the FCA [2]. The AKS is violated when someone (1) knowingly and willfully, (2) offers, pay, receives or solicits money, directly or indirectly, (3) to induce referrals for the furnishing of medical services, (4) paid for by federal healthcare programs [3, 4].

2   Discussion

2.1   Elements of AKS Violation

Knowingly and Willfully to Induce Referrals

To satisfy the "knowingly and willfully" element, neither specific intent to violate the AKS, nor even actual knowledge of the AKS are required. Rather, "knowingly and willfully," often referred to as the scienter element, is a question of the state of mind or intent behind the making of the payment, and it "may be inferred when a person, like an ostrich burying its head in the sand, takes deliberate steps in order to avoid learning the truth" [5]. However, mere negligence is insufficient. For example, deliberate ignorance is *not* demonstrated merely where "a reasonable person in the defendant's position should have been strongly suspicious, or should have been aware" of misconduct [5]. "Willfully" means that a defendant knew that his conduct was generally unlawful, not that he specifically knew that his actions violated the statute [6].

The AKS prohibits payments if *any* purpose for that remuneration is to induce referrals, even if inducement is not the only purpose—even mixed-motive situations where the offeror may also have legitimate business purposes for the payment arrangement are unlawful under the AKS [7–9]. As equipment and space leases are common place in the industry, we note that courts "can reasonably infer that a landlord would not enter into a lease agreement for a price that fell below the fair market rate if some other consideration were not involved" [10]. But, in any event, even a "fair market value payment will not legitimize a payment if there is also an illegal purpose" [7].

If the arrangement was not entered into for the purpose of inducing referrals, however, it does not violate the AKS, even if there is a collateral hope or expectation that referrals may result from the arrangement [7, 11]. Similarly, where there is no strong evidence of a purpose to induce referrals, evidence that arrangements are "compatible with legitimate business purposes" or even for an "arbitrary" purpose or no purpose at all, can disprove intent [5, 12]. Nonetheless, drawing the line between legitimate business purposes and unlawful intent to induce is very difficult for jurors (and courts) [11]. In matters where the government or relators prevail at trial without smoking gun evidence of a purpose to induce referrals, they often instead put forth proof "in such instances [that] tends to meaningfully exclude a legitimate (or negligent) explanation for the defendants' conduct—consistent with the heightened scienter requirement imposed by a 'knowing and willfully' standard" [5].

2.2   Collective Knowledge

For hospitals and large clinician or prescriber practices, it is important to note that corporate knowledge (and thus corporate liability) may be assessed based on the conduct and knowledge of individual employees who are not necessarily in positions of great authority and responsibility. Under the FCA, "the knowledge of an employee is imputed to the corporation when the employee acts for the benefit of the corporation and within the scope of his employment" [13, 14].

However, many courts have rejected the "collective knowledge theory," which "allows 'a plaintiff to prove scienter by piecing together scraps of 'innocent' knowledge held by various corporate officials, even if those officials never had contact with each other or knew what others were doing in connection with a claim seeking government funds" [15]. Instead, typically, the government attempts to prove a company's knowledge by "comment[s], email[s], memo[s], or other indication[s] by any relevant agent that suggests [persons] individually or collectively were knowing and willful participants in any kickback scheme," in conjunction with "evidence of [a company's] corporate practices and pressure, and that [the company] knew those practices likely caused" violations [5, 16].

Renumeration
"Remuneration" under the AKS is broadly defined as "anything of value" and includes leased space and personal services provided for free or less than fair market value [5, 17]. Where the item or service is part and parcel of arguably lawful and legitimate conduct, however, the ancillary benefit of the service may be found to have no independent value separate from the underlying conduct [18]. For example, courts have held that patient support services provided by pharmaceutical companies to patients did not constitute remuneration to prescribers because the support services did not provide "substantial independent value" [19]. On the contrary, co-pay and deductible waivers providing an immediate financial benefit to patients are closely scrutinized [20]. Such waiver can constitute AKS violations in several ways.

First, a provider directly forgiving co-pay or deductible charges for patients can be viewed as an inducement to use the clinician's services, Second, forgiveness of co-pays or deductibles by an upstream provider such as a clinical laboratory are often regarded as inducements to clinicians to use the laboratory services. In such cases, the clinicians get the benefit of telling their patients they will not receive co-payment or deductible bills for laboratory or other work provided by the third-party provider [20].

Safe Harbor Exceptions to the Anti-Kickback Statute

The AKS establishes certain exceptions to the broad definition of remuneration. These exceptions are known as safe harbors. Safe harbor provisions frequently relevant to clinicians are the space rental, equipment rental, and personal services and management contracts and outcomes-based payment arrangements exceptions.

These safe harbors are subject to multiple specific requirements each of which must be fulfilled to qualify for the safe harbors [21]. The safe harbors, however, have certain common core requirements that are intended to mitigate the risk that the arrangement is actually an inducement scheme dressed up as a legitimate business transaction. For example, leases of office space are common for medical device manufacturers who may rent a closet to store products, or for laboratories that might rent space for phlebotomists to draw blood or otherwise obtain specimens on behalf of the lab. Such a space may also have computers that integrate with the laboratories laboratory information systems. The government recognizes that under such arrangements it might be reasonable for the device company or laboratory to rent the space needed for its equipment or personnel to work in the practice. Such arrangements, however, also carry risk that the payments to the clinician may also induce purchase of products or services from or refer patients to these healthcare companies. The requirements of the safe harbor are intended to mitigate such risk. Thus, in addition to other technical requirements, any payments must be pursuant to a written arrangement made in advance of the referrals for the term of at least 1 year and the payment must be fair market value, not take into account the volume or value of the devices purchased or referrals made, and otherwise commercially reasonable.

Personal services and management contracts are also common transactions between healthcare providers and clinicians who may be referral sources. For example, it is not uncommon for institutional healthcare providers to employ medical directors or contract clinicians who may also be referral sources to the institution. Like the safe harbor for office leases and equipment rental, these arrangements must be under agreements set out in writing and in advance for the term of at least a year, and the compensation must be fair market value, not take into account the volume and value of referrals and commercially reasonable in the absence of any referrals. The key inquiry is whether the payments are for the services rendered under the agreement.].

For the new outcomes-based payment exception to apply, in short, a clinician must meet several complicated requirements relating to, among other things, the legitimacy of the outcome measures, the quality of patient care to be provided, and the reduction of costs associated with the arrangement. Like many of the other exceptions, an outcomes-based agreement is subject to multiple detailed requirements. Clinicians considering availing themselves of AKS safe harbors must carefully parse each requirement of the safe harbors and ensure compliance with each [21].

In litigation, clinicians will bear the burden of proving that an exception is satisfied, and application of a safe harbor is primarily addressed at summary judgment at the earliest, and more often, trial [22, 23]. Some cases and regulatory guidance imply that a clinician's reasonable belief that they were in compliance with a safe harbor—even if they were not—can serve as proof of a motive other than for the purpose of inducing referrals. However, to invoke any such evidence, a clinician must have "actually come to that reasonable but incorrect conclusion," because "[t]here is no support for the proposition that a defendant may disregard its obligations under the FCA and then argue *ex post facto* that a reasonable interpretation of applicable law supports its prior position" [24, 25]. Occasionally, high profile settlements and criminal cases brought by the government, some of which are discussed below, involve allegations that scheme participants funneled funds through shell companies and laundered bribes through real estate and yacht purchases, thereby demonstrating by their conduct that they knew the illegitimate purpose of the payments at issue [26].

3   Application

3.1   Speaker Programs

On November 16, 2020, OIG issued a Special Fraud Alert specifically highlighting speaker programs, defined as events, typically sponsored by a drug or device manufacturing company, at which prescribers speak or present to other healthcare professionals about drugs (or disease states targeted by these drugs) and devices manufactured by the sponsoring company, as a high priority area for enforcement [27]. OIG issued the alert only months after the United States Attorney's Office for the Southern District of New York reached a $678 million settlement with Novartis Pharmaceuticals Corporation alleging that it held sham educational events designed to increase prescriptions of cardiovascular and diabetes drugs [28].

While, on their face, speaker programs purport to promote honorable objectives—education and the sharing of information among clinician peers about the benefits, risks, and appropriate uses of devices and medication—OIG has expressly declared that it "is skeptical about the educational value of such programs" [28]. To avoid becoming investigation targets, prescribers should be mindful of their involvement in programs sponsored by third-parties and scrutinize the attendees, the venue and incentives associated therewith, the amount of any honorarium compared to the scope and complexity of the work, whether the disease state or products are new and/or evolving, and the frequency with which programs occur despite limited or no changes to knowledge about the condition and treatments.

3.2   Electronic Health Records

The ease with which health information technology developers may influence clinicians also has not gone unnoticed by the DOJ. To comply with the Health Information Technology for Economic and Clinical Health ("HITECH") Act and to facilitate the provision of patient care, clinicians make use of sophisticated electronic health record ("EHR") technologies and services. The need for superior, ever evolving technology has spurred competition among developers for clinician attention, sometimes through means which violate the AKS.

In January 2020, the DOJ settled with EHR developer Practice Fusion Inc. for $145 million (including civil penalties relating to both data portability certifications and criminal fines/forfeiture) based on allegations that the developer accepted bribes in the form of sponsorship payments from an opioid company to influence prescribers [29]. According to the government, the developer allowed the opioid company to design alerts, which were then implemented within Practice Fusion's EHR software, to prompt clinicians to prescribe more of the sponsored drugs. Although the physicians themselves were not part of the large dollar Practice Fusion settlement, clinicians can and do become embroiled in these types of kickback investigations as witnesses.

In 2021, DOJ has maintained its focus on EHR with two multimillion dollar settlements relating to kickbacks in the form of cash, cash equivalent credits, trips with complimentary travel, accommodations, meals, and alcohol paid to prescribers through "Champions" and "Lead Generation" programs designed to influence clinicians to recommend EHR products and related services to other clinicians. The government takes the position that such arrangements result in the submission of false claims by clinicians who certify and attest to adoption and meaningful use of appropriate EHR technology [30, 31].

3.3   Telemedicine, Durable Medical Equipment, Genetic Testing, and Compound Pharmacies

Although telehealth services seem poised to become a healthcare staple as a result of the COVID-19 pandemic, participation in telehealth services is not without risk to clinicians, as the government has expended significant resources investigating telefraud in recent years.

"Operation Brace Yourself" first made national news in 2019 when prescribers and telemedicine companies were criminally charged with a $1 billion fraudulent scheme relating to the prescription of unnecessary back, neck, and knee braces to elderly and disabled patients [32, 33]. Brace manufacturers paid kickbacks to telemedicine clinicians in exchange for referrals for the durable medical equipment ("DME"). The investigation involved multiple FBI field offices and several federal districts from New Jersey to Florida, and its effects continue even to-date, resulting in FCA whistleblower suits and multimillion-dollar settlements, as well as guilty pleas, convictions, and potential multiyear prison sentences [34–36].

Building on the momentum of Operation Brace Yourself, in September 2020, federal agencies announced a record shattering telemedicine takedown involving $4.5 billion in allegedly false claims submitted by more than 86 criminal defendants in 19 judicial districts [37]. In exchange for bribes from DME, testing, and drug companies, telemedicine companies purportedly paid clinicians for referrals for products manufactured by the companies paying bribes. The clinicians prescribing the products had no patient interaction or prescribed the products after only a brief telephonic conversation with patients they had never met or seen. Beyond the clinicians caught up in criminal charges, CMS revoked Medicare billing privileges for an additional 265 clinicians. DOJ formed a National Rapid Response Strike Force within its Criminal Division Fraud Section to spearhead the telemedicine initiative and other similar investigations across the country [37].

Subsequently, on May 26, 2021, DOJ announced charges against clinicians and telehealth executives relating to purported abuse of the COVID-19 emergency declaration which broadened telehealth access for Medicare beneficiaries. In exchange for bribes from testing companies, clinicians allegedly ordered medically unnecessary genetic and laboratory testing for drive-through patients (among others), resulting in false billing in excess of $143 million [38]. Additionally, more than 50 clinicians faced administrative actions and potential exclusion from participation in the Medicare program.

Last, but not least, another enforcement priority in the AKS context is compounded pharmaceuticals. Though not a new trend, having been firmly within the government's crosshairs for years, compounded pharmaceuticals remain an area of risk of which clinicians should be cognizant. As is the case with telemedicine actions, although the AKS does not require a referral or prescription be medically unnecessary to be illegal—the mere fact payment was intended to induce referrals is enough—the common sense purpose behind the rule is that renumeration biases clinician judgment and results in medically unnecessary billings to Medicare.

In September 18, 2019, the government secured a settlement under the FCA of more than $21 million from a compounding pharmacy, Patient Care America ("PCA"), and its executives relating to allegations of AKS fraud against the military [39]. PCA purportedly targeted military service members and families for prescriptions for compounded creams and vitamins designed to trigger high reimbursement from TRICARE, the federal health insurance program for the military. To obtain the prescriptions, marketers for PCA paid telemedicine clinicians to order the products without seeing (and occasionally, without even speaking with) the patients. Criminal convictions are also common in compounding pharmacy investigations. The co-owner of several compounding pharmacies and distributors was recently sentenced to 18 years in prison, in addition to being ordered to make restitution and to forfeit millions as a result of his kickback scheme [40]. Notably, one form of remuneration at issue in these investigations, in addition to direct payments to prescribers, is the routine and systematic waiver and reduction of copays by the pharmacies [39]. The government often considers the forgiveness of patient co-payments as integral to these schemes. If the patient received and had to pay a large co-payment due to the exorbitant cost of the products, the patients would likely complain to authorities about the bills. By not charging co-payments, the participants in the scheme remove the most likely source of complaints to enforcement authorities.

On February 26, 2021, OIG Principal Deputy Inspector General Christi Grimm released an open letter acknowledging "the promise that telehealth and other digital health technologies have for improving care coordination and health outcomes," while nonetheless defending telefraud investigations [41]. As mentioned above, the government focuses on medical necessity and the purpose behind prescriptions, regardless of the in-person or remote manner by which a clinician assessed the patient. Nonetheless, as evidenced by the foregoing examples of enforcement actions, a telemedicine clinician's judgment and motivations will be closely scrutinized.

4 Summary

The AKS applies to any person that pays, offer, solicits, or receives remuneration in to induce, of in exchange for, the ordering of any service or products paid for by federal healthcare programs. Clinicians should be aware that remuneration is broadly construed under the statute is any benefit to the referring or purchasing party. Although the AKS may broadly apply to many arrangements, violation of the statute requires that the parties "willfully and knowingly" enter into the transaction. Generally, that mean the parties have to know they are doing something wrongful or illegal. Unlike civil fraud and abuse laws, recklessness is not sufficient for a violation of the AKS.

Any arrangement the involves remuneration between a provider and referral source or a provider and company selling healthcare services must be carefully vetted for compliance with AKS safe harbors. Importantly, however, failure to qualify for the designated safe harbors does not result in an AKS violation. The government still must show that the parties to the transaction acted wrongfully in entering the transaction to establish violation of the AKS.

References

1. *Bingham v. HCA, Inc.*, 783 F. App'x 868 (11th Cir. 2019)
2. 42 U.S.C. § 1320a-7b
3. *U.S. ex rel. Mastej v. Health Mgmt. Assocs., Inc.*, 591 F. App'x 693 (11th Cir. 2014)
4. *U.S. ex rel. Gohil v. Sanofi U.S. Servs. Inc.*, No. CV 02-2964, 2020 WL 4260797 (E.D. Pa. July 24, 2020)
5. *Klaczak v. Consol. Med. Transp.*, 458 F. Supp. 2d 622 (N.D. Ill. 2006)
6. *United States v. Starks*, 157 F.3d 833 (11th Cir. 1998)
7. *U.S. ex rel. Ruscher v. Omnicare, Inc.*, No. 4:08-CV-3396, 2015 WL 5178074 (S.D. Tex. Sept. 3, 2015), aff'd sub nom. *U.S. ex rel. Ruscher v. Omnicare, Inc.*, 663 F. App'x 368 (5th Cir. 2016)
8. *United States v. Mallory*, 988 F.3d 730 (4th Cir. 2021)
9. *United States v. Hill*, 745 F. App'x 806 (11th Cir. 2018)

10. *U.S. ex rel. Osheroff v. Tenet Healthcare Corp.*, No. 09-22253-CIV, 2013 WL 1289260 (S.D. Fla. Mar. 27, 2013)
11. *United States v. McClatchey,* 217 F.3d 823 (10th Cir. 2000)
12. *U.S. ex rel. McDonough v. Symphony Diagnostic Servs., Inc.*, 36 F. Supp. 3d 773 (S.D. Ohio 2014)
13. *Grand Union Co. v. United States*, 696 F.2d 888 (11th Cir. 1983)
14. *U.S. ex rel. Silva v. VICI Mktg., LLC*, 361 F. Supp. 3d 1245 (M.D. Fla. 2019)
15. *United States v. Sci. Applications Int'l Corp.*, 626 F.3d 1257 (D.C. Cir. 2010)
16. *U.S. ex rel. Martin v. Life Care Centers of Am., Inc.*, 114 F. Supp. 3d 549 (E.D. Tenn. 2014)
17. *Miller v. Abbott Lab'ys*, 648 F. App'x 555 (6th Cir. 2016)
18. *U.S. ex rel. Westmoreland v. Amgen, Inc.*, 812 F. Supp. 2d 39 (D. Mass. 2011)
19. *U.S. ex rel. Suarez v. AbbVie Inc.*, No. 15 C 8928, 2019 WL 4749967 (N.D. Ill. Sept. 30, 2019)
20. *U.S. ex rel. STF, LLC v. Vibrant Am., LLC*, No. 16-CV-02487-JCS, 2020 WL 4818706 (N.D. Cal. Aug. 19, 2020)
21. 42 C.F.R. § 1001.952
22. *United States v. Rogan*, No. 02 C 3310, 2006 WL 8427270 (N.D. Ill. Oct. 2, 2006), aff'd, 517 F.3d 449 (7th Cir. 2008)
23. *United States v. Millennium*, No. 1:11cv825, 2014 WL 4908275 (S.D. Ohio Sept. 30, 2014)
24. *Waldmann v. Fulp*, 259 F. Supp. 3d 579 (S.D. Tex. 2016)
25. *U.S. ex rel. Phalp v. Lincare Holdings, Inc.*, 857 F.3d 1148 (11th Cir. 2017)
26. U.S. Dept. of Justice, Telemedicine Company Owner Charged in Superseding Indictment for $784 Million Health Care Fraud, Illegal Kickback and Tax Evasion Scheme (Aug. 10, 2021), https://www.justice.gov/opa/pr/telemedicine-company-owner-charged-superseding-indictment-784-million-health-care-fraud
27. Office of Inspector General, Special Fraud Alert: Speaker Programs (Nov. 16, 2020), available at https://oig.hhs.gov/compliance/alerts/index.asp
28. U.S. Dept. of Justice, U.S Attorney's Office Southern District of New York, Acting Manhattan U.S. Attorney Announces $678 Million Settlement Of Fraud Lawsuit Against Novartis Pharmaceuticals For Operating Sham Speaker Programs Through Which It Paid Over $100 Million To Doctors To Unlawfully Induce Them To Prescribe Novartis Drugs (Jul. 1, 2020), https://www.justice.gov/usao-sdny/pr/acting-manhattan-us-attorney-announces-678-million-settlement-fraud-lawsuit-against
29. U.S. Dept. of Justice, Electronic Health Records Vendor to Pay $145 Million to Resolve Criminal and Civil Investigations (Jan. 27, 2020), https://www.justice.gov/opa/pr/electronic-health-records-vendor-pay-145-million-resolve-criminal-and-civil-investigations-0
30. U.S. Dept. of Justice, U.S Attorney's Office District of Massachusetts, Athenahealth Agrees to Pay $18.25 Million to Resolve Allegations that It Paid Illegal Kickbacks (Jan. 28, 2021), https://www.justice.gov/usao-ma/pr/athenahealth-agrees-pay-1825-million-resolve-allegations-it-paid-illegal-kickbacks
31. U.S. Dept. of Justice, U.S Attorney's Office Southern District of Florida, Miami-Based CareCloud Health, Inc. Agrees to Pay $3.8 Million to Resolve Allegations that it Paid Illegal Kickbacks (Apr. 30, 2021), https://www.justice.gov/usao-sdfl/pr/miami-based-carecloud-health-inc-agrees-pay-38-million-resolve-allegations-it-paid
32. NBC News, A. Kaplan, J. Blackman, T. Costello & S. Ploss, Feds take down $1 billion Medicare fraud scheme in 'Operation Brace Yourself' (Apr. 9, 2019), https://www.nbcnews.com/politics/justice-department/feds-take-down-1-billion-fraud-scheme-operation-brace-yourself-n992481 (last accessed Aug. 15, 2021)
33. Federal Bureau of Investigations, Billion-Dollar Medicare Fraud Bust FBI Announces Results of Operation Brace Yourself (Apr. 9, 2019), https://www.fbi.gov/news/stories/billion-dollar-medicare-fraud-bust-040919 (last accessed Aug. 15, 2021)
34. U.S. Dept. of Justice, U.S Attorney's Office Middle District of Florida, Local Businesswoman Pleads Guilty To Criminal Healthcare And Tax Fraud Charges And Agrees To $20.3 Million Civil Settlement (Feb. 4, 2021), https://www.justice.gov/usao-mdfl/pr/local-businesswoman-pleads-guilty-criminal-healthcare-and-tax-fraud-charges-and-agrees

35. U.S. Dept. of Justice, Five Individuals Charged for Roles in $65 Million Nationwide Conspiracy to Defraud Federal Health Care Programs (Apr. 22, 2021), https://www.justice.gov/opa/pr/five-individuals-charged-roles-65-million-nationwide-conspiracy-defraud-federal-health-care
36. U.S. Department of Justice, Jury Convicts Medical Equipment Company Owners of $27 Million Fraud (Jul. 9, 2021), https://www.justice.gov/opa/pr/jury-convicts-medical-equipment-company-owners-27-million-fraud
37. U.S. Dept. of Justice, National Health Care Fraud and Opioid Takedown Results in Charges Against 345 Defendants Responsible for More than $6 Billion in Alleged Fraud Losses (Sept. 30, 2020), https://www.justice.gov/criminal-fraud/hcf-2020-takedown/press-release
38. U.S. Dept. of Justice, DOJ Announces Coordinated Law Enforcement Action to Combat Health Care Fraud Related to COVID-19 (May 26, 2021), https://www.justice.gov/opa/pr/doj-announces-coordinated-law-enforcement-action-combat-health-care-fraud-related-covid-19
39. U.S. Dept. of Justice, Compounding Pharmacy, Two of Its Executives, and Private Equity Firm Agree to Pay $21.36 Million to Resolve False Claims Act Allegations (Sept. 18, 2019), https://www.justice.gov/opa/pr/compounding-pharmacy-two-its-executives-and-private-equity-firm-agree-pay-2136-million
40. U.S. Dept. of Justice, Compounding Pharmacy Mogul Sentenced for Multimillion-Dollar Health Care Fraud Scheme (Jan. 15, 2021) https://www.justice.gov/opa/pr/compounding-pharmacy-mogul-sentenced-multimillion-dollar-health-care-fraud-scheme
41. U.S. Dept. of Health and Human Services Office of Inspector General, Principal Deputy Inspector General Grimm on Telehealth (Feb. 26, 2021), https://oig.hhs.gov/coronavirus/letter-grimm-02262021.asp

The Stark Law

Jonathan H. Ferry and Lyndsay E. Medlin

Contents

Key Points
- The Stark Law prohibits referrals of DHS.
- Services personally performed by a physician are not subject to the Stark Law.
- The Stark Law broadly prohibits financial relationships between clinicians and entities to which clinicians refer DHS.
- Safe harbors are extremely important because they allow many essential business arrangements.
- Strict adherence to safe harbors is essential.
- Violation of the Stark Law does not require any intent on the part of the entities involved.
- Self-Disclosure of Stark Law violations is an option.

Legal Concepts
- Financial relationship between referring clinician and entity providing medical services is central to Stark Law
- Definition of Designated Health Services ("DHS")
- Stark Law violations are strict liability

J. H. Ferry (✉) · L. E. Medlin
Bradley Arant Boult Cummings LLP, Charlotte, NC, USA
e-mail: jferry@bradley.com; lmedlin@bradley.com

© The Author(s), under exclusive license to Springer Nature Switzerland AG 2022   303
A. S. Pasha (ed.), *Laws of Medicine*,
https://doi.org/10.1007/978-3-031-08162-0_19

- Safe Harbors are central to Stark Law compliance because many common health-care relationships fall under definition of financial relationship
- Safe Harbors require strict compliance with all requirements
- Self-disclosure protocols important to mitigating potential damages from inadvertent Stark Law violations

1   Introduction

Originally passed in 1989 to rectify perceived abuses in the clinical laboratory industry, the Physician Self-Referral Law, commonly known as the Stark Law, prohibits physicians from referring certain "designated health services" ("DHS") to entities with which the physician or a family member has a financial relationship. The Stark Law prohibits entities from submitting claims to Medicare and Medicaid for services referred by such physicians.

When originally passed, the Stark Law covered only diagnostic laboratories, but the scope of the statute was expanded in 1992 and now covers 10 broad categories of DHS [1]. Conceptually, the Stark Law is broadly drafted to prohibit any financial relationship between a referring physician and the entity providing DHS (the "DHS Entity"). This broad prohibition, however, affects many common business arrangements. Thus, statutory and regulatory safe harbors, akin to Anti-Kickback Statute ("AKS") safe harbors discussed herein, have been established to allow many of these arrangements as long as the arrangements are structured to strictly comply with the requirements.

Unlike the AKS, which is a criminal statute, the Stark Law, with exclusively civil consequences, is a strict liability statute, meaning the government need not make any showing of a defendant's intent to violate the statute to demonstrate that a healthcare provider or company violated the law. Nonetheless, violations of the Stark Law are routinely the basis of FCA cases when the government can show that the entities involved acted with the requisite FCA scienter in submitting claims involving conduct violative of the Stark Law. Such cases have resulted in enormous liability for healthcare providers [2–5].

2   Discussion

The Stark Law forbids submitting claims to Medicare for "designated health services," provided under a "referral" made by a doctor with whom the entity has a "financial relationship" [6, 7]. Understanding the application of the three quoted terms is critical to understanding the Stark Act.

2.1   Elements of Stark Violation

Designated Health Services

The Stark Act covers only medical services specified to be DHS under regulation. DHS includes:

1. Clinical laboratory services;
2. Physical therapy, occupational therapy, and outpatient speech-language pathology services;
3. Radiology and certain other imaging services;
4. Radiation therapy service and supplies;
5. Durable medical equipment and supplies;
6. Parenteral and enteral nutrients, equipment, and supplies;
7. Prosthetics, orthotics, and prosthetic devices and supplies;
8. Home health services;
9. Outpatient prescription drugs
10. Inpatient and outpatient hospital services [8].

More specifically, each year the government publishes the specific CPT codes associated with DHS in the Federal Register [1].

Financial Relationship

The Stark Law's definition of the various prohibited financial relationships is designed to identify and focus on those relationships pursuant to which a physician (or an immediate family member) would gain financial benefit from referring DHS to another person or entity. As the Third Circuit Court of Appeals explains "[a] referral is ripe for abuse only when the doctor who made it has a financial relationship with the provider. Only then can a doctor profit from his own referral" [6].

The Stark Law broadly defines a financial relationship as (1) an ownership or investment interest in an entity by the physician or family member; or (2) a compensation arrangement between the physician or family member and the entity performing DHS [7, 9]. Family members are defined as: (1) husband or wife; (2) birth or adoptive parent, child, or sibling; (3) stepparent; (4) stepchild; (5) stepbrother or stepsister; (6) father-in-law; (7) mother-in-law; (8) son-in-law; (9) daughter-in-law; (10) brother-in-law; (11) sister-in-law; (12) grandparent or grandchild; and (13) spouse of a grandparent or grandchild.

The Stark Law further defines "ownership or investment interest" and "compensation arrangement" [9]. An ownership or investment interest includes equity, debt, or other means of being invested in a company, and includes an interest in an entity that holds an ownership or investment interest in the entity that furnishes DHS. This can include ownership of stocks, partnership interests, limited liability memberships, and loans, bonds or other financial instruments secured with the DHS entity's property or revenue [9]. Compensation arrangements include any arrangement involving remuneration, direct or indirect, between a physician (or a member of the physician's immediate family) and the entity providing DHS.

Financial relationships between a physician or family members and the DHS entity may be either direct or indirect [9]. An indirect ownership or investment interest exists if there is an "unbroken chain of any number (but no fewer than one) of persons or entities having ownership or investment interest" between the referring physician (or immediate family member) and the entity providing DHS [9]. A direct compensation relationship involves remuneration paid directly by the DHS providing entity to the physician (or immediate family members). An indirect compensation arrangement exists only if several criteria are met including that the referring physician (or immediate family member) receives aggregate compensation from the person or entity with whom the physician has a direct financial relationship that varies with the volume or value of referrals generated by the physician for the entity providing DHS.

Referrals

Whether or not a physician's action constitutes a referral is often a straightforward application of the regulatory definitions. At times, however, whether or not certain activities constitute a referral are less clear. Generally speaking, a referral is "the request by a physician, or ordering of, or the certifying or recertifying the need for, any designated health service for which payment may be made under Medicare Part B..." and "a request by a physician that includes the provision of any designated health service for which payment may be made under Medicare, the establishment of a plan of care by a physician that includes the provision of such designated health service, or the certifying or recertifying of the need for such designated health service..." [8]. However, a referral does not include any service personally performed by the physician [8].

This rather complex definition of referral is meant to reach the various activities a physician may engage in that would trigger the provision of DHS. For example, it is clear that if a physician sends a patient to a particular radiology practice or diagnostic testing lab for imaging or testing, a referral by that physician to those entities providing DHS has occurred. But, less clear is the fact that the definition of referral also includes a clinician sending a patient for establishment of a plan of care or certification for home health. Referrals may also include facility fees charged by hospitals when a physician performs inpatient or outpatient services in the hospital [10].

2.2   Stark Safe Harbors

The drafters of the Stark Law chose to broadly prohibit referrals between any physician and the entity with which she or an immediate family member has a financial relationship. From the beginning, however, it was clear that certain financial relationships, either by their very nature, or if structured with the appropriate safeguards, do not raise concerns of over-utilization and financial influence on medical decision-making. The statute itself thus includes several safe harbor provisions, supplemented by multiple regulatory safe harbors, allowing certain relationships between DHS entities and their physician referral sources [7, 11–13]. The safe harbors are vitally important, as they allow a number of common and beneficial relationships to exist. Reliance on the safe harbors, however, carries risk. Failure to comply with every requirement of the applicable safe harbor negates its protections, resulting in Stark Law violations. Several of the more significant safe harbors are summarized below.

Rental of Office Space [13]

Clinicians frequently rent office space from larger healthcare providers such as hospitals or large group practices. Compliance with this safe harbor allows a physician to continue to refer patients to the entity from which she rents the office space. To comply with this safe harbor, the terms of lease agreement must meet several criteria including that the agreement be in writing and signed, specify the premises covered, be for a duration of at least 1 year, and commercially reasonable and consistent with fair market value for the legitimate business purposes set forth therein, without consideration of referrals.

Bona Fide Employment Relationships [13]

The bona fide employment relationship safe harbor is one of the most important, as it allows health systems, hospitals, and other large healthcare providers to employ physicians that refer patients for services to the provider. Health systems generally dedicate significant resources to ensuring that physician employment relationships comply with this safe harbor. For employment to be considered a bona fide exception, it must meet the following criteria:

1. The employment is for identifiable services;
2. The amount of remuneration under the employment is –
   (a) Consistent with fair market value for the services and
   (b) Not determined in a manner that takes into account the volume or value of referrals by the referring physician; and
3. The compensation is provided under an arrangement that would be commercially reasonable even if no referrals were made to the employer.

The safe harbor also allows for physician employee productivity bonuses based on services personally performed by the physician. Failure to comply with this safe harbor has been the basis of multiple FCA cases against hospital and major health systems [14–16].

Personal Services Arrangements [13]

The personal services safe harbor is critical to the healthcare industry because it allows for physicians to provide contract services even when the clinician cannot take on full employment. It is the exception utilized for most forms of medical director relationships when the physician is not a full-time employee of the DHS entity. Part-time medical director relationships are common for hospitals, skilled nursing facilities, hospice providers, and home health providers. The personal services exemption requires an arrangement meet generally the same criteria as the space rental safe harbor, with the added requirement that the services furnished under the arrangement do not involve the counseling or promotion of a business arrangement or other activity that violates federal or state law.

The use of medical directors is also a commonly abused practice, and as such, the subject of enforcement authority scrutiny and investigation for potential Stark Law violations. Such arrangements are often deemed to run afoul of the AKS as well when enforcement authorities determine that the arrangement is a mere artifice to disguise bribes to physicians for referrals [17]. Enforcement authorities have levied substantial penalties on companies and physicians who allegedly entered into sham professional services agreements to disguise payments for referrals [18]. In one settlement announcement, DOJ alleged that, although the defendant had entered into numerous physician-services contracts "ostensibly to retain physicians as medical directors…in reality the company's payments under these contracts were intended to induce the physician to refer patients" to the company's facilities [18].

In-Office Ancillary Services [11]

This exemption allows clinicians and medical practices to engage in referrals within the practice under certain circumstances. Practice groups rely on this exemption to permit clinicians to refer services to other practitioners within the group, including internal laboratory, imaging, and other diagnostic services. Group practice members may thus benefit from the profits earned on ancillary services that the clinicians do not personally provide. Absent this exemption, such activity would meet the definition of a prohibited referral for DHS to an entity with which the clinician has a financial relationship in violation of the Stark Law.

This exception has several highly technical requirements related to who must perform the services, where the services must be performed, how the services must be billed, how the group practice is organized, the range of care provided by the physicians within the group, and how physicians within the group are compensated [11]. Failure to properly organize the group practice and adhere to various requirements of this exception can result in all claims for ancillary services in the group practice being non-payable, resulting in significant liability.

2.3   Self-Disclosure Protocols

Recognizing that the Stark Law can potentially result in significant liability for unintentional violations, the Patient Protection and Affordable Care Act of 2010 required HHS to establish a self-disclosure protocol setting forth a process to enable providers of services and suppliers to self-disclose actual or potential violations of the Stark Law [19]. Importantly, CMS is allowed to decrease penalties otherwise applicable to Stark Law violations when an entity self discloses under the protocol. CMS has broad statutory discretion in deciding what factors to apply in determining appropriate penalties under the self-disclosure protocols including:

1. The nature and extent of the improper or illegal practice;
2. The timeliness of such self-disclosure;
3. The amount of the defendant's cooperation in providing additional information related to the disclosure; and
4. Such other factors as the Secretary considers appropriate [19].

The self-disclosure protocol has proven a valuable resource for companies that discover Stark Law violations through internal reviews or other compliance measures. From 2011 to 2020, CMS settled 369 matters with entities under the voluntary disclosure protocols [20].

3   Application

Although the Stark Law applies to all physicians, enforcement has often focused on large organizations that employ large numbers of physicians, and is based on compensation agreements that do not comply with the employment safe harbor, thereby creating the incentive for overutilization or otherwise distorting medical decision-making.

In *U.S. ex rel. Drakeford v. Tuomey Healthcare System*, the question of what constituted a referral for the purposes of the Stark Law was at issue. Tuomey was a hospital system in South Carolina [10]. At the time of the case, most of the clinicians who performed surgeries at the hospital were members of private practices. Many of these clinicians that had previously performed surgeries at Tuomey, started performing surgeries in outpatient surgery centers, resulting in a significant loss of revenue for Tuomey, as Tuomey billed Medicare a significant facility fee every time a surgeon conducted an operation in the hospital. Tuomey entered into part-time employment contracts with surgeons establishing a financial relationship under the Stark Law. The question was whether or not the physicians made prohibited referrals to Tuomey. Because the physicians conducted the surgery at the hospital and personally performed the services, Tuomey argued that there was no referral. But, the court found that, when the surgeon conducted the surgery at the hospital, there was a referral of the facility fee that the hospital billed to Medicare [21]. Thus, a prohibited referral did exist. At trial, the jury determined that Tuomey's contracts with the physicians provided compensation that varied with the volume and value of those referrals in violation of the Stark Law. The jury also determined that Tuomey violated the FCA by submitting the claims to Medicare despite knowing that they violated the Stark Law and were not payable. Tuomey was found liable for over $237 million in treble damages and penalties even though Medicare paid Tuomey only $39 million for the improper claims.

4   Summary

The Stark Law is violated if a physician refers DHS to an entity with which the physician as a financial relationship under the statute. Medicare will not pay claims for services referred in violation of the Stark Law. Financial relationships are broadly defined encompassing many common and beneficial arrangements. Multiple statutory and regulatory safe harbors are therefore established exempting certain arrangements from violating the Stark Law, but only if the parties strictly adhere to the safe harbor requirements. Unlike the AKS, or even the FCA, violation of the Stark Law requires no intent whatsoever. It is a strict liability statute. Thus honest mistakes still result in violations and potential liability.

Recognizing the potentially enormous liability that may result from even inadvertent violations of the Stark Law, CMS established self-disclosure protocols that allow entities to self report potential Stark Law violations and seek resolution on favorable terms. Such terms may often be significantly below that amount of the claims submitted in violation of the Stark Act.

References

1. Centers for Medicare & Medicaid Services, Code List for Certain Designated Health Services (DHS) (Dec. 23, 2020), https://www.cms.gov/Medicare/Fraud-and-Abuse/PhysicianSelfReferral/List_of_Codes
2. U.S. Dept. of Justice, West Virginia Hospital Agrees To Pay $50 Million To Settle Allegations Concerning Improper Compensation To Referring Physicians (Sept. 9, 2020), https://www.justice.gov/opa/pr/west-virginia-hospital-agrees-pay-50-million-settle-allegations-concerning-improper
3. U.S. Dept. of Justice, Home Health Agency and Former Owner to Pay $5.8 Million to Settle False Claims Act Allegations (Nov. 20, 2020), https://www.justice.gov/opa/pr/home-health-agency-and-former-owner-pay-58-million-settle-false-claims-act-allegations
4. U.S. Dept. of Justice, California Health System and Surgical Group Agree to Settle Claims Arising from Improper Compensation Arrangements (Nov. 15, 2019), https://www.justice.gov/opa/pr/california-health-system-and-surgical-group-agree-settle-claims-arising-improper-compensation
5. U.S. Dept. of Justice, United States Resolves $237 Million False Claims Act Judgment against South Carolina Hospital that Made Illegal Payments to Referring Physicians (Oct. 16, 2015), https://www.justice.gov/opa/pr/united-states-resolves-237-million-false-claims-act-judgment-against-south-carolina-hospital
6. *U.S. ex rel. Bookwalter v. Univ. Pittsburgh Med. Ctr.*, 946 F.3d 162 (3rd Cir. 2019)
7. 42 U.S.C. § 1395nn
8. 42 C.F.R. § 411.351
9. 42 C.F.R. § 411.354
10. *U.S. ex rel. Drakeford vs. Tuomey Healthcare System*, 675 F.3d 364 (4th Cir. 2012)
11. 42 C.F.R. § 411.355
12. 42 C.F.R. § 411.356
13. 42 C.F.R. § 411.357
14. U.S. Dept. of Justice, West Virginia Hospital Agrees To Pay $50 Million To Settle Allegations Concerning Improper Compensation To Referring Physicians (Sept. 9, 2020), https://www.justice.gov/opa/pr/west-virginia-hospital-agrees-pay-50-million-settle-allegations-concerning-improper

15. U.S. Dept. of Justice, Adventist Health System Agrees to Pay $115 Million to Settle False Claims Act Allegations (Sept. 21, 2015), https://www.justice.gov/opa/pr/adventist-health-system-agrees-pay-115-million-settle-false-claims-act-allegations

16. U.S. Dept. of Justice, Florida Hospital System Agrees to Pay the Government $85 Million to Settle Allegations of Improper Financial Relationships with Referring Physicians (Mar. 11, 2014), https://www.justice.gov/opa/pr/florida-hospital-system-agrees-pay-government-85-million-settle-allegations-improper

17. U.S. Dept. of Justice, Home Health Agency and Former Owner to Pay $5.8 Million to Settle False Claims Act Allegations, (Nov. 20, 2020), https://www.justice.gov/opa/pr/home-health-agency-and-former-owner-pay-58-million-settle-false-claims-act-allegations

18. U.S. Dept. of Justice, Post Acute Medical Agrees to Pay More Than $13 Million to Settle Allegations of Kickbacks and Improper Physician Relationships, Department of Justice Press Release (Aug.15, 2018), https://www.justice.gov/opa/pr/post-acute-medical-agrees-pay-more-13-million-settle-allegations-kickbacks-and-improper

19. 42 U.S.C. § 1395nn note (Affordable Care Act Section 6409)

20. Centers for Medicare and Medicaid Services, https://www.cms.gov/Medicare/Fraud-and-Abuse/PhysicianSelfReferral/Self-Referral-Disclosure-Protocol-Settlements (visited August 17, 2021)

21. U.S. ex rel. Drafeford vs. Tuomey Healthcare System, 675 F.3d 364, 374 (4th Cir. 2015).

Office of Inspector General Administrative Authorities

Jonathan H. Ferry and Lyndsay E. Medlin

Contents

Key Points
- OIG has authority to exclude individuals and business entities from federal healthcare programs.
- OIG is required to exclude persons or entities for various felony offenses involving healthcare.
- OIG is permitted to exclude persons or entities for a number of other activities detrimental to federal healthcare programs.
- Exclusion can be appealed to an administrative law judge.
- OIG may level monetary penalties on persons or companies for a number of activities detrimental to federal healthcare programs.
- Civil penalties and exclusion are subject to review by administrative process within HHS.
- Final administrative determinations are appealable to the federal courts.

Legal Concepts
- OIG has both permissive exclusion authority and mandatory exclusion responsibility
- Civil monetary penalties designed to address past violations and deter future violations

J. H. Ferry (✉) · L. E. Medlin
Bradley Arant Boult Cummings LLP, Charlotte, NC, USA
e-mail: jferry@bradley.com; lmedlin@bradley.com

A. S. Pasha (ed.), *Laws of Medicine*,
https://doi.org/10.1007/978-3-031-08162-0_20

- Exclusion is forward looking remedy focused on protecting federal health programs from providers that have a high risk of fraudulent behavior
- Administrative process with ultimate review by federal courts
- OIG may forgo exclusion in exchange for a corporate integrity agreement that mitigates the risk of future fraudulent

1 Introduction

OIG has responsibility for protecting federal healthcare programs from fraud and abuse. Due to their conduct, certain individuals and entities present an unacceptable risk to the programs or their beneficiaries. Such individuals are candidates for exclusion under the OIG's mandatory and permissive exclusion powers and may be subject to civil monetary penalties. Excluded individuals and entities cannot receive payment from federal healthcare programs for any services they perform. Entities submitting claims to federal healthcare programs can also be sanctioned for employing an excluded individual. OIG maintains a list of excluded individuals and entities that clinicians must check before hiring a healthcare employee.

Exclusion lasts for at least a specified period of time, after which a person or entity can petition OIG for reinstatement in the programs. OIG's decision to exclude an individual or entity is subject to appeal procedures within the Department of Health and Social Services, and ultimately, review by a United States Circuit Court of Appeals. OIG often negotiates voluntary exclusion with individuals or entities as part of civil FCA settlements or criminal plea agreements.

In addition to exclusion authority, OIG may levy monetary penalties on individuals or entities pursuant to the CMPS. The CMPS allows OIG to impose penalties for 10 different categories of offenses and specifies a specific penalty amount for each offense. Penalties under the CMPS may be appealed within HHS up to an administrative law judge and ultimately to a U.S. Circuit Court of Appeals.

2 Discussion

The administrative penalties that OIG can implement to protect federal healthcare programs include exclusion and civil monetary penalties [1]. Although not exactly parallel, the circumstances under which OIG may impose monetary penalties and exclude individuals or entities from participation in federal healthcare programs greatly overlap. Exclusion is often imposed in conjunction with the resolution of criminal and civil actions by the DOJ in healthcare related criminal cases or FCA cases.

This section addresses the circumstances under which OIG seeks civil penalties or exclusions and the bases upon which OIG will assess the amount of penalties and determine the length of exclusion. We also address procedures for imposition of penalties and exclusion and the rights of individuals or entities to challenge efforts by OIG to impose these administrative sanctions.

2.1   Bases for Imposition of Civil Monetary Penalties and Exclusion

Although OIG often seeks exclusion in the context of criminal and civil FCA cases brought by the DOJ, it has independent authority to impose these sanctions for a wide variety of violations that do not rise to criminal activity or claims under the FCA. Although OIG may impose civil monetary penalties or exclusion for many regulatory offenses, this section focuses on those offenses that relate to fraud and abuse in government healthcare programs.

OIG may impose civil monetary penalties on, and exclude, individuals involved in the submission of false or fraudulent claims to federal healthcare programs, as well as those who violate the AKS or Stark Law [1–3]. With respect to false and fraudulent claims, the scope of OIG's authorities broadly parallels the FCA, but OIG's authorities are in many cases more specific or encompass matters that may not be actionable under criminal statutes or the FCA. For example, OIG may impose civil monetary penalties or exercise exclusion on individuals or entities that:

1. Submit claims for services provided by a person who was not a licensed physician;
2. Submit claims for services by a physician whose license lapsed or represented to a patient that the physician was board certified when she was not;
3. Submit claims in violation of provider agreements with federal health programs;
4. Provide false or misleading information to hospital patients that could influence the decision of when to discharge the patient;
5. Are excluded individuals who retain interest or control of an entity that participates in federal healthcare programs, or are officers or managing employees of such entity;
6. Contract with an excluded individual for services that may be paid by a federal healthcare program;
7. Fail to report an overpayment; and
8. Fail to grant timely access to records to the OIG.

The maximum amount of civil monetary penalties and period of exclusion vary with the specific violation being addressed. OIG must apply certain considerations in determining the actual penalty or length of exclusion including: (1) the nature and circumstances of the violation; (2) the degree of culpability of the person or entity; (3) the history of prior offenses; and (4) other wrongful conduct [4]. Some exclusion periods, however, are dictated by statute. For example, in the case of felony criminal violations related to healthcare fraud, patient abuse, or controlled substances, OIG is required to exclude an the individual or entity for a *minimum* of 5 years for the first offense, 10 years for the second offense, and permanently for third offense [5].

Aside from the criminal offenses addressed above, OIG may, but is not required to, exclude individuals or entities for a variety of other offenses, including criminal misdemeanor convictions for fraud, other types of healthcare fraud, or controlled substances violations, and certain types of civil healthcare fraud. In cases of permissive exclusion, the period of exclusion may be established by statute, but OIG may increase or decrease the period depending on the factors detailed in the paragraph above. For example, criminal misdemeanor offenses related to fraud, healthcare fraud, patient abuse or neglect, or substance have a 3 year exclusion baseline "unless the secretary determines in accordance with published regulations that a shorter period is appropriate because of mitigating circumstances or a longer period is appropriate because of aggravating circumstances" [5]. Circumstances that may favor mitigation of culpability include whether the person took appropriate and timely corrective action such as notification of OIG of the issue and fully cooperating with OIG in any investigation. Aggravating factors include "actual" knowledge of the violations at issue [4].

2.2   OIG Administrative Procedures

When OIG intends to impose monetary penalties, it must send notice to the affected entities [6]. Generally, OIG's notice must include the statutory basis for the penalties, and the details of the violation and proposed penalty, including the what, why, and how the penalty was determined. The notice will also include instructions for responding and an explanation of the right to a hearing if requested in 60 days.

Similarly, OIG must provide notice regarding exclusion. OIG's process is slightly different depending on the basis for exclusion. For exclusions related to fraud and abuse, OIG issues a Notice of Proposal to Exclude [7]. The notice must include the basis, length, and effect of the exclusion, as well as the earliest date on which O IG will consider reinstatement and the process for seeking reinstatement and appealing the exclusion [7].

The exclusion will be effective 60 days after the notice is received, unless the excluded entity files a written request for hearing [7]. The request for a hearing can include an explanation of why the health and safety of individuals receiving services under Medicare or any of the state healthcare programs does not warrant exclusion going into effect prior to the completion of an administrative law judge ("ALJ") proceeding [7, 8]. If OIG determines that the health or safety of program beneficiaries does not require exclusion before a hearing, then exclusion will be stayed pending the hearing before the ALJ [7]. An appeal to the ALJ of a permissive exclusion is limited to only two issues: (1) the basis for the imposition of the exclusion; and (2) whether the length of the exclusion is reasonable [9]. If the appeal is of a mandatory exclusion, however, no such limitation exists [9]. Parties before the ALJ may be represented by counsel, conduct document discovery, present evidence, present and cross-examine witnesses, present oral arguments, and submit written briefs and proposed findings of fact and conclusions of law after the hearing [10].

An ALJ is empowered to affirm, reduce, or increase penalties or the length of exclusion after the hearing. Either the defendant or the OIG may then appeal the ALJ's decision to the Departmental Appeals Board ("DAB") within 30 days of the decision. The notice of appeal to the DAB must fully describe the appellant's argument against the initial decision by the ALJ. The DAB may decline to hear the case, affirm, reduce, or increase sanctions previously in place. Appeal to the DAB stays the ALJ decision in CMPS cases. After the DAB decision, the respondent case request a stay of the DAB decision pending appeal to a federal court.

2.3 Exclusion in Context of DOJ Civil and Criminal Cases

Exclusion is often a factor in criminal healthcare and FCA cases. In the context of criminal healthcare cases, defendants often agree to a period of exclusion in the context of plea negotiations. Thus, OIG need not engage in formal exclusion proceedings because the defendant agreed to voluntary exclusions. Similarly, in FCA investigations, OIG is engaged in all settlements, and in fact, is a signatory to FCA settlements between DOJ and defendants. As part of the settlement agreement, OIG may negotiate a period of voluntary exclusion, which is then included in the agreement terms. Often, in lieu of exclusion, OIG will agree to a corporate integrity agreement pursuant to which the activities of the defendant remain subject to monitoring by OIG, and the defendant company is subject to mandatory reporting obligations to OIG.

3 Summary

OIG authorities to level civil monetary penalties and exclude individuals and organization from federal healthcare programs are potent tools in protecting these programs from abuse. Unlike many other healthcare fraud and abuse remedies that seek to rectify past wrongs, exclusion is forward looking and prevents high risky players from continuing to damage federal healthcare programs. OIG maintains a list of excluded entities and persons, and healthcare providers are held responsible for checking this list before employing individuals in jobs that involve federal heathcare programs.

Because OIG's exlucsion authorities are meant to address future risk the inquiry is quite different and past bad actors can rehabilitate themselves with mitigation, remediation and revamped compliance measures going forward. In many case, notwithstanding past bad conduct and in non-criminal healthcare fraud cases, OIG will agree to corporate integrity agreements pursuant to which the entity is monitored by OIG on an ongoing basis to ensure improved compliance with federal programs rules and regulations.

References

1. 42 C.F.R. § 1003.100
2. 42 C.F.R. § 901
3. 42 C.F.R. § 942
4. 42 C.F.R. § 1003.140(a)

5. 42 U.S.C. § 1320a-7(a)-(c)
6. 42 C.F.R. § 1003.1500
7. 42 C.F.R. § 1001.2003
8. 42 C.F.R. 1001.2002(c)
9. 42 C.F.R. § 1001.2007(a)
10. 42 C.F.R. § 1005.3(a)

Criminal Healthcare Fraud

Jonathan H. Ferry and Lyndsay E. Medlin

Contents

Key Points
- The HCF Statute is a criminal statute with maximum penalty of 10-year imprisonment.
- It is applicable to fraud against federal healthcare programs and private insurance companies.
- The requirement of criminal intent distinguishes these cases from actions pursuant to the FCA and other civil enforcement statutes.

Legal Concepts
- Parallel investigations include both criminal and civil investigations proceeding together
- Criminal scienter requires defendant acted with knowledge that it was engaged in illegal conduct
- Criminal cases involve multiple potential criminal statutes in addition to the HCF Statute.
- Prosecutions under the HCF Statute are subject to "beyond a reasonable doubt" burden of proof, which is significantly higher than "preponderance of evidence" standard applied in civil enforcement actions.

J. H. Ferry (✉) · L. E. Medlin
Bradley Arant Boult Cummings LLP, Charlotte, NC, USA
e-mail: jferry@bradley.com; lmedlin@bradley.com

© The Author(s), under exclusive license to Springer Nature Switzerland AG 2022   319
A. S. Pasha (ed.), *Laws of Medicine*,
https://doi.org/10.1007/978-3-031-08162-0_21

1   Introduction

The HCF Statute penalizes anyone who knowingly and willfully executes or attempts to execute a scheme to defraud any healthcare benefit program or to obtain by false or fraudulent pretenses, representations, or promises, any of money or property owned by, or under the custody or control of, any healthcare benefit program, in connection with the delivery of or payment for health care benefits, items, or services. A person need not have actual knowledge or specific intent to violate the HCF Statute, but courts strictly interpret scienter with respect to this criminal statute, to the benefit of defendants.

Typical schemes involved in cases under the HCF Statute include false claims submitted to obtain payment from federal programs and private insurers [1, 2]. These cases often involve billing for services not rendered or products not delivered. Occasionally, however, the government will pursue cases that do not involve false billing or monetary loss [3]. Courts have also found that the execution of a provider enrollment form stating that the individual would comply with regulations when the clinician had no intent to do so may constitute a sufficient false statement in violation of the statute [4].

Evidence of the intent of the defendants is the key difference between cases in which the government pursues civil claims under the FCA and those where it brings criminal charges under the HCF Statute. Clinicians should also be aware, that while government civil enforcement tools are focused on monetary harm to federal programs, the HCF Statute requires no monetary damage to the government and is equally applicable to fraudulent schemes against private insurers [5].

2   Discussion

Cases prosecuted under the HCF Statute often involve broad schemes encompassing several different illegal activities that may be charged as one or more counts against defendants. Importantly for clinicians, however, courts rigorously police the intent requirements under this statute so that the unwitting, careless, or even negligent are not swept into criminal charges in the context of this complex regulatory environment.

The Department of Justice dedicates substantial resources to investigating and prosecuting criminal healthcare fraud. The Healthcare Fraud Unit within the Criminal Division claims over 80 prosecutors organized in Washington, D.C. and throughout the country in strike forces in areas particularly beset with fraud. DOJ identifies potential targets through a variety of means.

First, whistleblowers who bring False Claims Act cases under the *qui tam* provisions described in Chap. 17, often identify potentially criminal conduct. DOJ policy requires close coordination between civil and criminal healthcare prosecutors to identify appropriate cases for parallel civil and criminal investigations [6]. *Qui tam* complaints are reviewed by criminal healthcare prosecutors to evaluate the cases for potential criminal liability. Importantly for practitioners advising healthcare providers or for clinicians that find themselves in an investigation, contact from the government that implies a focus on civil liability does not preclude the potential that criminal investigation is either possible in the future or currently underway. For example, receipt of a civil investigative demand under the False Claims Act means there is certainly a civil investigation. But, civil attorneys at DOJ can share any information obtained from such demands with criminal prosecutors. Even if a criminal investigation is not underway at the beginning of such an inquiry, information developed during the civil investigation can lead to criminal charges or investigation. Clinicians and those advising them often find it advisable to clarify the scope (civil, criminal, or both) at the beginning of any investigation, and check in on this status periodically. The government may not mislead an investigatory target about the nature of its investigation if asked [7].

Additionally, DOJ's criminal division has become increasingly proficient in the use of data analytics to identify outlier billing patterns and financial relationships that can lead to criminal investigations. In 2019, the Deputy Assistant Attorney General of DOJ's Criminal Division noted the importance of data analytics stating, "our health care fraud prosecutors are able to leverage data analytics to identify fraud indicators within Medicare claims data. This use of data analytics has allowed for greater efficiency in identifying investigation targets, which expedites case development, saves resources, makes the overall program of enforcement more targeted and effective" [8]. Applications to the Medicare program and claims data contain a wealth of information about healthcare providers and the services they provide. In conjunction with numerous contractors, DOJ can isolate billing trends, identify suspect codes, services, diagnoses, and pinpoint the healthcare providers that fit the patterns of potential fraud. Clinicians may find themselves on target lists due to nothing more than outlier billing patterns in suspect services. One lesson from DOJ's activity in this area is the need for compliance programs to similarly analyze data to identify areas of potential concerns before enforcement authorities come to investigate.

Criminal healthcare fraud cases are varied in scope and complexity. The most basic cases of a clinician submitting claims for services never performed are based on the simple allegation that the clinician lied to the payor to obtain payment to which he or she was not entitled. More complex cases may involve multiple co-conspirators and the violation of numerous statutes including the Anti-Kickback Statute and the Healthcare Fraud Statute. An example of such cases currently being pursued by DOJ involve billing Medicare for durable medical equipment that was never properly ordered by a physician and not needed by the patients [9]. In one case, the government alleged, that call centers, some based outside the United States, found Medicare beneficiaries and convinced them to take durable medical equipment. The call centers then partnered with telemedicine companies that employed physicians to write fraudulent orders for the equipment without examining or even in many cases speaking to the patient. The orders were then sold to DME companies who shipped the equipment to Medicare beneficiaries and billed Medicare for the equipment [9].

3   Application

In *United States v. Merino*, the defendant was convicted by jury trial of eight counts of healthcare fraud [10]. The government charged Merino with healthcare fraud and conspiracy to commit healthcare fraud for acting as a patient recruiter for a medical clinic. The government alleged that she conspired with others to fraudulently bill Medicare in violation of the HCF Statute. The court noted, however that the evidence indicated only that "Merino knew she was accepting kickbacks—which is a violation of the Title 42 anti-kickback statutes—for recruiting patients to Glazer's clinic, not that she knew Glazer was billing Medicare fraudulently." Violation of the AKS was not charged in the case. The government further argued that Merino's efforts to hide her conduct implied knowledge that she had violated the law, but the court nonetheless surmised that this indicated only that she knew she violated the AKS, not that she had engaged in violations of the HCF Statute.

In *United States v. Nora*, defendant Jonathan Nora was convicted of several charges including aiding and abetting healthcare fraud in violation of the HCF Statute [2]. Nora was the office manager at a home health company and was convicted for involvement in several schemes to submit false claims to Medicare for medically unnecessary home healthcare. The Fifth Circuit Court of Appeals overturned his convictions, however, finding that the government had presented insufficient proof that Nora acted knowingly and willfully as required by the HCF Statute. The court first noted that the term willfully in the statute requires that the government prove that the defendant "acted with knowledge that his conduct was unlawful." The court went on to find the government's proof lacking as to this element.

First, the court found that the government did not establish that the training Nora attended would have established knowledge that the activity he later engaged in was unlawful. Second, the court found vague statements that Nora was briefed on the compliance program and was told to report compliance issues to the CEO to be too general to establish knowledge. Third, the court found that staff meetings generally addressing Medicare regulations, testimony from Nora's co-workers describing his knowledge of the activities, and testimony that "everyone in the office knew" were insufficient. The court noted that there was no testimony that Nora actually understood the activities to be illegal. The court even said that, unless there was actually evidence Nora knew it was illegal, it was no more than negligence if Nora did not recognize the illegal nature of the activity. The *Nora* case exemplifies the rigorous application of the criminal scienter requirement to prevent conviction of those who may be no more than ignorant or even merely negligent.

4   Summary

One of the many weapons in the government's arsenal to combat healthcare fraud, prosecutions under the HCF Statute are reserved for the most egregious schemes to defraud. Unlike the FCA and other civil enforcement remedies, the HCF Statute is subject to criminal scienter requirements that defendants knew they were doing something unlawful. Additionally, as a criminal prosecution the government must meet its burden of proof of "beyond a reasonable doubt," as opposed to the "preponderance of the evidence" burden applicable in civil cases. Thus, criminal cases are significantly different in character from the civil case counterparts.

Importantly for both clinicians and healthcare law practitioners, the elements of violations are very similar for both civil and criminal healthcare fraud. DOJ policy is to pursue "parallel" civil and criminal investigations and determine if both civil and criminal sanctions are appropriate. Thus, even when an investigation appears wholly civil, criminal prosecutors can be unseen participants raising the risk of criminal fines and potential incarceration.

References

1. *United States v. Boesen*, 491 F.3d 852 (8th Cir. 2007)
2. *United States v. Nora*, 988 F.3d 823 (5th Cir. 2021)
3. *United States v. DuPont*, 672 F.3d 580 (8th Cir. 2012)
4. *United States v. Medina*, 485 F.3d 1291 (11th Cir. 2007)
5. 18 U.S.C. § 24(b)
6. Department of Justice, Justice Manual, Title 9 Section 27, available at https://www.justice.gov/jm/organization-and-functions-manual-27-parallel-proceedings
7. *United States of America v. Stringer*, 408 F.Supp. 2d 1083 (Dist. Or. 2006)
8. Remarks of Deputy Assistant Attorney General Matthew S. Miner at the 6th Annual Government Enforcement Institute, September 12, 2019; available at, https://www.justice.gov/opa/speech/deputy-assistant-attorney-general-matthew-s-miner-delivers-remarks-6th-annual-government
9. Department of Justice Press Release, April 9, 2019, "Federal Indictments & Law Enforcement Actions in One of the Largest Health Care Fraud Schemes Involving Telemedicine and Durable Medical Equipment Marketing Executives Results in Charges Against 24 Individuals Responsible or Over $1.2 Billion in Losses," available at: https://www.justice.gov/opa/pr/federal-indictments-and-law-enforcement-actions-one-largest-health-care-fraud-schemes
10. *United States v. Merino*, 2021 WL 754589 (9th Cir. 2021)

Part VIII

Antitrust Law

Antitrust Law

E. John Steren

Contents

Key Points

- The Antitrust Division of the U.S. Department of Justice, the Federal Trade Commission, State Attorneys General, and private parties can all enforce the antitrust laws.
- The Sherman Act, passed in 1890 and codified at 15 U.S.C. §§ 1–38, is a federal statute that prohibits activities that unreasonably restrain competition in the marketplace.
- Section 1 of the Sherman Act addresses "agreements" among two or more persons that unreasonably restrain trade or commerce.
- Most conduct is analyzed under the rule of reason, requiring a detailed factual analysis of the markets at issue, the impact of the conduct, and countervailing procompetitive justifications of the conduct.
- Section 2 of the Sherman Act addresses single firm conduct and prohibits efforts to monopolize through the use of anticompetitive means.

E. J. Steren (✉)
Epstein Becker & Green, P.C., Washington, DC, USA
e-mail: esteren@ebglaw.com

A. S. Pasha (ed.), *Laws of Medicine*,
https://doi.org/10.1007/978-3-031-08162-0_22

- The Clayton Act, passed in 1914 and codified at 15 U.S.C. §§ 12–27, amended the Sherman Act to specifically address competitive effects from price discrimination, exclusive dealing, mergers and acquisitions, and overlapping officers and directors.
- Section 7 of the Clayton Act seeks to prevent anticompetitive harm before it occurs by prohibiting transactions that might lessen competition substantially in any relevant market.
- Joint ventures raise potential issues relating to both the market share of the venture, as well as the legality of the operational agreements between the venture participants.
- Exclusive contracts can raise issues under section 1 of the Sherman Act, and serve as the basis for a claim of unlawful monopolization under section 2 of the Sherman Act.

Legal Concepts
- **Agreements among competitors that result in an unreasonable restraint on trade violate section 1 of the Sherman Act.**
- **Unilateral efforts to monopolize or attempt to monopolize through anticompetitive means can result in a violation of section 2 of the Sherman Act.**
- **Mergers and acquisitions that result in significant market share can violate the Clayton Act.**
- **Competitor collaborations including joint ventures raise issues at both the formation and operational stages of the collaboration.**

1   Introduction to Antitrust Law

The antitrust laws of the United States are premised on the belief that competition among market participants provides the most effective means of ensuring that markets work efficiently to produce the highest quality, lowest cost, and most innovative products and services possible for the ultimate consumers. As a result, these laws aim to ensure, at least in theory, that markets are characterized by active and robust competition. While individual competitors may benefit from active enforcement of the antitrust laws, these laws are generally intended to protect competition, and not individual competitors.

For many years there was doubt as to whether the antitrust laws should be applied to the "learned professions" including healthcare professionals. However, the Supreme Court, made clear that the antitrust laws apply with equal force and effect to healthcare professionals and the provision of healthcare products and services [1].

Enforcement of the antitrust laws comes from four (4) sources. The Antitrust Division of the U.S. Department of Justice has jurisdiction to enforce the federal antitrust laws including, as described further below, the Sherman Act and the Clayton Act. This jurisdiction includes the right to investigate potential violations of the federal antitrust laws as well as to bring civil enforcement actions and, when warranted, criminal enforcement actions.

The Federal Trade Commission, like the Antitrust Division of the U.S. Department of Justice, also is authorized to enforce the antitrust laws. Its main antitrust authority stems from section 5 of the Federal Trade Commission Act which, in relevant parts, declares that "[u]nfair methods of competition in or affecting commerce, and unfair or deceptive acts or practices in or affecting commerce, are hereby declared unlawful." [2] Unfair methods of competition include any conduct that would violate the Sherman Act (including sections 1 and 2 of the Sherman Act) or the Clayton Act.

The Federal Trade Commission, unlike the Antitrust Division, has the ability to enforce the antitrust laws through its administrative process by filing administrative complaints heard initially by administrative law judges, then ultimately by the five Commissioners that preside over the FTC. Appeals from these decisions are taken directly to the United States circuit courts of appeal. The FTC however does not have criminal enforcement authority. As a result, when it determines that a potential violation warrants criminal enforcement, it will refer the matter to the Antitrust Division.

Antitrust enforcement actions may also be brought by the individual states acting through their respective State Attorneys General. Most states have statutes that are either based, or interpreted in conformity with the federal antitrust statutes. These state antitrust statues are commonly referred to as "mini" Sherman Acts. States Attorneys General will frequently work in tandem with the federal enforcers to bring antitrust actions against suspected violators.

Finally, many of the federal and state antitrust laws can be enforced by private parties through civil litigation in the courts. These actions can take many years to work their way through the court systems, are typically very fact intensive, and often very expensive to both pursue and defend. A successful plaintiff in a private antitrust action can also be awarded treble damages and attorneys' fees raising the stakes considerably for any defendant to such an action [3].

Considerable guidance prepared by the federal enforcers is available to help explain both the general enforcement approach of these agencies as well as application of the antitrust laws to specific factual scenarios. Much of this guidance is applicable to, and in fact some is specifically tailored to, the healthcare industry. This guidance includes the following:

- U.S. Department of Justice and Federal Trade Commission *Horizontal Merger Guidelines* (August 19, 2010) [4]
- U.S. Department of Justice and Federal Trade Commission *Vertical merger Guidelines* (June 30, 2020) [5]
- U.S. Department of Justice and Federal Trade Commission *Antitrust Guidelines for Collaborations Among Competitors* (April 2000) [6]
- U.S. Department of Justice and Federal Trade Commission *Statements of Antitrust Enforcement Policy in Health Care* (August 1, 1996) [7]
- *Enforcement Policy Regarding Accountable Care Organizations Participating in the Medicare Shared Savings Program* (October 28, 2011) [8]

In addition, both the Antitrust Division through its business review process and the Federal Trade Commission through its advisory opinion process can provide specific guidance on proposed courses of conduct that will describe the enforcement intentions of these agencies [9].

2 Price Fixing and Monopolization

Much of today's antitrust law is based on judicial interpretation of what seems to be rather straight forward, albeit somewhat vague, statutory language contained in the Sherman Act. The most commonly enforced provisions of the Sherman Act are section 1 (the "anti-conspiracy" statute) and section 2 (the "anti-monopolization" statute).

2.1 Section 1 of the Sherman Act

Section 1 of the Sherman Act mandates that:

Every contract, combination in the form of trust or otherwise, or conspiracy, in restraint of trade or commerce among the several States, or with foreign nations, is declared to be illegal. Every person who shall make any contract or engage in any combination or conspiracy hereby declared to be illegal shall be deemed guilty of a felony, and, on conviction thereof, shall be punished by fine not exceeding $100,000,000 if a corporation, or, if any other person, $1,000,000, or by imprisonment not exceeding 10 years, or both said punishments, in the discretion of the court [10].

The focus of section 1 is on "agreements" between two or more persons (defined as any corporation, company, partnership, firm, association or society, as well as a natural person). Without the requisite agreement, section 1 cannot apply. Note, however, that an agreement for purposes of section 1 can be oral as well as written, and expressed as well as implied. The existence of the requisite agreement must be established by evidence that can be as direct as the existence of a contract, or as subtle and circumstantial as a wink and a nod.

The requisite agreement can be either horizontal or vertical in nature. Agreements among persons at the same level of competition are deemed to be horizontal in nature and tend to attract the most attention from the antitrust enforcers. However, vertical agreements, i.e., agreements among persons at different levels of the supply chain, can also raise antitrust issues.

In addition, the agreement must be between two or more persons that are "legally capable of conspiring." The interests of certain parties are deemed to be so aligned and so intertwined that for antitrust purposes, they are viewed as a single entity and deemed incapable of conspiring [11]. Examples of these types of relationships include: (1) a parent corporation and its wholly-owned subsidiary; (2) employers and their employees; and (3) certain agency relationships.

Despite the broad and sweeping language contained in the above-quoted statute, judicial interpretation of this statute has made clear that only "unreasonable restraints" of trade are deemed to be unlawful [12]. Case law has developed two frameworks for testing whether any particular conduct creates an "unreasonable restraint" of trade. Certain conduct that over time and with experience has been found to always, or almost always restrict competition or decrease output is deemed to be *per se* unlawful [13]. If an agreement relating to this type of conduct is found, no opportunity is provided to study the reasonableness of the restraint in light of the real market forces at work, including proffered business purposes, procompetitive benefits, anticompetitive harms, or overall competitive effects. "As a consequence, the *per se* rule is appropriate only after courts have had considerable experience with the type of restraint at issue ... and only if courts can predict with confidence that it would be invalidated in all or almost all instances under the rule of reason." [14] Conduct that falls into this category includes agreements among competitors to fix prices or output, rig bids, or share or divide markets by allocating customers, suppliers, territories or lines of commerce. For example, an agreement among competing providers to fix the rates for their services would likely amount to a *per se* violation of section 1 of the Sherman Act.

Most conduct, however, is tested under the rule of reason. Under this analytical framework, "the factfinder weighs all of the circumstances of a case in deciding whether a restrictive practice should be prohibited as imposing an unreasonable restraint on competition" [15]. Factors to be considered include: "specific information about the relevant business" and "the restraint's history, nature, and effect" on a relevant antitrust market [14]. Relevant antitrust markets include both the product market, i.e., the product or services (including all reasonable substitutes) at issue, and the geographic market, i.e., the area where customers would likely turn to buy the goods or service in the relevant product market. Market shares also become relevant in determining whether the conduct is reasonable [13]. In the end, this analysis seeks to weigh the real or potential anticompetitive effects of the restraint against potential benefits of the conduct to determine whether on balance, the restrain is unreasonable.

2.2   Application

Case 1: Arizona v. Maricopa County Med. Soc'y, 457 U.S. 332 (1982) The Maricopa Foundation for Medical Care was a nonprofit Arizona corporation composed of approximately 1750 physicians (or approximately 70% of the clinicians) in Maricopa County. It was organized for the purpose of promoting fee-for-service medicine and to provide the community with a competitive alternative to existing health insurance plans. A second organization, the Pima Foundation for Medical Care, included approximately 400 clinicians, and was created for a similar purpose.

At the time the lawsuit was filed by the State of Arizona, both foundations were utilizing fee schedules that included "relative values" and "conversion factors" used to determine fees for members of each participating medical specialty. The foundations believed that the use of these fee schedules was beneficial because the schedules purportedly set "maximum" limits on clinician charges and, therefore, provided a cost-containment mechanism to save patients money. The clinician members of these foundations shared no other forms of meaningful integration, neither financially nor clinically.

The State of Arizona sued both foundations claiming that their members, in establishing and agreeing to fee-for-service schedules, were engaged in unlawful price fixing constituting *per se* violations of section 1 of the Sherman Act, and the Supreme Court agreed. In disbanding the arguments raised by the foundations, the Supreme Court ruled that maximum-price-fixing agreements were equally as unlawful as minimum-price-fixing agreements. The Supreme Court stated, among other things, that the rule prohibiting price fixing "is violated by a price restraint that tends to provide the same economic rewards to all practitioners regardless of their skill, their experience, their training, or their willingness to employ innovative and difficult procedures in individual cases. Such a restraint also may discourage entry into the market and may deter experimentation and new developments by individual entrepreneurs." The Supreme Court also made clear that it was "unpersuaded by the argument that we should not apply the *per se* rule in this case because the judiciary has little antitrust experience in the health care industry." "Nor does the fact that doctors – rather than nonprofessionals – are the parties to the price-fixing agreement support the respondents' position."

Case 2: <u>United States v. Alston, 974 F.2d 1206 (9th Cir. 1992)</u> Defendants were three dentist that provided dental services to members of prepaid dental plans in Tucson, Arizona. Payments to the defendants included capitated amounts paid pursuant to contract by the plans, and co-pays paid directly by the patients. Purportedly, the fees paid by the plans had not risen in approximately 10 years and were no longer sufficient to cover the cost of commonly performed services such as porcelain crowning.

Over time, a number of dentists (including the defendants) had individually approached the plans seeking higher reimbursements. These efforts, however, proved unsuccessful. Subsequently, defendants met with approximately fifty other local dentist at defendant Dr. Alston's office to discuss the fees. After the meeting, many of those in attendance mailed in letters to the plans requesting higher fees. In the face of these letters, the plans revised their fee schedules upwards resulting in higher costs to plan members for some services.

The Justice Department quickly obtained criminal indictments against all three defendants for conspiring to fix prices in violation of section 1 of the Sherman Act, and a jury convicted all three defendants of conspiring to fix prices.

The trial court however, granted two defendants complete acquittals notwithstanding the guilty verdicts (i.e., a finding of not guilty), and granted a new trial to one defendant. On appeal, however, the Ninth Circuit Court of Appeals threw out the acquittals of the two defendants and remanded the matter back to the district court to allow for new trials of all three defendants. The Ninth Circuit found that in a criminal prosecution of a section 1 violation under the Sherman Act, the government need not prove specific intent to produce anticompetitive effects. Instead, the jury need only find that the defendants "knowingly conspired" to fix and raise rates. In other words, evidence of the agreement alone could suffice to justify finding of a *per se* violation of section 1 of the Sherman Act.

Case 3: Federal Trade Commission v. Indiana Federation of Dentists, 476 U.S. 447 (1986)

In a purported effort to save cost of dental care for plan members, dental health insurers in the State of Indiana implemented a policy of limiting payment of benefits to the cost of the "least expensive yet adequate treatment" suitable to the needs of the patients. In order to effectuate this policy, the plans required evaluation of the diagnosis and recommendations of treating dentist either in advance or following the provision of care. This typically required a review of patient X-rays by claims examiners who would review the claims for a determination as to whether the treatment option satisfied the policy and was appropriate for payment.

In response to this new policy, the Indiana Dental Association, comprising approximately 85% of the practicing dentist in the State of Indiana, organized an aggressive effort to hinder implementation of the cost containment plans. As a result, a large number of dentists signed a pledge not to provide the dental insurers with copies of the X-rays hindering the ability of claims examiners to review the proposed care for compliance with the new cost-containment goals.

The FTC subsequently challenged the concerted action. In response to this challenge, the Indiana Dental Association abandoned its efforts to prevent insurers from obtaining X-rays. However, a large group of dentist formed a new group known as the Indiana Federation of Dentists and implemented a policy forbidding its members from submitting X-rays to dental insurers.

The Federal Trade Commission then issued a complaint against the Federation alleging that the collective efforts to prevent compliance with the requests for X-rays amounted to a violation of section 1 of the Sherman Act, and consequently an unfair method of competition in violation of section 5 of the Federal Trade Commission Act. After a lengthy administrative hearing, the Commission ruled that the actions of the Federation in requiring its members to withhold X-rays from insurers amounted to a conspiracy and an unreasonable restraint of trade in violation of section 1 of the Sherman Act.

Although the Court of Appeals for the Seventh Circuit reversed the decision of the Commission, the Supreme Court determined that the Commission had correctly found that under the rule of reason, the actions of the Federation in mandating that its members withhold X-rays amounted to a violation of section 1 of the Sherman Act. In its findings, the Supreme Court noted that the actions of the Federation resembled those of a "group boycott", or a concerted refusal to deal on particular terms with patients covered by group dental insurance. The Supreme Court went on to find that application of the rule of reason to the actions of the Federation were fairly straight forward and the policy "takes the form of a horizontal agreement among participating dentists to withhold from their customers a particular service that they desire—the forwarding of X-rays to insurance companies along with claim forms." Noting further that "[w]hile not price-fixing as such, no elaborate industry analysis is required to demonstrate the anticompetitive character of such an agreement."

2.3   Section 2 of the Sherman Act

Section 2 of the Sherman Act states that:

Every person who shall monopolize, or attempt to monopolize, or combine or conspire with any other person or persons, to monopolize any part of the trade or commerce among the several states, or with foreign nations, shall be deemed guilty of a felony, and, on conviction thereof, shall be punished by fine not exceeding $100,000,000 if a corporation, or, if any other person, $1,000,000, or by imprisonment not exceeding 10 years, or by both said punishments, in the discretion of the court [16].

Unlike section 1 of the Sherman Act, the focus of section 2 of the Sherman Act is on the conduct of a single entity. The two primary claims under section 2 of the Sherman Act are: (1) monopolization, and (2) attempted monopolization. A third claim under section 2—conspiracy to monopolize, largely overlaps the claims of monopolization and attempted monopolization, albeit with an element of concerted conduct, and is beyond the scope of this chapter.

A claim of monopolization under section 2, requires: (1) possession of monopoly power in a relevant market; and (2) acquisition or maintenance of that monopoly power that is distinguished from "natural growth or development as a consequence of a superior product, business acumen, or historic accident" [17]. Importantly, being a monopoly, a characteristic that many businesses aspire to but few achieve, is not itself unlawful, and success in the market through legitimate means is generally applauded. However, obtaining a monopoly, or maintaining a monopoly through anticompetitive conduct as opposed to natural growth is prohibited by section 2.

Monopoly power is defined as the ability to raise prices profitably above those found in a competitive marketplace, and to exclude competition from that market. The concept/definition of "monopoly power" is similar that of "market power" but differs marginally in that monopoly power is considered to be greater and more sustained than market power.

A primary factor in determining whether a firm possesses monopoly power is that firm's market share. Although there is no bright-line test, market shares in excess of 90% have been shown to constitute a monopoly, while a firm with market share below 50% would almost, by definition, not possess monopoly power [18]. Whether the market is characterized by high entry barriers (such as Certificate of Need restrictions, high cost of entry) is also relevant in assessing whether a firm possesses monopoly power.

A successful claim under section 2 requires a showing that the defendant engaged in "anticompetitive" or "predatory" conduct. The term "anticompetive conduct" is somewhat vague and open-ended, but is intended in the first place to distinguish between conduct that has a legitimate business justification from conduct that is merely exclusionary. Conduct that is economically irrational but for its ability to harm competition could be deemed anticompetive [19]. Business torts, and other types of conduct that act to impair the opportunities of rivals could all amount to exclusionary and anticompetitive conduct [20].

A claim of "attempted monopolization" requires a showing of: (1) the use of anticompetitive conduct; (2) for the specific intent of monopolizing a market; (3) and, a dangerous probability that unless the actions are stopped, monopolization will be achieved [21].

The element of "specific intent" is satisfied by a showing that not only did the defendant engage in anticompetitive conduct, but also that it did so for the purpose of monopolizing. Demonstrating a dangerous probability of success typically rests on a showing that the defendant had achieved a certain level of market share and was likely to achieve more.

2.4   Application

Case 1: Advanced Health-Care Servs. v. Radford Cmty. Hosp., 910 F.2d 139 (4th Cir. 1990) Plaintiff, Advanced Health-Care Services ("AHCS") is a supplier of durable medical equipment. AHCS alleged that Radford Community Hospital possessed a 75% market share and monopoly power in the provision of inpatient general acute care services. Radford was accused of monopolizing and attempting to monopolize the durable medical equipment (DME) market by using its employees to steer hospital patients that needed DME to its corporate subsidiary, Southwest Virginia Health Enterprises, Inc., ("Southwest"), and away from AHCS.

The District Court rejected AHCS' arguments at the preliminary motions phase finding that AHCS had failed to allege "predatory conduct". On appeal, however, The Fourth Circuit found that if "the plaintiff can prove that the DME now provided to patients in the relevant areas is inferior in quality and/or more expensive than AHCS's, it will have shown harm to competitors, short-term sacrifices by the hospitals, and adverse effects on competition that injure DME consumers, all as a result of the hospitals' entry into the DME markets. From this, a finder of fact may be able to infer that their motives were anti-competitive (*i.e.,* that these were predatory acts stemming from an illegal specific intent to monopolize)" [19].

Case 2: Cascade Health Solutions v. PeaceHealth, 502 F.3d 895, 904 (9th Cir. 2007) Cascade Health Solutions (f/k/a McKenzie-Willamette Hospital) and PeaceHealth are the only two providers of hospital care in Lane County, Oregon. The relevant market, for purposes of the case was defined as the primary and secondary acute care hospital services in Lane County, and PeaceHealth was alleged to possess 75% of that market.

Cascade accused PeaceHealth of engaging in anticompetitive conduct by offering insurers "bundled" or "package" discounts of 35% to 40% on tertiary services if the insurers made PeaceHealth their sole preferred provider for all services—primary, secondary, and tertiary. Cascade introduced evidence of a few specific instances of PeaceHealth's bundled discounting practices and accused PeaceHealth of attempting to monopolize the primary and secondary acute care hospital services markets in Lane County in violation of section 2 of the Sherman Act.

The case eventually made its way to the Ninth Circuit Court of Appeals which concluded that it was possible, at least in theory, for a firm to use a bundled discount to exclude an equally or more efficient competitor and thereby reduce consumer welfare in the long run. It referred to, as an example, the case of a competitor who sells only a single product in the bundle, and who produces that single product at a lower cost than the defendant, but is unable to match profitably the price created by the multi-product bundled discount.

The Court then struggled to define when a bundled discount could be deemed anticompetitive or predatory. The Court ultimately held that the exclusionary conduct element of a claim arising under section 2 of the Sherman Act could be satisfied if discounts result in prices that are below an appropriate measure of the defendant's costs. The Court then adopted a "discount attribution" standard pursuant to which the full amount of the discounts given by the defendant on the bundle are allocated to the competitive product or products. If the resulting price of the competitive product or products is below the defendant's incremental cost to produce them, then the bundled discount could be deemed anticompetitive or exclusionary for the purpose of section 2 of the Sherman Act. As a result, the price of each product offered within the bundle must each separately be above the cost to produce that product. In the case of PeaceHealth, that would have meant that the cost of providing its tertiary services would have been analyzed alone and compared with the price of those services to determine whether the bundle was predatory.

Case 3: <u>Omni Healthcare, Inc. v. Health First, Inc.: Case No. 6:13-cv-1509-Orl-37DAB (M.D. Fla. Jan. 20, 2015)</u> Plaintiffs in this case comprised several physicians and physician practice groups, a physician assistant (PA) and his PA practice located in Southern Brevard County in Florida. Defendants included Health First, Inc., a fully integrated healthcare corporation, along with three wholly owned subsidiaries: Holmes Regional Medical Center, Inc.; Health First Health Plans, Inc.; and Health First Physicians, Inc.

Plaintiffs alleged that Defendants possessed market power, and engaged in an anticompetitive scheme to monopolize Southern Brevard County's interrelated healthcare markets. Among other accusations, Defendants were alleged to have forced independent physicians to join Health First Physicians, Inc., or to enter into exclusive arrangements with Health First. Those that refused were driven out of the market by the revocation of hospital privileges and the refusal of the integrated practice groups to refer patients to the independent doctors.

Defendants filed a preliminary motion to dismiss identifying various purported deficiencies in the complaint, including a failure to properly define the market, or otherwise allege facts that might support a section 2 claim. The District Court, however, denied Defendants' efforts to have the case dismissed. The District Court found that the allegations of the complaint were sufficient to define the relevant markets at issue, and that the alleged facts if true, could constitute an unlawful monopolization of the relevant market. The case proceeded to trial but was settled after the first day of court hearing.

3   Exclusive Contracts

Exclusive contracts can simplistically be defined as an agreement by one party to buy all inputs or services from one provider to the exclusion of all other providers. The exclusivity may, but does not have to, go both ways such that the provider or supplier is also prohibited from providing the input or services to any other buyer.

In the healthcare industry, these arrangements come in many forms. Frequently, a hospital may designate one clinician, or group of clinicians, to the exclusion of other clinicians, the right to provide a service to hospital patients. These arrangements are most prevalent in hospital-based specialties including radiology, anesthesiology, pathology, and emergency-room medical services. Similar tie ups might involve the provision of DME, ambulance or other ancillary services.

Exclusive arrangements, including the hospital/clinician arrangement, are typically "vertical" in nature, i.e., the buyer and supplier are at different levels of the supply chain and not horizontal competitors of each other. As with many types of vertical arrangements, these relationships can generate efficiencies and can be pro-competitive. Frequently, such as in rural areas, the hospital's patient base may only support one group of clinicians making it necessary to provide an exclusive arrangement to entice a group to provide coverage. Exclusive arrangements can also promote scheduling, coordination and teamwork.

However, by their very nature, exclusive contracts foreclose competitors and thereby can raise competitive concerns. This foreclosure leaves patients, referring clinicians and health plans without competing alternatives. In addition, because options for the foreclosed competitors are restricted, it is not uncommon for clinicians that are foreclosed as a result of the exclusive contract to bring an antitrust claim to challenge the relationship.

The analysis of exclusive contracts focuses on the nature and impact of the foreclosure caused by the arrangement. Generally speaking, the question centers on whether the foreclosure affects a sufficiently large percentage of the market for a sufficiently long period of time that it results or enables the group that maintains the contract to obtain market power. Once market power is obtained, it may result in prices being raised above the competitive level, or a decrease in quality.

Claims attacking exclusive contracts have been raised under both section 1 and section 2 of the Sherman Act. The exclusive contract itself may suffice to satisfy the "agreement" element of a section 1 claim. However, some courts in analyzing this element have focused on whether the decision to grant the exclusive arrangement resulted from an agreement between the contracting parties, or was a unilateral decision made by the hospital. If, for example, the decision to enter the contract was the result of action by the Board, some courts have held that it was a unilateral decision and could not serve as the basis for a section 1 claim [22].

Clinicians that have been foreclosed from providing services as a result of an exclusive arrangement have also sought to challenge the conduct under a theory of unlawful tying. A tying claim is based on the premise that one party, in this case the hospital, has market power in the services it provides (the tying product), and forces the buyer to buy a second product from its preferred supplier, in this case the party with the exclusive contract (the tied product). In most of these cases, however, the claims typically fail either because of the inability to demonstrate the existence of two separate products/services, or because the seller of the tied product lacks the requisite market power in the tied product, and, therefore, is unable to coerce the sale of the tying product.

3.1   Application

Case 1: <u>White v. Rockingham Radiologists, Ltd., 820 F.2d 98 (4th Cir. 1987)</u> The plaintiff, Dr. White, was a neurologist with staff privileges at Rockingham Memorial Hospital, the only hospital located in Rockingham County, Virginia. The defendant, Rockingham Radiologists, Ltd., was a professional corporation made up of six radiologists that furnished all X-ray services at Rockingham Memorial. Before Rockingham Memorial obtained a CT scanner, its patients needing CT scanning services were sent to Charlottesville for scans that were interpreted by Dr. White.

Rockingham Memorial, in conjunction with two other hospitals, established a steering committee to assess the feasibility of acquiring CT equipment for the hospitals. That committee concluded that a CT scanner should be acquired for use at all three hospitals.

After acquisition of the CT scanner, the radiologist from Rockingham Radiologists, Ltd., made all official interpretations of head scans. Dr. White sought permission to interpret the scans from his patients, but ultimately the board chose the radiologists as the official interpreters of all CT scans on the grounds that one entity should be responsible for the entire CT operation.

Dr. White brought suit raising various claims including a claim under section 1 of the Sherman Act alleging a conspiracy between Rockingham Memorial Hospital and the radiologists. The district court granted the defendant's motion for summary judgment and the Fourth Circuit Court of Appeals agreed. In its decision, the Fourth Circuit found that:

> The district court properly held that Dr. White's direct and circumstantial evidence was insufficient for a jury to find a conspiracy among the radiologists... the participating hospitals, and their officials. The evidence conclusively establishes that the hospital's board acted unilaterally. It decided that assigning full responsibility for scans to the radiologists, rather than fragmenting accountability, would best serve the interests of patients and promote efficient use of the scanner. There is no evidence that "reasonably tends to prove... 'a conscious commitment to a common scheme designed to achieve an unlawful objective.' [citation omitted] [23]

Case 2: <u>Jefferson Parish Hosp. Dist. v. Hyde, 466 U.S. 2 (1984)</u> Dr. Hyde, a board-certified anesthesiologist, applied for medical staff privileges at East Jefferson Hospital. The credentials committee and the medical staff executive committee recommended approval of Dr. Hyde's application for privileges, but the hospital board denied the application because it was a party to an exclusive contract with a group of anesthesiologist to fulfill all hospital based anesthesiology needs. Dr. Hyde then commenced action against the hospital claiming that the exclusive arrangement was unlawful. The lower court rejected Dr. Hyde's claims, but the Court of Appeals overturned that decision finding that the exclusive agreement involved a "tying arrangement" because the "users of the hospital's operating rooms (the tying product) are also compelled to purchase the hospital's chosen anesthesia service (the tied product)."

The Supreme Court, however, reversed basing its determination on a finding that East Jefferson lacked market power in the hospital-based services and, therefore, could not force patients to use its services and force the use of the related anesthesiology services. Furthermore, the Supreme Court determined that there was insufficient evidence that the exclusive contract had an adverse effect on competition as there were at least 20 hospitals in the New Orleans metropolitan area for Dr. Hyde to seek privileges.

4   Mergers and Acquisitions

The healthcare industry has seen accelerating rates of integration through, among other transaction vehicles, mergers and acquisitions. The Sherman Act, despite its application to many forms of anticompetitive conduct, proved insufficient to handle the range of competitive issues associated with mergers and acquisitions. As a result, in 1914, and in an effort to clarify, strengthen and fill certain gaps of the Sherman Act, the Clayton Act was passed.

Section 7 of the Clayton Act, in particular, seeks to prohibit mergers and acquisitions where the effect "may be substantially to lessen competition, or tend to create a monopoly" [24]. Section 7 is commonly referred to as an "incipiency" statute in that it seeks to prohibit unlawful conduct before it occurs and involves a detailed analysis of the likelihood of competitive harms that "may" occur as a result of the proposed conduct. Section 7 was bolstered in 1976 by the addition of the Hart-Scott-Rodino Antitrust Improvements Act that requires companies planning larger mergers or acquisitions to notify the federal antitrust enforcement agencies of their transaction plans in advance to allow the federal agencies an opportunity to investigate the potential competitive harms that might result [25].

Enforcement of section 7 of the Clayton Act involves a burden-shifting framework. Initially, the government (or private plaintiff) must establish a prima facie case by showing that the transaction in question will significantly increase market concentration, thereby creating a presumption that the transaction is likely to substantially lessen competition. Once the government establishes the prima facie case, the respondent may rebut it by producing evidence to cast doubt on the accuracy of the governments' evidence as predictive of future anti-competitive effects. Finally, if the respondent successfully rebuts the prima facie case, the burden of production shifts back to the government and merges with the ultimate burden of persuasion which is incumbent on the government at all times [26].

Demonstrating that the transaction will result in a significant increase in market concentration, thereby establishing a prima facie case, requires the government to define the relevant antitrust markets affected by the transaction, and then demonstrating that within that market(s), concentration levels will increase substantially. A properly defined antitrust market includes an appropriate description of the relevant product market at issue, and the geographic area affected by the transaction. In defining the relevant market, the enforcers utilize what is referred to as the "hypothetical monopolist test". Under this test, the government identifies a product or service, and a geographic location, and asks whether a hypothetical monopolist providing that product or service in the designed area could profitably impose a small but significant and non-transitory increase in price. If the answer is yes, then the market has been properly defined. If the answer is no because of the presence of viable substitutes that could defeat the price increase, then those substitute products or services (or suppliers of the products and services) are added to the market and the hypothetical monopolist test is repeated to determine whether a hypothetical monopolist could profitably impose a small but significant and non-transitory increase in price [4].

Once the market is properly defined, the active market participants are then identified and their respective market shares are calculated. The market concentration is then measured using the Herfindahl-Hirschman Index ("HHI"). This is computed by squaring the market share of each firm competing in the market and then summing the results of those numbers. According to the *Horizontal Merger Guidelines*, the agencies generally consider markets in which the HHI is between 1500 and 2500 points to be moderately concentrated, and consider markets in which the HHI is in excess of 2500 points to be highly concentrated. Transactions that increase the HHI by more than 200 points in highly concentrated markets are presumed likely to enhance market power [4].

If the government is able to establish the prima facie case, then the burden shifts to the transacting parties to bring forth evidence that demonstrates that despite the prima facie showing, the transaction is unlikely to generate the concerned harms. This can be rebutted by showing, for example, that entry into the market is relatively easy and that in the face of an attempted price increase new entrants would enter the market and drive the prices back down to the competitive level. Parties can also attempt to demonstrate that expected efficiencies from the transaction will outweigh any potential competitive harm. According to the Merger Guidelines, efficiencies must be well supported and specifically generated as a result of the transaction, and could not be achieved in the absence of the transaction [4].

Case 1: Federal Trade Commission, et al. v. St. Luke's Health System, et al., 778 F3d 775 (9th Cir. 2015) Saltzer Medical Group, P.A. ("Saltzer") was the largest independent multi-specialty physician group in Idaho, with 34 physicians practicing in Nampa (the second largest city in Idaho), approximately 20 miles from Boise, Idaho. Saltzer was also the largest employer of primary care physicians ("PCPs") in Nampa with 16. St. Luke's Health Systems, Inc. ("St. Luke's") operated an emergency clinic in Nampa, and employed eight PCPs.

In 2012, St. Luke's acquired Saltzer's assets and entered into a 5-year professional service agreement (described by the Court as a "merger" or "acquisition") with the Saltzer physicians. The merger did not require the Saltzer physicians to send patients to St. Luke's hospital located in Boise, or to use St. Luke's facilities for any type of ancillary services.

The only hospital located in Nampa was operated by Saint Alphonsus Health System ("Saint Alphonsus"). Saint Alphonsus also employed nine PCPs. Saint Alphonsus sought unsuccessfully to enjoin the merger under section 7 of the Clayton Act. The FTC and the state of Idaho subsequently filed their own complaint also seeking to enjoin the merger under section 7 of the Clayton Act claiming that the merger would have anticompetitive effects in the adult PCP market. The District Court granted the FTC's and the State of Idaho's request to enjoin the transaction.

On review by the Court of Appeals for the Ninth Circuit, the Ninth Circuit noted the extensive procompetitive efficiencies that the merged parties expected to generate as a result of the transaction. Nevertheless, the Ninth Circuit was persuaded that the plaintiffs had appropriately defined the market and that the expected efficiencies could not overcome the potential adverse effects that would be created by the significant concentration in the adult PCP market (HHI of 6219 with a change in HHI of 1607, creating a presumption that the transaction would result in anticompetitive effects).

5   Joint Ventures

The term "joint venture" includes any number of collaborations between separate entities designed to pull together resources and achieve a common objective [27]. Joint ventures are typically distinguished, at least for antitrust purposes, from a merger or acquisition by the lack of total integration among joint venture participants. Generally, a merger or acquisition will end all competition among the merged or acquired entities, where as a joint venture may preserve some form of competition among the participants.

Joint ventures are generally procompetitive. Joint ventures often make it possible to provide new products and services, or achieve other efficiencies that might not be possible by participants acting alone. However, antitrust concerns can arise particularly when participants are themselves competitors. Among other things, joint ventures among competitors may "facilitate explicit or tacit collusion through facilitating practices such as exchange or disclosure of competitively sensitive information or through increased market concentration" [6]. As a result, the analysis of joint ventures looks at both the formation of the enterprise, as well as the collective conduct of the joint venture participants.

The *Antitrust Guidelines for Collaborations Among Competitors* describe various types of joint ventures. These include: production collaborations (defined to include "agreements jointly to produce a product sold to others or used by the participants as an input"); marketing collaborations ("agreements jointly to sell, distribute, or promote goods or services that are either jointly or individually produced"); buying collaborations ("agreements jointly to purchase necessary inputs"); and research and development collaborations ("agreements to engage in joint research and development") [6].

Clinician networks, including individual practice associations ("IPAs") and preferred provider organizations ("PPOs"), are a form of joint venture designed to market clinician services to health plans. These joint ventures (which might fall somewhere between a "marketing" and "production" collaboration) can include mechanisms to encourage clinicians to collaborate in practicing efficiently, thereby controlling costs while increasing the quality of care for the benefit of consumers of healthcare services.

These joint ventures can raise antitrust issues both relating to the formation of the network as well as operations. From a formation perspective, the primary issue is one of size—that is whether the network represents so large a percentage of clinicians in the relevant area that as a result the network obtains or could obtain market power—i.e., the power to raise prices above the competitive level. From the operational perspective, the issue becomes whether the participants of the network have sufficiently integrated such that joint decisions by the network on behalf of its members do not raise *per se* antitrust violations.

A rule of reason analysis is used to determine whether the size of a network joint venture gives rise to an issue of market power. Under this framework, the reviewing agency looks at the characteristics of the network, and the competitive environment in which it operates. In doing so, the agency first defines the relevant market at issue in a fashion similar to that undertaken under section 7 of the Clayton Act. Often each separate physician specialty represents a separate product market, however, the agencies will look for reasonably available substitutes (including services provided by non-physicians) and include those in the market as there may be overlaps in the services provided. Once the market is defined, the agencies then look at the potential competitive effects of the joint venture including the ability and incentives to raise prices. While there is no bright line when it comes to market share as market dynamics dictate the ultimate analysis, joint ventures that possess market shares in excess of 30% are worth looking at carefully.

A significant factor in determining whether a network may gain market power, is whether its members participate in the network on an exclusive basis. In considering the issue of exclusivity, Statement 8 of the *Statements of Antitrust Enforcement Policy in Health Care* sets forth the following relevant criteria:

1. whether there currently exist in the market viable competing networks or managed care plans with adequate physician participation;
2. whether physicians in the network actually individually participate in, or contract with, other networks or managed care plans, or there is other evidence of their willingness and incentive to do so;

3. whether physicians in the network earn substantial revenue from other networks or through individual contract with managed care plans;
4. the absence of any indicators of significant de-participation from other networks or managed care plans in the market; and.
5. the absence of any indications of coordination among the physicians in the network regarding price or other competitively significant terms of participation in other networks or managed care plans [7].

Statement 8 also sets forth a safety zone related to market shares. According to Statement 8, the U.S. Department of Justice and the Federal Trade Commission have indicated that they "will not challenge, absent extraordinary circumstances, an exclusive physician network joint venture whose participating participants share substantial financial risk and constitute 20% or less of the physicians in each physician specialty with active hospital staff privileges who practice in the relevant geographic market" [7]. This safety zone is consistent with the safety zone applicable more broadly to any competitor collaboration as described in the *Antitrust Guidelines for Collaborations Among Competitors* [6].

In addition, the Agencies have indicated that they "will not challenge, absent extraordinary circumstances, a non-exclusive physician network join venture whose physician participants share substantial financial risk and constitute 30% or less of the physicians in each physician specialty with active hospital staff privileges who practice in the relevant geographic market" [7]. It is also worth noting that non-exclusive networks with considerably more participation (in fact as much as 50% participation) have been permitted on the theory that the non-exclusivity of its members will protect against the exercise of market power [7].

As reference above, in order to qualify for either safety zone, network members must share substantial financial risk. Statement 8 describes various forms of financial risk that will qualify including:

1. an agreement by the venture to provide services to a health plan at a "capitated" rate;
2. an agreement by the venture to provide designated services or classes of services to a health plan for a predetermined percentage of premium or revenue from the plan;
3. the use by the venture of significant financial incentives for its physician participants, as a group, to achieve specified cost-containment goals. Two methods by which the venture can accomplish this are:
   (a) withholding from all physician participants in the network a substantial amount of the compensation due to them, with distribution of that amount to the physician participants based on group performance in meeting the cost-containment goals of the network as a whole; or
   (b) establishing overall cost or utilization targets for the network as a whole, with the network's physician participants subject to subsequent substantial financial rewards or penalties based on group performance in meeting the targets; and

4. agreement by the venture to provide a complex or extended course of treatment that requires the substantial coordination of care by physicians in different specialties offering a complementary mix of services, for a fixed, predetermined payment, where the costs of that course of treatment for any individual patient can vary greatly due to the individual patient's condition, the choice, complexity, or length of treatment, or other factors [7].

Analysis of the operations of the joint venture, and in particular the joint contracting on behalf of network members, depends on the level of integration, if any, among the network's members. Agreements among members to a network that lack any type of integration are deemed to be "naked" and may constitute *per se* violations of the antitrust laws. For example, an agreement to use the network as a vehicle for joint contracting, in a situation where the network members lack any form of integration, amounts to a price-fixing agreement and a *per se* violation of section 1 of the Sherman Act.

On the other hand, if members of the network are integrated, and the agreements of the network are reasonably necessary for the operations of the network, then they will be analyzed under the rule of reason. Network integration generally takes one of two forms. As described above, network members can share financial risk through one of the arrangements described above and deemed to be economically integrated for purposes of those risk sharing arrangements.

In addition, network members can seek clinical integration generally "evidenced by the network implementing an active and ongoing program to evaluate and modify practice patterns by the network's physician participants and create a high degree of interdependence and cooperation among the physicians to control costs and ensure quality. This program may include: (1) establishing mechanisms to monitor and control utilization of health care services that are designed to control costs and assure quality of care; (2) selectively choosing network physicians who are likely to further these efficiency objectives; and (3) the significant investment of capital, both monetary and human, in the necessary infrastructure and capability to realize the claimed efficiencies" [7].

5.1   Application

Case 1: <u>FTC Advisory Opinion Concerning MedSouth, Inc.</u> <u>(02-19-2020)</u> MedSouth is a physician independent practice association located in Denver, Colorado. It formed, in part, to enter into contracts with third-party payers for the provision of physician services on a fee-for-service basis. Without some form of integration, the collective negotiations on behalf of the independent members would likely have amounted to a *per se* price-fixing agreement in violation of section 1 of the Sherman Act.

MedSouth sought to integrate its members clinically to avoid *per se* treatment. As part of its program, MedSouth implemented a web-based electronic clinical data record system to permit its physicians access and to share clinical information relating to MedSouth patients. Through the electronic data record system, MedSouth members were able to rapidly access and exchange clinical information relating to patients, including lab and radiological reports, transcribed patient records and office visit information, treatment plans, and prescription information. The doctors were also able to order prescriptions on-line, and determine whether the patients filled the prescription. The system is able to aggregate data from multiple doctors to show, for example, the trend of results on tests done at different times and places. Each practice was required to acquire the hardware necessary to use the system.

In addition, it adopted and implemented clinical practice guidelines and performance goals relating to the quality and appropriate use of services provided by MedSouth physicians. At the time of the issuance of the Advisory Opinion, MedSouth was developing: (1) clinical protocols covering the majority of MedSouth physicians' patient population; and (2) measurable performance goals relating to the quality and appropriate utilization of services that are linked to those protocols.

MedSouth acted to secure members' commitment to adhere to those protocols in their office and hospital practices; review the performance of MedSouth physicians individually and collectively with respect to those goals; assist members in meeting the goals; and, if necessary, expel physicians who cannot or will not meet the goals. The physician participation agreement mandated the physicians' commitment to participate in all the network's programs; to adhere to the IPA's standards and protocols; and to implement the technology that permits MedSouth to report performance information to members and to third parties.

The FTC explained that in the absence of integration, the joint negotiation of price terms by non-integrated, competing physicians would constitute an agreement among the physicians not to compete on price, and would be illegal *per se*. This is consistent with the *Antitrust Guidelines for Collaborations Among Competitors* that states that if "participants in an efficiency-enhancing integration of economic activity enter into an agreement that is reasonably related to the integration and reasonably necessary to achieve its procompetitive benefits, [it will be analyzed] under the rule of reason, even if it is of a type that might otherwise be considered *per se* illegal" [6].

The FTC ultimately concluded that the program MedSouth proposed to engage in appears capable of creating substantial partial integration of the participating physicians' practices, and to have the potential to produce efficiencies in the form of higher quality or reduced costs for patient care services rendered by the network physicians. As a result, rule of reason analysis was warranted. It then concluded that if the doctors implemented the program as explained, the FTC would not challenge the arrangement or the joint contracting on behalf of its physician participants. Importantly, the FTC based its opinion on the fact that MedSouth operated on a "non-exclusive" basis, and that its members were free to contract, and did contract, outside of the network. The FTC warned that exclusive behavior could result in unlawful refusals to deal, particularly in light of the significant membership percentages (as high as 100% of certain specialists operating at nearby hospitals).

Case 2: __FTC Advisory Opinion Concerning Suburban Health Organization,__ __Inc. (03-28-2006)__ Suburban Health Organization, Inc. ("SHO") operated as a "super physician-hospital organization" ("PHO") with seven local PHOs each affiliated with a local hospital, and one multi-facility health system all located in the Indianapolis area. Together, these eight SHO members employed 192 primary care physicians. SHO was originally established to undertake risk-based contracts with health plans and other payers of health care services such as self-insured employers. It now wanted to collectively engage in fee-for-service contracting on behalf of all of its members with these same payers.

The FTC noted the efforts to achieve clinical integration at each of the PHOs, and efforts to collaborate in the development of quality management programs, outcomes measurement, and professional peer review among the members. However, apart from that, SHO provided no explanation of how the separate PHOs and their respective members were integrated. Specifically, the FTC noted that "SHO does not explain … how the physicians will or can work collaboratively to attain program's goals, since the program does not appear to involve collaborative provision of physician services, or direct involvement by participating physicians from any SHO hospital in the delivery of services by physician at any other SHO hospital." In other words, there was no apparent cross-integration among the physicians at the separate PHOs. As a result, the FTC concluded that it could not provide a favorable opinion and might seek to challenge the collective contracting of SHO.

6   Summary

In summary, the implications and impact of the antitrust laws of the United States should be considered before engaging in any conduct with a competitor, or in any conduct that could potentially have an adverse effect on competition. Conduct that creates an unreasonable restraint of trade could result in time-consuming and costly civil, and potentially criminal, prosecution. Actions by clinicians and other stakeholders in the healthcare industry are targeted equally, if not more so, than any other industry, and the antitrust laws apply with equal force an effect.

References

1. Goldfarb v. Virginia State Bar, 421 U.S. 773 (1975).
2. The Federal Trade Commission Act, Public Law No. 108-82, 15 U.S.C. §45 (1914).
3. The Clayton Antitrust Act, Pub. L. 63-212, 38 Stat. 730, 15 U.S.C. §15 (1914).
4. U.S. Department of Justice and Federal Trade Commission Horizontal Merger Guidelines (August 19, 2010). Available from: https://www.justice.gov/atr/horizontal-merger-guidelines-08192010.
5. U.S. Department of Justice and Federal Trade Commission Vertical merger Guidelines (June 30, 2020). Available from: https://www.justice.gov/opa/pr/department-justice-and-federal-trade-commission-issue-new-vertical-merger-guidelines.

6. U.S. Department of Justice and Federal Trade Commission Antitrust Guidelines for Collaborations Among Competitors (April 2000). Available from: https://www.justice.gov/atr/guidelines-and-policy-statements-0.

7. U.S. Department of Justice and Federal Trade Commission Statements of Antitrust Enforcement Policy in Health Care (August 1, 1996). Available from: https://www.ftc.gov/public-statements/1996/08/statements-antitrust-enforcement-policy-health-care.

8. Enforcement Policy Regarding Accountable Care Organizations Participating in the Medicare Shared Savings Program (October 28, 2011). Available from: https://www.ftc.gov/policy/federal-register-notices/ftc-doj-enforcement-policy-statement-regarding-accountable-care.

9. Code of Federal Regulations Chapter 28, §50.6 and Chapter 16 §1.1-1.4 (1977).

10. The Sherman Antitrust Act, 26 Stat. 209, 15 U.S.C. §1 (1890).

11. Copperweld Corp. v. Independence Tube Corp., 467 U.S. 752 (1984).

12. State Oil Co. v. Khan, 522 U.S. 3, 10 (1997).

13. Business Electronics Corp. v. Sharp Electronics Corp., 485 U.S. 717, 723 (1988).

14. Arizona v. Maricopa County Medical Soc, 457 U.S. 332, 344 (1982).

15. Contiental T.V., Inc. v. GTE Sylvania Inc., 433 U.S. 36, 49 (1977).

16. The Sherman Antitrust Act, 26 Stat. 209, 15 U.S.C. §2 (1890).

17. United States v. Ginnell Corp., 384 U.S. 563, 570 – 71 (1966).

18. American Tobacco Co. v. United States, 328 U.S. 781, 813-814 (1946).

19. Advanced Health-Care Servs. v. Radford Cmty. Hosp., 910 F.2d 139, 148 (4th Cir. 1990).

20. Cascade Health Solutions v. PeaceHealth, 502 F.3d 895, 904 (9th Cir. 2007).

21. Spectrum Sports, Inc. v. McQuillan, 506 U.S. 447, 456 (1993).

22. Todorov v. DCH Healthcare Authority, 921 F.2d 1438 (11th Cir. 1991).

23. White v. Rockingham Radiologists, Ltd., 820 F.2d 98, 103 (4th Cir. 1987)

24. The Clayton Antitrust Act, Pub. L. 63-212, 38 Stat. 730, 15 U.S.C. §18 (1914).

25. Hart-Scott-Rodino Antitrust Improvements Act, Pub. L. 94-435, 90 Stat. 1383, 15 U.S.C. §18a (1976).

26. FTC v. Univ. Health, Inc., 938 F.2d 1206 (11th Cir. 1991).

27. ABA Section of Antitrust Law, Joint Ventures: Antitrust Analysis of Collaborations Among Competitors, Second Edition (2014).

Part IX

FDA Law

Drug and Medical Device Law

Amirala S. Pasha

Contents

Key Concepts
- Regulation of drugs and devices in the US has been shaped by a patchwork of laws that have developed over the past 150 years.
- FDA does not regulate the practice of medicine.
- The Federal Food, Drug, and Cosmetic Act (FDCA), establishing the legal framework within which the FDA operates, can be found in the United States Code (USC) beginning at 21 USC 301.
- The FDA regulations are in Title 21 of the Code of Federal Regulations (CFR).

A. S. Pasha (✉)
Division of General Internal Medicine, Mayo Clinic, Rochester, MN, USA
e-mail: pasha.amirala@mayo.edu

© The Author(s), under exclusive license to Springer Nature Switzerland AG 2022
A. S. Pasha (ed.), *Laws of Medicine*,
https://doi.org/10.1007/978-3-031-08162-0_23

- The FDA has several pathways to expedite the review and approval of eligible new drugs.
- Off-label use is permitted.
- Off-label promotion is allowed under commercial speech but subject to scrutiny.
- Expanded Access Program does not provide any malpractice liability protection but is available for both drugs and devices.
- Right-to-Try Act provides sweeping malpractice liability protection but is only an option for drugs.

Legal Concepts
- Drugs: articles defined in the "official compendium" or those used in the diagnosis, cure, mitigation, treatment, or prevention of disease in man or other animals, or articles (other than food) intended to affect the structure or any function of the body of man or other animals or articles intended for use as a component of any article specified above [1].
- Device: "an instrument, apparatus, implement, machine, contrivance, implant, in vitro reagent, or other similar or related article, including any component, part, or accessory, which is (A) recognized in the official National Formulary, or the United States Pharmacopeia, or any supplement to them, (B) intended for use in the diagnosis of disease or other conditions, or in the cure, mitigation, treatment, or prevention of disease, in man or other animals, or (C) intended to affect the structure or any function of the body of man or other animals, and which does not achieve its primary intended purposes through chemical action within or on the body of man or other animals and which is not dependent upon being metabolized for the achievement of its primary intended purposes" [1].
- Off-Label Use: Use for indication, dosage form, dose regimen, population or other use parameter not mentioned in the approved labeling" [2].
- Off-Label Promotion: Promotion of a drug or medication by manufacturers for a purpose other than what the Food and Drug Administration (FDA) has approved.
- Preapproval Access: various legal schemes outside of a clinical trial, for patients to access drugs and medical devices that have not been fully and formally approved (or cleared) by a country's regulatory system.

1   Introduction

The earliest predecessor to the U.S. Food and Drug Administration (FDA) was the Bureau of Chemistry, established in the 1860s and housed within the U.S. Department of Agriculture (USDA). Although the federal government initially played a very small role in regulating drugs, that would change over the next 150 years due to numerous new laws. The federal government's role expanded greatly to include regulating drugs, biologics, medical devices, and other areas now regulated by the FDA. Consequently, to better address this expanded role, the agency that ultimately became the FDA was transferred out of the USDA and into other agencies beginning in 1940. Finally, in 1988, the FDA became part of the Department of Health and Human Services (HHS), which continues to be the case today [3].

The FDA is responsible for protecting public health by ensuring the safety and effectiveness of human and veterinary drugs, vaccines, other biological products, and medical devices; protecting the public from electronic product radiation; ensuring the safety and proper labeling of cosmetics and dietary supplements; regulating tobacco products; and assuring food safety and proper labeling. In part due to the FDA's origins within the USDA, meat from livestock, poultry, and some egg products continue to be regulated by the USDA [4].

As evident from the above description, the FDA is involved in a large portion of the economy. Based on 2019 figures, the FDA is responsible for overseeing more than $2.6 trillion in consumption, which accounts for about 20 cents of every dollar spent by U.S. consumers [5]. Multiple volumes of books have been written on FDA law, analyzing the regulatory aspects of its role in detail. However, in this chapter, the discussion will be limited to regulations of drugs, biologics, and medical devices. In subsequent chapters, tobacco regulation and artificial intelligence are discussed. In keeping with this book's overall theme, the focus will be on clinical applications of relevant FDA law. One recurring concept will be that the FDA is not involved in regulating the practice of medicine.

The Federal Food, Drug, and Cosmetic Act (FDCA), establishing the legal framework within which the FDA operates, can be found in the United States Code (USC) beginning at 21 USC 301. Like other federal agencies, the FDA also develops regulations based on various laws under which it operates. This process follows the typical procedures required by the Administrative Procedure Act discussed in an earlier chapter in this book. FDA regulations have the full force of law and are found in Title 21 of the Code of Federal Regulations (CFR). FDA also provides formal guidance which is not legally binding on the FDA or the public but describes the agency's current thinking on regulatory issues [6, 7].

As the FDA has expanded throughout the years, it now houses multiple specialized units, with nine Center-level organizations named below. Each unit has a high degree of specialized expertise and is responsible for the regulatory oversight of their relevant products within the FDA's wide-ranging jurisdiction. Each unit also has multiple offices within its structure with a narrower focus, but those details are beyond the scope of this chapter.

- Center for Food Safety and Applied Nutrition (CFSAN, includes Cosmetics)
- Center for Drug Evaluation and Research (CDER)
- Center for Devices and Radiological Health (CDRH)
- Center for Biologics Evaluation and Research (CBER)
- Center for Veterinary Medicine (CVM)
- Center for Tobacco Products
- National Center for Toxicological Research (NCTR)
- Office of Regulatory Affairs (ORA)
- Office of Operations

2   History of Drug and Device Regulation

To better understand the current scheme of drug and device regulation in the U.S., it is helpful to briefly review the history. This review highlights the passage of major laws within the drug and device regulatory arena and, in some cases, major events that sparked the passage of these laws. Although this historical review is not exhaustive, it does provide the background knowledge needed to better understand and appreciate the complexities of the current regulatory framework, which is based on a patchwork of laws and amendments that greatly expanded the minor role of the federal government in regulating drugs and medical devices over the past century to what it is today.

In 1906, the original Food and Drugs Act was passed, prohibiting interstate commerce of misbranded and adulterated foods, drinks, and drugs. However, there was no requirement to demonstrate safety and efficacy of drugs before marketing, and the act did not regulate cosmetics or medical devices. Furthermore, in a 1911 landmark case, *U.S. v. Johnson*, the Supreme Court ruled that the 1906 Food and Drugs Act only prohibited misleading statements about the ingredients or identity of a drug but not false therapeutic claims. In response to this ruling, Congress passed the Sherley Amendment, which prohibited the labeling of medicines with false therapeutic claims intended to defraud the purchaser. However, the required intent of the fraud element made proving this claim difficult [3].

In 1938, Congress passed the Federal Food, Drug, and Cosmetic Act (FDCA), replacing the 1906 Food and Drugs Act. For the first time, this major overhaul of the drug-regulation scheme in the U.S. required that new drugs be shown to be safe prior to marketing. Since then, before marketing, all new prescription drugs have been required to have an approved New Drug Application (NDA), discussed in more detail later in this chapter. However, all drugs marketed prior to the passage of the FDCA in 1938 were considered to be "pre-1938 grandfathered drugs" and exempt from the need for an approved NDA. To date, many continue to be marketed without an NDA. In other words, many prescription drugs commercially available in the U.S. market today (e.g., phenobarbital) have never been evaluated or approved formally by the FDA to ensure safety and efficacy. Among other notable provisions, the FDCA eliminated the Sherley Amendment's intent requirement, authorized factory inspections, and added court injunctions as an additional remedy to seizures and prosecutions for violators. In that same year, Congress passed the Wheeler-Lea Act, charging the Federal Trade Commission (FTC) with overseeing the advertising of FDA-regulated products. The FDCA, though amended numerous times since 1938, continues to be the principal law regulating drugs and medical devices in the U.S. [3].

Under the 1938 version of the law, there was still no requirement to demonstrate effectiveness before marketing, and the law did not cover medical devices. In the 1940s—initially with diabetic medications, including insulin, and then extending to antibiotics—amendments were passed to require FDA scrutiny beyond safety and establish potency and/or effectiveness requirements for drugs within those specific categories. It was not until the 1960s that the law required proof of effectiveness for all drugs [3].

In 1962, in response to the thalidomide tragedy in Europe that resulted in thousands of birth defects, Congress passed the Kefauver-Harris Drug Amendments, which, for the first time, required proof of effectiveness before marketing new drugs. This was followed by the FDA contracting with the National Academy of Science/National Research Council to evaluate the effectiveness of 4000 drugs approved based on safety alone between 1938 and 1962, before the Kefauver-Harris Drug Amendments were passed [3].

In 1944, the Public Health Service Act (PHSA) was passed, which included regulating biological products and controlling communicable diseases. The Act was followed in 1955 by the establishment of the Division of Biological Control, an independent agency within the National Institutes of Health (NIH), in response to a polio outbreak thought to be linked to the polio vaccine. The division was housed within the NIH until its transfer to the FDA in 1972 [3].

In 1976, Congress passed the Medical Device Amendments (MDA) to ensure safety and effectiveness of medical devices, giving the FDA jurisdiction over the regulation of medical devices before marketing; requiring some medical devices to undergo the FDA's premarket approval and others to meet specific standards. The amendments were followed by the passage of the Safe Medical Devices Act of 1990, which required certain facilities to report serious adverse events associated with medical devices and for medical device manufacturers to conduct post-market surveillance and tracing on permanently implanted medical devices. The 1990 Act also authorized the FDA to order device product recalls and other actions [3].

In the 1980s, the Orphan Drug Act, involving the research and marketing of drugs for rare diseases, and the Drug Price Competition and Patent Term Restoration Act, regarding generic drugs, were passed [3]. The specifics of these acts are discussed later in this chapter.

In 1994, Congress passed the Dietary Supplement Health and Education Act, defining "dietary supplements" and "dietary ingredients" and classifying them as foods rather than drugs. However, the Act authorizes the FDA to promulgate good manufacturing practice regulations for dietary supplements and establishes specific labeling requirements. In that same year, the FDA announced that it could consider regulating nicotine. The following year, the FDA declared cigarettes to be "drug delivery devices," proposing restrictions on marketing and sale. However, these efforts were thwarted. The agency withdrew its regulations in response to the landmark Supreme Court case, *FDA v. Brown & Williamson Tobacco Corp. et al.*, holding that the FDA does not have the authority to regulate tobacco as a drug. The FDA continued to stay clear of tobacco regulations until the Family Smoking Prevention and Tobacco Control Act was passed in 2009 [3]. Tobacco regulations are discussed in more detail in a later chapter.

In 1997, Congress passed the Food and Drug Administration Modernization Act, generally viewed as the most wide-ranging reform to the agency since the 1938 FDCA. Although the fine details of the act are beyond the scope of this chapter, the act did provide a pathway to accelerate the review of devices and regulate the advertising of unapproved uses of approved drugs and devices. In the 1990s and early 2000s, several other laws were passed to facilitate and encourage the testing of relevant new drugs in the pediatric population [3].

3   Drugs

The U.S. Code defines "drugs" as articles defined in the "official compendium" or those used in the diagnosis, cure, mitigation, treatment, or prevention of disease in man or other animals, or articles (other than food) intended to affect the structure or any function of the body of man or other animals or articles intended for use as a component of any article specified above [1].

The United States Pharmacopeial (USP) Convention, an independent nonprofit organization, publishes what the FDCA defines as the "official compendium," comprising the United States Pharmacopeia, the official Homoeopathic Pharmacopeia of the United States, the official National Formulary, and any supplement to any of them. USP also develops and publishes standards for drug substances, drug products, excipients, and dietary supplements in the United States Pharmacopeia–National Formulary (USP-NF). However, enforcing these provisions is the FDA's responsibility. USP's goal is to have monographs for all FDA-approved drugs. Furthermore, USP develops monographs for therapeutic products not approved by the FDA (e.g., pre-1938 drugs, dietary supplements, and compounded preparations). Once a monograph is available, compliance with nonproprietary name, identity, strength, quality, purity, packaging, and labeling is mandatory. Additionally, USP, in association with the United States Adopted Names (USAN) Council, creates the nonproprietary names for drugs and biologics under the authority derived from section 502(e) of the FDCA. Proprietary names (including brand names) are designated by the FDA while working with the NDA applicant [8].

Unlike small-molecule drugs with well-defined chemical structures that can be thoroughly characterized, biologic products are generally derived from living material (e.g., human, animal, or microorganism) and are complex structures that are not usually fully characterizable [9]. Biologics, with some exceptions, are licensed under section 351 of the Public Health Services Act (PHSA). However, since biologics meet the definition of "drugs" under the FDCA, they are also subject to regulation under the FDCA provisions. The U.S. Code defines "biologic product" as a "virus, therapeutic serum, toxin, antitoxin, vaccine, blood, blood component or derivative, allergenic product, protein, or analogous product, or arsphenamine or derivative of arsphenamine (or any other trivalent organic arsenic compound), applicable to the prevention, treatment, or cure of a disease or condition of human beings" [10]. This includes monoclonal antibodies for in-vivo use, cytokines, growth factors, enzymes, immunomodulators, thrombolytics, and non-vaccine therapeutics [9].

Drugs and biologics require premarketing approval by the FDA before being transported or distributed across state lines. To obtain approval, typically, one of the following applications needs to be filed with the FDA. These applications are filed by the sponsor.

- Investigational New Drug (IND): This application pertains to investigational drugs or biologics before entering the clinical investigation phase. By submitting this application, the sponsor of an investigational drug or biologic is essentially seeking an exemption from prohibition to transport and/or distribute an unapproved drug or biologic across state lines. The IND must contain information about animal pharmacology and toxicology studies, manufacturing information, clinical protocols, and investigator information. Upon submitting an IND, the sponsor must wait at least 30 calendar days before initiating any clinical trials. During this time, the FDA reviews the application to ensure research subjects are not subjected to unreasonable risk.
- New Drug Application (NDA): Before commercializing a drug in the U.S., an approved NDA is required. This applies to both new drugs and drugs that have been approved in other countries but are seeking commercialization in the U.S. The NDA can be submitted once the sponsor believes sufficient evidence exists to show the drug's safety and effectiveness. In addition to information on safety and effectiveness from pre-clinical and clinical trials, the NDA provides information on the proposed labeling and manufacturing.
- Abbreviated New Drug Application (ANDA): As the name suggests, this is an abbreviated application, specifically for the approval of generic drugs. The application is abbreviated as it does not require any new preclinical or clinical data for the generic drug while deriving the safety and effectiveness data from the approved NDA of the pioneer drug. Rather, the application is focused on establishing bioequivalence between the proposed generic drug and the pioneer drug. For approval, the sponsor must demonstrate that the generic delivers the same amount of active ingredients into a patient's bloodstream in the same amount of time as the pioneer drug.
- Biologic License Application (BLA): Necessary for biological products, the BLA requires the submission of information on the manufacturing processes, chemistry, pharmacology, clinical pharmacology, and the medical effects of the biologic product. If the submitted information meets the criteria, the FDA issues a license to allow marketing of the product.

An FDA-approved drug is a drug product that has an approved NDA and is thereby determined by the FDA to have been shown to be *safe* and *effective* for its intended use in humans. "Safe" in this context does not equate to absolute safety or absence of risk but rather means that the benefits outweigh known potential risks for the intended population. This balancing of risks versus benefits is a somewhat subjective standard and shifts based on intended conditions as well as currently available treatments for the intended conditions. For instance, the FDA may consider a higher risk tolerance for drugs that treat life-threatening conditions with no other available treatments than for so-called "me-too" drugs for relatively benign conditions. Safety data is obtained from clinical studies provided by the sponsor [11]. "Effectiveness" must be established through "substantial evidence." Adequate and well-controlled investigations must demonstrate that the drug "will have the effect it purports or is represented to have under the conditions of use prescribed, recommended, or suggested in the labeling or proposed labeling thereof" [12]. Typically, the FDA requires at least two well-designed clinical trials but sometimes allows a single trial in exceptional cases such as rare disease treatments [11].

An FDA-licensed biologic is a biologic product that has been issued a license under an approved BLA; the FDA has determined that the biologic is shown to be *safe*, *pure*, and *potent* for its intended use in humans. Safety is defined similarly to drugs as discussed above, while purity means relative freedom from extraneous matter regardless of its potential for harm or deleterious effect [13]. The FDA has included effectiveness in the "potency" requirement [14]. Effectiveness is determined under the same standard of "adequate and well-controlled studies" discussed earlier, although this can be waived for biologic products when effectiveness can be substantiated through an alternative method such as serological response. Of note, Congress has directed the FDA to minimize the differences in the review and approval of products regulated under BLAs and products regulated under NDAs [15].

4   Expedited Programs

Congress and the FDA recognized that certain new drugs that significantly decrease morbidity and mortality associated with serious conditions require expedited review to hasten availability of those drugs for patients. Consequently, a number of programs have been devised for such drugs and are briefly introduced below. Each program has detailed requirements that are beyond the scope of this chapter.

4.1   Fast Track

The Fast Track process is designed to facilitate the development and expedite the review of drugs that treat serious conditions and fill an unmet medical need. The conditions that can qualify a drug for Fast Track designation are broad and include AIDS, Alzheimer's, cancer, diabetes, and many others. These conditions are selected based on various factors, including survival, day-to-day functioning, or the likelihood of progression without treatment. Receiving Fast Track designation allows for expedited review, including Accelerated Approval and Priority Review if relevant criteria are met (discussed below). Furthermore, Fast Track designation allows for more frequent communication with the FDA and, in some aspects, a more flexible review process [16].

4.2   Accelerated Approval

The Accelerated Approval pathway was established in 1992. It allows for expedited approval of new drugs for promising therapies treating serious or life-threatening conditions. Under this pathway, new drugs can be approved using a surrogate or an intermediate clinical endpoint instead of the traditional clinical endpoints. For instance, tumor shrinkage could be used as satisfying evidence of a new cancer drug's effectiveness; this allows for its approval while awaiting confirmatory trials that confirm mortality benefit. Approval may be withdrawn, or labeling changed if confirmatory trials fail to verify sufficient clinical benefits [17].

4.3   Priority Review

Priority Review designation indicates that the FDA intends to take action on an application for a new drug within 6 months compared to 10 months under standard review. Priority Review designation is reserved for drugs that significantly improve the treatment's safety or effectiveness, the diagnosis, or the prevention of serious conditions. Of note, Priority Review designation does not change scientific/medical standards, or the type/quality of evidence needed for approval; rather, it simply expedites the review process [18].

4.4   Breakthrough Therapy

Breakthrough Therapy designation is given to drugs that are intended to treat a serious condition with preliminary evidence indicating substantial improvement over available therapy on the clinically significant endpoint(s). Like the Accelerated Approval pathway, a "clinically significant endpoint" may differ from the traditional clinical endpoint. Such drugs qualify for all benefits of Fast Track designation in addition to more intensive guidance and involvement from the FDA to assist and support an efficient FDA approval process [19].

5   Orphan Drugs

Orphan drugs are those drugs, biologics, or vaccines intended to treat, prevent, or diagnose rare diseases. The FDA defines rare diseases as those that affect fewer than 200,000 patients in the U.S.. It is estimated that more than 7000 rare diseases currently exist. In the past, due to limitations associated with drug development for rare diseases, such drugs rarely progressed beyond the initial IND phase. They would tactically remain on the market as an investigational drug without ever undergoing a full investigation of their safety and efficacy. Particularly concerning for these drugs' sponsors was that by definition, rare diseases have a small patient population, making it difficult to find an adequate number of research subjects to satisfy the FDA's traditional clinical trial standards. Additional concerns included the required large research and development investment with a limited number of potential patients to recuperate the cost. In response, the Orphan Drug Act was passed in 1983 to facilitate drug development for rare diseases, affording significant advantages to orphan drugs' sponsors. These advantages include considerable financial incentives, such as the Orphan Drug Tax Credit (ODTC), providing a tax credit equal to 25% of the cost of qualified trials, exemption from the application fee of the Prescription Drug User Fee Act (PDUFA), and a generous 7-year market exclusivity for the first approved NDA, independent of the drug's patentability, thereby barring the FDA from approving any other application for the same drug for the same orphan disease or condition during this period. Additional benefits of orphan drug designation include funding, such as FDA funding through the Orphan Products Grants Program, and more relaxed requirements for clinical trials, which, in many cases, include waiving the requirement for the placebo arm and Phase 1 study, and combining Phase 2 and Phase 3 trials. A 2014 study showed that most clinical trials for therapeutics involving rare diseases between 2006 and 2012 were likely single-arm, nonrandomized, and open-label compared to non-rare disease clinical trials [20, 21]. Although this scheme has enabled the development of many drugs that treat actual rare diseases, it has also been ripe for abuse. Pharmaceutical companies seek approval for rare indications of a given drug to take advantage of the many benefits afforded to orphan drugs. For instance, in 2009, one colchicine producer secured orphan drug designation for colchicine, a readily and cheaply available drug, for treating Familial Mediterranean Fever (FMF), considered a rare disease, even though the medication had been used for centuries to treat acute gout attacks, a very common disease. Consequently, the manufacturer secured a 7-year market exclusivity for the brand-name formulation of colchicine, Colcrys, resulting in a price hike from nine cents to five dollars per tablet, more than a 50-fold increase in price! The story of colchicine is not unique, and there are many other examples of unintended consequences of the Orphan Drug Act in the literature [22]. Given the tremendous benefits afforded to the development of orphan drugs, it is no surprise that in 2020, 58% of all novel drug approvals were for orphan drugs [23].

6   Controlled Substances

Federal control of narcotics dates back to the passage of the Harrison Narcotic Act of 1914, which required prescriptions for products exceeding the allowable limit of narcotics and mandated increased record-keeping for dispensers [3]. Subsequently, a patchwork of other federal laws were enacted to regulate narcotics. After the passage of the Drug Abuse Control Amendments of 1965, the FDA became responsible for its enforcement to prevent abuse of depressant and stimulant drugs. However, it was not until 1970 that congress enacted a new comprehensive law, the Controlled

Substances Act (CSA), which continues to be the primary legislation regulating controlled substances. The enforcement of the CSA was shifted to the Drug Enforcement Agency (DEA), an agency within the Department of Justice. The CSA establishes five schedules of controlled substances, with Schedule I drugs being substances with high abuse potential and no currently accepted medical use (e.g., cocaine) that are thereby prohibited from domestic distribution. Schedule II–V drugs have a progressively lower potential for abuse and accepted medical uses (e.g., tramadol). Under the statute, the DEA must consult the FDA on the scheduling of controlled substances. The FDA's recommendations are binding on scientific and medical matters, barring the DEA from scheduling a drug if the FDA recommends against it. Of note, the CSA does not prohibit the FDA from approving an NDA for any controlled substance with legitimate medical use. The inclusion of a drug as a scheduled controlled substance imposes a significant additional burden on all involved parties, but those details are beyond the scope of this chapter. Regarding the use of Schedule I drugs for religious purposes, with few minor exceptions, the courts have been relatively unsympathetic to plaintiffs and have generally afforded the DEA wide latitude on scheduling substances and enforcing bans [24].

7   Prescription vs Non-prescription

Since the passage of the FDCA in 1938, the FDA has claimed the regulatory authority, backed by the courts, to mandate some drugs to be only available by prescription instead of being available as over-the-counter (OTC) drugs. The FDA claimed this regulatory authority based on the labeling requirements of the FDCA. In 1951, by enacting the Durham-Humphrey Amendments, Congress codified the FDA's regulations distinguishing between prescription and nonprescription drugs in section 503(b) of the FDCA. Accordingly, drugs held for sale or dispensed in contravention of section 503(b) are deemed to be misbranded, a violation of the FDCA. The 503(b) section referenced above, requires drugs that are too toxic or have a harmful effect that is not safe except under the supervision of a licensed clinician or as otherwise determined by the FDA to be available only by prescription. Originally, the law also required habit-forming drugs to be available by prescription only, but that requirement has been eliminated. Most new chemical entities are initially approved as prescription drugs. Subsequently, usually after at least 5 years, there can be a status change from prescription drug to OTC drug (e.g., proton pump inhibitors such as omeprazole were initially only available by prescription but now are OTC). The change in status from prescription to OTC can be prompted by the FDA, the sponsor, or any other person in the case of off-patent prescription drugs. Although states do not have the authority to switch prescription drugs to OTC drugs, they do have the authority within their jurisdictions to mandate a switch from OTC to prescription. For instance, Oregon mandated that ephedrine-containing drugs, considered an OTC drug by the FDA, to require a prescription [25].

Authority to prescribe and requirements to obtain such authority, as well as any limitation to such authority, are derived from state law. The FDA defers that determination to the states, as the FDA and the FDCA do not regulate the practice of medicine. Although in all states, both allopathic and osteopathic physicians have full prescribing powers, different states impose various limitations on other types of clinicians. In addition to needing the authority to prescribe, there is both federal statutory and case law precedent that the prescription must have been issued in the course of professional practice, as part of a bona fide prescriber/patient relationship, and after a good faith determination that the prescription is for a "legitimate medical purpose." The pharmacist is also prohibited from dispensing a prescription medication when the pharmacist knows that there is no valid prescription [25].

8   Over-The-Counter (OTC) Drugs

Prior to marketing, OTC products must meet the same effectiveness standards as prescription drugs and, in many cases, need to clear a higher bar for safety given their availability without a prescription. Most OTC products reach the market through the OTC drug monograph system. The OTC monograph serves as a rule book for various therapeutic categories of OTC drugs. It establishes several parameters (e.g., active ingredients, indications, doses, testing, and labeling) under which an OTC drug is generally recognized as safe and effective (GRASE), and can be marketed without an FDA pre-market review or an approved NDA. Monographs can be added, removed, or changed through an administrative process, which can be initiated by the industry or the FDA. Final monographs are published as a rule in the Code of Federal Regulations [26]. As discussed above, many OTC drugs initially start as prescription drugs going through the FDA approval process and then, based on post-marketing information regarding their safety, they are switched to OTC drugs years later. Others get direct OTC approval, although this route is rare and usually pertains to drugs that have been approved and used for some time in foreign markets [27].

9   Generic Drugs

To facilitate generic drug development and manufacturing, The Drug Price Competition and Patent Term Restoration Act (aka the Hatch-Waxman Act) was passed in 1984. The Act allowed generic drug companies to use an abbreviated application process (i.e., ANDA) to obtain FDA approval for marketing. Essentially, ANDA only requires a demonstration of therapeutic equivalence between the generic and pioneer drugs, also known as the Reference Listed Drug (RLD), thus bypassing the other costly studies and processes associated with the approval of a new drug that are usually required under the New Drug Application (NDA) [28]. The Act also addressed issues related to pioneer drug exclusivity, generic drug exclusivity, patent term, and patent adjudication [29]. The Hatch-Waxman Act extended the patent term to account for the time the patented product is under review by the FDA and provided certain periods of marketing exclusivity for pioneer drugs. Also, the first generic drug company to obtain approval for an ANDA is entitled to a 180-day generic drug exclusivity [28].

Before the passage of the Act, only 35% of top-selling drugs no longer under patent protection had generic equivalents, while by the late 1990s, almost all such drugs had generic equivalents [30]. However, from a cost perspective, more notable is the exponential growth in the generic drug market share; now, generic drugs constitute over 85% of U.S. prescriptions, resulting in a significant reduction in pharmaceutical spending [31]. Although the Act has introduced generics into the market by providing various incentives to generic drug companies, it alone cannot explain the market share growth of generic drugs. A main driver of such growth in the market share has been state-level generic substitution laws and promotion of such laws by various payers [30]. By the early 1980s, every state had effectively adopted some form of generic substitution law.

Without state-level generic substitution laws, there are no legal mechanisms that allow the pharmacist to dispense any medication other than what is exactly stated on the prescription. In essence, these laws provide a legal pathway for the dispensing pharmacist to substitute a cheaper, typically generic drug for a more expensive, typically pioneer drug if the two are therapeutically equivalent without having to obtain an affirmative authorization from the prescriber. For instance, fluoxetine can be dispensed in place of Prozac if the prescriber has prescribed Prozac without the need for an affirmative authorization from the prescriber. Such laws come in different variations; some states make substitutions mandatory while others only make it permissible; some require patient consent while others do not. Some states use a positive formulary approach (i.e., limiting drug substitution to a specific list); others use a negative formulary approach (i.e., allowing substitution unless prohibited by a specific list) [32]. Nonetheless, all states prohibit such substitution if the prohibition is specifically indicated by the prescriber or the patient declines the substitution.

In the late 1970s, due to the number of requests from various states regarding therapeutic equivalence determination and development of formularies to facilitate generic substitution, the FDA developed the *Approved Drug Products with Therapeutic Equivalence Evaluations* (commonly known as the Orange Book and, more recently with the online version, the Electronic Orange Book). In addition to other eligible drug products, it lists every prescription drug that has been approved by the FDA as a new drug (both pioneer and generic drugs), which, with a few exceptions of older drugs, includes the vast majority of drug products available in the U.S.. Additionally, the Orange Book notates therapeutic equivalence evaluation based on the FDA's specific criteria for prescription drugs [33], thereby providing a single master list for the vast majority of pioneer drugs and all generics of pioneer drugs approved under an ANDA. Though the FDA clearly states that the Orange Book's content is neither an official FDA action nor a legally binding regulation, many states have incorporated the Orange Book's therapeutic equivalence determination into their generic substitution laws and regulations, elevating its status to a legally binding source [32]. Some jurisdictions only allow substitution if the generic drug is recognized as a therapeutic equivalent in the Orange Book. Other jurisdictions allow some flexibility and defer to the pharmacist's professional judgment, although the scope of such flexibility is usually very limited.

As opposed to small-molecule drugs where generic drugs are identical in structure to their pioneer drugs, due to the nature of biologics, it is practically impossible to require that level of similarity for generic versions of biologics. Rather, the FDA allows *biosimilars* where the biologic product is "highly similar" to and has "no clinically meaningful difference" from an existing FDA-approved reference product. The biosimilar's manufacturer must establish through extensive analysis of the structure and function that the product is "highly similar" to the reference product. Furthermore, the biosimilar has "no clinically meaningful difference" if the sponsor can demonstrate through human pharmacokinetic (exposure) and pharmacodynamic (response) studies, an assessment of clinical immunogenicity and, if needed, additional clinical studies that the product is as safe, pure, and potent as the reference product. *Interchangeable* products are biosimilar products that further establish they can be expected to produce the same clinical result as the reference product in any given patient and that their safety or efficacy is not adversely affected by alternating or switching between the products [34]. The FDA considers interchangeable products to be an acceptable substitute for the reference product without the need for prescriber involvement. However, as discussed above, permissibility of pharmacy-level substitution is a state-level decision, and the FDA lacks jurisdiction in the matter. State-level substitution laws for biosimilars are not as robust and universal as they are for small-molecule generic drugs. Similar to the Orange Book, the FDA publishes the *Lists of Licensed Biological Products with Reference Product Exclusivity and Biosimilarity or Interchangeability Evaluations*, known as the Purple Book, with information on biologics, including whether a specific product is a reference, a biosimilar, or an interchangeable product [35].

10 Devices

In reference to medical devices, the U.S. Code defines a "device" as "an instrument, apparatus, implement, machine, contrivance, implant, in vitro reagent, or other similar or related article, including any component, part, or accessory, which is (A) recognized in the official National Formulary, or the United States Pharmacopeia, or any supplement to them, (B) intended for use in the diagnosis of disease or other conditions, or in the cure, mitigation, treatment, or prevention of disease, in man or other animals, or (C) intended to affect the structure or any function of the body of man or other animals, and which does not achieve its primary intended purposes through chemical action within or on the body of man or other animals and which is not dependent upon being metabolized for the achievement of its primary intended purposes" [1]. Of note, the term "device" does *not* include software functions excluded pursuant to 21 U.S.C. § 360j(o), including but not limited to software functions intended for administrative support, maintaining, or encouraging a healthy lifestyle, electronic patient records, or medical decision support systems for health-care professionals. These exceptions are discussed in more detail in the Artificial Intelligence chapter of this book.

Medical devices have been subject to FDA oversight since the passage of FDCA in 1938. However, initially, the main concern was fraudulent devices. Therefore, the 1938 provisions did not afford the FDA authority to review medical devices for safety and effectiveness before marketing or to set and enforce performance standards. The FDA's authority was limited to misbranding and adulteration provisions, essentially limiting the FDA's enforcement actions to postmarket judicial remedies. In 1976, Congress enacted the Medical Device Amendments (MDA), creating a complex and novel regulatory regime for medical device development and marketing [36]. Several subsequent laws and amendments have been enacted since; many are summarized in this section of the book's introductory chapter. Medical devices are regulated by the CDRH.

The 1976 MDA established the current 3-tier classification scheme that creates a system to require a varying degree of assurance and scrutiny before a device can be introduced into the market based on its potential risk (see Table 1).

- Class I: Includes devices for which neither special controls nor premarket approval is warranted because general regulatory controls, as provided under the FDCA, are sufficient. These include devices with the lowest risk that present minimal potential for harm. A manual stethoscope is an example of a Class I medical device.
- Class II: Includes devices for which general controls are insufficient but sufficient information exists about the device to devise special controls such as special labeling, design characteristics or specifications, performance standards, premarket data requirements, and guidance documents. These include devices at higher risk than Class I devices. An electronic stethoscope is an example of a Class II medical device.
- Class III: Includes all devices introduced after the 1976 MDA's enactment that are not substantially equivalent to a device marketed before the 1976

Table 1 Medical Device Classification [39]

Class	Risk	Potential Harms	Regulatory Controls	Submission Type	Percent of Devices* (%)
I	Lowest	Minimal	General	510(k) 510(k) – exempt (~93%)	35
II	Moderate	Higher than Class I	General and Special (some are exempt)	510(k) 510(k) – exempt	53
III	Highest	Sustain or support life, are implanted or present potential unreasonable risk of illness of injury	General and PMA	PMA	9

*3% of devices are unclassified

amendments or other devices already classified as Class I or II, unless they are down-classified under a "de novo" procedure. These include but are not limited to devices that sustain or support life, are implanted, or present potentially unreasonable risk of illness or injury. Examples include mechanical heart valves, artificial knee joints, and intraocular lenses.

All Class III medical devices require premarket approval prior to marketing. Premarket approval is obtained through an approved Premarket Approval (PMA) application, which is analogous to the NDA for new drugs and effectively serves as a private license for marketing a particular device. Devices that fail to meet PMA requirements are adulterated and cannot be marketed. Alternatively, unless otherwise exempt from this requirement, devices that are not subject to a PMA, namely Class I and Class II devices, can be marketed through premarket notification (instead of approval) under section 510(k) of the FDCA (aka 510(k) clearance).

In a 510(k) submission, manufacturers must demonstrate that their "new device" is "substantially equivalent" to another legally marketed device, known as the predicate, and therefore exempt from PMA. For a device to be considered substantially equivalent, the device must have the same intended use as the predicate and the same technological characteristics or, if different, should not raise questions of safety and effectiveness. For instance, the manufacturer of a new at-home urine pregnancy test, classified as a Class II medical device, can satisfy the premarket notification requirements through a 510(k) submission by demonstrating that its test has the same intended use and technological characteristics as another legally marketed at-home urine pregnancy test, without needing to prove safety and effectiveness through clinical studies as part of a PMA.

Although safety and effectiveness data are required for most 510(k) submissions, the overall process is significantly less costly and burdensome, and the requirements are not nearly as rigorous as those for a PMA submission. Most Class I devices (over 90%) and even some Class II devices are exempted from the 510(k) clearance requirements and can be marketed directly without the FDA's involvement (e.g., manual stethoscopes). To conduct clinical studies on a medical device to collect safety and effectiveness data in support of a PMA or 510(k) clearance, an approved Investigational Device Exemption (IDE) is required allowing the investigational device to be shipped lawfully. Although the purpose of an IDE is similar to an IND, the details of the process are quite different and beyond the scope of this chapter [37, 38].

As previously defined, Class III medical devices include any device that is not substantially equivalent to a device marketed prior to the 1976 amendment or other medical devices already classified as Class I or Class II. Under this definition, any "new device" would automatically be classified as Class III, requiring a costly and lengthy PMA before marketing, irrespective of its potential risk. As this burdensome process may not be needed for low and moderate risk devices, the De Novo classification request allows manufacturers to request an otherwise Class III device to be reclassified as a Class I or a Class II device. If granted, the device can be marketed, and this can create a new category of devices serving as a predicate for future substantially equivalent devices under the 510(k) process.

11   Off-Label Use

Once a drug or device is approved (or cleared) for any specific use, with few exceptions, it can be prescribed by prescribers for any patient and any use. The FDA defines off-label use as "use for indication, dosage form, dose regimen, population or other use parameter not mentioned in the approved labeling." [2]. As long as the off-label use is done in the regular course of practicing medicine, it is outside of the scope of the FDCA and thereby not within the FDA's purview. However, if the off-label use is part of performing additional research on that drug, then the use will be regulated by the FDA [40].

Clinical off-label use is customary and some of the most commonly prescribed medications are prescribed for off-label use. Although the FDA does not regulate this practice, clinicians must be careful to avoid malpractice liability associated with off-label use, typically brought under negligence (including informed consent) cause of action [41, 42]. The specific legal elements to succeed in such cases and likewise, available defenses are discussed in earlier chapters of this book. Usually, the legal question with regards to off-label use in such cases is about disclosure of the off-label nature of the use. Tort law is a creature of state law and therefore, rules differ from state to state. However, many courts have held that there is no affirmative duty on behalf of the clinician to disclose that the proposed use of the drug or device is considered off-label but rather the focus is on the adequacy of disclosure of material risks. The FDA approved labeling, any applicable warnings (e.g., black box warning) and prescribing guidelines in the Physician's Desk Reference are examples of legitimate sources of such information [43]. Even though off-label use alone is not sufficient to establish presumption of breach of duty, off-label nature of the use can be introduced as evidence in such cases [40].

12   Off-Label Promotion

Off-label promotion involves promotion of an approved (or cleared) medical product for a use other than the ones approved by the FDA. This is to be differentiated from promotion of an unapproved drug or device which is treated differently under the law and is beyond the scope of this chapter. Unlike off-label use, federal law does limit off-label promotion. Legal liability may arise under False Claims Act which is outside of the FDA's jurisdiction, and it can also lead to misbranding actions by the FDA, including criminal prosecution under the FDCA. In some instances, state consumer protection laws may also be implicated [44]. It is common for clinicians to become involved with off-label promotion; for example, depending on the circumstance, discussing uses other than the ones approved by the FDA for a given drug while presenting at a conference, could be considered off-label promotion. Therefore, it is of vital importance to ensure the off-label promotion falls within the legal framework [40].

Generally, the FDA is mainly concerned with off-label promotion by manufacturers. Therefore, activities that are not supported by the industry are exempt from FDA regulations. For activities that are in some way supported by the industry, the FDA differentiates between (1) activities performed by, or on behalf of, the companies that market the products which is subject to the FDA regulations; and (2) activities, supported by companies, that are otherwise independent from the promotional influence of the supporting company which are exempt from the FDA regulations [40].

For activities that fall under the FDA's jurisdiction, rules and regulations on off-label promotion have been in flux for many years. Traditionally, the FDA took a more rigid approach and largely prohibited off-label promotion. However, in recent years, in part as a result of evolving case law, the FDA has changed its stance and it is more permissive in allowing off-label promotion as long as it is truthful and non-misleading. One of the challenges in this area is that limitation on off-label promotion involves restricting commercial speech which is afforded qualified protection under the First Amendment of the US Constitution. In such cases, if the speech involves an unlawful activity or is false or misleading, then it can be banned. Otherwise, a more nuanced analysis must be taken by the court to see if prohibition is justified. Specifically, with regards to off-label promotion, the courts have held that since ultimately the goal is to promote a lawful activity (i.e., off-label use), as long as it is not false or misleading, it is constitutionally protected commercial speech and cannot be prohibited [45–47]. Since courts have not clarified the threshold for determining what is considered truthful and non-misleading, clinicians should take extra care to ensure that their promotion meets such standard even under extra scrutiny [40].

13 Preapproval Access

Preapproval access describes various legal schemes outside of a clinical trial, for patients to access drugs and medical devices that have not been fully and formally approved (or cleared) by a country's regulatory system. "Compassionate use," "named patient programs," and "managed access" are some other commonly used terms to describe such access. In the US, outside of clinical trials, the law provides three pathways for preapproval access, (1) Emergency Use Authorization (EUA), (2) Expanded Access Program (EAP), and (3) Right-to-Try (RTT). Unlike off-label use discussed earlier, these pathways provide access to regulated products that have not been approved (or cleared) by the FDA for any purpose. While EUA and EAP can potentially provide preapproval access to both drugs and medical devices, the RTT pathway only provides preapproval access to drugs. Although these pathways provide a legal mechanism for manufacturers to provide access to unapproved drugs and medical devices, none of the mentioned pathways entitles the patient or the clinician to be provided with the desired products. Furthermore, in *Abigail Alliance v. von Eschenbach*, the U.S. Court of Appeals for the District of Columbia held that patients have no legal right to demand "a potentially toxic drug with no proven therapeutic benefit" and further that "there is no fundamental right… to experimental drugs for the terminally ill." The Supreme Court declined to review the case [40, 48].

13.1 Emergency Use Authorization (EUA)

The EUA pathway was established under the Project BioShield Act amending the FDCA in 2004 to permit "the FDA Commissioner to authorize the use of an unapproved medical product or an unapproved use of an approved medical product during a declared emergency" [49]. Before the FDA can issue an EUA, the HHS Secretary must declare that circumstances exist justifying the authorization (aka EUA declaration). The requirements for the basis of such declaration are detailed in the law and can be based on domestic, military or a public health emergency [50]. A drug or medical device may be issued an EUA if the totality of the evidence makes it reasonable to believe that the product *may be* effective, and *relatively* safe [51]. This is in contrast to the much higher bar for full FDA approval, requiring *substantial evidence* of efficacy and *proof* of safety [52]. The EUA may subsequently be revoked by the FDA if it is determined that the required criteria are no longer met, or such revocation is appropriate to protect public health or safety, or when the EUA declaration is terminated [40, 53, 54].

The EUA process was in widespread use during the COVID-19 pandemic where numerous medical products including drugs, vaccines and medical devices became available on the market at least initially through an EUA. Furthermore, a number of previously approved drugs and medical devices also received new indications under an EUA. Since off-label use allows clinicians to prescribe previously approved drugs and medical devices for any purpose in the regular course of the practice of medicine, it may not seem as important, however, as discussed below, this will have tremendous implications regarding immunity from tort liability when using or prescribing such products. There may be other potentially important implications for approving new indications for previously approved drugs and medical devices including insurance coverage and governmental actions to increase availability and accessibility, but they are beyond the scope of this chapter.

Clinicians seldom become a sponsor of an EUA medical product. But are usually affected by the EUA regulatory scheme when they are involved in administering or prescribing an EUA medical product specifically in relation to potential liability. Therefore, the focus will be on legal liability associated with EUA products rather than the details of the regulatory process of obtaining an EUA [40].

Although the EUA pathway does not provide any liability protection, Public Readiness and Emergency Preparedness Act ("PREP Act") enacted in 2005 does provide a mechanism to limit tort liability associated with EUA medical products or as referred to in the PREP Act, covered countermeasures. Technically, this liability protection does not automatically attach to EUA products. Instead, the liability protection only applies after the HHS secretary has issued a separate declaration. Such a declaration will identify covered individuals and entities, covered counter-measures, any limitations on distribution, category (or categories) of diseases, health conditions or threats to health, population, geographic area, and the declaration's effective time period. Even though the immunity from tort liability is fairly sweeping and generally preempts state law, it does not cover willful misconduct. Of note, by law, acts consistent with applicable directions, guidelines, or recommendations of the HHS Secretary are not to be considered "willful misconduct" so long as a federal, state or local health authority is notified regarding the suffered loss associated with the administration or use of the covered countermeasure within 7 days of actual discovery of such information [55]. Finally, the immunity against tort liability is only provided for harms caused by the use of a covered countermeasure; harms caused by failure to use a covered countermeasure do not qualify for immunity under this scheme. The Countermeasures Injury Compensation Program (CICP) provides a separate compensation mechanism for eligible individuals injured because of administration or use of a covered countermeasure [40, 56].

Given the nature of EUA products and settings in which they are typically used, clinicians should take extra precaution to ensure that there is a separate declaration by the HHS Secretary triggering the immunity under the PREP Act. Clinicians should also ensure that they are covered under the declaration and their use and administration of the EUA medical product is also consistent with such declaration. Doing so, ensures sweeping tort liability protection [40].

13.2   Expanded Access Program (EAP)

Although since the 1970s, the FDA has facilitated access to investigational drugs, Early Access Program (EAP) as a distinct pathway became available in 1987 for drugs and biologics and in 1996 for devices. EAP was codified in law in 1997 followed by significant revision in 2009. EAP provides a potential pathway for patients with an immediately life-threatening condition or serious disease or condition to gain access to an investigational medical product for treatment outside of clinical trials in absence of alternative options. Additionally, potential patient benefits must justify the potential risks of the treatment [57].

Licensed physicians are authorized to become a sponsor to request access to an investigational medical product under EAP. Sponsors must submit necessary forms to the FDA, secure IRB approval, obtain patient's informed consent and coordinate access with the manufacturer. Unless otherwise advised by the FDA, the requested investigational medical product can be shipped, and treatment commenced 30 days after the application is received by the FDA. In an emergency, when treatment must begin before a written submission can be made, for drugs and biologics, a licensed physician can obtain authorization by contacting the FDA by telephone (or other rapid means of communication) while devices can be used without authorization. In all cases, the FDA must be notified, and appropriate reports must be filed within a specified time period. Accessing investigational medical products through EAP is time consuming, with one study estimating administrative burden to be approximately 30 hours inclusive of physician and staff time to support an EAP application for an individual patient [40, 57].

Given the nature of EAP and clinical settings in which EAP is utilized, clinicians should at least be aware of liability risk which is likely to be higher than what is encountered in the typical course of clinical practice. Unlike other preapproval access programs, there is no immunity or other liability protections associated with EAP. Best mitigation strategy for clinicians when utilizing EAP is to obtain and document a thorough informed consent. Lastly, even though it is theoretically possible, it is highly improbable that declining to become a sponsor under EAP would lead to additional liability [40].

13.3   Right-to-Try Act (RTT)

The Right-to-Try Act (RTT) (aka the Trickett Wendler, Frank Mongiello, Jordan McLinn, and Matthew Bellina Right to Try Act of 2017) was signed into law on May 30, 2018. The RTT allows a patient with a life-threatening disease or condition that has exhausted approved treatments and is unable to participate in a clinical trial involving the eligible investigational drug, access to that eligible investigational drug. Eligibility of the patient must be certified by a licensed physician who is not being compensated directly by the manufacturer [58, 59]. Unlike the EUA and EAP, RTT is limited to investigational drugs and does not allow access to investigational devices.

An eligible investigational drug is a drug that has completed Phase 1 clinical trial but has not been approved or licensed by the FDA for any use. Furthermore, the drug must have an application filed with the FDA or must be under investigation in a clinical trial in support of the FDA approval and be subject to an active IND. Finally, the drug's active development or production must be ongoing and not been discontinued by the manufacturer or placed on clinical hold by the FDA [40].

Once the RTT eligibility requirements for patient and the investigational drug are met, access can be obtained directly from the manufacturer through the treating physician without FDA involvement. The FDA's role is limited to receiving information from the drug sponsor and posting of certain information as required by law [40].

RTT explicitly shields sponsors, manufacturers, prescribers, dispensers and other individual entities from most liabilities in the absence of reckless or willful misconduct, or gross negligence. However, RTT does not provide any liability protections for causes of action arising from intentional torts. As discussed elsewhere in this book, some jurisdictions construe absent and in rarer cases inadequate informed consent as "battery," an intentional tort, in which case RTT does not provide any liability protection. However, malpractice suits brought under negligence theory (absent gross negligence, reckless or willful misconduct) including, in most cases, inadequate informed consent cases are preempted by the Act. Finally, the Act protects the above-mentioned entities against liability if electing not to participate [40, 60].

14 Application

14.1 *U.S. v. Caronia* 703 F.3d 149 (2d Cir. 2012)

Alfred Caronia, a pharmaceutical sales consultant, was criminally charged and convicted under the FDCA for conspiracy to introduce a misbranded drug into interstate commerce for his off-label promotion of Xyrem. The Second Circuit vacated his conviction and remanded the case on First Amendment grounds. The Court noted that off-label promotion in its essence is a form of commercial speech. Therefore, the restrictions should be analyzed under the *Central Hudson* test. If the speech involves lawful activity, it is truthful, and not misleading, the government can restrict it only if (1) the government has asserted a substantial interest; (2) the regulations must directly and materially advance the government's substantial interest and (3) the regulation must be narrowly tailored [40, 61, 62].

In this case, by applying the *Central Hudson* test, the court found that Off-label promotion involved a legal activity (i.e., off-label use) and Caronia's statements were neither false nor misleading. Therefore, he met the threshold prong of the test and was afforded qualified protection under the First Amendment. Subsequently the court proceeded with its analysis under the other three prongs of the test and found that the government's interest in preventing unsafe prescribing and usage of drugs were a substantial interest. However, since the prohibition aimed to prevent an outcome (i.e., off-label use) which remained legal, prohibition of off-label promotion was not found to be advancing the government's substantial interest. Lastly, it was found that a complete and criminal ban on off-label promotion was not narrowly tailored. Thus, a complete and criminal ban of off-label promotion was held to be unconstitutional holding that the "the government cannot prosecute pharmaceutical manufacturers and their representatives under the FDCA for speech promoting the lawful, off-label use of an FDA-approved drug." The FDA opted not to appeal this decision [40, 61].

15 Summary

Drugs and devices are regulated in the US based on a patchwork of laws and regulations developed over the past 150 years. The predecessor of the FDA initially started with a very limited role within the Department of Agriculture. Over the years, the role of the FDA has expanded to now include regulating consumption amounting to a fifth of US consumer spending. Initially, drug and device regulations were limited to preventing false advertising. Over time, this evolved to require review prior to marketing of drugs and devices including the need to prove safety and effectiveness. Then, post-market monitoring became part of the regulatory framework. However, as regulatory complexity grew, additional pathways were developed to expedite and decrease the regulatory burden on life-saving drugs as well as drugs for rare diseases although at times these were accompanied by unintended consequences.

As the FDA is not involved in regulating the practice of medicine, off-label use in the regular course of the practice of medicine is outside of the FDA's purview, while off-label promotion continues to be regulated although with evolving jurisprudence. Lastly, recognizing that there are times that patients may benefit from accessing drugs and devices prior to full approval, there are regulatory pathways to provide preapproval access.

References

1. 21 USC § 321.
2. Wittich CM, Burkle CM, Lanier WL. Ten common questions (and their answers) about off-label drug use. Mayo Clinic Proceedings. 2012; 87:982–90.
3. U.S. Food and Drug Administration [Internet]. Milestones in U.S. Food and Drug Law. 2018 Jan 31 [cited 2021 Dec 28]. Available from: https://www.fda.gov/about-fda/fda-history/milestones-us-food-and-drug-law.
4. U.S. Food and Drug Administration? [Internet]. What Does FDA Do. 2021 Jun 06. [cited 2021 Dec 28]. Available from: https://www.fda.gov/about-fda/fda-basics/what-does-fda-do.
5. U.S. Food and Drug Administration [Internet]. Regulated Products and Facilities. 2019 Oct. [cited 2021 Dec 28]. Available from: https://www.fda.gov/media/131874/download
6. U.S. Food and Drug Administration [Internet]. What is the difference between the Federal Food, Drug, and Cosmetic Act (FD & C Act), FDA regulations, and FDA guidance?. 2018 Mar 28. [cited 2021 Dec 28]. Available from: https://www.fda.gov/about-fda/fda-basics/what-difference-between-federal-food-drug-and-cosmetic-act-fdc-act-fda-regulations-and-fda-guidance
7. 21 CFR §10.115.
8. USP [Internet]. Legal Recognition – Standards Categories. [cited 2021 Dec 28]. Available from: https://www.usp.org/about/legal-recognition/standard-categories.
9. U.S. Food and Drug Administration [Internet]. Therapeutic Biologics Applications (BLA). 2020 Feb 24. [cited 2021 Dec 28]. Available from: https://www.fda.gov/drugs/types-applications/therapeutic-biologics-applications-bla.
10. 42 USC § 262.
11. U.S. Food and Drug Administration [Internet]. Development & Approval Process – Drugs. 2019 Oct 28. [cited 2021 Dec 28]. Available from: https://www.fda.gov/drugs/development-approval-process-drugs.
12. 21 USC § 355.
13. U.S. Food and Drug Administration [Internet]. Frequently Asked Questions About Therapeutic Biological Products. 2015 Jul 07. [cited 2021 Dec 28]. Available from: https://www.fda.gov/drugs/therapeutic-biologics-applications-bla/frequently-asked-questions-about-therapeutic-biological-products.
14. 21 CFR § 600.3(s).
15. U.S. Food and Drug Administration [Internet]. Providing Clinical Evidence of Effectiveness for Human Drug and Biological Products. 1998 May. [cited 2021 Dec 28]. Available from: https://www.fda.gov/files/drugs/published/Providing-Clinical-Evidence-of-Effectiveness-for-Human-Drug-and-Biological-Products.pdf.
16. U.S. Food and Drug Administration [Internet]. Fast Track. 2018 Jan 04. [cited 2021 Dec 28]. Available from: https://www.fda.gov/patients/fast-track-breakthrough-therapy-accelerated-approval-priority-review/fast-track.
17. U.S. Food and Drug Administration [Internet]. Accelerated Approval. 2018 Jan 04. [cited 2021 Dec 28]. Available from: https://www.fda.gov/patients/fast-track-breakthrough-therapy-accelerated-approval-priority-review/accelerated-approval.

18. U.S. Food and Drug Administration [Internet]. Priority Review. 2018 Jan 04. [cited 2021 Dec 28]. Available from: https://www.fda.gov/patients/fast-track-breakthrough-therapy-accelerated-approval-priority-review/priority-review.

19. U.S. Food and Drug Administration [Internet]. Breakthrough Therapy. 2018 Jan 04. [cited 2021 Dec 28]. Available from: https://www.fda.gov/patients/fast-track-breakthrough-therapy-accelerated-approval-priority-review/breakthrough-therapy.

20. Srivastava G, Winslow A. Orphan Drugs: Understanding the FD Orphan Drugs: Understanding the FDA Approval Process. Academic Entrepreneurship for Medical and Health Scientists. 2019; 1:1–19.

21. U.S. Food and Drug Administration [Internet]. Patents and Exclusivity. 2015 May 19. [cited 2021 Dec 28]. Available from: https://www.fda.gov/media/92548/download.

22. Murphy SM, Puwanant A, Griggs RC. Unintended Effects of Orphan Product Designation for Rare Neurological Diseases. Annals of Neurology. 2012; 72:481–90.

23. U.S. Food and Drug Administration [Internet]. New Drug Therapy Approvals 2020. 2021 Jan 08. [cited 2021 Dec 28]. Available from: https://www.fda.gov/drugs/new-drugs-fda-cders-new-molecular-entities-and-new-therapeutic-biological-products/new-drug-therapy-approvals-2020.

24. Hutt PB, Merrill RA, Grossman LA. Food and Drug Law: Cases and Materials. 4th ed. St. Paul: Foundation Press; 2014. p. 810-14.

25. Hutt PB, Merrill RA, Grossman LA. Food and Drug Law: Cases and Materials. 4th ed. St. Paul: Foundation Press; 2014. p. 802-10.

26. U.S. Food and Drug Administration [Internet]. Over-the-Counter (OTC) Drug Monograph Process. 2020 Sep 03. [cited 2021 Dec 28]. Available from: https://www.fda.gov/drugs/over-counter-otc-drug-monograph-process.

27. U.S. Food and Drug Administration [Internet]. How FDA strives to ensure the safety of OTC products. 2016 Mar 10. [cited 2021 Dec 28]. Available from: https://www.fda.gov/drugs/special-features/how-fda-strives-ensure-safety-otc-products.

28. U.S. Food and Drug Administration [Internet]. Abbreviated New Drug Application (ANDA). 2022 Jan 14. [cited 2022 Jan 25]. Available from: https://www.fda.gov/drugs/types-applications/abbreviated-new-drug-application-anda.

29. U.S. Department of Health and Human Services, Office of the Assistant Secretary for Planning and Evaluation [Internet]. Expanding the Use of Generic Drugs. 2010 Nov 30. [cited 2022 Jan 25]. Available from: https://aspe.hhs.gov/basic-report/expanding-use-generic-drugs.

30. Congressional Budget Office [Internet]. How Increased Competition from Generic Drugs Has Affected Prices and Returns in the Pharmaceutical Industry. 1998 Jul. [cited 2021 Dec 28]. Available from: https://www.cbo.gov/sites/default/files/105th-congress-1997-1998/reports/pharm.pdf.

31. Gupta R, Kesselheim AS, Downing N, Greene J, Ross JS. Generic Drug Approvals Since the 1984 Hatch-Waxman Act. JAMA Internal Medicine. 2016; 176:1391-1393.

32. Vivian JC. Generic-Substitution Laws. US Pharmacist. 2008; 33:30-34.

33. U.S. Food and Drug Administration [Internet]. Orange Book Preface. 2022 Jan 19. [cited 2022 Jan 25]. Available from: https://www.fda.gov/drugs/development-approval-process-drugs/orange-book-preface.

34. U.S. Food and Drug Administration [Internet]. Biosimilar and Interchangeable Products. 2017 Oct 23. [cited 2022 Jan 25]. Available from: https://www.fda.gov/drugs/biosimilars/biosimilar-and-interchangeable-products.

35. U.S. Food and Drug Administration [Internet]. Prescribing Biosimilar and Interchangeable Products. 2017 Oct 23. [cited 2022 Jan 25]. Available from: https://www.fda.gov/drugs/biosimilars/prescribing-biosimilar-and-interchangeable-products.

36. Hutt PB, Merrill RA, Grossman LA. Food and Drug Law: Cases and Materials. 4th ed. St. Paul: Foundation Press; 2014. p. 1202-11.

37. U.S. Food and Drug Administration [Internet]. Investigational Device Exemption (IDE). 2019 Dec 12. [cited 2021 Nov 30]. Available from: https://www.fda.gov/medical-devices/premarket-submissions-selecting-and-preparing-correct-submission/investigational-device-exemption-ide.

38. U.S. Food and Drug Administration [Internet]. Premarket Notification 510(k). 2020 Mar 03. [cited 2021 Nov 30]. Available from: https://www.fda.gov/medical-devices/premarket-submissions-selecting-and-preparing-correct-submission/premarket-notification-510k.

39. U.S. Food and Drug Administration [Internet]. How is My Medical Device Classified?. [cited 2021 Nov 30]. Available from: https://fda.yorkcast.com/webcast/Play/17792840509f4 9f0875806b6e9a1be471d.

40. Quang TS, Taft MS, Beriwal S. Understanding the Principles and Practice of Legal Oncology. 1st ed. New York: McGraw Hill; 2022. p. 305–22.

41. Riley JB Jr, Basilius PA. Physicians' liability for Off-label Prescriptions. Nephrology News Issues. 2007; 21:43-47.

42. *Femrite v. Abbott Northwestern Hospital*, 568 N.W.2d 535 (Minn. App. 1997).

43. Edersheim JG, Stern TA. Liability Associated with Prescribing Medications. Primary Care Companion Journal Clinical Psychiatry. 2009; 11:115-119.

44. U.S. Food and Drug Administration [Internet]. Industry Supported Scientific and Educational Activities. 1997 Nov. [cited 2021 Nov 30]. Available from: https://www.fda.gov/regulatory-information/search-fda-guidance-documents/industry-supported-scientific-and-educational-activities.

45. Mazer D, Curfman GD. FDA Sanctions Off-Label Drug Promotion. Health Affairs Blog. 2016 Jul 19. [cited 2021 Nov 30]. Available from: https://www.healthaffairs.org/do/10.1377/hblog20160719.055881/full/.

46. Jacobson L. Don't Fix What Ain't Broken—Off-Label Marketing, the FDA's Regulatory Regime, and the First Amendment. Emory Corporate Governance and Accountability Review. 2018; 5:19–60.

47. *Central Hudson Gas & Elec. Corp. v. Public Serv. Comm'n of N.Y.*, 447 U.S. 557 (1980).

48. *Abigail Alliance v. von Eschenbach*, 495 F.3d 695 (2007).

49. Institute of Medicine (US) Forum on Medical and Public Health Preparedness for Catastrophic Events [Internet]. Medical Countermeasures Dispensing: Emergency Use Authorization and the Postal Model, Workshop Summary. Washington (DC): National Academies Press (US); 2010. Available from: https://www.ncbi.nlm.nih.gov/books/NBK53122/.

50. FDCA § 564 (b).

51. 21 USC § 360bbb-3.

52. 21CFR § 314.126.

53. DCA § 564 (f).

54. U.S. Food and Drug Administration [Internet]. Emergency Use Authorization of Medical Products and Related Authorities. 2017 Jan. [cited 2021 Nov 30]. Available from: https://www.fda.gov/media/97321/download.

55. 42 U.S. Code § 247d–6d.

56. 42 CFR § 110.

57. U.S. Food and Drug Administration [Internet]. Expanded Access Program. 2018 May. [cited 2021 Nov 30]. Available from: https://www.fda.gov/media/119971/download.

58. Public Law 115-176.

59. 21 CFR § 312.81.

60. Public Law 115-176(b).

61. *United States v. Caronia*, 7 03 F.3d 149 (2d Cir. 2012).

62. *Central HudsonGas & Electric Corp. v. Public Service Commission*, 447 U.S. 557 (1980).

Tobacco Regulations

Amirala S. Pasha and Richard Silbert

Contents

Key Points

- The FDA requires new tobacco products to be approved and tobacco product health claims to be justified through a special approval process. Previous tobacco modifiers such as "light" or "mild" are banned.
- Manufacturers must report harmful ingredients as well as levels of tar and nicotine in their products. The TCA allows the FDA to limit but not eliminate nicotine levels in products.
- Under the TCA, additive flavors in cigarettes other than menthol were banned. A similar but narrower flavor ban was later extended to E-cigarettes.
- The TCA set the nationwide floor for tobacco purchases to 18 years of age which was subsequently raised to 21 years of age. State and local authorities are allowed to set higher limits.

A. S. Pasha (✉) · R. Silbert
Division of General Internal Medicine, Mayo Clinic, Rochester, MN, USA
e-mail: pasha.amirala@mayo.edu; silbert.richard@mayo.edu

© The Author(s), under exclusive license to Springer Nature Switzerland AG 2022
A. S. Pasha (ed.), *Laws of Medicine*,
https://doi.org/10.1007/978-3-031-08162-0_24

Legal Concepts

- Tobacco Master Settlement Agreement (MSA): An agreement entered in 1998, originally between the four largest tobacco companies in the United States and the attorneys general of 46 states, whereby, states settled their Medicaid lawsuits against the tobacco industry. In exchange the companies agreed to limit or eliminate certain tobacco marketing practices, and pay, in perpetuity, to the states to compensate them for some of the medical costs of caring for persons with smoking-related illnesses. Since then, other tobacco companies have joined the MSA.
- Section 907 of the TCA pertains to constraints and disclosure requirements for tobacco product ingredients under which the FDA can limit the amount of nicotine in products but crucially cannot decrease the amount to zero. This section also bans all added flavors other than menthol.

1   Introduction

Since a link between smoking and lung cancer was published by Doll and Hill in the British Medical Journal in 1954, there have been myriad federal, state, and local efforts to regulate tobacco products [1]. A landmark 1964 report by the surgeon general recognizing the health impacts of smoking was followed 2 years later by congressionally mandated health warnings on cigarette packaging. In 1970, congress passed the Public Health Cigarette Smoking Act, strengthening health warnings labels and banning cigarette marketing on TV and Radio [2]. On the state level, Minnesota was the first of many states to limit smoking in indoor spaces such as restaurants in 1975. Local municipalities such as Aspen, Colorado and San Luis Obispo, California took this a step further by requiring completely smoke-free restaurants. The next 20 years saw additional actions on tobacco including restricting smoking on domestic flights, regional increases in tobacco taxes, and the first FDA approval of a tobacco cessation aid (i.e. nicotine gum) [1].

Despite these varied efforts, an attempt at comprehensive regulation of tobacco did not occur until the 1990's. Federal efforts centralized in 1994 when the FDA announced it was discussing regulating nicotine as a drug; ultimately culminating in the 1995 decision to classify nicotine as a drug and cigarettes as drug delivery devices [3]. Prior to this, nicotine was exempt from FDA drug classification and regulation of tobacco products was not within the purview of the FDA except if the products made specific health claims [4]. Following the FDA's 1995 classification, then President Clinton directed the agency in 1996 to regulate the advertising, sale, and distribution of tobacco products to minors [1, 5, 6].

However, these efforts at formal tobacco oversight were overturned in 2000 in a 5–4 decision in *FDA v. Brown & Williamson Tobacco Corp*, where the Court ruled that the FDA could not regulate tobacco without authorization from congress [1]. Writing for the majority, Justice O'Connor acknowledged the significant public health threat caused by tobacco but argued that Congress "has created a distinct regulatory scheme for tobacco products, squarely rejected proposals to give the

FDA jurisdiction over tobacco, and repeatedly acted to preclude any agency from exercising significant policymaking authority in the area" [7].

The Court's decision catalyzed the development and eventual passage of the Family Smoking Prevention and Tobacco Control Act (TCA) in 2009. The passage of the TCA brought broad new regulation of tobacco including new approval processes for tobacco products, requiring disclosure of ingredients, banning many flavors of tobacco, and adding additional restrictions on the sale of tobacco to minors. The FDA was placed in charge of administering these regulations. The act represented the first major comprehensive congressional action to mitigate the harmful effects of tobacco use and provide basic oversight of tobacco.

There have been significant additional regulations on tobacco since 2009. Provisions in the 2010 Affordable Care Act required insurance companies to increase their coverage of tobacco cessation interventions. In the years following the TCA, the use of e-cigarettes skyrocketed, especially among youth, leading to the FDA's 2016 ruling giving the agency oversight of all tobacco products including e-cigarettes [1].

In 2019 congress raised the national age requirement for tobacco purchases to 21; an effort previously passed in 19 states and the District of Columbia [1]. In September 2019, the Trump Administration outlined plans to "clear the market of flavored e-cigarettes" promising a formal announcement a few weeks later [8]. Final regulation did not arrive until January 2020, when the FDA issued guidance that banned unauthorized e-cigarette cartridge flavors but contained loopholes that permitted the continued sale of thousands of flavors at vape shops [9]. At the time of this printing, additional regulation on e-cigarettes is being considered. Similarly, discussions on the fate of menthol in conventional cigarettes are ongoing.

It is worth mentioning the key role in tobacco regulation of the major 1998 Master Settlement Agreement (MSA) reached between the largest cigarette manufacturers and the attorneys general of a majority of US states and territories. The settlement was an amalgamation of individual lawsuits brought by states during the 1990's to recoup healthcare costs attributable to smoking. The MSA required initial payments by manufacturers at the time of settlement and increasing annual payments ($9.2 billion in 2018) to be paid in perpetuity. It also restricted key tobacco industry marketing tactics including targeting youth, providing free samples, most outdoor advertising, and product placements in media. Importantly, the agreement prohibited tobacco industry misrepresentation and suppression of research related to the harms of smoking. Of note, these terms only apply to the manufacturers' party to the agreement. The number of manufacturers party to the MSA has increased from 4 in the initial agreement to 50 as of 2018 as additional manufacturers sought to avoid tobacco related litigation [10].

While states vowed to use a sizable portion of the agreement's payments for public health and prevention, data suggests they have only used a small fraction of the estimated $126 billion paid out by the agreement on these measures, levels far below CDC recommendations [10, 11]. The tobacco industry continues to significantly outspend national, state, and private marketing efforts aimed at reducing smoking [11]. Despite these shortfalls, cigarette smoking rates in the United States have declined since the MSA, with youth rates decreasing from 36.4% to 8.8% from 1997 to 2017 and adult rates decreasing from 24.1% to 13.7% from 1998 to 2018. Among those who do smoke, the average number of cigarettes smoked per day has steadily decreased over the same period [12]. While the rates of cigarette smoking have decreased, the rates of e-cigarette use over the past decade has eroded much of these public health gains.

2   Discussion

The main tenets of Tobacco Control Act include adding warning labels to smokeless tobacco products, requiring scientific evidence to support claims of "low risk" products, compulsory reporting of product ingredients, and imposing stricter limits on the sale and marketing to youth. The act does not restrict state and local authorities from imposing additional restrictions [13].

The Tobacco Control Act does impose limitations on the FDA's regulatory authority. For instance, the agency can reduce the amount of nicotine in products to a "non-addicting" level but cannot reduce the amount to zero. While the act does allow the agency to restrict most flavors of cigarettes, the FDA is not allowed to outright ban conventional tobacco products such as chewing tobacco or cigarettes.

The TCA originally applied FDA regulation to cigarettes, roll-your-own tobacco, smokeless tobacco, and additional products that the agency "by regulation deems to be subject to" the new law [14]. In 2016, the FDA issued new guidance expanding oversight to include additional products such as pipe tobacco, cigars, hookah tobacco, and electronic nicotine delivery systems (ENDS) partially in response to skyrocketing e-cigarette use [15]. Unlike the agency's previous actions in the 1990's labeling nicotine a drug, the TCA does not classify tobacco products as a drug, a device, or a combination product.

3   Ingredients

A main component of the TCA is section 907 which pertains to constraints and disclosure requirements for tobacco product ingredients. Manufacturers must now limit the amount of pesticide residues detectable in their products and report the levels of harmful and potentially harmful constituents present. These reporting requirements extend both to the tobacco itself and any part of the tobacco product. Additionally, the amount of nicotine and tar in tobacco products must now be provided to the FDA. Under the act, the FDA can limit the amount of nicotine in products but crucially cannot decrease the amount to zero. Section 907 of the act also bans all added flavors other than menthol in cigarettes to curtail use among minors. The FDA has reserved the right to ban menthol in the future and the issue is under active debate given the disproportionate use of menthol in communities of color most harmed by smoking. Later regulatory action in 2020 extended the ban on most flavors to e-cigarettes [14].

4   Approval Process

Following the TCA's passage, new tobacco products come to market in a manner similar to the FDA's regulation of medical devices allowing different pathways for products that are and are not substantially equivalent. Substantially equivalent products need only justify their similarity to products marketed prior to the passage of the act. Such products marketed prior to 2007 were designated as "Grandfathered Tobacco Products". Products that are not substantially equivalent to a grandfathered product are subject to approval through a new tobacco product application. During this process, the product must demonstrate that its approval "would be appropriate for the protection of the public health" [14].

Under the act, marketing modifiers such as "light" or "mild" used to promote cigarettes as "lower risk" were prohibited. Instead, tobacco companies can only release products with "modified risk" claims if approved by a new FDA process requiring demonstration that the product will lead to a significant public health benefit. This process involves multiple rounds of review including a period for public comments prior to final approval. One factor the FDA considers during the approval process is how the risks and benefits of a "modified risk" product compare to already approved medical products for nicotine cessation [14].

5   Labeling/Marketing

Warning labels on tobacco packaging have been required since the Federal Cigarette Labeling and Advertising Act of 1965 [2]. Afterwards, the Public Health Cigarette Smoking Act in 1970 required sterner warning labels, prohibited television and radio advertising, and limited advertising in print and billboard media. These restrictions were challenged in the case of *Capital Broadcasting Company v. Mitchell* where 6 radio companies claimed the act violated their first amendment right of freedom of speech. The district court disagreed with the plaintiffs, holding that "product advertising is less vigorously protected than other forms of speech" [16]. The Supreme Court denied certiorari on appeal.

The TCA placed more stringent restrictions on tobacco product labeling and marketing. For instance, a greater portion of tobacco packaging must now display text and images depicting the harms from smoking. Though the TCA outlines how tobacco products come to market, producers are forbidden from using this process to claim their products are safe or "FDA approved" [14]. Similarly, the FDA limits how the health claims of "modified risk" products can be marketed.

The TCA further prohibited tobacco advertising through free samples, event sponsorship, loyalty programs, and tobacco-branded merchandise. These restrictions were challenged by tobacco companies in *Discount Tobacco City & Lottery, Inc v. United States* on first amendment grounds (see below) [17]. The case upheld the TCA's provisions to require significant packaging space be devoted to health warnings including color graphics and non-graphics on tobacco warning labels and that tobacco advertising be limited to black and white text. However, the case struck down the TCA's ban on tobacco loyalty (aka continuity) programs. FDA regulations pertaining to the previously upheld TCA's graphic warning label requirements were later successfully challenged in *R.J. Reynolds Tobacco Co. V. FDA* in 2012 [18]. In this case, the court held that the FDA had failed to meet its burden under *Central Hudson* [19] to show that its regulations on graphic warnings directly and materially advanced government's goals of encouraging tobacco cessation and dissuading others from picking up the habit. Of note, the ruling in *R.J. Reynolds Tobacco Co. V. FDA* only applied to the FDA regulations that implemented the graphic warning provisions of the TCA and not to the underlying provisions of the TCA which was previously held to be constitutional. At the time of this writing, the FDA has once again promulgated new regulations on required graphic warnings, but their implementation have been delayed by courts due to pending litigation.

6   Tobacco in Minors

Prevention of youth tobacco use has a long history in this country, dating back to at least 1883 when New Jersey set a minimum age for tobacco purchases to 16. By 1920, 46 of 48 states had set a minimum age for tobacco purchases with age requirements as high as 21 years of age. However, due to tobacco industry lobbying, many states lowered their age limits in the 1950's leading to a proliferation of tobacco marketing and use among youth. Tobacco companies have long viewed age restrictions as a serious threat to their business model as most smokers (66%) start before age 18 and the 17–20 age group was viewed by manufacturers as a key revenue demographic [7].

Limiting tobacco use among minors has been a central aim of both the MSA and the TCA. The MSA prohibited direct and indirect marketing to youth; specifically banning use of cartoon characters, gift rewards to youth for tobacco purchases, and free tobacco samples except in adult only facilities (13). The TCA went further by banning sports and event sponsorship, restricting vending machines to adult-only facilities, and placing further limits on free samples and giveaways that had been common youth advertising tactics. As discussed above, the TCA banned flavored tobacco products more favored among youth. Most significantly, the TCA established the first national minimum age for tobacco sales to 18 years. The act strengthened this provision by requiring retailers to check photo ID to verify age and enforcing strict penalties on retailers who were caught selling to minors [4]. In 2019, congress went further and raised the national age requirement for tobacco purchases to 21.

7   Application

7.1   *Discount Tobacco City & Lottery, Inc. v. United States,* 674 F.3d 509 (6th Cir. 2012)

As discussed previously, the TCA had a number of provisions limiting and restricting tobacco advertisement. This case was a challenge to a number of these requirements, namely (1) reserving significant portion of tobacco packaging for health warnings including graphic images; (2) restriction on commercial marketing of "modified risk tobacco products;" (3) ban of statements to convey the impression that tobacco products are approved or safe due to now being regulated by the FDA; (4) restricting the advertising of tobacco products to black text on a white background; and (5) bar on distribution of free samples of tobacco products, tobacco brand sponsorship of events, branded merchandising of tobacco products, and distribution of free items in consideration of a tobacco purchase (i.e., "continuity programs").

These challenged provisions all involved restriction on commercial speech. The First Amendment to the US Constitution does afford protection to commercial speech, albeit in a more limited fashion than protection afforded to non-commercial speech. These challenges are typically reviewed under the *Central Hudson* test, a form of intermediate scrutiny. Under this test, commercial speech is only protected by the First Amendment if it involves a lawful activity and is truthful and not misleading. However, such speech can be restricted if (1) the government has asserted a substantial interest, (2) the restrictions directly and materially advance the government's substantial interest, and (3) the restrictions are narrowly tailored.

In this case, the appellate court held that applying the *Central Hudson* test, the challenged restrictions, except for the limiting tobacco advertising to black text on white background and prohibition on continuity programs, satisfy the requirements of the First Amendment and are therefore constitutional. With regards to the provisions held to be unconstitutional, the court reasoned that limiting advertising to black text on white background was overly broad (i.e., not narrowly tailored) while on the prohibition on continuity programs, government had failed to meet its burden on establishing that such prohibition advances the government's stated interest and therefore, neither could be sustained under the *Central Hudson* test.

8   Summary

The net effect of the TCA and prior tobacco regulation was to forge a multifaceted approach to limit the public health costs of smoking while handicapping many of the tobacco industry's long-held marketing methods. The regulation of tobacco today encompasses a holistic, end to end approach that supplants prior piecemeal regulation. Requirements that tobacco companies disclose ingredients, justify health claims, and seek approval for new products have increased public transparency and aid the clinician in effectively communicating the harms of smoking. Prohibitions on marketing and sales to youth have led to a generational decline in smoking rates that will hopefully be accompanied by a similar decline in smoking-related disease in the decades to come [20]. Likewise, increased marketing and labeling requirements have occurred hand in hand with declining usage rates among those who do smoke [12]. That bodes well for the burden of smoking-related disease going forward.

Challenges still loom. The rising prevalence of e-cigarettes threatens progress made on youth smoking rates. While smoking rates have declined, as of 2018 an estimated 19.7% of adults still reported tobacco and smoking endures as a leading cause preventable death in the United States. Finally, the prevalence of smoking among those with mental illness, lower socioeconomic status, and less formal education remains disproportionately high and a worthy target of future efforts [20].

References

1. American Lung Association [Internet]. Tobacco Control Milestones. [cited 2020 Dec 20]. Available from: https://www.lung.org/research/sotc/tobacco-timeline.
2. Miles D. Public Health Cigarette Smoking Act of 1969 [Internet]. The First Amendment Encyclopedia. 2009 [cited 2021 Jan 15]. Available from: https://www.mtsu.edu/first-amendment/article/1089/public-health-cigarette-smoking-act-of-1969.
3. U.S. Food and Drug Administration [Internet]. Milestones in U.S. Food and Drug Law. 2018 [cited 2020 Dec 24]. Available from: https://www.fda.gov/about-fda/fdas-evolving-regulatory-powers/milestones-us-food-and-drug-law-history.
4. Public Health Law Center [Internet]. Litigation update: Recent cases that implicate the family smoking prevention and tobacco control act. 2014 [cited 2021 Jan 15]. Available from: https://www.publichealthlawcenter.org/sites/default/files/resources/tclc-fs-litigationupdate-TobaccoControlAct-Sept.2014.pdf.
5. U.S. Food and Drug Administration [Internet]. Nicotine in cigarettes and smokeless tobacco is a drug and these products are nicotine delivery devices under the federal food, drug, and cosmetic act: Jurisdictional determination. 1996 [cited 2021 Jan 15]. Available from: https://www.govinfo.gov/content/pkg/FR-1996-08-28/pdf/X96-20828.pdf
6. Statements by President Bill Clinton and the US Food and Drug Administration on regulations to restrict the marketing, sale, and distribution of tobacco to children. Tob Control [Internet]. 1995;4(3):299–309. Available from: http://www.jstor.org/stable/20747440.
7. Apollonio, Dorie E., and Stanton A. Glantz. "Minimum ages of legal access for tobacco in the United States from 1863 to 2015." American journal of public health 106.7 (2016): 1200-1207.
8. U.S. Food and Drug Administration [Internet]. Trump administration combating epidemic of youth E-cigarette use with plan to clear market of unauthorized, non-tobacco-flavored E-cigarette products. 2019 [cited 2020 Dec 28]. Available from: https://www.fda.gov/news-events/press-announcements/trump-administration-combating-epidemic-youth-e-cigarette-use-plan-clear-market-unauthorized-non.
9. U.S. Food and Drug Administration [Internet]. FDA finalizes enforcement policy on unauthorized flavored cartridge-based e-cigarettes that appeal to children, including fruit and mint. 2020 [cited 2020 Dec 28]. Available from: https://www.fda.gov/news-events/press-announcements/fda-finalizes-enforcement-policy-unauthorized-flavored-cartridge-based-e-cigarettes-appeal-children
10. Public Health Law Center [Internet]. Tobacco Control Legal Consortium. The Master Settlement Agreement: An Overview. 2019 [cited 2022 Jan 31]. Available from: https://www.publichealthlawcenter.org/sites/default/files/resources/MSA-Overview-2019.pdf.
11. Campaign for Tobacco-Free Kids [Internet]. Broken promises to our children. 2017 [cited 2021 Jan 5]. Available from: https://www.tobaccofreekids.org/what-we-do/us/statereport.
12. American Lung Association [Internet]. Overall Tobacco Trends. [cited 2021 Jan 10]. Available from: https://www.lung.org/research/trends-in-lung-disease/tobacco-trends-brief/overall-tobacco-trends.
13. U.S. Food and Drug Administration [Internet]. Family Smoking Prevention and Tobacco Control Act - An Overview. 2020 [cited 2021Jan31]. Available from: https://www.fda.gov/tobacco-products/rules-regulations-and-guidance/family-smoking-prevention-and-tobacco-control-act-overview.
14. Hutt PB, Merrill RA, Grossman LA. Food and Drug Law: Cases and Materials. 4th ed. St. Paul: Foundation Press; 2014. p. 1351-72.

15. Center For Tobacco Products [Internet]. FDA Requirements for Newly Regulated Tobacco Products. [cited 2021 Jan 5]. Available from: https://www.fda.gov/media/110022/download.
16. Capital Broadcasting Company v. Mitchell, 333 F. Supp. 582 (D.D.C. 1971).
17. Discount Tobacco City & Lottery, Inc. v. US, 674 F.3d 509 (6th Cir. 2012).
18. R.J. Reynolds Tobacco Co. v. Food & Drug Admin 696 F.3d 1205 (D.C. Cir. 2012).
19. Central Hudson Gas & Electric Corp. v. Public Service Commission, 447 U.S. 557 (1980).
20. Center for Disease Control and Prevention [Internet]. Cigarette Smoking Among U.S. Adults Hits All-Time Low. 2019 [cited 2021 Jan 10]. Available from: https://www.cdc.gov/media/releases/2019/p1114-smoking-low.html.

Artificial Intelligence: A Legal Landscape

Ashleigh Giovannini and Amirala S. Pasha

Contents

Key Points

- Artificial Intelligence (AI) uses algorithmic units to create intelligent machines that have the potential to evolve over time.
- Regulations pertaining to software programs used for medical purposes in the United States are well-suited to technological advancement.
- The Food and Drug Administration typically regulates software used for the treatment, diagnosis, cure, mitigation, or prevention of disease, known as Software as a Medical Device, through various approval processes at the beginning of a product's lifecycle.
- The Food and Drug Administration has proposed a Total Product Life Cycle regulatory approach to monitor the safety and efficacy of complex AI technology throughout the product's circulation in an effort to protect patients and promote technological advancement and transparency.

A. Giovannini (✉)
Nixon Gwilt Law, Memphis, TN, USA
e-mail: ashleigh.giovannini@nixongwiltlaw.com

A. S. Pasha
Division of General Internal Medicine, Mayo Clinic, Rochester, MN, USA
e-mail: pasha.amirala@mayo.edu

387

A. S. Pasha (ed.), *Laws of Medicine*,
https://doi.org/10.1007/978-3-031-08162-0_25

- The 21st Century Cures Act excluded certain software from FDA regulation, allowing certain AI to go unregulated.
- The Federal Trade Commission enforces Section 5 of the FTC Act which prohibits unfair or deceptive practices in or affecting commerce.
- In recent years, the Federal Trade Commission has indicated it will approach the transparency and algorithmic bias of applications and devices developed using consumer health data with increased scrutiny and attention toward unfair and deceptive practices.
- The tort law concepts of negligence and products liability have not yet adapted to provide redress for plaintiffs injured by or through the use of AI.
- Algorithms are not generally patentable because they are mathematical concepts that fall under the category of abstract ideas.
- To determine if AI, as made up by algorithms, is patentable, an additional inquiry is required to determine whether the ineligible concept (e.g., abstract idea) is integrated into a practical application whereby the patent claim in practice becomes more than the ineligible concept.
- An AI machine would be ineligible to be named as the author of the "original work" for purposes of copyright protection, but the developer of the AI may be eligible for authorship and copyright protection for the AI system's creation since the output can be viewed as the "fruit" of the initial work by the developer.

Legal Concepts
- Software as a Medical Device: a regulatory vehicle used by the Food and Drug Administration to classify software used to treat, diagnose, cure, mitigate, or prevent disease without the need for hardware.

1 Introduction

Think for a moment about the data footprint of the average patient. With the proliferation of technologies like health and wellness applications, connected devices, and electronic medical records, large volumes of data are accumulated with previously unfathomable swiftness. This data can be critical to the clinician. Information derived from these technologies can influence patient treatment and care plans, create efficiencies, and facilitate practice management. Indeed, some technologies can even generate predictions about clinical outcomes or suggest diagnoses using Artificial Intelligence (AI).

AI is the science and engineering of making intelligent machines using basic operational units called algorithms [1, 2]. Algorithms are often simple, carefully crafted programs with detailed instructions to solve narrow and well-defined problems. AI programs are often similarly simple. However, certain algorithmic networks can resemble human intellectual abilities through a process called Machine Learning (ML), a subset of AI. ML is a culmination of algorithms trained to identify patterns in data and generate a new algorithmic model used to make predictions [3].

AI and ML technologies are emerging in the clinical setting. Currently, though, advancements in AI are only clinically beneficial in certain instances, like in the detection of seizures, hypoglycemia, or atrial fibrillation, or the diagnosis of certain conditions using histopathological examination or medical imaging [4]. As these tools become increasingly sophisticated, however, there is potential for large-scale transformation in the delivery of clinical care.

While AI technology is rapidly evolving, the same cannot be said for the regulatory and legal landscape surrounding the development and use of AI. Today, AI is governed by a patchwork of regulations, guidance documents, and occasional case law. Unlike technological advancements of the past, AI may present insurmountable challenges to the law's ability to adapt. However, appropriately regulating AI is critical to protecting consumers and developers of these potentially groundbreaking technologies. This chapter explores these challenges and potential solutions.

2   Regulation of AI

To date, the US federal government has taken a relatively hands-off approach to centralized regulation of AI, instead choosing to develop a regulatory framework on an agency-by-agency basis. This chapter will focus on the regulatory efforts of the Food and Drug Administration (FDA) and the Federal Trade Commission (FTC).

2.1   FDA Regulation of AI

The FDA and the International Medical Device Regulators Forum (IMDFR) classify software programs that do not include hardware intended to be used for medical purposes as Software as a Medical Device (SaMD) [1, 2]. This classification implicates the traditional regulatory pathway for medical devices. The definition of a medical device comes from 21 U.S.C. § 321(h) of the Federal Food, Drug, and Cosmetic Act (FDCA), which classifies any instrument, apparatus, implement, machine, appliance, implant, reagent for in vitro use, *software*, material, or other similar or related article that is intended for the specific medical purpose of: (1) diagnosis, prevention, monitoring, treatment or alleviation of disease; (2) diagnosing, monitoring, treatment, alleviation of or compensation for an injury; (3) investigation, replacement, modification, or support of the anatomy or of a physiological process; (4) supporting or sustaining life; (5) control of conception; (6) disinfection of medical devices; or (7) providing information by means of in vitro examination of specimens derived from the human body [4]. As discussed in more detail in the earlier chapters of this book, medical devices are generally regulated by the FDA through the 510(k) premarket notification, De Novo classification, and premarket approval processes [1].

However, these established procedures for the regulation of typical SaMD are not well-suited to address technological advancements in AI-based SaMD because the devices are constantly evolving. The current regulatory pathway requires developers of software that has received clearance from the FDA to submit a 510(k) for renewed clinical evaluation each time a change is made with the intent to significantly affect safety or effectiveness. This includes changes made to significantly improve clinical outcomes, to mitigate a known risk, or to respond to adverse events within the software [5, 6]. This is relatively simple for traditional SaMD that do not change when populated with additional data. However, the same cannot be said for SaMD that utilize AI to evolve as data inputs are processed.

Within the category of AI, the FDA classifies machine learning into "locked" or "adaptive" categories. "Locked" machine learning creates a fixed behavior that does not change with additional data. Because locked algorithms produce the same results with the same input, regulatory review is relatively simple. However, as treatments and clinical practices evolve, the training data used to develop a locked algorithm may begin to yield less accurate results with the input of real-world data. In a clinical setting, locked algorithms may be referred to as "rules-based AI," which uses validated information, such as published studies or clinical guidelines, to develop a logic of individual decisions that make up a final recommendation [7].

"Adaptive" machine learning allows for the behavior of the algorithm to change using a defined learning process [1]. The adaptation of algorithms occurs in two stages: learning and updating. Learning can take place through increasingly complex methods (e.g., linear and logistic regression, decision trees, support vector machines, and deep learning or neural networks), depending on the nature of the problem the system must solve. Generally, by inputting data and labeling whether the algorithm's recommendations are correct, the system begins to develop an internal logic [8]. With this development comes an increased risk for distributional shift, false-positives or -negatives, "black box" decision-making, automation bias, and outmoded practices that outweigh the benefits of the algorithm [8]. This can be particularly problematic where the algorithm serves a clinical function.

Often touted as a major impediment to the regulation of AI, the term "black box" decision-making is used to describe the opacity of learning algorithmic systems as they become more complex. There are two types of opacity: literal and practical. Literal opacity exists when the algorithmic relationships are completely hidden while the machine learning process is known. With literal algorithms, the mechanisms used to classify information remain unknown to everyone, including the person who built the program. Practical opacity occurs when numerous parameters interact with numerous decision trees to classify observations. While the relationship may not be hidden from view, the logic used by the system could be so complex as to defy understanding [9].

Not all algorithms are "black box," but, as algorithms become increasingly sophisticated, the lack of transparency in development has affected the establishment of a meaningful regulatory framework for assuring safety and efficacy [10]. If it is difficult to determine the predictability of AI-based SaMD, the FDA's safety requirements could be impossible to satisfy. Something that is unpredictable has the potential to be dangerous. Similarly, if an AI-centered SaMD lacks explainability, the inability to properly address problematic outputs could jeopardize the technology's ability to meet FDA efficacy standards.

Proposed Regulatory Pathway

To date, the FDA has approved devices using "locked" algorithms. However, because adaptive algorithms change over time, the FDA has proposed a new regulatory pathway to provide ongoing safety and efficacy assurances to consumers [8]. Unveiled by the FDA in 2019, the Total Product Life Cycle (TPLC) approach is modeled after the Software Precertification (Pre-Cert) Program, which provides a voluntary regulatory pathway for SaMD products that are not well-suited for the hardware-focused pathway used for traditional medical devices [11]. The TPLC would provide reasonable assurances about safety and effectiveness throughout the life of AI-driven SaMD products. This approach would create regulatory evaluation and monitoring processes that span from premarket development of an AI SaMD product through post market performance. The organization responsible for the development of the AI-based SaMD would also be required to demonstrate continued excellence to the FDA [12].

The proposed TPLC approach would involve four general functions: (1) to establish expectations for quality systems and good machine learning practices (GMLP); (2) to conduct premarket review; (3) to set expectations among manufacturers to monitor AI-driven devices and incorporate risk management in developing, validating, and executing algorithmic changes; and (4) to implement post market performance reporting to increase transparency and maintain assurances of safety and effectiveness [12].

As with all medical device manufacturers, the FDA would require AI-driven SaMD device developers to establish a quality system to ensure compliance with regulatory standards and continued development, delivery, and maintenance of high-quality products. For premarket review, AI device manufacturers would be expected to demonstrate valid clinical association, analytical validation, and clinical validation. The amount and depth of each of these data types required to assure safety and effectiveness would depend on the function of the AI device, the risk to users, and the intended use. GMLP are the AI analogue to good software engineering practices. With the function of SaMD in mind, GMLP would include best practices relating to the relevance of available data to the clinical problem and practice, the acquisition of data that aligns with the SaMD's intended use, adequate separation between test, tuning, and training datasets for the algorithm, and transparency about the algorithm and its outputs geared toward the user [12].

The proposed approach to modifications of AI-based SaMDs would rely on a "predetermined change control plan." This plan would detail the types of anticipated changes an SaMD might undergo and the relevant protocol to implement those changes while managing the risk to patients. SaMD Pre-Specifications (SPS) describe the changes the manufacturer will make to the algorithm as it continues to learn. The Algorithm Change Protocol (ACP) describes the method by which a manufacturer would effectuate these changes with additional details pertaining to risk mitigation [12].

The TPLC approach would incorporate continued monitoring of the risks to patients while allowing the manufacturer to improve the technology's performance. The current SaMD framework requires manufacturers to perform risk assessments to determine whether an algorithmic modification should result in the submission of a new 510(k) for premarket review or whether placing documentation and analysis of the modification in the risk management and 510(k) files is sufficient. The proposed rule would allow AI SaMD manufacturers to document modifications that fit the parameters of the approved SPS and ACP. In instances where the modifications exceed the bounds of the SPS and ACP, the FDA proposes a "focused review" of the SPS and ACP. During the focused review, manufacturers would be able to request a concurrence from the agency that the proposed modification fits the current SPS and ACP, enter a pre-submission document discussing how the modification fits the current parameters, or submit a premarket application to modify the SPS or ACP [12].

Finally, through the TPLC approach, the FDA would expect AI SaMD manufacturers to commit to transparency and provide frequent reporting about updates made to the algorithm and performance metrics for the SaMD. Transparency would include updates to the FDA, appropriate labeling changes that completely describe and justify any modifications to the SaMD, notice about changes in inputs, and updated reports pertaining to the performance of the SaMD. The FDA would also encourage innovation in transparency measures by manufacturers. The frequency and reporting type of the real-world performance monitoring requirement would be tailored with consideration for the risk of the device, the number of anticipated modifications, and the age of the device compared to the number of previous modifications [12].

FDA's 2021 Action Plan Update

In January 2021, the FDA published an action plan in response to the agency's 2019 call for feedback on its proposed total lifecycle approach to AI regulation. Going forward, the FDA intends to issue draft guidance on what must be included as part of SaMD Pre-Specifications (SPS) and Algorithm Change Protocol (ACP) to demonstrate the safety and effectiveness of AI. The FDA also plans to encourage Good Machine Learning Practices, such as AI best practices in data management, training, interpretability, evaluation, and documentation. The action plan includes the FDA's commitment to supporting regulatory efforts to develop a methodology for the evaluation and improvement of AI, including the identification and elimination of bias, and for evaluation and promotion of algorithm robustness.

In addition to the Action Plan, the has FDA indicated an evolving agency approach to regulation and guidance in the rapidly progressing field of AI-based SaMD [13]. In October 2021, the FDA held a public workshop on how a patient-centered approach to labeling devices with information about algorithm training, the relevance of data inputs, and the logic employed by the algorithm can promote transparency in AI-based medical technologies [14]. As part of the Action Plan, the FDA has begun engaging with stakeholders who are piloting a real-world performance monitoring process for AI-based SaMD to allow manufacturers to understand how their products are being used, identify opportunities for improvement, and respond proactively to safety or usability concerns [13].

Exclusion from FDA Regulation

While the proposed TPLC approach put forth by the FDA includes some important parameters surrounding the development of AI, any finalized regulations may not apply to certain types of AI-based technologies because they have been excluded from FDA jurisdiction. In 2016, federal legislators added Section 520(o) to the Food, Drug & Cosmetic Act to exclude certain health-related software from the Section 201(h) definition of a device for purposes of FDA review through the 21st Century Cures Act (Cures Act) [15]. Exemptions include software that is intended:

1. **For administrative support of a health care facility.** This includes software functions used for the processing and maintenance of financial records, billing, admissions, business analytics, and practice and inventory management and analysis.
2. **For maintaining or encouraging a healthy lifestyle unrelated to diagnosis, cure, mitigation, prevention, or treatment of a disease or condition.** This includes mobile applications (apps) that calculate calories, log, record, track, evaluate or make decisions or behavioral suggestions for general fitness, health, and wellness.
3. **To serve as electronic patient records, including patient-provided information, with a function not intended to interpret or analyze patient records for the purpose of diagnosis, cure, mitigation, prevention, or treatment.** This includes an Electronic Health Records (EHR) or Electronic Medical Records (EMR). Additionally, an app that collects patient-reported health data simply for incorporation into the medical record would not be considered a device requiring FDA regulation.
4. **For transferring, storing, converting formats, or displaying data or results.** This includes hardware or software products that store data, such as blood pressure readings, to be viewed at a later time by the patient's clinician, convert health data into a format that can be printed, or display previously stored health data [16].
5. *Not* **to acquire, process, or analyze data from scanning or diagnostic devices such as MRIs or** *in vitro* **clinical tests,** *and* **is used for the purpose of:**
   (a) **Displaying, analyzing, or printing medical data, such as patient information or peer-reviewed clinical studies; or**
   (b) **Supporting or providing medical recommendations to a health care professional, on the condition that the software allows the provider to independently review how those recommendations were made; or**
   (c) **Enabling a health care professional to independently review the basis for such recommendations that such software presents so that it is not the intent that such health care professional rely primarily on any of such recommendations to make a clinical diagnosis or treatment decision regarding an individual patient** [13].

Of particular importance is the final exemption, which can encompass Clinical Decision Support (CDS) software. CDS software can provide clinicians with "knowledge and person-specific information, intelligently filtered or presented at appropriate times to enhance health and healthcare." The FDA's ability to regulate these systems hinges on two considerations: (1) the type of intended user, and (2) the ability for the user to independently review the basis for the system's decision. Notably, the Cures Act excluded from FDA oversight all CDS software used by clinicians where the clinician can independently review the basis for the decision, regardless of the gravity of the healthcare situation or condition [17]. In response to this exclusion, the FDA warned that too many AI-driven products may now evade review under the Cures Act and expressed concerns that dependence on CDS software that is not subject to FDA oversight to provide diagnoses and treatment could harm patients [16].

Alternative to FDA Regulation

Some legal scholars believe that the FDA will be unable to adequately protect consumers as technology advances [18]. Indeed, FDA leaders have indicated that the regulation of AI-based SaMD presents significant challenges for the agency [13]. However, many scholars expressing skepticism about the FDA regulatory regime for algorithms also doubt that the statutory authority or expertise of any existing governmental agency would suffice [8].

Some critics believe that an algorithmic regulatory regime would hamper market incentives for companies to develop new and innovative products [6]. Others believe that a dedicated federal agency with extensive expertise on AI technologies could impart safety and efficacy standards without impeding innovation through standard-setting authority and responsibility for the coordination and development of classifications, design standards, and best practices for algorithmic technology. Such an agency might use a nuanced rubric to assign regulatory scrutiny based on an algorithm's complexity. Subjecting only the most complex and opaque algorithms to regulatory scrutiny would come closer to developing flexible guidelines without stifling innovation [18].

Other suggestions to address the risks of quickly developing AI technologies include federal agency certification with limited tort liability as an incentive, licensure of technology developers, and "nudging" companies by proposing policy that "fosters the beneficial and benign development of AI" [19, 20]. Regardless of the method, it is clear that an innovative regulatory approach is necessary to address the revolutionary developments in algorithmic technology, especially those that influence consumer health.

2.2   FTC Regulation of AI

The FDA is not alone in its AI regulatory endeavors. The Federal Trade Commission (FTC) enforces Section 5 of the Federal Trade Commission Act (FTC Act), which prohibits unfair or deceptive practices in or affecting commerce, in virtually every industry. The FTC routinely uses its subpoena power, granted to the agency through

Section 6 of the FTC Act, to investigate and regulate uses of information and AI in the private sector. Under the FTC Act, an act or practice is considered unfair if it is likely to cause substantial harm to consumers without any larger benefit. A deceptive practice involves a statement, omission, or other practice that is likely to mislead a consumer acting reasonably under the circumstances [21]. The FTC Act gives the FTC the ability to seek relief on behalf of consumers through injunctions, restitution, and, sometimes, civil penalties [22].

In recent years, the FTC has issued guidance on the development of AI, algorithms, and predictive analytics [23]. As a baseline, the FTC urges organizations to embrace transparency and fairness at all stages of AI development and implementation [24]. In practice, AI developers and manufacturers may face enforcement action if the entity overpromises regarding the AI's performance, is dishonest about how data is used or processed, or trains the algorithm with biased or unfair data [25].

Overpromising from developers may come in the form of exaggeration or statements not backed by evidence about the AI system. For example, if a developer markets a product as providing "100% accuracy in the diagnosis of multiple myeloma," but the technology was built using testing data that lacked racial diversity, the FTC may deem this statement to be deceptive to consumers. While this example is oversimplified, it illustrates the importance of paying particular attention to how AI-based products are described.

As of late, the FTC has also begun doubling down on companies that do not tell the truth about how data is used or processed. Often the FTC learns about such practices from consumer complaints. Consumer complaints about the way user information is collected for AI development often focus on the statements made by companies about how they process or use data in their privacy notices. However, the FTC may also discover impropriety through close examination of published data [25]. The FTC has taken a punitive approach to AI that is developed dishonestly. In a recent case against a photo application, the FTC ordered the developer to delete data it collected deceptively, and the algorithms created with that data [26].

The FTC is also particularly concerned with algorithmic bias. Algorithmic bias arises when the data used for training is imbalanced or misrepresents various populations. Bias may also arise when the algorithm is influenced by human subjectivity or prejudice, a surprisingly common occurrence [27]. It is important to note that the incorporation of bias is not always intentional [28]. However, when particular groups are underrepresented in data training sets, the algorithmic outcomes may be of limited predictive value [27]. The FTC views the outcomes associated with biased datasets as harmful to consumers because the model may yield results that are unfair or inequitable to certain groups [25]. Additionally, in 2021, the FTC issued guidance as part of its enforcement of the Health Breach Notification Rule, which includes notification obligations for vendors of personal health records in the event of any breach of unsecured information, stating that it would begin exercising close regulatory scrutiny for applications and devices that collect or use consumer health information in their development [29].

3   Artificial Intelligence and Tort Law

The widespread use of AI in the clinical setting presents special considerations for tort law theories. The basics of tort law are covered in earlier chapters of this book. Instead, this chapter provides an overview of tort law as applicable to AI. It is therefore assumed that the reader has an understanding of tort law principles.

AI can be a useful tool for providing patient care, but it also raises complex questions about clinician liability. Predictive algorithms can guide patient care through risk identification, diagnoses and drug recommendations, and crisis care management that replace human decision-making. However, because much of AI processing is developed or conducted in a "black box," it is often impossible to know or discover the path to an algorithm's conclusion. Who, then, is liable when an error is made?

Liability is traditionally governed by principles of tort law [30]. Clinician liability for medical errors is often adjudicated as part of a malpractice or negligence action. The purpose of a malpractice or negligence action is to determine whether the clinician engaged in "conduct which falls below the standard established by law for the protection of others against unreasonable risk of harm" [31]. Healthcare organizations can also be found liable under the legal doctrine of *respondeat superior,* or for negligent hiring or supervision of the clinician [32, 33].

Each of these causes of action stems from the action or inaction of a human clinician. When a clinician uses a CDS system to make decisions about a patient's care, we expect that the clinician understands the use of this tool well enough to satisfy the duty of care in using it. If the clinician understands the tool, relies on it, and makes an error, there is a clear cause of the harm. However, this does not account for AI-powered devices that are intended to reach conclusions about patient care independently through complicated processes that are not revealed to the clinician. In these instances, a plaintiff may need to forgo redress for negligence and instead pursue a claim under the theory of products liability.

Plaintiffs in products liability cases are entitled to recovery when they are harmed by products which, due to defective design, manufacture, or warning are not reasonably safe. A product is defectively designed if the foreseeable risks of harm created by the product sufficiently outweigh the foreseeable benefits such that a reasonable clinician would not prescribe the product to any class of patients [34]. A manufacturing defect is a deviation from design specifications that occurs during production, resulting in a product's defect, frailty, or shortcoming [35]. Defective warning, also known as the failure to warn, occurs when warnings or instructions included with the product fail to reasonably disclose foreseeable risks to clinicians who are in a position to reduce the risks of harm to patients [36]. Manufacturers of medical devices are often insulated from liability under these theories because of clinician intervention.

Under the *learned intermediary doctrine*, patients are prevented from directly pursuing claims against medical device manufacturers if the device was suggested or prescribed by a clinician because the manufacturer has no duty to the patient. In this instance, the clinician is considered the consumer of the medical device. Clinicians are expected to weigh the risks and benefits of using a device and pass this information along to patients. If the risks of the medical device are not properly disclosed to the patient, it is the clinician who may face liability in a products liability action [32]. This theory becomes muddied, though, if the developer is unable to pass to the clinician and the clinician is unable to understand even a foundational exposition of the constantly changing AI processes that make up the device. Put simply, the incomprehensible and unpredictable nature of particular AI fundamentally changes the ability to satisfy the element of foreseeability in products liability cases.

It is unsurprising, given the methods of proving liability in tort law claims against humans, that plaintiffs would find proving the liability of AI quite onerous. After all, because the goal of development for AI products is often to go beyond the capabilities of humans to process large and complex sets of data, there is an inherent amount of unpredictability that prohibits total control by software developers, making it nearly impossible to describe why some AI-based products make certain decisions. To date, modern US legal theories do not account for this level of uncertainty. However, as the use of AI-based technologies becomes more prolific, there is potential for rapid development of a legal precedent surrounding negligence and products liability.

4   Artificial Intelligence and Intellectual Property

Introductory chapters on intellectual property (IP) law have been provided later in this book. For readers new to the field of IP law, it is recommended to read those chapters first as very little time is dedicated to discussion of basic IP law principles here. Rather, this chapter is focused on the applicability of IP law principles, in particular patent law and copyright law, to the field of AI and it is assumed that readers possess a basic understanding of IP law and terminology.

Patenting AI algorithms creates unique challenges. Generally, for an invention to be patentable, it must be useful and must fall into one of the following categories: machine, process, manufacture or composition matter. Of importance is what is not patentable, including abstract ideas (e.g., mathematical concepts), natural phenomena, and laws of nature. Because algorithms are at their core essentially mathematical concepts and fall under the abstract idea category, they are not patentable. However, algorithms definable as a series of steps to solve real-world problems are transformed from a non-patentable abstract idea to a patentable category of "process." The determination of whether an AI algorithm or for that matter any ineligible concept is patentable relies on the "Alice/Mayo test" named after two supreme court cases on the topic, Mayo Collaborative Servs. v. Prometheus Labs., Inc. [37] and Alice Corporation v. CLS Bank International [38]. The US Supreme Court ruled that if a patent claim is directed to an ineligible concept such as an abstract idea including software and AI algorithms, then additional inquiry must be made to determine subject matter eligibility before a patent can be issued. The additional inquiry focuses on whether the claim has additional element(s) where the ineligible concept (e.g., abstract idea) is integrated into a practical application whereby the patent claim in practice becomes more than the ineligible concept. If this integration is present, the invention is patentable. Stated differently: If there is more to the algorithm than the algorithm, it may be patentable [39].

An illustration may be useful to better understand this complex concept. Imagine a hypothetical system that can predict a patient's mortality risk based on their ECG tracing using an ML algorithm. This hypothetical system is patentable since the algorithm, a mathematical concept, provides a practical application. Although such a system is patentable, the source code for such a system is not. However, the source code may be eligible for copyright protection. In reality, the analysis and the Alice/Mayo test is more intricate than presented here, but the above discussion serves as an introduction to complexities of IP protection for AI systems.

While AI systems are potentially patentable, inventions produced by such systems are not. Under current US law, patents can only be awarded to the inventor(s). Although the inventor(s) can later assign their rights to a third party, the patent can only be awarded to the inventor(s). To qualify as an inventor, the individual must have contributed to the conception of the invention (i.e., the mental aspect of inventing). Given the immense potential of AI, this raises the question whether an "invention" conceived solely by an AI machine can be patented. This very question was subject of litigation in Thaler v. Hirshfeld [40]. The plaintiff in this case filed two US patent applications with the US Patent and Trademark Office (USPTO) naming his previously created AI system, DABUS, as the inventor of the two new "inventions." The plaintiff conceded that he did not meet the definition of an inventor since conception, the cornerstone of any invention, was done solely by DABUS and not by the plaintiff. In this case, the court held that an AI system cannot be an inventor and therefore, an "invention" solely conceived by an AI system is not patentable in the US. The court reasoned that that by statute, only one or more "individual(s)" are eligible to be named as inventor(s) and the term "individual" can only mean a natural person. Since a US patent can only be awarded to the inventor(s), at least one natural person must have been involved in the conception of each claim. This is not to assert that computers, including AI systems, cannot contribute to a natural person's conception of a patentable invention, but an "invention" without conception by a natural person is not patentable in the US. Of note, and as discussed in the Patent chapter later in this book, patent law is a creature of national law; consequently, different countries may approach patentability differently as illustrated by this case. Although the US, United Kingdom and Australia ultimately denied patent applications naming DABUS as the inventor, a patent listing DABUS as the inventor was issued in South Africa [40]. Finally, even though this case explains the current patentability of AI inventions in the US, the approach can be changed by Congress or case law as this, or other cases work their way through the judicial process in the coming years.

Similar to the above discussion regarding patentability of AI inventions, an analogous inquiry must be made in availability of copyright protection for AI-created content. Given the promises of AI, it is easy to envision an AI system creating "original work" that would ordinarily qualify for copyright protection had it been produced by a natural person. At the time of this writing, the law and guidance from the Copyright Office are silent on this exact question. However, based on related guidance from the Copyright Office, current statutory language and case law, it can be deduced that at least at this time in the US, authorship requires creation by a human being. Thereby, the AI machine would be ineligible to be named as the author of the "original work." However, potentially, the developer of the AI system may be eligible for authorship and copyright protection for the AI system's creation since what the AI "creates" can be viewed as the "fruit" of the initial work by the AI system's developer.

Another, more immediate inquiry with regards to AI and copyright is the potential for copyright infringement by the AI system development. For instance, there may be infringement when a ML algorithm "learns" from scanning copyrighted works (e.g., images). The law is somewhat clearer in instances such as this where traditional copyright principles apply. If the AI machine's work is sufficiently transformative, fair use doctrine can absolve the AI machine (or its developer) from infringement. Nonetheless, this does not completely remove the risk of potential copyright infringement claims depending on the extent of utilization of the copyrighted material and the final product. Consequently, it is recommended that developers consult with an experienced attorney in the field to avoid potential legal liability in the future [41].

5   Application (Examples)

5.1   Example 1: FDA Regulation

Happy Clinic, Inc. (Happy Clinic) is a data analytics company that provides AI-powered software to clinicians who want to better understand claims data, and increase practice and care management efficiencies. Because this software is not intended for a medical purpose, it is not regarded as a SaMD and therefore, not regulated by the FDA.

Healthy Cycle, Inc. (Health Cycle) has also created a mobile application that processes user inputs about menstruation, ovulation, and other personal details to provide predictive analytics regarding the user's fertility and reproductive health. The app uses algorithmic reasoning to provide suggested educational materials based on these analytical outputs. The intended use of the app is to inform and to provide resources to the person tracking their menstrual cycle. The app includes specific disclaimers about how the software is not intended to "diagnose, prevent, monitor, treat, or alleviate disease" and instructs the user to consult with a clinician for medical advice. Under the FDA's regulatory definition, the Healthy Cycle fertility and reproductive health app is not a SaMD and therefore, not regulated by the FDA.

Risk Disk is an AI-powered software that uses electronic health records coupled with patient-generated data to predict the risk of developing a stroke or heart disease. Because this software is designed to aid the clinician in developing prevention and interventional strategies, Risk Disk is a classic example of SaMD subject to FDA regulation.

FluIQ is an ML-based software that is intended to diagnose seasonal influenza based on location and symptoms using data from patients with suspected or confirmed cases. The purpose of this algorithm is to aid clinicians in the differentiation of seasonal flu cases and other viral illnesses. FluIQ does not provide the clinician with information or explanation about the logic used to reach a certain output. Therefore, FluIQ is classified as a device (specifically, SaMD) Clinical Decision Support system. FluIQ will be subject to FDA oversight.

5.2   Example 2: Dermatology—Convolutional Neural Network for Skin Lesion Image Classification

Healthy Skin, Inc. (Healthy Skin) has developed a mobile application that purports to be superior to dermatologists in the classification of melanoma. The deep learning algorithm calculates the fractal dimension of a lesion and builds a neural network to determine whether the lesion is malignant or benign and provides a suggested diagnosis. When presenting this result, the software displays and explains the logic and inputs for the ML algorithm to the clinician. Because the clinician is able to independently evaluate the basis for the software's recommendations, it is therefore exempt from FDA regulatory oversight because it is classified as a non-device Clinical Decision Support (CDS) system under the 21st Century Cures Act.

The product has experienced booming success since entering the market. Recently, however, the training images used for further development of the Healthy Skin algorithm have depicted predominantly white patients, therefore significantly decreasing the program's accuracy in detecting melanoma in patients of color. This is particularly troublesome because Black patients have the highest mortality rate for melanoma [42].

Healthy Skin recently published a study with a well-known dermatologist, Clinician A, about the promise of AI-based tools in the detection of melanoma. Upon reading this study, Clinician B, a dermatologist with experience in predictive algorithm development and significant involvement in the care of Black patients, notes that the data indicates that the algorithm may be trained with racially biased images that could create particular harm for Black consumers. Because the Healthy Skin application has received widespread praise, Clinician B feels that it is best to file a complaint with the FTC.

Following Clinician B's complaint, the FTC opens an investigation into the Healthy Skin, Inc. app. After subpoenaing the source code and training data used to develop the app's algorithm, the FTC discovers that less than 5% of the skin lesion sample images are of Black patients. In addition, the FTC discovers that the Healthy Skin, Inc. privacy notice fails to notify users that their data is being used to further train the algorithm and that many patient images were involved in a breach of unsecured information.

The FTC issues an order against Healthy Skin requiring them to (1) provide immediate notice to individuals whose data was breached by notifying the media and the patients individually, and (2) to delete the biased training data that causes more harm than benefit to consumers. Eventually, complying the with FTC order became too costly for Healthy Skin and the app was removed from the market.

Coinciding with the FTC's investigation, the family of a Black male patient, John Doe, brings a tort action against the developer of the Healthy Skin app and the patient's dermatologist, Clinician C. Clinician C owns and operates an independent dermatology clinic. In their complaint, John Doe's family alleges that Clinician C was negligent when he relied on the defective Healthy Skin app to provide an incorrect diagnosis. As a result, John Doe did not seek additional treatment for melanoma that quickly progressed. John Doe is now deceased.

As a defense, Healthy Skin plans to argue that it had no duty to John Doe because the app was intended to assume the role of a Clinical Decision Support system and because of Clinician C's involvement under the learned intermediary doctrine. Indeed, the Healthy Skin app provides a detailed description of the logical processes used to reach a recommendation. Just prior to the trial, the FTC issued the order detailing Healthy Skin's use of biased training data. Clinician C intends to rely on this information to make the argument that Healthy Skin did not provide adequate information to allow Clinician C to describe to John Doe the unreasonable danger of using the Healthy Skin app for melanoma detection, and that the product was therefore defective when it was put into the market by Healthy Skin.

6   Summary

In summary, AI used in the clinical setting presents special challenges for the current legal and regulatory regimes in the US. The classification by the FDA of many AI-based tools as SaMD evokes regulatory oversight that is important for the protection of patients, but there are notable shortcomings in the ability to regulate tools qualified by constant evolution. In recent years, other federal agencies, like the FTC, have begun to fill in some regulatory gaps by paying particular attention to fairness in the use and development of AI software in the healthcare setting. The continuous internal progression of AI also poses problems for establishing causation and foreseeability, key tenets of tort theory in the US, when AI-based tools are used by a clinician to aid in patient care. Finally, due to the nature of AI technology, the current considerations for obtaining intellectual property rights could stifle AI innovation as the ability for developers to obtain a patent or copyrights for their creations is unsettled. While not particularly robust at the moment, intense focus on the development of AI-based technology in the clinical setting may force significant legal and regulatory expansion.

References

1. Food and Drug Administration [Internet], Artificial Intelligence and Machine Learning in Software as a Medical Device. 2021 Jan [cited 2021 Oct 11]. Available from: https://www.fda.gov/medical-devices/software-medical-device-samd/artificial-intelligence-and-machine-learning-software-medical-device.
2. McCarthy J, What Is Artificial Intelligence?. Stanford University [Internet], 2007 Nov 12 [cited 2021 Oct 9]. Available from: http://jmc.stanford.edu/articles/whatisai/whatisai.pdf.
3. Benjamens S, Dhunnoo P, Mesko. The state of artificial intelligence-based FDA-approved medical devices and algorithms: an online database. npj Dig. Med. 2020; 3:18. doi: https://doi.org/10.1038/s41746-020-00324-0.
4. Food, Drug, and Cosmetic Act, 21 U.S.C. § 321(h) (2021).
5. Food and Drug Administration [Internet]. Deciding When to Submit a 510(k) for a Software Change to an Existing Device. 2017 Oct 25 [cited 2022 Jan 27]. Available from: https://www.fda.gov/media/99785/download.
6. Food and Drug Administration [Internet], Proposed Regulatory Framework for Modifications to Artificial Intelligence/Machine Learning (AI/ML)-Based Software as a Medical Device (SaMD). 2019 Apr [cited 2021 Oct 9]. Available from: https://www.fda.gov/media/122535/download.
7. Nicholson Price II W. Black-Box Medicine. Harv. J. L. Tech. 2015; 28:419-67.
8. Toews R. Here Is How The United States Should Regulate Artificial Intelligence. Forbes [Internet]. 2020 Jun [cited 2021 Oct 9]. Available from: https://www.forbes.com/sites/robtoews/2020/06/28/here-is-how-the-united-states-should-regulate-artificial-intelligence/?sh=2a5087ba7821.
9. Challen R, Denny J, Pitt M, Gompels L, Edwards T, Tsaneva-Atanasova K. Artificial intelligence, bias and clinical safety. BMJ Quality & Safety. 2019; 28:231-237.
10. International Medical Device Regulators Forum SaMD Working Group [Internet], Software as a Medical Device (SaMD): Key Definitions. 2013 Dec [cited 2021 Oct 9]. Available from: http://www.imdrf.org/docs/imdrf/final/technical/imdrf-tech-131209-samd-key-definitions-140901.pdf.
11. U.S. Food and Drug Administration [Internet]. Medical Device Data Systems. 2019 Sep 26 [cited 2021 Nov 28]. Available from: https://www.fda.gov/medical-devices/general-hospital-devices-and-supplies/medical-device-data-systems
12. Daniel G, Silcox C, Sharma I, Wright M.B. Current State and Near-Term Priorities for AI-Enabled Diagnostic Support Software in Health Care. 2019 Jun 6 [cited 2021 Nov 19]. Available from: https://healthpolicy.duke.edu/sites/default/files/2019-11/dukemargolisaienableddxss.pdf.
13. U.S. Food and Drug Administration [Internet]. Virtual Public Workshop – Transparency of Artificial Intelligence/Machine Learning-enabled Medical Devices. 2021 Oct 14 [cited 2021 Nov 28]. Available from: https://www.fda.gov/medical-devices/workshops-conferences-medical-devices/virtual-public-workshop-transparency-artificial-intelligencemachine-learning-enabled-medical-devices.
14. Pew Trusts [Internet]. How FDA Regulates Artificial Intelligence in Medical Products. 2021 Aug [cited 2021 Oct 15]. Available from: https://www.pewtrusts.org/en/research-and-analysis/issue-briefs/2021/08/how-fda-regulates-artificial-intelligence-in-medical-products.
15. 21st Century Cures Act, Pub. L. No. 114-225, 130 Stat. 1033 (2016).
16. U.S. Food and Drug Administration [Internet]. Artificial Intelligence/Machine Learning (AI/ML)-Based Software as a Medical Device Action (SaMD) Action Plan. 2021 Jan [cited 2021 Oct 15]. Available from: https://www.fda.gov/media/145022/download.

17. U.S. Food and Drug Administration [Internet]. Clinical Decision Support Software—Draft Guidance for Industry and Food and Drug Administration Staff. 2019 Sep [cited 2021 Oct 15]. Available from: https://www.fda.gov/media/109618/download.

18. Tutt A. An FDA for Algorithms. Admin. L. Rev. 2017; 69:83-122.

19. U.S. Food and Drug Administration [Internet], Mission Possible: How FDA Can Move at the Speed of Science. 2015 Sep [cited 2021 Oct 9]. Available from: https://www.fda.gov/files/about%20fda/published/Report—Mission-Possible—How-FDA-Can-Move-at-the-Speed-of-Science.pdf.

20. Guilhot M, Matthew A, Suzor N. Nudging Robots: Innovative Solutions to Regulate Artificial Intelligence. Vand. J. Ent. & Tech. L. 2017; 20:385-454.

21. Federal Trade Commission Act. 15 U.S.C. § 45.

22. U.S. Federal Trade Commission [Internet]. What the FTC Does. Cited 2021 Nov 1. Available from: https://www.ftc.gov/news-events/media-resources/what-ftc-does.

23. Gesser A, Gressel A, Eudy P. The Future of AI Regulation: The FTC's New Guidance on Using AI Truthfully, Fairly, and Equitably. Program on Corporate Compliance and Enforcement at the New York University School of Law [Internet]. 2021 May [cited 2021 Oct 15]. Available from: https://www.wp.nyu.edu/compliance_enforcement/2021/05/10/the-future-of-ai-regulation-the-ftcs-new-guidance-on-using-ai-truthfully-fairly-and-equitably/.

24. Jillson E. Aiming for truth, fairness, and equity in your company's use of AI. Federal Trade Commission [Internet]. 2021 Apr 19 [cited 2021 Nov 28]. Available from: https://www.ftc.gov/news-events/blogs/business-blog/2021/04/aiming-truth-fairness-equity-your-companys-use-ai.

25. Everalbum, Inc., Case No. 1923172 (Decision and Order May 7, 2021), https://www.ftc.gov/system/files/documents/cases/1923172_-_everalbum_decision_final.pdf.

26. Norori N, Hu Q, Aellen FM, Faraci FD, Tzovara. Addressing bias in big data and AI for health care: A call for open science. Cell Press Open Access [Internet]. Patterns. 2021; 2(10).

27. Turner Lee N, Resnick P, Barton G. Algorithmic bias detection and mitigation: Best practices and policies to reduce consumer harms. Brookings Institute [Internet]. 2019 May 22 [cited 2021 Dec 5]. Available from: https://www.brookings.edu/research/algorithmic-bias-detection-and-mitigation-best-practices-and-policies-to-reduce-consumer-harms/.

28. 46 C.F.R. § 318.3 (2009).

29. U.S. Federal Trade Commission [Internet]. FTC Warns Health Apps and Connected Device Companies to Comply with Health Breach Notification Rule. 2021 Sep 15 [cited 2021 Dec 4]. Available from: https://www.ftc.gov/news-events/press-releases/2021/09/ftc-warns-health-apps-connected-device-companies-comply-health.

30. Restatement (Second) of Torts: § 283 (1965).

31. Sullivan H, Schweikart S. Are Current Tort Liability Doctrines Adequate for Addressing Injury Caused by AI?. AMA J Ethics. 2019;21(2):E160-166. doi: https://doi.org/10.1001/amajethics.2019.160.

32. Allain J. From jeopardy to jaundice: the medical liability implications of Dr. Watson and other artificial intelligence systems. LA Law Rev. 2013;74(4):1049.

33. Levin H. Hospital Vicarious Liability for Negligence by Independent Contractor Physicians: A New Rule for New Times. U. Ill. L. Rev. 2005;2005:1291.

34. Restatement (Third) of Torts: § 6 (1998).

35. Scherer M. Regulating Artificial Intelligence Systems: Risks, Challenges, Competencies, and Strategies. Harv. J.L. & Tech. 2016; 29:353.

36. The Law Dictionary [Internet]. What is Manufacturing Defect?. [cited 2021 Nov 1]. Available from: https://www.thelawdictionary.org/manufacturing-defect/.

37. Mayo Collaborative Servs. V. Prometheus Labs, Inc., 566 U.S. 66, 71 (2012).

38. Alice Corporation v. CLS Bank International, 573 U.S. 208 (2014).

39. United States Patent and Trademark Office [Internet]. 2106 Patent Subject Matter Eligibility R-10.2019. [cited 2022 Jan 27]. Available from: Trahttps://www.uspto.gov/web/offices/pac/mpep/s2106.html#ch2100_d2e827_17196_6c.
40. Thaler v. Hirshfeld, 1:20-cv-903 (E.D. Va. Sep. 2, 2021).
41. Patel S, Rajeevan S. Copyright Considerations in Artificial Intelligence. Law Journal Newsletters [Internet]. 2020 Aug [cited 2022 Jan 27]. Available from: https://www.lawjournal-newsletters.com/2020/08/01/copyright-considerations-in-artificial-intelligence/.
42. American Cancer Society [Internet]. Cancer Facts & Figures 2020. [cited 2021 Dec 4]. Available from: https://www.cancer.org/content/dam/cancer-org/research/cancer-facts-and-statistics/annual-cancer-facts-and-figures/2020/cancer-facts-and-figures-2020.pdf.

Part X

Intellectual Property Law

Patent Law

Ana Santos Rutschman

Contents

Key Points
- In order to obtain a patent, an applicant must establish that the invention cumulatively meets the criteria of novelty, non-obviousness and utility.
- Certain subject matters are not eligible for patent protection (e.g. laws of nature).
- Patents are granted and enforced on a national basis.
- Applicants seeking patent protection outside the U.S. must file patent applications in the relevant foreign countries.
- U.S. law shields medical practitioners and related healthcare entities from liability for practicing a patented medical or surgical procedure on a human body without permission of the patent holder.

Legal Concepts
- Utility patent: a set of exclusive rights covering a technological "invention" meeting the cumulative criteria of novelty, non-obviousness (inventive step) and utility (practical application).
- Implicit knowledge: information about a product or process that is not captured by the patented disclosure and that is hard or impossible to acquire through use or reverse-engineering.

A. S. Rutschman (✉)
Charles Widger School of Law, Villanova University, Villanova, PA, USA
e-mail: ana.santosrutschman@law.villanova.edu

- Infringement: defined conduct in regard to the patented invention (or parts of it) that is not authorized by law or by the patent holder.
- Accused product or device: products and devices (or components thereof) alleged to infringe one or more claims of and asserted patent or patents.

1 Introduction

The focus of this chapter is on utility patents. In the United States, the constitutional intellectual property clause the Constitution establishes that patents and other types of intellectual property rights may be granted in order "[t]o promote the Progress of Science and useful Arts, by securing for limited Times to Authors and Inventors the exclusive Right to their respective Writings and Discoveries." [1] Patent protection is available to qualifying inventions that are novel, non-obvious and useful [2]. If granted, a patent is valid for 20 years counted from the filing date [3], but adjustments of the term are allowed in some cases to compensate for delays associated with the process of obtaining a patent [4]. In addition to utility patents, patent laws at the national level often recognize additional types of patents, such as design patents, which cover qualifying ornamental designs, and plant patents, which cover certain new varieties of plants.

Patents are negative rights, enabling the patent holder to prevent others from making, using, selling, offering for sale, or importing the protected components of the patented product or process (as well as any embodiments thereof). Patents do not give the patent holder any positive right or privilege, such as the right to use or commercialize the patented invention. As such, a separate analysis must be conducted to determine whether a patented medical or pharmaceutical product is subject to approval or authorization before entering the market. For example, most new drugs, whether covered by patents or not, must undergo review by the Food and Drug Administration (FDA) before entering the market [5, 6].

While U.S. patent law recognizes both direct and indirect infringement of patent rights, in 1996 an immunity from patent liability was created to cover "medical practitioners" and "related healthcare entities" who practice a patented medical or surgical procedure on a human body [7]. Non-covered infringers may nonetheless be held liable for patent infringement, even in cases of indirect infringement in which the direct infringer is covered by the immunity [8].

2 Discussion

Patent law is one of the main branches of intellectual property. Intellectual property consists of a set of laws protecting creations of the mind [9]. A utility patent (hereafter called a patent) refers to the bundle of rights covering an invention of scientific or technical nature [10]. The World Intellectual Property Organization defines a utility patent as "an exclusive right granted for an invention, which is a product or a process that provides, in general, a new way of doing something, or offers a new technical solution to a problem" [10]. Utility patents can thus be divided into product and process patents.

Patents are regulated by both international and domestic laws. The main legal framework at the international level is the Agreement on Trade-Related Aspects of Intellectual Property Rights [11], commonly known as the TRIPS Agreement, which came into force on January 1, 1995. TRIPS directs member-states to grant patents to eligible inventions across all fields of technology, without defining either key term [2]. However, TRIPS allows countries to elect not to award patent protection to a limited categories of inventions [12]. The most relevant of these categories in the context of healthcare is the one encompassing inventions that may be necessary to protect human health as well as diagnostic, therapeutic and surgical methods [13]. TRIPS further establishes three criteria that any patent-eligible invention must meet in order to qualify for TRIPS-required patent protection: novelty, inventive step (also known as non-obviousness) and capability of industrial application (also known as utility) [2], again without defining these key terms and thus leaving substantial flexibility to national laws to define the basic requirements for patent eligibility and patentability. TRIPS also establishes that patent protection must last for a minimum of 20 years generally counted from the filing date [3]. The minimum standards of protection set forth in the TRIPS Agreement must be adopted by member-countries in their national laws (although many least-developed countries have additional time to adopt such requirements), but the only penalties for failure to implement treaty requirements are to come into compliance or, in the event of recalcitrance, authorized retaliatory trade sanctions or compensation obligations. All members of the World Trade Organization—164 countries as of late 2021—are parties to TRIPS [14]. The quasi-global reach of TRIPS has meant that most national patent laws are relatively similar in their basic contours (while differing significantly in specifics), a phenomenon often alluded to as harmonization of patent law [15].

Even though there is strong harmonization of patent laws at the international level, patents are governed entirely by laws at the national level. For example, if an inventor applies for a patent in the United States, the grant and enforcement of that patent will be assessed according to United States law. If the same inventor wishes to apply for patent protection in a second country, e.g. Canada, the inventor will need to apply for a patent covering the same invention in Canada, where Canadian domestic patent rules will apply. There are no "international patents."

For an invention to qualify for patent protection, it must fall into one or more categories of eligible subject matter as statutorily defined [16]. Under U.S. law, a "process, machine, manufacture, or composition of matter, or any new and useful improvement thereof" may qualify as patentable subject matter [16]. In spite of the broad statutory language, caselaw has progressively limited the realm of patent-eligible subject matter. "Natural phenomena," "laws of nature" and "abstract ideas" are not deemed patent-eligible [17–19]. If a product or process falls to meet eligibility criteria, a patent cannot be granted. As recently articulated by the Supreme Court, limitations on subject matter eligibility have important consequences in the field of medicine, and particularly with regard to medical technologies. For example, the Supreme Court ruled in 2013 that isolated genes were not patentable, as further discussed in the Application section of this chapter [18]. It is also worth noting that current limitations on subject matter eligibility are also relevant in the development of medical technologies relying on computational methods. Many of these methods are not patentable under the 2014 Supreme Court's decision in *Alice*, which held that "merely requiring generic computer implementation" of an abstract idea does not render the process patent-eligible [19].

Pursuant to TRIPS, U.S. law requires that, in addition to being patent-eligible, an invention must meet three cumulative criteria: novelty [20], non-obviousness [21] and utility [16].

Under U.S. law, an eligible invention is new (not "anticipated") if there is no *single* source ("prior art") that already discloses *all* the components covered by the patent [22]. Current U.S. patent law establishes that a patent should not be granted if the claimed invention was described in "a printed publication, or in public use, on sale, or otherwise available to the public." [20]. Examples of a source that can be used to defeat novelty include but are not limited to a product that is publicly used, an advertisement for sale, a published article, or another patent. For instance, if a patent applicant seeks protection for a respirator and there is a pre-existing published article describing a respirator that possesses all the elements that the patent applicant has listed in the patent claim, that claim will fail for lack of novelty. While patent applications are evaluated according to domestic law, relevant prior art is assessed globally. For instance, in the respirator example provided above, even if the patent application is made in the United States, a pre-existing published article in another country could be used to defeat novelty [20]. Unlike the laws of several other countries, U.S. patent law provides a 1-year grace period during which an inventor may disclose an invention under certain conditions without barring a patent from issuing [20]. Under this rule, if the pre-existing publication describing a respirator was published by the inventor herself 1 year (or less than 1 year) before the effective filing date of the claimed invention, the publication would not be considered as integrating prior art [20].

The doctrine of non-obviousness prevents governments from issuing patents on products or processes that constitute merely trivial variations on existing inventions [21]. It also precludes the patenting of inventions that would otherwise become public without the inducement of the patent right [21]. Under U.S. law, it applies to situations in which, although an existing single source does not disclose all the components of the patent claim, the product or process in question results from a predictable combination of components disclosed in *more than one* pre-existing source [22]. The determination of whether this combination is obvious is highly technical [22] and is performed with reference to the knowledgebase and creative efforts expected of a person having ordinary skill and creativity in the technical and scientific fields to which the invention pertains [21]. For instance, in the case of the previously referenced respirator, the relevant fields may include chemistry (if the analysis involves components related to the filtration system) or materials engineering (if the analysis involves components related to the material from which parts of the respirator are made).

The doctrine of utility requires that an invention possess a "demonstrated real-world use" [22]. Under U.S. law, an invention must be useful at the time of the patent application [16, 23]. For instance, if a process patent application claims the use of a large molecule to treat a specific disease, the invention likely meets the utility requirement, provided that the claim is otherwise credible. If, however, a process patent application claims the use of the same molecule to treat an unspecified disease that may be caused by the emergence of a future pathogen in the coronavirus family, the invention will fail to meet the utility requirement. The utility requirement is commonly understood as a low-bar, easily satisfiable requirement [22]. The level of utility required under current law is much lower than what would be required for a showing of efficacy or safety for drugs and medical devices subject to review by medical regulatory authorities (e.g., the FDA). Patent applications for medical products and processes thus are usually filed much earlier than efficacy and safety testing, and often are based on *in vitro* rather than *in vivo* studies [24].

In addition to novelty, non-obviousness and utility, domestic patent laws typically require patent applicants to comply with additional requirement designed to promote the disclosure of information about the invention. Under U.S. law, a patent application must contain a written description of the invention [23], enable an average expert in the relevant scientific and technical field(s) to understand how to make and use the claimed invention without undue experimentation [23] and comply with the "best mode" requirement, which prompts inventors to disclose their preferred way of practicing the invention, if any [23]. These disclosure obligations prevent the applicant from overclaiming relative to the technological invention that the applicant discloses, but failure to disclose a subjectively understood best mode is no longer a basis for invalidating a patent after grant, and the requirement applies only to information known at the time of filing. Therefore, any trade secrets or best modes developed during subsequent testing or manufacturing need not be disclosed by a patent applicant.

If a patent application meets the necessary patentability requirements, the application matures into a patent. In the U.S., the patent term is currently set at 20 years from filing, in accordance with obligations imposed by the TRIPS Agreement [3], but a complex set of rules allows for adjustments of the term to compensate for delays associated with the process of obtaining a patent [4].

In the case of product patents, a valid patent gives the patent holder a set of rights that translate into the ability to prevent others from making, using, selling, offering for sale, or importing the protected components of the patented invention, as well as any embodiments thereof, without the authorization of the patent holder [25]. Additional provisions bar these acts in regard to substantial components of the invention that have no non-infringing uses or inducing others to make, use, sell or import the invention, or similarly export such components or export components to induce infringement overseas [26–28].

In the case of process patents, a valid patent gives the patent holder the ability to prevent others from performing the patented process or processes, without authorization from the patent holder [25]. Additional provisions bar importing of products directly made from using such processes abroad, even though performing such processes abroad does not infringe a U.S. patent [29].

Patents rights are thus negative rights, giving the patent holder exclusionary powers throughout the duration of the patent. Importantly, patents do not give the patent holder any right to use the patented invention. This is especially relevant in the context of healthcare and other heavily regulated industries that require prior approval to market products. The need for permission to bring a patented invention to market must be addressed separately from any patent-related issues. For instance, patent law establishes that an inventor that has obtained a utility patent on a medical device has a right to prevent others from making or using that device (e.g., the patent holder may elect not to license the patented invention to would-be competitors). But, if the patent holder wishes to commercialize that device, a separate inquiry must be conducted to determine the applicability of laws and regulations outside the realm of intellectual property that relate to such marketing approval. These requirements in the U.S. often entail some degree of pre-market review by the U.S. Food and Drug Administration and typically involve the submission of significant amounts of data or other information about the product, which in some cases has to be obtained through clinical trials.

Even though the patent application process is designed to foster the disclosure of substantial levels of information about the invention for which the applicant seeks patent protection, it is still possible that some information about the patented product or the process may remain secret. This is especially salient in the case of non-codified or implicit knowledge that cannot be easily apprehended or reverse-engineered through analysis of the patented product or process. Structurally complex pharmaceutical products and processes, and especially those in the area of biotechnology (such as monoclonal antibodies or vaccines), are routinely covered by both patents and non-patented knowledge that is kept secret by the patent owner [30]. An agreement to transfer patent rights, or to allow the use of a patented product or process, does not create any obligation to transfer secret knowledge, unless that transfer is expressly negotiated between the parties and incorporated in the contract.

The number of both patent applications filed, and patents granted has increased significantly over time across most fields of technology [31]. This trend is especially pronounced in the context of pharmaceutical products. Most patented conventional or small-molecule drugs are or have been covered by fewer than five patents [32, 33]. Most patented biologics are covered by dozens of patents, with that number sometimes exceeding 100 [34, 35]. For instance, a study on the HPV vaccines sold under the trade names *Gardasil* and *Cervarix* in the U.S. found that there were 81 patents issued in the United States covering these two vaccines [36]. And the biologic Humira, the world's best-selling drug since 2012, is covered by over 100 U.S. patents on its own [35]. One of the major implications of the large number of patents covering many existing and routinely prescribed drugs is that it is difficult, time-consuming and costly to challenge the validity of large drug patent estates [35]. The fact that most new drugs are covered by many patents is one of the contributing causes of high prescription drug prices in countries like the U.S., as patent rights allow pharmaceutical companies to preserve a monopoly-like position in the market. It should nonetheless be noted that patents are not the sole cause of high prescription drug prices [37]. For instance, other contributing factors in the U.S. include the lack of drug price regulation [38], the inability for centralized purchasers (such as the Medicare program) to negotiate drug prices [39], as well as the existence of regulatory exclusivities administered by non-patent actors such as the U.S. Food & Drug Administration [40] awarded separately from, and irrespective of, patent status of a drug.

Infringement of patent rights may be direct or indirect. U.S. law recognizes two different types of direct infringement: literal infringement and infringement under the doctrine of equivalents. Direct infringement occurs when a party makes, uses or sells a patented invention without consent of the patent owner [25]. Technically, literal infringement occurs when the accused product or method embodies or otherwise contains every element of a patent claim [22]. Infringement by equivalents occurs when the accused product or method does not literally infringe the patented invention, but is so similar the invention is treated as covering things that perform similarly [41]. The theory of protecting such equivalents is that a finding of non-infringement would frustrate the purpose of the patent system as applicants cannot envision how to draft claims for all future embodiments that rely on the same basic invention. Current caselaw directs courts to assess whether the two products or processes are substantially the same, or whether they employ substantially the same functions in substantially the same way to achieve substantially the same result [42, 43].

Indirect infringement occurs in cases in which a party that is not directly liable under the law has nonetheless facilitated infringement by the party that is directly liable [22]. U.S. law recognizes two different types of indirect infringement: inducement and contributory infringement [26, 27]. Inducement occurs when a party provides sufficient information or components with the intention to cause others to infringe, knowing of the patent (or at least not being willfully ignorant of it) [44]. Contributory infringement occurs when a party provides a component of the patented invention knowing that it has no substantial non-infringing use other than in an infringement (and thus similarly is seeking to induce an infringement) [45, 46].

In the U.S., there is an immunity from patent liability that is relevant for healthcare professionals [47]. Congress amended the Patent Act in 1996 to prevent patent holders from holding liable clinicians, hospitals and other entities in the healthcare space for practicing a patented medical or surgical procedure on a human body [7]. The creation of immunity from patent liability of healthcare providers was a direct response to ongoing litigation between two surgeons over a patent covering a new surgical method for use during cataract surgery [48]. More broadly, it responded to growing concerns among the medical community about the chilling effects of patents on medical procedures [8, 49]. The immunity from liability applies to "medical practitioners" and "related healthcare entities" [7].[1] The Patent Act defines a medical practitioner as "any natural person who is licensed by a State to provide" medical activities [50]. The Act defines related healthcare entities as those "with which a medical practitioner has a professional affiliation under which the medical practitioner performs the medical activity, including but not limited to a nursing home,

---

[1] same as 35 U.S.C. § 287(c).

hospital, university, medical school, health maintenance organization, group medical practice, or a medical clinic." [50] While patent law still allows for the patenting of medical procedures, in practice these patents are not enforceable against the actors covered by the immunity established in 35 U.S.C. § 287(c). However, immunity is only available to cases of direct infringement or inducement by such providers and does not limit claims of indirect infringement of patent rights by others not provided with such immunity from liability [8].

3   Application

Case 1: Mayo v. Prometheus, 566 U.S. 66 (2012) Prometheus Laboratories was the sole licensee of patents related to the use of thiopurine drugs in the treatment of autoimmune diseases. As these drugs metabolize, the body releases metabolites into the bloodstream at varying rates among patients. This variability had long caused difficulties in determining the precise dose of the drug that should be administered to individual patients, with lower doses being ineffective and high doses being harmful to recipients. Researchers found a way to establish correlations between metabolite levels and the likelihood of ineffectiveness or harm. Prometheus had licensed the patents covering the processes for establishing these correlations. These processes included three steps: (1) an "administering step" directing a doctor to administer the drug to a patient; (2) a "determining step" directing the doctor to measure the resulting metabolite levels; (3) a "wherein step" describing the metabolite concentrations below which the dosage is likely ineffective and above which the dosage will likely result in harmful side effects, and directing the doctor to adjust drug levels accordingly.

Mayo developed and intended to commercialize a diagnostic test similar but not identical to the one patented by Prometheus. Prometheus sued Mayo for patent infringement. One of the most salient legal issues became the issue of whether the claimed processes constituted patentable subject matter under 35 U.S.C. § 101. The District Court granted summary judgment for Mayo, holding that the steps claimed in the patents amounted to "natural laws." A longstanding tradition in patent law theory and caselaw holds that laws of nature are ineligible subject matter. The Court of Appeals for the Federal Circuit (CAFC) reversed after applying the "machine or transformation test," holding that the steps for administering the drug and determining metabolite levels elevated these steps beyond the realm of laws of nature, thus rendering them eligible subject matter. The Supreme Court reversed the CAFC, ruling in favor of Mayo, holding that these steps constituted insufficient applications of laws of nature and did not amount to "additional steps."

Case 2: Association for Molecular Pathology v. Myriad Genetics, Inc., 569 U.S. 576 (2013) Myriad Genetics identified the exact location and sequence of two human genes (BRCA1 and BRCA2). Mutations associated with these genes can substantially increase the risk of breast and ovarian cancer. Following this scientific breakthrough, Myriad obtained several patents, including the ones litigated here, which covered the isolated BRCA1 and BRCA2 genes, as well as complementary DNA (cDNA). The Association for Molecular Pathology (AMP) sued Myriad for a declaratory judgment that Myriad's patents claiming complementary cDNA of the isolated genes were invalid. The District Court granted summary judgement in favor of AMP, holding that Myriad's patents were invalid because the claims were directed at products of nature, which do not constitute patentable subject matter under 35 U.S. Code § 101. The Court of Appeals for the Federal Circuit (CAFC) affirmed the District Court in part and reversed in part, with each member of the panel writing a separate opinion. Collectively, the Court held that both cDNA and the isolated genes were patent-eligible.

The Supreme Court agreed that the isolated BRCA1 and BRCA2 genes were not patentable. The Court reasoned that the claimed products—the isolated genes— were not sufficiently distinct from the patent-ineligible naturally occurring products from which they were derived. By contrast, the Court held that cDNA constituted patentable subject matter. Noting that "cDNA retains the naturally occurring exons of DNA, but it is distinct from the DNA from which it was derived," the Court reasoned that cDNA was not a "product of nature" and thus was patent-eligible.

Case 3: Johns Hopkins Univ. v. Alcon Labs., Inc., No.15-525-SLR-SRF, 2018 U.S. Dist. LEXIS 70403, at *3 (D. Del. Apr. 5, 2018) Johns Hopkins University obtained a patent covering "sutureless ocular surgical methods" and the instruments used to practice those methods. Johns Hopkins sued Alcon, a manufacturer of eyecare products, for contributory patent infringement. Having recognized that 35 U.S.C. § 287(c) shields "medical practitioners" and "related healthcare entities" from patent liability, the court found nonetheless that Alcon could be held liable for contributory infringement. Even though the direct infringers in this case (the physicians) were shielded from liability, § 287(c) immunity does not extend to manufacturers.

4   Summary

In summary, patents play a significant role in the context of healthcare, particularly with regard to pharmaceutical products and medical procedures. The making, using, selling, offering for sale, or importing the protected components of the patented product or process (or any embodiments thereof) is subject to authorization from the patent holder. Unauthorized acts constitute infringement of patent rights, except in the limited context of "medical practitioners" and "related healthcare entities" who practice a patented medical or surgical procedure on a human body.

Acknowledgements The author thanks Joshua Sarnoff for helpful feedback on the chapter.

References

1. U.S. Constitution, article I, section 8, clause 8.
2. TRIPS Agreement, article 27.1.
3. TRIPS Agreement, article 33.
4. 35 U.S.C. § 154.
5. 21 U.S.C. § 355(a)]
6. 42 U.S.C. § 262(a)(1)(A)
7. 35 U.S.C. § 287(c).
8. Anderson J, Nonexcludable Surgical Method Patents, 61 Wm. & Mary L. Rev. 637 (2020).
9. World Intell. Prop. Org., What is Intellectual Property?, https://www.wipo.int/about-ip/en/
10. World Intell. Prop. Org., Patents: What is a Patent?, https://www.wipo.int/patents/en/
11. TRIPS: Agreement on Trade-Related Aspects of Intellectual Property Rights, Apr. 15, 1994, Marrakesh Agreement Establishing the World Trade Organization, Annex 1C, 1869 U.N.T.S. 299, 33 I.L.M. 1197 (1994)
12. TRIPS Agreement, article 27.2.
13. TRIPS Agreement, article 27.3.a.
14. World Trade Org., Members and Observers, https://www.wto.org/english/thewto_e/whatis_e/tif_e/org6_e.htm
15. World Intell. Prop. Org., Patent Law Harmonization, https://www.wipo.int/patent-law/en/patent_law_harmonization.htm
16. 35 U.S.C. § 101.
17. Mayo v. Prometheus, 566 U.S. 66 (2012).
18. Association for Molecular Pathology v. Myriad Genetics, Inc., 569 U.S. 576 (2013).
19. Alice Corp. v. CLS Bank, 573 U.S. 208 (2014).
20. 35 U.S.C. § 102.
21. 35 U.S.C. § 103.
22. Masur JS & Ouellette LL, Patent Law: Cases, Problems, and Materials (2021).
23. 35 U.S.C. § 112(a).
24. Manual Pat. Exam. Proc. (10.2019), 2164.02 Working Example [R-11.2013].
25. 35 USC 271(a).
26. 35 USC 271(b).
27. 35 USC 271(c).
28. 35 USC 271(f).
29. 35 USC 271(g).
30. Price II WN & Rai AK, Manufacturing Barriers to Biologics Competition and Innovation, 101 Iowa L. Rev. 1023–1063 (2016).
31. U.S. Pat. Office, U.S. Patent Activity Calendar Years 1790 to the Present (2021), https://www.uspto.gov/web/offices/ac/ido/oeip/taf/h_counts.htm
32. Hemphill CS, Sampat BN., Evergreening, Patent Challenges, and Effective Market Life in Pharmaceuticals, 31. J. Health Econ. 327 (2012).
33. Ouellette LL, How Many Patents Does It Take to Make a Drug - Follow-On Pharmaceutical Patents and University Licensing, 17 Mich. Telecomm. & Tech. L. Rev. 299 (2010).
34. Storz U,Of patents and patent disputes: The TNFα patent files. Part 1: Humira, Hum. Antibodies 2017;25(1–2):1-16. doi: https://doi.org/10.3233/HAB-160300.
35. Rutschman AS, Regulatory Malfunctions in the Drug Patent Ecosystem, 70 Emory L. J. 347 (2020).
36. Padmanabhan S, et al., Intellectual Property, Technology Transfer and Developing Country Manufacture of Low-cost HPV vaccines - A Case Study of India, Nat. Biotechnol. 28(7): 671–678 (2010).
37. Rosenthal E, An American Sickness: How Healthcare Became Big Business and How You Can Take It Back, Penguin (2018).
38. Hirschler B, How the U.S. Pays 3 Times More for Drugs, Scientific American (Oct. 13, 2015).

39. Feldman R, Perverse Incentives: Why Everyone Prefers High Drug Prices -- Except for Those Who Pay the Bills, 57 Harv. J. on Legis. 303 (2020).
40. Heled Y, Patents v. Statutory Exclusivities in Biological Pharmaceuticals - Do We Really Need Both, 18 Mich. Telecomm. & Tech. L. Rev. 419 (2012).
41. Royal Typewriter Co. v. Remington Rand, Inc., 168 F.2d 691, 692 (2d. Cir. 1948).
42. Warner-Jenkinson Co. v. Hilton Davis Chem. Co., 520 U.S. 17 (1997).
43. Abbott Laboratories v. Sandoz, Inc., 566 F.3d 1282, 1296 (Fed. Cir. 2009).
44. Global-Tech Appliances, Inc. v. SEB S.A., 563 U.S. 754 (2011).
45. Aro Manufacturing Co. v. Convertible Top Replacement Co., 365 U.S. 336 (1961).
46. Aro Manufacturing v. Convertible Top Replacement ("Aro II"), 377 U.S. 476 (1964).
47. Ho C, Patents, Patients, and Public Policy: An Incomplete Intersection at 35 U.S.C. 287(c), 33 U.C. Davis L. Rev. 601 (2000).
48. Pallin v. Singer, No. 5:93-22, 1995 WL 608365, at *1 (D. Vt. May 1, 1995).
49. Glasson J, Reports of Council on Ethical and Judicial Affairs: Patenting of Medical Procedures, 144 AMA PROC. H. DEL. 200 (1995).
50. 35 U.S.C. § 287(c)(2).

Copyright and Trademark Law

Stephen McJohn

Contents

Key Points
- Copyright applies broadly to works, without the need for registration
- Registration is inexpensive and makes the copyright much stronger to enforce, especially making statutory damages and attorney's fees available for infringement
- Fair use authorizes many uses of copyrighted works, especially if there is no loss of licensing revenue to the copyright owner
- Trademark applies only to distinctive symbols used to indicate the source of goods or services
- Not every business name or term qualifies as a mark
- Generic terms and merely descriptive terms are not protected by trademark

S. McJohn (✉)
Suffolk University Law School, Boston, MA, USA
e-mail: smcjohn@suffolk.edu

© The Author(s), under exclusive license to Springer Nature Switzerland AG 2022   419
A. S. Pasha (ed.), *Laws of Medicine*,
https://doi.org/10.1007/978-3-031-08162-0_27

Legal Concepts

- Copyright—the exclusive right of a copyright owner to make copies, adapt, publicly distribute, publicly display, or publicly perform a copyrighted work.
- Copyright infringement—unauthorized copying, adaptation, public distribution, public display, or public performance of a copyrighted work.
- Fair use—uses of copyrighted material that does not infringe copyright.
- Actual damages—damages caused by infringement, such as lost sales or licensing revenue.
- Statutory damages—damages that are awarded without the need to show a loss.
- Trademark—a symbol used by a person in commerce to indicate the source of goods or services and to distinguish them from the goods sold or made by others.
- Generic term—the name for a product, not protectable by trademark.
- Merely descriptive symbol—terms that merely describe products or services, not protected by trademark.
- Trademark infringement—unauthorized commercial use of a mark or similar symbol that is likely to confuse or deceive consumers.

1 Introduction

Copyright applies to a very broad range of works. Healthcare professionals should be aware that they may hold copyright in things they create and be aware of the risks of infringement of others' copyrights. Fair use excuses many uses of works in healthcare that do not harm the copyright holder.

Trademarks are used widely in healthcare. Healthcare professionals should be aware that brand names and other business symbols may be protected marks. Healthcare professionals may wish to develop their own marks and should choose protectable symbols, not generic or merely descriptive terms.

2 Discussion

2.1 Copyright

Many things that a healthcare professional deals with are copyrighted works. For example, the following are all likely to be protected by copyright: books, articles, software, artwork, labels, photographs, music, floorplans, long email messages, etc. The copyright holder has a set of exclusive rights: [1]

1. Reproduce the copyrighted work in copies or phonorecords (often called the reproduction right or the right to make copies);
2. Prepare derivative works based on the copyrighted work (adaptation right);
3. Distribute copies or phonorecords of the copyrighted work to the public by sale or other transfer of ownership or by rental, lease, or lending (public distribution right);

4. Perform the copyrighted work publicly (public performance right); and
5. Display the copyrighted work publicly (public display right).

Assume a clinician writes a book about neurosurgery. The clinician would hold the book copyright (unless they wrote it as an employee or transferred the copyright to someone else). The copyright would be infringed if someone else, without permission,

- Made a copy or copied material into another work
- Made an adaptation, such as a translation, abridgment, or musical (unlikely!)
- Sold unauthorized copies
- Performed the work publicly, such as reading it to an audience, including playing a recording (less likely with nonfiction books)
- Displayed unauthorized copies (such as displaying unauthorized copies online)

What works are protected by copyright? Copyright applies to any original work of authorship, fixed in a tangible medium of expression [2]. A Broadway play, for example, would easily qualify as a copyrighted work. The work is original (as long as even a smidgen of creativity was used in writing it), qualifies as a work of authorship (as a literary work), and is fixed in a tangible medium of expression (as long as the work has been put in tangible form, such as the script being written down or the work being videotaped).

What works are not protected by copyright? The "public domain" includes material that are not under copyright: government works, works published before 1925 (as of 2021—increasing over time); works that were published without copyright notice before 1989; works published before 1964, if copyright was not renewed after 28 years; and, most importantly, works, or aspects of copyrighted works, that are not protected by copyright, such as non-original material, facts, ideas, and functional elements.

2.2   Original

To qualify for copyright, a work must be original, meaning it was created by the author (not just copied from someone else) and reflects at least a minimal level of creativity. That is an easy standard to meet. A New York white pages phone book is not original because there is no creativity in the selection or arrangement of material, and the author does not make up the names and phone numbers. But as long as there is even a little bit of creativity, copyright applies. A New York yellow pages phone book would qualify if some creativity was used in selecting the categories of business. Likewise, a photo of workers in a healthcare office or the notes taken by a clinician could have the minimal creativity standard required for copyright. Some works might lack originality: a healthcare form calling only for standard information or an X-ray image taken with a standard procedure might not be original. But if some creativity was used in drafting the form or in making the image, even those works could qualify for copyright. The bar is quite low. As the Supreme Court put it, "The vast majority of works make the grade quite easily, as they possess some creative spark, 'no matter how crude, humble or obvious' it might be." The vast majority of works make the grade quite easily, as they possess some creative spark, "no matter how crude, humble or obvious" it might be [3].

3   Works of Authorship

The categories of works of authorship are likewise broadly inclusive, comprising the following (with examples relevant to healthcare):

1. literary works (e.g., medical books, case histories, text for a medical office's web page);
2. musical works, including any accompanying words (e.g., music for advertising, music played in a medical office or during surgery);
3. dramatic works, including any accompanying music (e.g., a script for a pharmaceutical commercial);
4. pantomimes and choreographic works (less often relevant to healthcare but still might be in the case of, for example, a modern dance about surgery, and could include therapeutic dance);
5. pictorial, graphic, and sculptural works (e.g., pictures in a medical textbook, diagnostic images if made with creativity, blueprints for a medical device, artwork in a medical office);
6. motion pictures and other audiovisual works (e.g., videos, PowerPoint presentations);
7. sound recordings (e.g., a sound recording of a lecture, recordings of music played at a medical office); and
8. architectural works (e.g., architectural plans for a hospital building).

Two of those categories are especially broad. "Literary works" include anything done with letters, numbers, or other symbols. So literary works include not only novels, poetry, and short stories but medical journal articles, clinical notes, instruction manuals, computer programs, diary entries, and much more. "Pictorial, graphic, and sculptural works" include any applied or fine art in two or three dimensions and so go beyond paintings and sculpture to include such things as photos, doodles, and textbook illustrations.

Not everything qualifies as a work of authorship. Short phrases, everyday conversation and behavior, or a person's persona does not qualify as a work of authorship. Such material should not be protected by copyright, for it could bar others from discussing these topics. In addition, copyright does not apply to U.S. government works, works created by federal employees within the scope of their employment. Likewise, copyright does not apply to "government edicts," works with legal authority such as court opinions and state laws. Otherwise, it could be copyright infringement to quote a source of the law, which would raise real questions about whether copyright was consistent with the First Amendment.

3.1   Fixed in a Tangible Medium of Expression (Fixation)

Copyright protection is simple to obtain. A work is protected by copyright when it is put in tangible form. As soon as a picture is snapped and saved in the memory of a phone, or a painting is painted (it need not even dry), or a song is written down, copyright applies. Copyright applies automatically, without any further action by the author and without the author even needing to be aware of copyright.

Some works may be created without being put in tangible form. A choreographer might make up a dance on their feet, or a clinician might give a lecture at a conference, or a jazz musician might make up a solo. If the work is not put in tangible form, copyright does not apply. The rule makes sense because it would be hard to show someone else copied without a copy to compare the accused work to. But each author could easily get copyright: the choreographer could use notation or simply a video, as could the lecturer or jazz musician. Copyright law is flexible in permitting works to be protected.

An author may register their copyright with the United States Copyright Office. Registration is not required to have copyright, but registration does make the copyright stronger to enforce. In particular, if a work has been registered, then anyone who infringes is liable for statutory damages, amounting to a minimum of $750 to a maximum of $150,000 per work infringed, even if there were no actual losses to the copyright holder. Registration is not onerous: it can be done online at the Copyright Office website, the form is not too complicated, and the fee is around $45.

3.2   Case Example

A grad student in Boston uploaded some thirty copyrighted songs without permission. Actual damages would have been about $30. But the copyrights were registered, allowing statutory damages of $750 to a maximum of $150,000 per work infringed. The jury awarded $22,500 per song because the infringement was willful, for a judgment of $675,000 [4].

3.3   Copyright Does Not Protect Nonoriginal Material, Ideas, or Functional Aspects

We have seen it is easy for a work to qualify for copyright. Many things in the healthcare industry are automatically copyrighted: books, articles, websites, photographs, forms (if made with a little creativity), music, notes, drawings, X-ray or MRI images (as always, only if made with a little creativity). But copyright law does provide some balance. Even though all those things are under copyright, many elements in them may be copied without infringing copyright. Copyright protects only original creative expression; it does not apply to nonoriginal material, ideas, or functional aspects.

Nonoriginal material is not copyrighted. The biggest example is factual material. Facts are not created by the author but rather are discovered or copied from other material. Healthcare often involves lots of data: patient information, test results, clinical trial data, and so on. That data, being factual, is not protected by copyright, so it may be copied without infringing on copyright. Other laws may apply (from HIPPA to trade secret law), but copying data is not copyright infringement. If an entire database were arranged in a creative fashion, then copying the entire database could infringe, but copying data from the database would not. Other, nonoriginal material is also not protected by copyright. Suppose an author compiled a set of photographs in a creative way by selecting and arranging the photographs. We'll assume the photographs themselves are not copyrighted. They could be government works or images produced automatically by a security camera. The compilation would be protected, but only its original selection and arrangement. Others could still copy the individual photographs without infringing on the copyright in the compilation.

Ideas are not protected by copyright—another large limit on copyright protection. Copyright exists to encourage creativity, and rule against copyright protection for ideas serves the same goal. An author has copyright protection of their creative expression but not of any ideas expressed in the work. So, the ideas in a medical journal article, an inspired email, or a clinician's autobiography may all be copied without infringing copyright (although plagiarism can be an ethical violation if the source of the idea is not attributed).

Copyright also protects only creative expression, not functional aspects. So, for example, medical devices fall under the category of "pictorial, graphical or sculptural works" but generally have no copyright protection. The functional aspects of a medical device would not be separable from its creative aspects, so copyright would not apply. Anyone could copy the device without infringing *copyright*. But, of course, the device might be patented. Patents exist to protect functional inventions, and the rule against the copyright protection of functional aspects serves to draw the line between the domains of copyright and patent.

3.4 Case Example

A great illustration of the nonprotection of ideas and facts is *Hoehling v. Universal City Studios*. [5] Hoehling's book *Who Destroyed the Hindenburg?* laid out Hoehling's theory that a crew member had sabotaged the airship. A disaster film was made (without securing rights from Hoehling) that copied many elements, such as historical facts (the age and birthplace of the alleged saboteur, information about the airship and crew, a letter from Germany's ambassador downplaying risks). There was no copyright infringement. The film copied only nonprotected material: facts (nonprotected nonoriginal material) and theories (nonprotected ideas) from his research. As this case illustrates, the scope of protection depends on the degree of creative expression in the work. A work conveying facts or theories, like a medical text, has more limited protection than a creative work like an epic poem. A book of history, then, may be copied more closely than a book of fiction. This principle has broad applicability in health care, where much material in copyrighted works is not protected by copyright.

4   Copyright Ownership

Who owns the copyright to a work? The author of a work owns the copyright (unless they have signed their rights away, as discussed below). There are several possibilities.

Individual author: If the author is an individual, they initially own the copyright.

Work made for hire: The employer owns the copyright for works by employees within the scope of their employment or for certain works specially ordered, where the parties expressly agree in writing that it is a work made for hire. Note that work made for hire is not as broad as it sounds. If the work is made by an independent contractor, as opposed to an employee, the independent contractor owns the copyright unless the parties have otherwise agreed. This applies even though the contractor has been paid to create the work. This is a trap for the unwary. Anyone hiring someone to create a work (a book chapter, a photograph, a manual, etc.) should have a clause in the contract stating who owns the copyright.

Joint authors: Joint authors are co-owners with equal, undivided interests in the copyright and an obligation to share profits with the other joint authors.

Collective work (such as a collection of articles): The author of the collective work will have a thin copyright in the work as a whole, but authors of individual portions own their separate copyrights.

These ownership rules are only default rules. If the parties involved sign a contract assigning ownership, the contract will govern. Whenever multiple people are involved in the creation of a work, it is sound practice to sign an agreement governing copyright ownership to clarify the matter and avoid subsequent disputes about ownership of the copyright.

Ownership of copyright is distinct from ownership of a copy of the work or even the original of a work. Suppose a hospital purchases a painting to hang in the lobby. The hospital would own the painting. But it would not own the copyright because it did not purchase the copyright, and copyright is separate from ownership of a physical object. The hospital would infringe the copyright of the painting if it made posters of the painting and sold them, because that would be copying (infringement) and distributing copies (also infringement) of a copyrighted work, even though it owned the original.

There is a rule called first sale that does give some rights to the owner of an authorized copy, including the original. The owner of an authorized copy may display that copy and distribute that copy to the public. So the hospital, under first sale, could publicly display the painting and sell the painting itself. But first sale does not apply to the other exclusive rights, the rights to make copies, to adapt the work, and to perform the work. The hospital could therefore infringe if it made copies of or adapted the painting. It cannot perform a painting, so the public performance right is moot in that case. But for musical works and sound recordings, the public performance right is important. If the hospital purchased copies of music, it would own those copies. It would not own the copyrights. It could display or distribute those copies under first sale. But first sale does not apply to the public performance right. The hospital could infringe copyright if it played the music to the public, such as playing it over the sound system in the hospital building.

The copyright owner may also use copyright notices on copies of the work, such as "©1962 James Baldwin." Until 1989, copyright notice was required. If a work was published without a copyright notice, the copyright was lost, a harsh forfeiture. In 1989, the United States dropped this requirement in order to join the Berne Convention, the leading international copyright treaty, which forbids formalities as a condition of copyright. Publishers still routinely use copyright notice, but notice is not required. One should not infer that a photo on the web is not under copyright simply because there is no copyright notice. On the contrary, copyright arises automatically, and no notice of copyright (or of registration, giving the threat of statutory damages) is required.

How long does copyright last? A long time. The copyright of a work by an individual lasts for the author's life plus 70 years. The copyright for works made for hire and for works made before 1978 lasts 95 years. Therefore, works from the 1920s, such as F. Scott Fitzgerald's *The Great Gatsby*, are only now reaching the end of their copyright. Contrast that with patent, with its shorter term of 20 years from the patent application. Inventions from the 1920s are not patented. Rather, patented inventions often go out of protection, such as when brand-name pharmaceuticals lose patent protection and generic manufacturers may enter the market.

The owners of copyrights for valuable works may enter into license agreements to allow others to use the works. There are many types of licenses, from commercial licenses, such as for software or music, to an author licensing their book to a publisher and free licenses that allow anyone to use the work.

5   Fair Use

Fair use is not copyright infringement, another key limitation of copyright. What is fair use? The Copyright Office regulations give some examples of fair use:

> [Q]uotation of excerpts in a review or criticism for purposes of illustration or comment; quotation of short passages in a scholarly or technical work, for illustration or clarification of the author's observations; use in a parody of some of the content of the work parodied; summary of an address or article, with brief quotations, in a news report; reproduction by a library of a portion of a work to replace part of a damaged copy; reproduction by a teacher or student of a small part of a work to illustrate a lesson; reproduction of a work in legislative or judicial proceedings or reports; incidental and fortuitous reproduction, in a newsreel or broadcast, of a work located in the scene of an event being reported [6].

Use of a work that does not cost the copyright holder licensing fees may be fair use, especially if done for nonprofit or educational purposes. In an age when the internet has us all copying, adapting, and distributing works on a daily basis, fair use is important. Sending a copy of a recent newspaper article to a fellow clinician is likely fair use. Copying a photo from a photographer's website to use on a medical office's website is likely infringement, not fair use. For cases in between, consulting an attorney may be helpful.

Copyright infringement (unauthorized copying, adaptation, public distribution, public display, or public performance of a copyrighted work) may lead to awards of damages and attorney's fees, seizure of infringing material, and an injunction against further infringement. Actual damages would include things like lost sales. Alternatively, statutory damages are available for registered copyrights, along with potential attorney's fees (the painful remedy of paying the lawyers who brought the action against you).

6 Trademark

A trademark is a symbol used by a person in commerce to indicate the source of goods and to distinguish them from the goods sold or made by others [7]. Some well-known trademarks in healthcare: United Healthcare® (for managed healthcare and insurance), Pfizer® (for pharmaceuticals), Medtronic® (for medical devices), Moderna® (for vaccines). Small businesses may have trademarks, such as The Local Bookie™ (a local bookstore).

The trademark owner has a much narrower set of rights than the owner of a copyright. Rather than a broad basket of exclusive rights, the trademark owner has the right to prevent others from using the mark or similar symbols in a way likely to confuse consumers about the source of goods or services. Putting Pfizer on bottles of drugs for sale without permission would be trademark infringement. Naming Pfizer in a critical journal article or discussing Pfizer at a public conference would not be trademark infringement.

Trademark law encourages the mark owner and allows her to invest resources in developing goodwill in the mark by forbidding deceptive practices. Trademark aids buyers by reducing search costs. A buyer may rely on their past experience and the reputation of the mark rather than searching for and testing the product or service.

6.1 Case Example

Would it infringe on Gucci®, for luxury clothing and accessories, if a local dog-washing business named itself the Gucci Poochie Wash? No. Although the local business is using the Gucci name, it is unlikely that potential customers would think the dog-washing business was owned or sponsored by Gucci®. Likewise, it did not infringe to use Chewy Vuitton on handbags. Consumers would likely recognize the parody and not think the bags came from Louis Vuitton [8]. By contrast, it could infringe to use Guci on fancy clothes and accessories without a license. The name is slightly different (Gucci v. Guci), but the similar symbol on similar goods would likely confuse potential buyers.

6.2   Acquiring Trademark Rights

Most trademarks are words, real or made-up. However, any symbol that acts as a source-identifier may be a trademark, including numbers, designs, logos, shapes, sounds, fragrances, and colors. The key is that potential buyers regard the symbol as an indication of the source. Generic terms cannot be marks. "Apple" could not be a mark for an apple seller because other apple sellers need to use the same term, and it does not indicate a single source. "Apple" can be a trademark for a computer firm because other computer firms do not need to use the word "Apple." For similar reasons, merely descriptive terms are not trademarks. "Fast" would not be a trademark for computers because it is merely descriptive. In some cases, a term may go from being merely descriptive to distinctive by becoming well-known. "Microsoft" is descriptive (referring to software for microcomputers, albeit now a little archaic), but Microsoft is so well-known that the term has become a trademark. The bar against protection for merely descriptive terms can cause conflict between lawyers and marketers, who want to choose a term that describes the products to increase their appeal to potential buyers. Many small businesses, with no need to market widely, pay little attention to trademark law and operate under trade names that are not protected as trademarks, such as Garden Store or Computer Repairs.

By contrast, pharmaceutical companies often choose trademarks that are not merely descriptive but are subtly suggestive of the product's desired qualities:

- Claritin and Flonase, allergies;
- Cardura, high blood pressure;
- Requip, against Parkinson's disease;
- Ambien, Stilnox, and Lunesta, as respites from insomnia;
- Provigil, for sleepiness.

Unlike copyright and patent, there is no fixed term of duration. A trademark remains a trademark as long as it is used as a mark. Thus, when a brand-name drug goes out of patent protection, others may sell the drug but cannot use its trademark.

As with copyright (and unlike patents), registration is not required to have trademark rights. Rather, rights exist as soon as the trademark is used as a mark, such as when marking a product or its container when the product is sold or transported. Registration is permitted, and registration in the United States Patent and Trademark Office (USPTO) does make the mark stronger to enforce. In particular, registration gives nationwide rights, whereas use gives rights only in the area of actual use. Registration also allows the trademark owner to use the ® symbol by their mark. Owners of unregistered marks may use "TM."

Trademark infringement (unauthorized commercial use of a mark or similar symbol that is likely to confuse consumers) may lead to awards of damages and attorney's fees, seizure of infringing material, and an injunction against further infringement.

6.3   Applications

1. Infox Inc. created forms with blanks for common information for visits to a clinician: name, date of birth, sex, and ailment. For each ailment, the forms provide the following information to be obtained: history of the present illness, a review of systems, medical and social history, physical exam, medical decision-making, clinical impressions, and finally, consultation, disposition, and instructions. The forms also list frequent answers, which speeds up the process of completing the forms. A medical office copies the form and makes copies for its own use. Copyright infringement?

   The forms lacked even the minimal creativity to be copyrighted. Rather, they simply called for the same information that a clinician would typically ask. There is no creativity in selecting, arranging, or coordinating the information on the forms, just as a white pages phone book lists all the phone holders in alphabetical order [9].

2. Browsing online, the social media editor for a medical clinic finds a recent photo of patients getting vaccinated. The photo does not bear a copyright notice or copyright warning. The editor uses the photo in the clinic's online and print marketing. Copyright infringement?

   Yes. If the photo was made after 1989, it has copyright automatically and no copyright notice is necessary for protection. Using it without permission will infringe that copyright. Even if the editor thinks in good faith that copyright only applies to works with notice, the use infringes. Nor does fair use apply. The use is commercial, the use is not transformative, and no license has been sought. It may be that the copyright owner never finds out or never complains. The worst-case scenario, however, is that the copyright is registered, the copyright owner sues, and the court awards statutory damages (in the range of $750 to $150,000) and attorney's fees.

3. A medical software company sells a software package for radiology. The software examines images and provides valuable diagnostic data. A competitor studies how the software works and copies all of its functionality into a competing software package developed by coders using the functionality specifications. Copyright infringement?

   Software is protected as a literary work (although a literature professor might cringe at that characterization). It would have been copyright infringement to simply copy all the code of a software package. But copyright does not protect functional aspects of works, and the competitor here copied only functional elements and wrote its own code. So there is no infringement. Even if the competitor had made copies of the software in the process of reverse engineering (studying a product to determine how it works), that would have qualified as fair use and not infringement. Copying of nonprotected elements, such as functionality, would not cost the copyright owner relevant licensing revenue and so would likely be fair use.

4. A clinic in Manhattan opens for business under the name Midtown Rapid Urgent Care. A few months later, another clinic appears a few blocks away using the name: Rapid Urgent Care Midtown. Trademark infringement?

There is no trademark infringement because the first clinic does not have a trademark on Midtown Rapid Urgent Care. That is not because its name is not registered as a trademark. Registration is not required for trademark protection, although it does make a trademark's protection stronger. But a symbol is protected by trademark only if it is distinctive, in the sense that it is distinguished from others in the marketplace. Toward that end, there is no protection for merely descriptive terms (e.g., "Rapid"), geographically descriptive terms (e.g., "Midtown"), or generic terms (e.g., "Urgent Care"). Nor is the name as a whole distinctive, such as a play on words using merely descriptive terms. If the business becomes well-known, that may qualify a descriptive term as distinctive. A safer approach, if a business wants trademark protection, is to incorporate a nondescriptive term in its name, perhaps one that indirectly suggests the desirable qualities (in this case, speed: "Cheetah Midtown Rapid Urgent Care").

7   Summary

We may all both own copyrights and use copyrighted works. Copyright is ubiquitous because almost everything made with even a smidgen of creativity is covered by copyright. We use the internet a lot, which often entails copying and distributing works. Much copying from copyrighted works is permitted under fair use along with the nonprotection of nonoriginal material, ideas, and functional matter.

Trademark is likewise everywhere in commercial life. Trademark protection is narrow, however, so many uses of trademarks do not raise issues of infringement. A healthcare professional with a business may wish to own a protectable trademark. That is not hard to do, provided they choose a symbol that is not merely descriptive or likely to cause confusion with an existing mark.

References

1. See Copyright Act, 17 U.S.C. §§ 101 et seq.
2. 17 U.S. Code § 105.
3. Feist Publ'ns, Inc. v. Rural Tel. Serv. Co., 499 U.S. 340, 345, 111 S. Ct. 1282, 1287 (1991).
4. Sony BMG Music Entm't v. Tenenbaum, 719 F.3d 67 (1st Cir. 2013).
5. Hoehling v. Universal City Studios, 618 F.2d 972 (2d Cir. 1980).
6. Copyright Office Circular FL 102.
7. See Lanham Act [the federal trademark statute], 15 U.S.C. §§ 1051 et seq.
8. Louis Vuitton Malletier S.A. v. Haute Diggity Dog, 507 F.3d 252 (4th Cir. 2007).
9. See Utopia Provider Sys., Inc. v. Pro-Med Clinical Sys., LLC, 596 F. 3d 1313 (11th Cir. 2010).

Part XI

Mental Health Law

Psychiatric Hospitalization and Civil Commitment

Jacqueline Landess, Ashley VanDercar, and Brian Holoyda

Contents

Key Points
- Involuntary psychiatric hospitalization balances an individual's constitutional right to liberty with society's interest in personal and public safety.
- An individual must be mentally ill and dangerous to self or others (or, in some states, gravely disabled) to meet statutory criteria for civil commitment.
- Because of the significant liberty deprivation at stake, involuntary commitment involves specific procedural protections, which include a neutral fact finder, such as a judge, and the right to an attorney.
- Patients who are voluntarily hospitalized to a psychiatric unit must have capacity to consent to hospitalization; even voluntary hospitalization may infringe on certain rights of hospitalized patients.

J. Landess (✉)
Medical College of Wisconsin, Milwaukee, WI, USA
e-mail: Jacqueline.landess@dhs.wisconsin.gov

A. VanDercar
Case Western Reserve University, Northcoast Behavioral Healthcare, Northfield, OH, USA
e-mail: axv253@case.edu

B. Holoyda
Martinez Detention Facility, Martinez, CA, USA

© The Author(s), under exclusive license to Springer Nature Switzerland AG 2022   433
A. S. Pasha (ed.), *Laws of Medicine*,
https://doi.org/10.1007/978-3-031-08162-0_28

Legal Concepts

- *Parens patriae*: A doctrine that allows the state government to stand as a guardian to individuals who are unable to care for themselves.
- Police power: The state's power to regulate and pass laws on issues dealing with public health, safety, and morals.
- Due process rights: The rights granted by the United States Constitution to receive due process of law before the government can take away someone's life, liberty (such as in civil commitment), or property. Due process rights include both procedural and substantive rights.
- Clear and convincing evidence: The standard of proof in most civil commitment proceedings, which is lower than a beyond a reasonable doubt standard, but higher than a preponderance of the evidence standard.
- Least restrictive treatment setting: A standard describing that when a patient is civilly committed, they are entitled to only have their liberty restrained to the extent necessary.

1   Introduction

When a patient with mental health symptoms presents to a clinician and is deemed dangerous to self or others and in need of psychiatric hospitalization, the clinician may initiate the process of emergency detention, which could result in civil commitment. Depending on the jurisdiction, the police, patient's family, mental health clinicians, and/or other physicians are involved in initiating the petition for emergency detention and must follow statutory procedures in doing so. If an individual continues to need treatment beyond the emergency detention period, they may be civilly committed through a more formal, often judicial, process. Civil commitment is a form of involuntary psychiatric hospitalization. Patients can also be voluntarily hospitalized. The avenues to civil commitment, and the interplay between voluntary and involuntary psychiatric hospitalization, are described in Fig. 1. As discussed in this chapter, numerous legal regulations protect the patient's individual liberty interests but recognize the need to counterbalance those interests with the need to protect the public and the patient from imminent harm.

Broadly speaking, civil commitment laws require the following: the individual being committed must have a mental illness; the individual must pose imminent danger to self or others (this usually includes if the individual is gravely disabled and unable to care for basic needs); and the commitment must occur in the least restrictive environment, which may be inpatient or outpatient. This chapter will first discuss the legal basis for, and history of, civil commitment, followed by an overview of the statutory process for civil commitment, legal definitions of civil commitment concepts, which vary by state, and a brief discussion of voluntary psychiatric hospitalization.

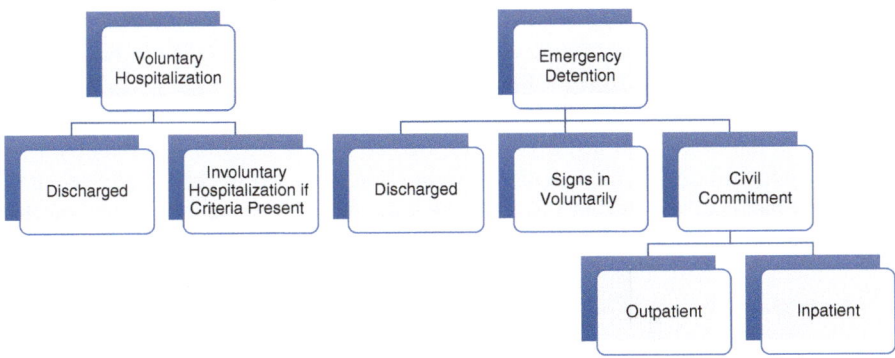

Fig. 1 Voluntary Hospitalization Versus Emergency Detention

2 Discussion

2.1 Brief History and Legal Basis for Civil Commitment in the United States

The government's authority to civilly commit a mentally ill individual is based on the police power and the *parens patriae* doctrine [1]. The police power is the state's power to regulate and pass laws on issues dealing with public health, safety, and morals [2]. When a state statute authorizes the civil commitment of a patient with mental illness, whose symptoms create an imminent threat of violence, this is an exercise of the police power. It is a power used to protect the public.

The *parens patriae* doctrine is a remnant from English law; it means "parent of his or her country" [3]. This doctrine allows the state government to stand as a guardian to individuals who are unable to care for themselves. When a state statute authorizes the civil commitment of a patient with mental illness, whose symptoms make them unable to care for their own needs, this is an exercise of the *parens patriae* doctrine. It is similar to how parents may be able to sign their children with mental illness into a psychiatric hospital without the child's consent [4].

When a clinician initiates the commitment of a person with mental illness, the clinician is acting based on a grant of power from the government under state law. These state laws, and the manner in which they are implemented by clinicians, must comply with the state constitution and United States Constitution. Americans with mental illness have a legally protected right to their liberty [5]. They can be deprived of their liberty, if necessary, by clinicians acting under state laws, but this deprivation of rights must be done in a legally permissible fashion, both substantively and procedurally.

During the late 1800s and 1900s, although asylums increased in number, the procedural protections associated with civil commitment remained quite lax [6]. The commitment process was left open for abuse by unscrupulous family members.

A good example of a case demonstrating abuse of civil commitment—and the importance of the procedural protections now in place—can be seen in the case of Elizabeth Packard, from 1864 [7, 8].

At the time that Elizabeth Packard's husband, Theophilus Packard, had her committed, they had been married 21 years and had six children together. Civil commitment laws in Illinois were particularly permissive at that time, especially when it came to a husband's ability to institutionalize his wife. A woman who was, in the opinion of the superintendent of the hospital, "insane or distracted" could be committed to a psychiatric hospital at her husband's behest.

Mr. Packard was a Calvinist minister. Mr. and Mrs. Packards' religious views had diverged. For example, amongst other things, Mrs. Packard believed slavery was a sin, and viewed the pulpit as the "proper place to combat [that] sin." Mr. Packard described Mrs. Packard's beliefs as "emanations from the devil," barred her from the bible class and sabbath school and began describing her as insane. Mr. Packard had a physician come to the family house in disguise to evaluate Mrs. Packard. The doctor told Mrs. Packard that he was selling sewing machines. Based on her supposed religious delusions, Mrs. Packard was declared insane. She was committed to an insane asylum for three years.

After Mrs. Packard was released, her husband continued her commitment at home. Mrs. Packard then filed a petition for *habeas corpus* (a judicial process that allows someone who has been involuntary detained or locked up to seek their freedom). After a jury trial, during which the jury deliberated for only seven minutes, Mrs. Packard was finally declared sane.

Cases like Elizabeth Packard's are helpful in understanding the push for greater procedural protections for individuals who were civilly committed prior to the twentieth century. In fact, before the 1960s and 1970s, individuals in need of psychiatric inpatient treatment were rarely voluntarily hospitalized; in 1949, only 10% of admissions were voluntary [9]. It was not until the 1960s, that a shift began from a focus on a patient's need for treatment to a patient's right to remain free unless he or she was imminently dangerous [10, 11]. Many of the modern substantive and procedural protections can be traced back to a series of seminal court cases which took place during this era in the wake of deinstitutionalization [12–15].

Deinstitutionalization refers to the societal and political movement to transition psychiatric patients out of long-term psychiatric hospitals and into community-based housing and treatment. Deinstitutionalization occurred in response to a variety of factors. By the 1950s, several journalists had written exposés on the poor conditions of some psychiatric facilities, reporting overcrowding, abuse, mistreatment, and neglect of inpatients [16].[1] Other factors which led to deinstitutionalization included the advent of anti-psychotic medication, such as chlor-

---

[1] Perhaps one of the more famous accounts was that of Nelly Bly, the investigative journalist who pretended to be mentally ill in order to be committed to Blackwell's Island in New York City in 1887. Bernard D. She went undercover to expose an insane asylum's horrors. Now Nellie Bly is getting her due. Washington Post (Internet). 2019 July 28. https://www.washingtonpost.com/history.2019/07/28/she-went-undercover-expose-an-insane-asylums-horrors-now-nellie-bly-is-getting-her-due/. Accessed 18 Sept 2021.

promazine, which allowed psychiatrists to treat, rather than merely confine, patients with mental illness, and the passage of increasingly strict civil commitment laws.

The seminal court cases of this era determined that if an individual's constitutional right to liberty is compromised by involuntary hospitalization, then appropriate safeguards must be in place to ensure a fair and just process. Most of these court cases were decided on Fourteenth Amendment grounds. The Fourteenth Amendment contains the Due Process Clause, which prohibits the states from depriving "any person of life, liberty, or property, without due process of law" [17]. This clause limits the reasons the government can deprive a citizen of their life, liberty, or property; it also requires that adequate procedures be provided when those rights are taken away [18]. These cases determined that patients undergoing civil commitment proceedings should be allowed many of the same protections afforded to individuals undergoing criminal proceedings, such as the right to an attorney, due to the liberty interests at stake. Further discussion of these procedural protections is provided later in this chapter.

2.2   Emergency Detention

A psychiatric emergency hold can begin the process of involuntary civil commitment in all jurisdictions in the United States. Initiating a hold involves a requestor completing a form or affidavit, or appearing before a judge to testify that an individual has a mental illness and meets the state's criteria for a hold [19]. In most jurisdictions, the statutory criteria require that the individual must be mentally ill and, because of mental illness, pose a danger to self or others or be unable to meet his or her basic needs (a condition often referred to as "grave disability"). The standard of proof a requestor must meet to demonstrate that a detainee satisfies the hold criteria is probable cause. Probable cause is a low standard of proof to meet; it simply requires a reasonable belief. If the standard is met, the individual is then typically transferred to a psychiatric facility for a brief period of assessment, observation, and stabilization. Stabilization may entail psychiatric treatment, if it is indicated, and if the patient is willing to accept treatment [12].

Psychiatric emergency holds differ across jurisdictions in important ways, including their permitted duration, who can initiate a hold, the extent of judicial oversight, and the rights afforded to patients placed on holds [19]. The most common maximum permitted duration of an emergency hold is 72 hours, though some states allow holds up to 10 days and some states prohibit holds lasting more than 23 hours. There is wide variability in who can initiate an emergency hold and what level of judicial review is required prior to a hold. Police officers in all states have the authority to detain someone who poses a risk of danger but may or may not have the ability to initiate an emergency psychiatric hold. Police officers are the only individuals allowed to initiate psychiatric holds in several states. In most jurisdictions, mental health practitioners, such as psychiatrists, psychologists, and other mental health professionals, can initiate a psychiatric emergency hold, whereas less than half the states allow non-psychiatric clinicians to do so.

2.3   Civil Commitment Laws

Prior to the expiration of an emergency hold, a detainee's mental health clinician determines whether he or she meets the jurisdiction's civil commitment criteria. If the detainee meets the criteria and is unable or unwilling to consent to voluntary hospitalization, then the clinician may petition to civilly commit the detainee for a lengthier period of assessment and treatment. Most state civil commitment statutes contain similar elements defining conditions necessary to commit an individual, including presence of mental illness or substance use disorder and dangerousness or grave disability.

Definitions of Mental Illness

State statutes governing civil commitment all require that the individual undergoing commitment has a mental illness. The definition of mental illness differs among states, but most state statutes defining mental illness list a variety of symptoms or related impairments that may qualify and identify conditions that are exclusionary. For example, New Jersey's civil commitment statute defines a mental illness as:

> .... a current, substantial disturbance of thought, mood, perception or orientation which significantly impairs judgment, capacity to control behavior or capacity to recognize reality, but does not include simple alcohol intoxication, transitory reaction to drug ingestion, organic brain syndrome or developmental disability unless it results in the severity of impairment described herein. The term mental illness is not limited to "psychosis" or "active psychosis," but shall include all conditions that result in the severity of impairment described herein [20].

New Jersey's statute suggests that individuals presenting with simple substance intoxication, a neurocognitive disorder, or an intellectual disability would not meet criteria for involuntary civil commitment, unless they are experiencing other mental health symptoms causing significant impairment. West Virginia's statute, in contrast, provides a broader and less clear definition of mental illness:

> "Mental illness" means a manifestation in a person of significantly impaired capacity to maintain acceptable levels of functioning in the areas of intellect, emotion and physical well-being [21].

This vague definition of "mental illness" could enable the civil commitment of individuals with a variety of psychiatric problems in West Virginia. Other states, such as Wisconsin, specifically allow a clinician to seek civil commitment on the basis of conditions such as a developmental or intellectual disability [22].

Recently, states have increasingly established civil commitment procedures specifically for individuals with substance use disorders (SUDs). Some jurisdictions provide for such commitment in the same statutes as their general mental illness commitment statutes. For example, West Virginia's emergency hold statute and civil commitment statute allow for the commitment of individuals identified as "addicted" [23, 24] However, other states have established civil commitment procedures whereby family members and others may petition the courts to obtain involuntary hospitalization and treatment for individuals with SUDs. One of the first of these commitment laws, the Marchman Act of Florida, was passed in 1993 and established mandatory commitment treatment programs so that individuals with SUDs could receive assessment, stabilization, and subsequent encouragement to enter longer-term voluntary treatment of their substance abuse [25].

Definitions of Dangerousness

Beginning in the 1960s, states began modifying the statutory requirements for civil commitment from a need-for-treatment model to a dangerousness model [12]. Like emergency hold statutes, most involuntary civil commitment statutes provide for the detainment of individuals who, due to mental illness, are at imminent risk of harm to self or others or are "gravely disabled," meaning they are incapable of providing for their basic needs. Thus, the treatment provider responsible for applying for civil commitment should do so based on concerns regarding suicide risk, homicide, or other physical violence risk, and the individual's inability to care for basic needs, such as obtaining food, clothing, and shelter in the community. There are only a few jurisdictions in the United States that do not follow these general standards. Some states such as Delaware, Kansas, Maryland, and Maine, for example, do not have statutory provisions allowing civil commitment on the basis of grave disability [26]. Additional criteria to consider in petitioning for civil commitment may include the potential for an individual's mental status to deteriorate further without treatment and the lack of competence to make decisions about treatment [27].

Determining dangerousness is a complex and difficult process for clinicians. Civil commitment laws define dangerousness in a variety of ways, some more specifically than others, with a focus on recent behaviors, acts, and/or statements which convey dangerous intent. What constitutes recency is often not defined or clear. For instance, in Wisconsin, an individual must "evidence a substantial probability of physical harm to himself or herself by evidence of recent threats of or attempts at suicide or serious bodily harm" [28]. Furthermore, when it comes to a risk of physical harm to others, the Wisconsin law requires evidence of a "recent overt act, attempt or threat to do serious physical harm" [28]. Several states have an "overt act" requirement. A minority of states define the time period during which dangerous behaviors must have occurred. For example, in Pennsylvania, a person must present a "clear and present danger" to others and have "inflicted or attempted serious bodily harm on another" in the past 30 days [29].

Legally, those people that act under state authority (e.g., clinicians who are using a state statute of civil commitment to authorize the detention of a patient), must ensure that patients who are detained and subsequently civilly committed meet the statutory criteria for dangerousness. If a person is civilly committed who is non-dangerous and able to care for himself or herself, this could be a violation of their constitutional rights, as the Supreme Court held in *O'Connor v. Donaldson*. Mr. Donaldson was an adult with schizophrenia who was involuntarily committed to a Florida state psychiatric hospital at the behest of his father; he was kept there for nearly 15 years. Mr. Donaldson was not dangerous to himself or others. The hospital refused to release him, despite friends and organizations that agreed to care for him if he'd be released. The legal issues in the case were complicated, and involved questions such as whether there was a constitutional right to treatment during civil commitment (the lower courts said that there was; the U.S. Supreme Court did not answer this question), and whether the state psychiatric hospital had qualified immunity (and thus whether O'Connor could be required to pay monetary damages). Notable for our purposes, the Court held that the commitment was a violation of Donaldson's constitutional rights, because "a State cannot constitutionally confine without more a non-dangerous individual who is capable of surviving safely in freedom by himself or with the help of willing and responsible family members or friends" [5].

Procedural Protections

After an emergency detention is initiated, the patient's clinicians may decide to recommend formal civil commitment, which would allow for a longer period of treatment and assessment. If granted, civil commitments can extend for months before they are reviewed and/or expire. Because civil commitment entails a more significant deprivation of liberty than that resulting from an emergency psychiatric hold, there are greater procedural protections in place for patients who are being civilly committed than for those subject to an emergency hold [12].

Patients facing civil commitment are entitled to a hearing before the civil (usually probate) court, where a judicial authority determines whether the involuntary commitment should commence. Patients have the right to be present for these hearings, to have legal assistance, and to present evidence in support of their case. The treating clinicians will usually be asked to testify at these hearings to explain their diagnosis and assessment of dangerousness. The government, which is usually represented by a county attorney, will argue for commitment. The patient's attorney will present evidence and cross-examine the patient's clinicians. The patient may also testify.

The standard of proof to certify a patient's civil commitment is higher than that for an emergency psychiatric hold, with most states adopting a clear and convincing standard. The clear and convincing standard is in between a preponderance of the evidence standard and a beyond a reasonable doubt standard. In essence, it means that something is highly probable.

The level of procedural protections afforded to potential civil commitment detainees has shifted, much like a pendulum, over the years. The case of *Lessard v. Schmidt,* from 1972, is often viewed as the high-water mark [27]. *Lessard* was a Wisconsin district court case involving Mrs. Lessard, who had schizophrenia. She had allegedly attempted suicide and was then picked up by police. She was civilly committed, and although she was then released for outpatient commitment, the process itself was problematic. She did not receive notice of her hearings, which were held without her being present. She had no right to legal representation and lacked other procedural protections. Mrs. Lessard enlisted the aid of an attorney at the Milwaukee Legal Services agency [30].

The trial court[2] found Mrs. Lessard's commitment to be unconstitutional. The court mandated notice of civil commitment hearings, a right to a jury trial, an initial hearing within 48 hours, representation by counsel, and consideration of less restrictive alternatives to commitment. The court also required that patients receive notice that any statements they make to a doctor during a psychiatric assessment could later be used against them, and required the exclusion of hearsay[3] at hearings. In so doing, the trial court attempted to apply criminal procedural protections in the context of civil commitment proceedings, based on their interpretation of the United States Constitution. Most notably, the court required proof beyond a reasonable doubt that the patient was both mentally ill and dangerous.

The *Lessard* court interpreted the constitutionality of the Wisconsin involuntary commitment statute in place at that time. Following the court's decision, the Wisconsin statute was then amended, incorporating much of the restrictive *Lessard* decision. Of note, Wisconsin now follows the clear and convincing standard of proof in commitment proceedings [31]. Though most jurisdictions did not adopt the *Lessard* court's procedural requirements, the case is emblematic of a time when the procedural protections associated with civil commitment were more similar to those given to individuals in criminal proceedings. At that time—and to a certain extent still today—commitment procedural requirements could be so stringent that patients were left without needed care.

States can choose to use a beyond reasonable doubt standard; however, in *Addington v. Texas*, the United States Supreme Court clarified it was not constitutionally required. This is a high bar and is meant for criminal cases. It is also a difficult standard to apply to civil commitment cases, where the necessary facts are amorphous and include concepts such as psychiatric diagnoses. From a constitutional perspective, the state must only show that a patient meets civil commitment criteria based on a "clear and convincing" evidence standard [1]. This is a less stringent requirement than proof beyond a reasonable doubt, but more stringent than a preponderance of the evidence standard (Table 1).

Clinicians and the court must also consider the setting in which an involuntarily committed patient is best treated. Involuntary psychiatric treatment is commonly thought of as hospitalization in an inpatient psychiatric facility. However, there are

---

[2] This was a federal trial court—the Eastern District Court of Wisconsin – because the case was based on § 1983, a federal statute.

[3] Hearsay is an out of court statement that is being offered, in court. For instance, when a doctor testifies as to what a patient's family member reported the patient as having said, as an indicator of the patient's dangerousness.

Table 1 Relevant Standards of Proof [3]

Standard of Proof	Definition	Examples of Use
Preponderance of the evidence	A lower standard, requiring only that the truth of the facts be proven as being more likely than not.	Civil cases, such as in the tort of medical malpractice.
Clear and convincing	An intermediate standard, between a preponderance of the evidence standard, and a beyond a reasonable doubt standard; the "truth of the facts asserted is highly probable."	Cases which involve more than money damages, but less than criminal incarceration; e.g., civil commitment, deportation.
Beyond a reasonable doubt	The highest standard, used in criminal cases; the truth of the facts is "clear, precise and indubitable."	Criminal cases, which typically involve a potential deprivation of liberty (incarceration).

other forms of civil commitment as well. In almost all states, patients can be civilly committed to outpatient psychiatric treatment [32].[4] Outpatient civil commitment is less restrictive than inpatient civil commitment.

The case of *Lake v. Cameron*, from 1966, is emblematic of what is known as the least restrictive alternative requirement. It is an assessment of whether a patient's inpatient commitment is in fact necessary, or the least restrictive alternative to manage their acute needs [14, 33]. *Lake v. Cameron* involved Mrs. Lake, a 60-year-old woman who had been found wandering in the community. The police filed a petition for involuntary hospitalization. She was committed to a hospital, and two weeks later filed a *habeas corpus* petition asking the court to release her. The court refused and did not hold a hearing. After a series of legal proceedings, an appellate court in the District of Columbia issued an opinion on the case.

Mrs. Lake's attorneys had been attempting to have her released from the hospital to be cared for by her family in the community. The trial court had declined to permit that, citing Mrs. Lake's mental state and the family's lack of funds to provide an appropriate level of care. The trial court placed the burden of showing the availability of less restrictive alternatives to hospitalization on Mrs. Lake. On appeal, the appellate court noted that the government was required to bear the burden of showing that inpatient commitment was the least restrictive setting for Mrs. Lake's commitment.

Lake v. Cameron was based on the District of Columbia's statute at the time. It is not binding law for other states. Yet, the views expressed in the case are reflective of a shift toward an increased focus on patient rights. Language originating in that case has been incorporated into many civil commitment statutes across the country [33].

---

[4]There are also civil commitments associated with the criminal justice system, such as those used for certain types of criminals with prior sexual offenses in need of mental health treatment.

2.4   Civil Commitment Outcomes

Though civil commitment laws exist in every jurisdiction in the United States, there is relatively little data regarding the utilization and outcomes of civil commitment in different states or nationally [34]. One exception is Virginia, which collects and annually publishes data on the frequency of emergency custody placements and the outcomes of civil commitment hearings. In 2019, 18.5% of initial commitment hearings in Virginia resulted in dismissal, whereas 59.9% of patients remained on involuntary status and 20.9% transitioned to voluntary status. An additional 0.7% were mandated to outpatient treatment [35]. Data tracking of civil commitment could assist states in identifying trends in emergency holds and civil commitment over time and coordinating appropriate legislative responses to changing mental health needs in their jurisdictions.

2.5   Voluntary Psychiatric Hospitalization

By the 1980s, more than 70% of psychiatric hospitalizations in the U.S. were voluntary, in stark contrast to the early part of the twentieth century [36]. Voluntary hospitalizations offer a route to treatment without court involvement and are offered to those patients who are seeking, and consent to, treatment. However, even a voluntary psychiatric admission implicates a patient's liberty rights. For example, a voluntarily admitted patient cannot simply leave the hospital at any time. In nearly every state, a voluntary psychiatric patient must request discharge, usually in writing, and the patient's clinicians must respond to the request within a certain amount of time. State statutes specify these time frames, which range from immediately up to five days, with the most common time frame being 72 hours [37]. In addition to the required waiting period before discharge, the treating clinician could also determine that the patient is not safe for discharge and initiate involuntary hospitalization proceedings.

Further, a voluntary patient on a psychiatric unit will experience other rights restrictions as dictated by institutional policy. For instance, psychiatric units may not have patient areas which allow smoking, patients may not be allowed to wear their own clothing, use their cell phones, or make regular phone calls, and/or patients could be admitted to a unit with unpredictable peers. Thus, a patient being admitted to a psychiatric unit—even voluntarily—potentially faces a greater deprivation of liberty than a patient admitted to a medical service.

The case of *Zinermon v. Burch* highlights the concerns around deprivation of patient rights in the context of voluntary psychiatric hospitalization. Mr. Burch was found wandering along a Florida highway in a disoriented state. He was assessed and deemed to be suffering from paranoid schizophrenia. He subsequently signed voluntary admission forms believing he was "in heaven." He was sent to a Florida psychiatric hospital where he was hospitalized for approximately six months. After discharge, Mr. Burch filed a lawsuit claiming that the hospital and staff violated his constitutional rights because he had lacked capacity to consent to the hospitalization. He also alleged violation of a state law, which required that patients give express and informed consent to hospitalization. Mr. Burch argued that staff should have pursued involuntary commitment procedures, which would have allowed him greater liberty protections, due to the legal procedures required.

The lower court dismissed Mr. Burch's suit, finding that his complaint had failed to state a valid claim for federal court. On appeal, the Supreme Court, interpreting the intersection of state laws, federal laws, and their prior case holdings, held that Mr. Burch's case must be allowed to proceed. In so doing, the Court noted the importance of hospital staff making an "inquiry" into a patient's capacity to consent to voluntary hospitalization. The Court did not, however, specify the knowledge a patient must possess to consent to hospitalization and/or how staff should conduct such an inquiry.

The *Zinermon* case produced a palpable response within the psychiatric community. Many patients presenting for psychiatric treatment are often suffering such severe symptoms that they cannot understand detailed information about psychiatric hospitalization. These patients might not meet civil commitment criteria either, effectively eliminating avenues for intensive treatment, even though they are assenting to such treatment.

Recognizing the concerns prompted by *Zinermon*, the American Psychiatric Association convened a task force on consent to voluntary hospitalization and generated a position statement on the issue in 1992 [36]. Position statements are informative and reflective of a general consensus. However, they are not binding. More important in any particular situation is the applicable state law, and how the courts in that jurisdiction have interpreted the concept of consent, as it pertains to voluntary psychiatric hospitalization.

The American Psychiatric Association task force recommended that a patient first should be able to communicate a choice and understand relevant information in order to have the capacity to sign in voluntarily. If the patient could not communicate a choice, then alternative mechanisms, such as a substitute decision-maker (including limited guardianship), could be pursued. Some states allow next of kin to make medical decisions for family members when they are incapacitated, while other states require temporary or permanent guardians. Guardians are designated by a judge following a court (civil) hearing. The individual usually is evaluated by a qualified clinician, such as a psychiatrist or psychologist, and this report is made available to the court. Relevant information would include, at a minimum, that the patient understand they are being admitted to a psychiatric hospital for treatment and that release from the hospital may not be automatic [38]. Other experts have recommended that the patient understand more extensive information in order to consent to voluntary hospitalization, including alternatives to hospitalization, the possibility of involuntary commitment, the institution's ability to file for guardianship, and related information [37].

3   Application

3.1   Clinical Vignette A: Determining Dangerousness

Ms. B is brought into the emergency room by police for a psychiatric evaluation. In Ms. B's jurisdiction, clinicians must formally initiate an emergency detention. Police state that Ms. B is homeless and well-known to them. Earlier that day, she was allegedly approaching passers-by and demanding money from them. When they did not comply, she began to yell aggressively and swear at them and spit in their direction. However, she did not directly touch the pedestrians. One pedestrian stated that Ms. B told him that she was a federal agent and would find him and "set the record straight." Ms. B has a history of schizophrenia and does not take medication or receive treatment for this condition. On examination, she is not agitated but refuses to answer questions, stating it is her "constitutional right to remain silent."

This case highlights the complexity of assessing dangerousness. Ms. B appears to meet statutory criteria for mental illness, though the clinician who evaluated her would need to seek collateral information, such as previous treatment records, to confirm her history of schizophrenia, since she is not cooperative during the exam. Since nearly all state laws require recent evidence of behaviors, acts, or statements for a finding of dangerousness, it appears that Ms. B's alleged acts and statements probably meet this standard. The standard of proof at the emergency detention stage is only probable cause. However, subsequently, Ms. B may not meet the involuntary commitment criteria, depending on her behaviors while detained and collateral information received (e.g., the frequency and severity of recent harm and imminence of future harm). If Ms. B lives in a jurisdiction where grave disability may be grounds for civil commitment, this may be an alternative avenue for commitment, assuming that she is unable to care for her basic needs.

3.2   Clinical Vignette B: Least Restrictive Treatment Setting

Mr. X is emergently detained after a suicide attempt by hanging. He is diagnosed with severe depression. While hospitalized, he continues to make statements about wanting to die and is refusing treatment. His clinicians pursue an involuntary commitment, which is granted. At the hearing, the judge considers whether Mr. X should continue to be hospitalized. Mr. X's attorney argues that Mr. X should receive treatment as an outpatient, due to the restrictive nature of an inpatient psychiatric unit. Because of Mr. X's recent suicidal statements, serious suicide attempt, refusal of treatment and lack of insight into his depression, the judge commits Mr. X to inpatient involuntary treatment. Mr. X is then mandated to take medication for his illness and he subsequently recovers within three weeks. He is then transitioned to outpatient civil commitment because his risk of suicide has been sufficiently reduced.

3.3   Clinical Vignette C: Informed Consent to Hospitalization

Mr. Jones is a 35-year-old man with a documented history of schizophrenia and alcohol use disorder. He self-presents to the emergency room requesting admission to the psychiatric unit. He reports worsening hallucinations and is observed to have disorganized thoughts and persecutory delusions. He apparently ran out of his pre-scribed anti-psychotic medication four days ago. He has a slightly elevated blood alcohol level. Mr. Jones is not exhibiting or reporting dangerous thoughts or urges, such as a desire to kill himself or harm others. Mr. Jones has one previous voluntary admission to a psychiatric unit. The clinician agrees that Mr. Jones would benefit from inpatient hospitalization while his anti-psychotic medication is restarted. Mr. Jones expresses a clear choice to be admitted. However, after the clinician reviews the rights that may be restricted while Mr. Jones is an inpatient, including a waiting period after request for discharge, Mr. Jones shrugs and states: "I don't care about any of that. The FBI is trying to find me and I just need to hide. I figured I would be safe here. If I need to leave, I'll leave." He is unable to restate the information the clinician provided and does not appear to appreciate this information.

This case highlights the classic dilemma and balance between obtaining ade-quate informed consent, protecting patient rights, and avoiding coercion, while ensuring adequate access to treatment. Mr. Jones may be able to assent to voluntary admission but is appears to be unable to provide proper informed consent to admis-sion due his psychotic symptoms. In this case, the clinician has several options. First, the clinician should review the law in her state regarding voluntary hospital-ization [37]. She could seek out a substitute decision-maker for Mr. Jones. Each state has differing criteria for how this is accomplished. She may give Mr. Jones a dose of his prescribed medication and revisit the issue later, though one dose of medication might not substantially change his capacity. She could also consider involuntary commitment, but Mr. Jones is not likely a proper candidate given that he is not acutely dangerous. Lastly, she could request that Mr. Jones be admitted to the medical floor for detoxification/withdrawal monitoring with ongoing assessment of his capacity to consent to voluntary hospitalization, if that continues to be recommended.

4   Summary

In summary, involuntary psychiatric hospitalization is permissible under legal doc-trines such as *parens patriae* and the police power, which recognize society's inter-est in caring for those who are incapacitated and/or preventing harm to the patient or others. Being detained and confined is a serious deprivation of liberty and thus clinicians and the government must follow specific statutory procedures in pursuing and defending the need for commitment. These procedures include obtaining an opinion that the person is mentally ill and dangerous or gravely disabled and identi-fying the least restrictive setting for treatment. The government must prove these conditions by clear and convincing evidence, which is a high bar. Finally, clinicians must also be aware of the restrictions on patient rights that can occur even with voluntary psychiatric hospitalizations and must ensure that patients seeking voluntary hospitalization are capable of consenting and understand the basics of hospitalization. If such patients are not capable of consenting, the clinician should consider alternative routes to treatment.

References

1. Addington v. Texas, 441 U.S. 418 (1979).
2. Barnes v. Glen Theatre, 501 U.S. 560 (1991).
3. Black's Law Dictionary (6th ed. 1990).
4. Parham v. J. R., 442 U.S. 584 (1979).
5. O'Connor v. Donaldson, 422 U.S. 563, 576–583 (1975).
6. H. J. J. Civil Commitment of the Mentally Ill. 107 U. Pa. L. Rev. 668, 668–684 (1959).
7. Packard, E. Marital power exemplified in Mrs. Packard's trial, and self-defense from the charge of insanity; or three years' imprisonment for religious belief, by the arbitrary will of husband, with an appeal to the government to so change the laws as to protect the rights of married women. Clark & Co., Publishers; 1870.
8. Hendrik Hartog. Mrs. Packard on dependency. 1 Yale J.L. & Human. 79, 79–103 (1988).
9. Appelbaum PS, Bateman AL. Competency to consent to voluntary psychiatric hospitalization: a theoretical approach. Bull Am Acad Psychiatry Law. 1979; 7:390.
10. Stone AA. Mental health and law: System in transition. Rockville, U.S. Dept. of Health, Education, and Welfare; 1975. p. 43–82.
11. Sara Gordon. The danger zone: How the dangerousness standard in civil commitment proceedings harms people with serious mental illness. 66 Case W. Res. L. Rev. 657, 657–700 (2016).
12. Testa M, West SG. Civil commitment in the United States. Psychiatry (Edgmont). 2010;7(10):30-40.
13. Jackson v. Indiana, 406 U.S. 715 (1972).
14. Lake v. Cameron, 364 F.2d 657 (D.C. Cir. 1966).
15. Lessard v. Schmidt, 349 F. Supp. 1078 (E.D. Wis. 1972).
16. Morrissey J, Goldman H. Care and treatment of the mentally ill in the United States: Historical developments and reforms. Ann Am Acad Pol Soc Sci. 1986 Mar;484:12-27.
17. U.S. Const. amend. XIV, § 1.
18. Erwin Chemerinsky. Substantive due process. 15 Touro L. Rev. 1501, 1501–1534 (1999).
19. Hedman LC, Petrila J, Fisher WH, Swanson JW, Dingman DA, Burris S. State laws on emergency holds for mental health stabilization. Psychiatr Serv. 2016 May 1;67(5):529-535. doi:https://doi.org/10.1176/appi.ps.20150025
20. N.J. Stat. Ann §30: 4-27.2 (2013).
21. W. Va. Code § 27-1-2 (2019).
22. Wis. Stat. §51.20-1-a-1 (2019).
23. W. Va. Code. § 27-5-2 (2019).
24. W. Va. Code § 27-5-4 (2019).
25. Cavaiola AA, Dolan D. Considerations in civil commitment of individuals with substance use disorders. Subst. Abus. 2016 Jan 20;37(1):181-7. doi:https://doi.org/10.1080/0889707 7.2015.1029207
26. Treatment Advocacy Center. State Standards for Civil Commitment. 2020 Sept. Arlington, VA.
27. Substance Abuse and Mental Health Services Administration. Civil commitment and the mental health care continuum: Historical trends and principles for law and practice. 2019. Rockville, MD.
28. Wis. Stat. §51.20-1-a-2-a (2019).
29. 50 Pa. C. Stat. § 7301(b) (2019).

30. Arnold H. Landis. Civil commitment of the mentally ill: Lessard v. Schmidt, 23 DePaul L. Rev. 1276, 1276-1297 (1974)
31. Steven K. Erickson, Michael J. Vitacco, Gregory J. Van Rybroek. Beyond overt violence: Wisconsin's progressive civil commitment statute as a marker of a new era in mental health law. 89 Marq. L. Rev. 359, 359-405 (2005).
32. Daily L, Gray M, Johnson B, Muhammad S, Sinclair E, Stettin B. Grading the states: An analysis of involuntary psychiatric treatment laws. Treatment Advocacy Center. 2020. https://www.treatmentadvocacycenter.org/storage/documents/grading-the-states.pdf. Accessed 16 May 2021.
33. Michael L. Perlin & Alison J. Lynch. Toiling in the danger and in the morals of despair: Risk, security, danger, the constitution, and the clinician's dilemma. 5 Ind. J. L. & Soc. Equality 409, 409-440 (2017).
34. Morris NP. Detention without data: Public tracking of civil commitment. Psychiatr Serv. 2020 Jul 1; 71(7):741-744. doi:https://doi.org/10.1176/appi.ps.202000212
35. Zelle H, Gwinn KMF. Adult civil commitment proceedings in Virginia: Annual statistical report FY19. 2020 June. University of Virginia Institute of Law, Psychiatry and Public Policy.
36. Cournos F, Faulkner LR, Fitzgerald L, Griffith E, Munetz MR, Winick B. Report of the Task Force on Consent to Voluntary Hospitalization. Bull Am Acad Psychiatry Law. 1993; 21(3):293-307.
37. Garakani A, Shalenberg E, Burstin S, Brendel R, Appel JM. Voluntary psychiatric hospitalization and patient-driven requests for discharge: a statutory review and analysis of implications for the capacity to consent to voluntary hospitalization. Harvard Review of Psychiatry. 2014 Jul-Aug; 22 (4):241-49.
38. Appelbaum BC, Appelbaum PS, Grisso T. Competence to consent to voluntary psychiatric hospitalization: a test of a standard proposed by APA. American Psychiatric Association. Psychiatr Serv. 1998 Sep; 49(9):1193-96.

Confidentiality and Privilege

Cristah Artrip Prost and Edmund G. Howe

Contents

Key Points
- The HIPAA Privacy Rule applies to patients receiving mental healthcare. Mental health clinicians should note exceptions which exist which would permit them to share information in good faith to a patient's family when the patient has and does not have capacity.
- Under the HIPAA Privacy Rule, psychotherapy notes are considered private, personal notes for the clinician and it is not required that they be shared with or accessed by the patient.
- Information received directly from the patient or that clinicians observe during the professional relationship between the clinician and patient is considered privileged information. A patient may exercise their right to prevent documentation from psychotherapy from being released in court, and the clinician cannot use privileged information to testify as a witness in court. Both specific federal and state definitions may apply and there are exceptions.

C. A. Prost
School of Medicine, Uniformed Services University, Bethesda, MD, USA

E. G. Howe (✉)
Department of Psychiatry, Uniformed Services University, Bethesda, MD, USA
e-mail: edmund.howe@usuhs.edu

© This is a U.S. government work and not under copyright protection in the U.S.; foreign   449
copyright protection may apply 2022
A. S. Pasha (ed.), *Laws of Medicine*,
https://doi.org/10.1007/978-3-031-08162-0_29

- Federal laws and state laws may restrict access to information regarding a patient's treatment for a substance use disorder. Confidentiality of these documents may also limit their use in criminal or civil legal proceedings such as those involving criminal charges, child custody, and employment.
- Clinicians' duty to warn others about patients who are imminently dangerous may be met legally if they make reasonable and timely efforts to reduce this risk. Permissible responses may be civilly committing the patient, arranging an appropriate treatment plan, informing legal authorities and warning the intended victim(s). State mandates vary widely in defining these requirements.

Legal Concepts
- Confidentiality: The professional ethical obligation of clinicians to keep all information of the patient's private, and not sharing this information with other parties unless prior consent is given.
- Privilege: The right of the patient to prevent a clinician from disclosing health information, particularly during legal proceedings.
- Privacy: The right of a person to control how their personal or health information is collected, used or disclosed.
- Duty to Warn: The obligation and sometimes legal requirement for a clinician, upon learning their patient may pose an imminent risk of harm to an identifiable other person, to inform third parties such as law enforcement personnel and/or potential victims of a serious threat of harm.

1   Introduction

The legal and professional obligations of clinicians in the mental healthcare sector pose unique challenges. In clinicians' attempt to honor the tradition of their medical profession's confidentiality, clinicians must weigh their legal duties with their ethical obligations to offer maximal protection to their patients while also optimizing health outcomes. The Hippocratic Oath and various current follow-on professional codes have set forth the ethical guideposts which define clinicians' chief considerations regarding confidentiality [1]. Secondarily, laws and statutes have offered a codified moral posturing within which providers can respond with optimal ethical resolutions, by outlining exceptions to confidentiality which compel clinicians to divulge patient information. Commonly, law will establish a minimum standard to reduce the conflict between mainstream practices. Ethically, however, there may be exceptions that clinicians should consider at the margins of these laws. When this is the case, they should generally err by doing what they believe is in the best interest of the patient and honoring patient desires as opposed to what the law, taken literally, may seem to require. Simply put, if questioned on the stand as to why the clinician did what he did, to knowingly give the patient sub-optimal care to conform to the law invites legal liability. Clinicians erring to meet patients' best interest or preference, on the other hand, and documenting their rationale is likely to affect the best legal outcome.

We will consider the diversity of legal definitions and exceptions at the federal and state level within which mental health clinicians navigate to arrive at general guidelines involving safe, yet ethical, clinical decision-making. Even when there are seemingly discrete laws that can readily be applied in most cases, this presupposition of black and white decision-making may be illusory. There is diversity in the contextual factors in many mental health cases as well as different interpretations regarding the legal definitions. Both may affect where confidentiality can and should be breached. At these margins, in some extreme scenarios, ethical dilemmas may exist that have higher stakes. Clinician behavior then may become more conflicted and at times, contested.

Laws and their definitions vary from state to state, so clinicians should ascertain their specific state's law whenever this is uncertain. Here, while making general points most important to clinicians, we will provide specific examples of Maryland state law to illustrate legal concerns and solutions likely to be predominant also in other states.

These core concepts will be first presented in a discussion of federal and state laws which address confidentiality, privilege, documentation, reporting, and their exceptions. Then, specific cases and clinical vignettes will elucidate the challenges and considerations clinicians should weigh as they seek to arrive at optimal legal and ethical decisions effecting the optimal medical treatment decisions made through a process of shared decision-making with their patients. This shared decision-making involves clinicians anticipating the medical and legal information patients must know to meaningfully decide what both should do. Then together, they may decide how to advance within the scope of applicable laws. The general aim should be then to make informed decisions together, taking fully into account the applicable laws.

The examples provided will involve general considerations regarding confidentiality and privilege, patients who may harm themselves, patients who may harm others, and practical recommendations. The last of these topics will fundamentally emphasize shared decision-making and considerations regarding documentation.

Supporting a shared decision-making approach between clinicians and patients will ultimately provide the strongest legal rationale for clinicians' decisions. When clinicians have articulated all pertinent risks considered and an optimized treatment plan for their patients that is supported by both their patients' preferences and the evidence-based medical literature, the legal grounds for their actions is most likely to withstand challenge and scrutiny. Conversely, clinicians may risk legal, ethical, and patient harm if they frame the clinical case solely in the interest of legal protection or impose isolated clinical judgment presupposing, perhaps, that a patient is insufficiently knowledgeable to participate. A case that may meet confidentiality and privilege requirements could still lend itself to sub-optimal care and may even unnecessarily harm patients. In clinical decision-making, even shared, it may also be necessary to document this decision-making process and obtain consent for information release up front when legally necessary and prescribed for confidentiality and privilege in mental health contexts.

After discussion of the basic legal concepts, cases and vignettes will elaborate the law's complexities and subtleties regarding how providers should best implement these laws.

2   Discussion

Confidentiality, privacy, and privilege as required specifically in mental health require consideration of the basic legal framework as well as its exceptions. Patients with mental illness are afforded and protected under the legal protections of the Health Insurance Portability and Accountability Act (HIPAA) Privacy Rule, a U.S. federal law covered previously in this book. Mental health has notable exceptions to some of the applications of privacy rule. Below we will provide a broad overview of confidentiality from the perspective of HIPAA, state legal variations in privilege and "duty to warn," which makes special exceptions as well for documentation and access to mental health related visits.

Confidentiality: Review of the Health Insurance Portability and Accountability Act of 1996.

(HIPAA)- Federal Privacy Law and Exceptions in Mental Health:

Regarding confidentiality and the HIPAA Privacy rule, it may be important to note that "confidentiality" as previously defined focuses on the obligation of the clinician, and "privacy" focuses on the legal right of the patient/individual. While not an exhaustive review of HIPAA, the emphasis below is placed on several exceptions which pertain directly to the complexities encountered in mental health. Often, due to the number of exceptions and complexities regarding safety and confidentiality which arise in HIPAA as a U.S. federal law, mental health clinicians receive additional advanced training on HIPAA.

In general, the HIPAA Privacy rules applies uniformly to all patients and all clinicians who transmit patient information, including mental healthcare professionals. When the patient is present and does not object, clinicians may disclose relevant health information including mental health information to family members or other people the patient has involved in their care or payment for care [2]. When the patient is not present or is incapacitated, a clinician may legally share the patient's information if the clinician professionally deems it in the best interest of the patient to do so. Still though, any disclosures to other people involved in the patient's care should be limited to only the protected health information most directly relevant to the person's involvement in the patient's care [3].

Within the HIPAA Privacy Rule, an important exception permits a clinician to share directly relevant health information with the patient's family, even when the patient does not have capacity or has objected if doing so aligns with state confidentiality statutes and the professional standard of ethical conduct. This may occur when clinician has a "good faith belief" that the patient poses a threat to the health or safety of themselves or others, and that disclosure to the family member may prevent and reduce this threat [4].

The most notable mental health exception in the HIPAA Privacy Rule distinguishes psychotherapy notes as a separate, private type of documentation. These notes are defined as the clinician's private documented analysis of conversations occurring during private and/or group sessions which are stored separately from the rest of the patient's medical record. These are used for personal purposes by the treating clinician, and the HIPAA Privacy Rule does not provide a right of access to these by any individual, including other clinicians unless required by state statutes in instances of abuse or an obligation of duty to warn, as outlined later in this chapter [5, 6]. Otherwise, the patient's treatment plan, including the assessment and medications prescribed, are not included in this exception and are subject to the same treatment as other health information in the HIPAA regulations.

Other state confidentiality and professional codes permitting, clinicians may disclose directly relevant health information about an adult patient with mental health disease to a family member if there is reason to believe there is imminent harm or danger to the patient. Additionally, "any information disclosed to the provider by another person who is not a healthcare provider that was given under a promise of confidentiality (such as that shared by a concerned family member), may be withheld from the patient if the disclosure would be reasonably likely to reveal the source of the information," thus protecting the family relationship while also aiding the clinician in optimizing safety assessment for the patient [3, 7].

In general, the HIPAA Privacy Rule allows clinicians to disclose health information to other clinicians and third parties in the instance of coordination of care or arrangement of services. Similarly, where a patient poses an imminent threat of harm to themselves or others, the clinician may disclose information necessary to inform family members, law enforcement or other officials to intervene and minimize this threat. There are limits to disclosure. For example, clinicians may not turn over certain protected health information such as DNA or dental records for identification to law enforcement while a patient is on psychiatric hold even though clinicians are permitted to disclose the timing of admission [3].

Beyond HIPAA at the federal level, states enact laws and regulations regarding confidentiality which may be more stringent than those defined in HIPAA. State issued board licensures also play a role in defining the legal definitions and enforcing the consequences of breaching confidentiality. One such example of Maryland state law compels clinicians to release confidential information if they become aware of child abuse or neglect, a mandatory reporting requirement [8]. The most notable federal confidentiality laws pertain to confidentiality for federally funded substance abuse programs, under which the release of treatment information for substance or alcohol abuse to certain entities like legal proceedings and employers is restricted [9].

2.1 Privilege and its Exceptions under Maryland State Law

Privilege refers to a person's right to prevent disclosure of certain information in legal proceedings which was learned during during specified relationship which was assumed to be confidential such a conversations with a professional or spouse. This concept is often referred to as a lawyer-client relationship, but in a landmark Supreme Court case in 1996, *Jaffee v. Redmond*, 518 U.S. 1, the Supreme Court also recognized the psychiatrist-patient privilege [10]. Since this case, many state and federal cases have further sought to define the extent of this protection in legal proceedings for additional types of mental health clinicians. Examples of privilege include the limitation on sharing health information, the mental health clinician being summoned as a witness, or the release of medical record to courts. Below, Maryland Law is utilized as an example outlining this scope and notable exceptions [11].

In defining mental health clinician and patient privilege, this reads:

Unless otherwise provided, in all judicial, legislative, or administrative proceedings, a patient or the patient's authorized representative has a privilege to refuse to disclose, and to prevent a witness from disclosing:

1. Communications relating to diagnosis or treatment of the patient; or
2. Any information that by its nature would show the existence of a medical record of the diagnosis or treatment [11].

States vary in the precision of their exact definitions, exceptions, and scope, but the principles remain similar. Clinicians may consult their local statutes, in addition to professional boards for legal guidance specific to their practice. The Maryland statute goes on to define exceptions to privilege as outlined below, stating:

There is no privilege if:

1. A disclosure is necessary for the purposes of placing the patient in a facility for mental illness;
2. A judge finds that the patient, after being informed there will be no privilege, makes communications in the course of an examination ordered by the court and the issue at trial involves his mental or emotional disorder;
3. In a civil or criminal proceeding: (1) The patient introduces his mental condition as an element of his claim or defense; or (2) After the patient's death, his mental condition is introduced by any party claiming or defending through or as a beneficiary of the patient;
4. The patient, an authorized representative of the patient, or the personal representative of the patient makes a claim against the psychiatrist or licensed psychologist for malpractice;
5. Related to civil or criminal proceedings under defective delinquency proceedings;
6. The patient expressly consents to waive the privilege, or in the case of death or disability, his personal or authorized representative waives the privilege for purpose of making claim or bringing suit on a policy of insurance on life, health, or physical condition;

7. In a criminal proceeding against a patient or former patient alleging that the patient or former patient has harassed or threatened or committed another criminal act against the psychiatrist or licensed psychologist, the disclosure is necessary to prove the charge; or.
8. In a peace order proceeding under Title 3, Subtitle 15 of this article in which the psychiatrist or licensed psychologist is a petitioner and a patient or former patient is a respondent, the disclosure is necessary to obtain relief [11].

As seen above, privilege is primarily a right afforded to the patient, protecting in court of law their prior disclosures made to their mental health clinician under an assumed confidential relationship. Exceptions above outline general principles seen elsewhere in healthcare law which coincide with confidentiality and consent laws where possible. They also generally limit the use of healthcare documentation in other civil sectors such as court proceedings or employment beyond immediate healthcare for the individual.

2.2   Duty to Warn, Maryland State Law

Duty to Warn, rising out of landmark cases *Tarasoff I* and *Tarasoff II* reviewed below, refers to the obligation for a clinician to warn law enforcement or potential victims of a serious threat of harm when they believe their patient discloses a credible or imminent threat of harm to another identifiable individual [12]. States vary widely in their requirements for clinicians and the legal definitions of an adequate warning or effort to mitigate the threat. These variations include differing mandates on when to warn or protect the potential victim, when to report the threat to law enforcement, and when to make attempts to treat or hospitalize the patient to protect others. Below, Maryland state law is utilized as an example of state laws which interpret the legal definitions of "duty to warn," rising out of the legal definition of Tarasoff [12, 13]. Some states, such as Texas do not have laws of this kind, potentially only placing the responsibility of harm to an individual on the perpetrator alone and not on the clinicians treating their mental health disorder [14].

2.3   Excerpt of Maryland's Duty to Warn Law

A cause of action or disciplinary action may not arise against any mental health care provider or administrator for failing to predict, warn of, or take precautions to provide protection from a patient's violent behavior unless the mental health care provider or administrator knew of the patient's propensity for violence and the patient indicated to the mental health care provider or administrator, by speech, conduct, or writing, of the patient's intention to inflict imminent physical injury upon a specified victim or group of victims.

The duty described under this section is deemed to have been discharged if the mental health care provider or administrator makes reasonable and timely efforts to:

Seek civil commitment of the patient;
Formulate a diagnostic impression and establish and undertake a documented treatment plan calculated to eliminate the possibility that the patient will carry out the threat;
Inform the appropriate law enforcement agency and, if feasible, the specified victim or victims of: (1) the nature of the threat; (2) the identity of the patient making the threat; and (3) the identity of the specified victim or victims [15].

In summary, the law above compels Maryland mental health clinicians to treat, warn, or protect and under these circumstances they will not be held liable for breaches of confidentiality. Here, the law outlines the order of operations for the Maryland clinician, prioritizing a first duty to treat the patient through inpatient admission or other forms of treatment as a means of protecting the potential victim. After this has failed, Maryland clinicians have the duty to warn and protect by notifying law enforcement and making a good faith effort to warn the intended victim or group of the threat [12, 13, 15].

Discussing the historical context of duty to warn laws below, this subset of mandates primarily falls on mental health clinicians, with wide variation in state practices. Confidentiality and duty to warn legislation often conflict. Texas, for example, favors strict confidentiality laws such that duty to warn may result in violating confidentiality laws [14]. Overall, duty to warn laws point to the complexity of confidentiality and its specific exceptions which touch almost all aspects of mental healthcare.

The laws above demonstrate the diversity, complexity and challenges which occur specifically in confidentiality and privilege in mental health. For many mental health clinicians, understanding the federal regulations, state legislation, and professional code of ethics is just one arc of the broader context which guides their clinical decision-making as they assess and treat patients with mental health disorders. However, without the knowledge of these laws, it may be unwittingly easy to violate their patient's rights even when acting in good faith.

Important questions to consider as a clinician in light of legal diversity in different jurisdiction:

• What are the laws in my state regarding confidentiality and duty to warn?
• To what degree is the warning sufficient according to the law?
• To what degree am I legally and ethically compelled to commit or psychiatrically treat the patient?
• To what degree am I legally and ethically compelled to contact third parties such as law enforcement or potential victims?

Below are some examples of ethical challenges and applications of the laws above.

3   Application

Case 1: Confidentiality in Mental Health   Clinical Vignette: A 36-year-old male with major depressive disorder who is prescribed an antidepressant medication he usually takes every day tells his psychologist during his regularly scheduled telehealth appointment that he ran out of medication 3 weeks ago and has been having increasing thoughts of suicide. He has a plan for suicide but no current means for how he would carry out his suicide plan. He is very distressed by his mental state and feels ashamed that he is letting his wife down. Later the same day, the patient's wife calls the psychologist concerned about the patient. She states that her husband has been acting strangely and has a habit of forgetting to take his medication. She asks the psychologist whether the patient is taking his medication and whether she should be concerned.

In the example above, the clinician knows that the patient is not at a therapeutic level of his antidepressant and may pose an imminent risk of harm to himself. Additionally, the patient has not indicated whether or not he has given prior consent for the clinician to release health information to his family. In this example, the HIPAA privacy rule does allow mental health exceptions for good faith breach of confidentiality to families in that the clinician believes that releasing certain information may mitigate the risk or increase the patient's safety to prevent a serious and imminent threat such as loss of life. However, had the patient expressly withheld consent, the clinician may consider prioritizing the patient's confidentiality and autonomous wishes. Alternatives to full disclosure would be suggesting the patient may be having a health emergency and needs to come into the emergency room without revealing the explicit details of his medication regimen or his mental health condition. Of note, confidentiality laws in some instances often include not indicating whether that patient is even under the care of a certain clinician. Thus any sort of conversation with that acknowledgement may be a breach of confidentiality in the absence of consent. Even in the instance above, the HIPAA legal precedent suggests only the minimum disclosure necessary to accomplish the desired end goal.

Other important factors to consider in the example above include the clinical context. While the clinician may seek to optimally treat the patient and maintain a maximally trusting relationship, potential harms may result from disclosing information about the patient. This may include irreparable damage to the rapport already established in the physician-patient relationship, especially if the patient has explicitly requested full confidentiality and non-disclosure to named family members. Resolving these conflicting concerns may be especially difficult when, legally, these patients are not considered adults or with full capacity. When this is the case, there are likely to be different start rules regarding the conditions under which minors may be viewed as having adult capacity and when their confidentiality and privacy must be respected. These rules involve, for example, when a patient may be deemed a mature minor. Different rules may be applied in certain circumstances as when adolescents request birth control medications and are using illegal substances.

Clinicians faced with these situations should particularly check their state's law prior to making decisions regarding minors that could violate their state's requirements.

The HIPAA Privacy Rule exceptions generously note clinicians' intentions of acting in good faith to avert serious adverse outcomes. Other factors related to patients having optimal outcomes may also guide the final decision on whether a clinician may disclose confidential information, even when legally they may do so [16–18].

Case 2: Tarasoff vs. Regents of the University of California (1976) [19–21] Tatiana Tarasoff attended folk dance lessons at the International House on campus at the University of California at Berkeley where she met Prosenjit Poddar, a graduate student from Bengal, India. They became friends, and Tarasoff later rejected Mr. Poddar's advances stating she was not interested in a relationship with Poddar. This led to an emotional crisis and psychological decline for Poddar. He became more withdrawn and ultimately sought treatment from a campus therapist, Dr. Lawrence Moore. Over the course of their sessions, Poddar stated that while Tarasoff was away, he intended to kill Tarasoff when she returned to California. Dr. Moore notified campus police who briefly detained, then released Poddar whom they deemed was "thinking rationally," and he promised to stay away from Tarasoff. Following direction from his superior, Dr. Moore did not seek further action to commit the patient and no action was taken to warn Tatiana Tarasoff or her family of Poddar's stated intention to kill Tarasoff. On October 27, 1969, Poddar went to Tarasoff's house requesting to speak with her. She was not home, and her mother requested Poddar leave. Poddar left and returned later the same evening when he again requested to speak with her. He then attacked her and shot her with a pellet gun. When she attempted to run away, he chased her and fatally stabbed her with a knife [21].

While on trial, Poddar claimed insanity attempting to lessen the charges and potential sentencing. Recall that legally, this claim is an exception allowing the release of privileged information between Dr. Moore and Poddar in a court of law. Additionally, Tarasoff's parents sued Dr. Moore, campus police, and their employer, The University of California, claiming that they, too, should be held accountable for the death of their daughter as the risk of harm to Tatiana was known but no one had warned Tatiana or her family. The case ultimately landed in the California Supreme Court, which put forth two decisions. Initially in 1974, the defendants were all considered immune from liability, but they still admonished that psychiatrists have a "duty to warn." After a significant outcry at the amount of responsibility placed on mental health clinicians, the court revised and republished their decision in 1976. The decision then was that if mental health clinicians in their professional judgment believe their patient is an imminent risk of harm to themselves or others, then the clinician bears a duty to exercise reasonable care to protect the foreseeable victim from that danger [19, 20]. It was out of this landmark case that individual states and the psychiatric community began the ethical and legal debates on the tension of confidentiality and duty to warn, treat and protect.

Answering the questions above may first begin with reviewing laws in the state or region of clinical practice. Beyond these definitions, a clinical impression of the patient's mental health status and dangerousness as well as ability to identity or reasonably discover the potential victim's identity may be critical. Many court decisions have changed the scope and definitions of duty to warn such that these vary by state. It may be of use to review the above discussion with the Maryland definitions for reference.

4   Summary

In this chapter, we have reviewed legal and ethical concerns in mental healthcare practice that involve confidentiality, privilege, and the duty to warn. We have used suicide and domestic violence as paradigmatic cases. We have emphasized the law's limitations, particularly at the margins of cases lying along the same legal spectrums where optimal outcomes may be less clear. Here, we have urged that there is a place, legally and ethically, for providers to exercise their best clinical judgement. We have also urged that in all cases, whenever possible, they share decision-making with their patients. This sharing ideally may involve not only the medical options patients can choose but the legal and ethical considerations that affect these decisions. Clinicians responding reflexively at these edges may give suboptimal and even harmful care if they rigidly apply literal interpretations of the law, as clinical contexts plays a large role in patient outcomes. Thus, clinicians, especially those in mental health, must assess their knowledge of the black letter laws for which they may be held accountable and extend their ethical considerations to the fullest extent as they weigh both the possible outcomes and ethical pros and cons. The above case example involving suicide demonstrates the possible benefit and risk of breaching confidentiality. Where the law almost always favors confidentiality, even the HIPAA Privacy Rule outlines exceptions pertaining to optimizing patient safety, thus making clinical judgement all the more important.

Patients may still have other needs most important to them, like keeping their present emotional distress confidential. The risk that they will take their lives can be sufficiently reduced to a preferable clinically and legally permissible extent in other ways [16, 18]. Clinicians may then, as opposed to hospitalizing these patients invariably thereby revealing health information to their family, consider alternative plans which may be more acceptable to both the clinician and the patient. For example, clinicians may arrange more frequent contact with the patient such that their safety may be at least almost as closely maintained as when they are an inpatient, while benefiting these patients more and respecting their autonomy by also preserving their confidentiality. Long term maintenance of clinician rapport in a therapeutic relationship may be the most significant factor in patients' long term outcomes, increasing their positive outcomes and minimizing their harm [18]. Patients who are chronically suicidal may do better over time if they are not so often hospitalized. Courts have indicated their understanding and have accepted the clinical wisdom in this clinical judgement not to hospitalize these patients to maximize their long-term outcome. Providers should, however, carefully document in the patient's chart their rationale for giving this long-term benefit priority over the patient's isolated short-term risk.

The Tarasoff Case demonstrates early challenges in domestic violence law touching subtly on the legal navigation of privilege as a right for patients in their professional relationship with mental health clinicians. However, exceptions to these state laws suggest that privilege no longer applies when the patient utilizes a mental health claim as part of their defense in legal proceedings, as Mr. Poddar did in his own criminal prosecution for killing Tatiana Tarasoff. Of note, privilege in the clinician-patient relationship came to be legally defined in 1996, long after the Tarasoff case.

Most notably, the landmark Tarasoff cases catapulted the mental health community and legal experts to define for clinicians the scope of their legal obligations to treat patients at risk of harming others as well as the clinician's obligation to warn and protect. To this day, these definitions and the extent to which clinicians are obligated to warn varies from state to state on a spectrum from strict confidence to the fullest extent of duty to treat the patient, warn local law enforcement, victim(s) and/ or family, and protect potential victim(s).

The thrust of this chapter illustrates that when patients have needs involving confidentiality, clinicians should do more than just look to the most applicable law. They must additionally consider key ethical nuances in cases in which laws may be at tension with one another or with treating their patient optimally. Ethically and legally, these decisions may have uncertain answers. The clinical rationale for care may be most likely to hold up in a court of law when the clinician, though knowledgeable about the legal limits of confidentiality in mental health that are at stake also discusses and attempts to optimize when possible the ethical pros and cons of the patient's health outcomes and safety.

Acknowledgements This work was prepared by a military or civilian employee of the US Government as part of the individual's official duties and therefore is in the public domain and does not possess copyright protection (public domain information may be freely distributed and copied; however, as a courtesy it is requested that the Uniformed Services University and the author be given an appropriate acknowledgement.

The opinions and assertions expressed herein are those of the author(s) and do not reflect the official policy or position of the Uniformed Services University of the Health Sciences or the Department of Defense.

References

1. Rothstein MA. The Hippocratic bargain and health information technology. J Law Med Ethics. 2010 Spring;38(1):7–13.
2. 45 CFR 164.510(b)
3. U.S. Department of Health and Human Services. HIPAA Privacy Rule and Sharing Information Related to Mental Health [Internet]. HHS.GOV. Available from: https://www.hhs.gov/sites/default/files/hipaa-privacy-rule-and-sharing-info-related-to-mental-health.pdf
4. 45 CFR 164.512(j)

5. 45 CFR 164.501.
6. 45 CFR 164.524(a)(1)(i)
7. 45 CFR 164.524(a)(2)(v)
8. Md. Code Ann., Fam. Law § 5-704
9. 42 CFR 2.11
10. 42 USC § 290dd–2
11. Md. Code Ann., Cts. & Jud. Proc. § 9-109
12. Gorshkalova O, Munakomi S. Duty to warn. In: StatPearls. Treasure Island (FL): StatPearls Publishing; 2020
13. Werth JL, Welfel ER, Benjamin GAH. The Duty to Protect: Ethical, Legal, and Professional Considerations for Mental Health Professionals. First. American Psychological Association; 2009.
14. Thapar v. Zezulka, 994 S.W.2d 635 (Tex. 1999)
15. Md. Code Ann., Cts. & Jud. Proc §5-609
16. Goldsmith SK, Pellmar TC, Kleinman AM, Bunney WE, editors. Reducing Suicide: A National Imperative. In: Society and Culture. Washington, D.C., DC: National Academies Press; 2002.
17. Fisher MA. The ethics of conditional confidentiality. Oxford University Press; 2015.
18. Garvey KA, Penn JV, Campbell AL, Esposito-Smythers C, Spirito A. Contracting for safety with patients: clinical practice and forensic implications. J Am Acad Psychiatry Law. 2009;37(3):363–70.
19. Tarasoff v. Regents of the U. California, 551 P. 2d 334, 1976
20. SCOCAL, Tarasoff v. Regents of University of California, 17 Cal.3d 425 available at: (https://scocal.stanford.edu/opinion/tarasoff-v-regents-university-california-30278)
21. Case study - The Case of Tarasoff [Internet]. Practicalbioethics.org. Available from: https://practicalbioethics.org/case-studies-study-guide-the-case-of-tarasoff.html

Informed Consent

Marc D. Ginsberg

Contents

Key Points
- Law of informed consent is an integral component of the physician-patient relationship.
- Informed consent focuses on patient autonomy—the patient's right to informed treatment decisions.
- Informed consent contemplates a required physician disclosure of the risks, benefits and complications of proposed treatment, and of alternatives to recommended treatment.
- Physicians may be obligated to also disclose personal matters which may increase patient risk.
- Physicians should make proper chart notations indicating that the required disclosure occurred and the patient agreed.

Legal Concepts
- Informed Consent
- Patient Autonomy
- Mandatory Disclosure
- Professional Based Model

M. D. Ginsberg (✉)
University of Illinois—Chicago School of Law, Chicago, IL, USA
e-mail: mgins@uic.edu

- Patient Based Model
- Evidence of Informed Consent

1   Introduction

The doctrine of informed consent is an integral component of the law of medicine. At its foundation is patient autonomy [1, 2]. Of course, unconsented treatment may result in tort liability [3]. Unconsented treatment constitutes a battery, which is an intentional tort. A claim based on lack of informed consent sounds in negligence, which implicates the applicable standard of care—what a reasonably well qualified physician would do or not do in the same or similar circumstances.

Informed consent may have originated in ancient times [4]. Insofar as American law is concerned, the concept was first used in a published judicial opinion by a California court in 1957 [5].

"The doctrine of informed consent requires physicians to disclose to patients (without having been asked by the patient) the risks and benefits of, and alternatives to, proposed treatment. The doctrine may be based on common law or a statute" [6]. That informed consent is best understood as a disclosure doctrine is evidenced by the critical footnote 36 contained in *Canterbury v. Spence* [7], which states:

> We discard the thought that the patient should ask for information before the physician is required to disclose. Caveat emptor is not the norm for the consumer of medical services. Duty to disclose is more than a call to speak merely on the patient's request, or merely to answer the patient's questions; it is a duty to volunteer, if necessary, the information the patient needs for intelligent decision, the patient may be ignorant, confused, overawed by the physician or frightened by the hospital, or even ashamed to inquire. Perhaps relatively few patients could in any event identify the relevant questions in the absence of prior explanation by the physician. Physicians and hospitals have patients of widely divergent socio-economic backgrounds, and a rule which presumes a high degree of sophistication which many members of society lack is likely to breed gross inequities.

Therefore, in contrast to shared decision making, even though medicine has also considered informed consent as an interactive, communicative process [2], unless so recognized by the common law or statute of a particular jurisdiction, it is not. It contemplates a one-way disclosure, from physician to patient. The disclosure is required even if the patient does not interact with the physician by questions or conversation.

2   Discussion

2.1   The Scope of Disclosure

The scope of the informed consent disclosure is jurisdictionally dependent. Certainly, the physician is obligated to disclose the risks, benefits, complications of, and alternatives to a proposed treatment but the actual disclosure contents may depend on the informed consent model adopted in a given jurisdiction. The physician-based or professional model requires the disclosure which would be made by a reasonably well qualified physician under the same or similar circumstances [2]. The patient-based model contemplates the disclosure of that information which a reasonable patient would want to know [2]. Under either model, it is unlikely that the physician would be required to disclose risks which are extraordinarily rare, as that disclosure could discourage a patient from accepting a recommended treatment, placing the patient at greater risk than would have existed had the recommended treatment been received. Variations on these models may exist in various jurisdictions [8].

2.2   Specifics of the Disclosure: To Disclose or Not to Disclose

In addition to the disclosure of risks, benefits and complications of, and alternatives to a proposed treatment, other topics for potential disclosure have been considered by courts. Again, it is necessary to become aware of the requirements in a specific jurisdiction.

1. Physician's Financial Research Interest In Patient Treatment: pursuant to *Moore v. Regents of Univ. of Calif.* [6], this must be disclosed. It constitutes a classic conflict of interest, potentially elevating the physician's financial interest over the patient's treatment interest.
2. The Differential Diagnosis: controversial, but should not be a required disclosure due to time involved and potential risk to patient [3, 9]. The disclosure should relate to the actual diagnosis arrived at by the physician.
3. Physician Experience: should be disclosed if inexperience creates risk to patient [3, 6].
4. Physician Disability & Health: should be disclosed if physician's condition creates risk to patient [3, 6].
5. Physician Disciplinary History: should be disclosed if staff privileges or medical license discipline relates to physician providing standard of care medicine [3, 6].
6. Physician Qualifications & Training: should be disclosed if creates risk to the patient [6].
7. Participation Of Trainees, Non-Physicians In Patient Care: participation of residents, fellows in academic medical centers should be assumed by patients. Other participants should be disclosed [3, 6].
8. FDA Status Of Drug or Medical Device: off-label use of FDA approved drug or device does not require disclosure [6].
9. Statistics Regarding Procedure Success Rate, Life Expectancy, Mortality: disclosure not required as these statistics do not relate to a specific patient [6].
10. Lack of Professional Liability Insurance: disclosure not required as no increased risk to patient. Issue is whether the insured physician will be more or less careful. Physician must check state law regarding required notice to all patients [3].

11. Detail Of Medical Procedure: general description should be sufficient. Patient cannot understand technical detail [3].

12. Sophisticated Care Facility: physician not required to disclose that there may be a better physician or facility for treatment. Theoretically, there is always a superior physician or facility elsewhere. Of course, if the patient needs to be referred due to a recognized risk of a non-referral, the disclosure should be required [3].

2.3   Delegability of Disclosure

It is of the utmost importance to understand that the physician's informed consent required disclosures are not delegable to other allied health professionals or administrative personnel with whom the physician works [3]. The treating physician is responsible for obtaining the patient's informed consent.

2.4   The Referring Physician & Informed Consent

There is some judicial authority suggesting that a referring physician may be obligated to obtain a patient's informed consent for treatment to be performed by another physician if the referring physician co-manages the patient [3]. Of course, this would not eliminate the need for the treating physician to obtain the patient's informed consent, as well. The problem here is the possibility of inconsistent disclosures as to the risks, benefits, complications of and alternatives to treatment.

2.5   Causation Requirement

"Causation is established in an informed consent case if the plaintiff can prove a link between the failure of a doctor to disclose and the patient's injury—first that the risk not disclosed in fact materialized, and second, that a patient would have declined treatment if [the patient] had received full information about that risk" [3]. The possibility of an informed consent claim in the absence of an injury has been addressed [3, 10] but is not conventional wisdom.

2.6   Consent Forms

Patients, particularly in a hospital setting, are routinely requested/required to sign consent forms. The executed consent form is _not_ informed consent. It is simply evidence of consent obtained following the required physician disclosure [2, 11].

2.7   Obtaining Informed Consent from Patients Incapable of Giving Informed Consent to Treatment

1. Pediatric Patients: Informed consent must be obtained from the patient's parent. The American Academy of Pediatrics has published an excellent paper on informed consent in the pediatric practice, including patient assent to treatment [12].
2. Patients with Legally Appointed Guardians: Legal guardianship is determined by state law. The physician should confirm that the guardian has the authority to consent to medical treatment on behalf of the patient [13].

3   Application

A brief hypothetical should serve to demonstrate the informed consent process. A patient requiring an appendectomy is seen by the general surgeon pre-operatively. The surgeon explains the patient's condition. Assume that the patient asks no questions and is not interested in a conversation with the surgeon. The surgeon is obligated to disclose the risks, benefits and complications of the planned appendectomy. The surgeon is similarly required to disclose any alternatives to surgery. Additionally, should there be any factors personal to the surgeon which would increase risk to the patient, those factors must be disclosed to the patient. An excellent example of a case concerning a required disclosure of a factor personal to a surgeon is *Goldberg v. Boone* [14]. Here, the Court of Appeals of Maryland focused on the inexperience of a surgeon, holding that a combination of the patient's condition and the surgeon's inexperience, would require the surgeon to disclose his lack of experience to the patient. The required disclosure recognizes that physician inexperience may constitute a risk to the patient about which the patient should be informed. The surgeon should be certain to enter a chart note confirming that the informed consent disclosure occurred and the patient understood.

4   Summary

The doctrine of informed consent is a critical component of the practice of medicine. Informed consent, by law, is a disclosure doctrine, unless the law of a particular jurisdiction provides that the informed consent process includes a dialog with the patient, parent or guardian. Patient understanding is difficult to gauge due to general literacy, health literacy and the fact that some patients simply prefer that the physician make the treatment choices based on the physician's experience and judgment.

In addition to the proper informed consent disclosure, the physician must be certain to memorialize that the patient's consent was obtained, through a chart notation. The detail of the disclosure need not be included in the chart notation.

The notation should, at least, state that the risks, benefits and complications of, and alternatives to, treatment were explained and the patient agreed with the treatment recommendation. The chart notation is evidence of informed consent.

Acknowledgements The author acknowledges the assistance of Ms. Christine Cotter, Executive Assistant for Faculty Administrative Support, who typed the manuscript for this chapter. The author also acknowledges Mr. Hudson Cross, his Research Assistant, for his proof reading and citation checking efforts.

References

1. Pope, Thaddeus M., *Certified Patient Decision Aids: Solving Persistent Problems With Informed Consent Law*, 45 Jour. Law, Med. & Ethics 12 (2017) (see references cited therein).
2. Ginsberg, Marc, *Beyond Canterbury: Can Medicine And Law Agree About Informed Consent? And Does It Matter?*, 45 Jour. Law, Med. & Ethics 106 (2017) (see references cited therein).
3. Ginsberg, Marc, *Informed Consent: No Longer Just What The Doctor Ordered? Revisited*, 52 Akron L. Rev. 49, 52 (2018) (see references cited therein).
4. Dalla-Vorgia, P. et al, *Is Consent In Medicine A Concept Only Of Modern Times?*, 27 J. Med. Ethics 59 (2001) (see references cited therein).
5. *Salgo v. Leland Stanford Jr. Univ.*, 317 P.2d 170 (Cal. Ct. App. 1957).
6. Ginsberg, Marc, *Informed Consent: No Longer Just What The Doctor Ordered? The "Contributions" Of Medical Associations And Courts To A More Patient Friendly Doctrine*, 15 MSU Jour. Med. & Law 17, 19–20 (2010) (see references cited therein).
7. 464 F.2d 772, 783 (D.C. Cir. 1972).
8. *See*, for example, *Scott v. Bradford*, 606 P.2d 554 (Okla. 1979) (adopts doctrine of informed consent through a model, apparently, more patient friendly than that announced in *Canterbury v. Spence*).
9. Ginsberg, Marc, *Informed Consent And The Differential Diagnosis: How The Law Can Overestimate Patient Autonomy And Compromise Health Care*, 60 Wayne L. Rev. 349 (2014) (see references cited therein).
10. *Looney v. Moore*, 861 F.3d 1303 (11th Cir. 2017).
11. Meisel, A. & Kuczewski, M., *Legal And Ethical Myths About Informed Consent*, 156 Arch. Int. Med. 2521, 2522 (1996) (see references cited therein).
12. Katz, A. et al, *Informed Consent In Decision-Making In Pediatric Practice*, 138 Pediatrics (2016) (see references cited therein).
13. Dreher, G., *Is This Patient Really Incompetent?*, 71 Am. Fam. Phys. 198 (2005) (see references cited therein).
14. 912 A.2d 698 (Md. 2006).

Right to Decisional Privacy

B. Jessie Hill

Contents

Key Points
- The U.S. Supreme Court has recognized a right to decisional privacy.
- The right to decisional privacy encompasses the freedom to make certain decisions without undue governmental interference, including some reproductive healthcare decisions and whether to refuse medical treatment.
- It is not clear whether the right to decisional privacy applies beyond the contexts specifically identified by the U.S. Supreme Court, and it is not always clear what test a court should apply in evaluating a decisional privacy claim.
- The right to decisional privacy is more limited for minors and mentally incompetent adults than for mentally competent adults. In addition, important government interests, such as public health needs, may override an individual's right to decisional privacy.

Legal Concepts
- Right to decisional privacy: The fundamental constitutional right to make certain personal decisions, including certain healthcare decisions, without undue governmental interference.

B. J. Hill (✉)
School of Law, Case Western Reserve University, Cleveland, OH, USA
e-mail: bjh11@case.edu

1   Introduction

The Fourteenth Amendment to the U.S. Constitution protects a right to decisional privacy. Though its contours are not clearly defined, the right to decisional privacy is understood to protect individuals' autonomy in making certain deeply personal and important decisions about their bodies and their intimate lives. In the healthcare context, this largely encompasses the right to make certain reproductive healthcare decisions without undue interference from the state, as those decisions touch directly upon the private realm of family life. The right has also been extended to other healthcare contexts, such as the right to refuse medical treatment. The application of the decisional privacy right to other healthcare contexts is less clear, however.

The right to decisional privacy in the healthcare context is subject to certain limitations. In particular, because the ability to make informed, autonomous decisions undergirds this right, minors and mentally incompetent individuals do not possess this right to the same extent as fully competent adults. In addition, public health needs may override the right to decisional privacy in some circumstances.

2   Discussion

The Fourteenth Amendment to the U.S. Constitution, which states that "[n]o State shall … deprive any person of life, liberty, or property, without due process of law," acts as a constraint on laws and other government actions [1]. The U.S. Supreme Court has recognized that the government deprives individuals of their liberty without due process of law when it interferes with their private decisionmaking without a sufficiently strong reason to do so [2]. This right to make private decisions without unwarranted government interference is referred to as the "right to privacy" [3]. Because the right to privacy (which is here referred to as the "right to decisional privacy") derives from the Constitution, it supersedes any state or federal laws that conflict with it.

The U.S. Supreme Court first recognized a constitutional right to privacy in the 1965 case *Griswold v. Connecticut*. In that case, the Court decided that a Connecticut law banning the use of contraceptive drugs and devices was unconstitutional. In so reasoning, the Court did not rely on the Fourteenth Amendment; rather, it found that a right to privacy could be inferred from numerous constitutional provisions, including the First Amendment protection for intimate personal associations; the Third and Fourth Amendment protections for the sanctity of the home; and the Fifth Amendment protection for information that individuals wish to keep private from the government [4]. In *Griswold*, unlike in later cases, the Court was particularly focused on the effects that the law would have on married couples and the "sacred" marital relationship, rather than the right of individuals to make private decisions.

Seven years later, in *Eisenstadt v. Baird*, the Supreme Court revisited the issue of access to contraception in a case involving distribution of contraceptive drugs and devices. The Court moved away from a right to privacy grounded in the marital relationship and instead proclaimed, "If the right of privacy means anything, it is the right of the *individual*, married or single, to be free from unwarranted governmental intrusion into matters so fundamentally affecting a person as the decision whether to bear or beget a child" [5]. Thus, the right to decisional privacy as enunciated in *Baird* probably includes not only the right to use contraception but also the right not to be sterilized by the government against one's will, which the Supreme Court had declared a fundamental human right in a much earlier case, *Skinner v. Oklahoma* [6]. But the full significance of the Court's statement in *Baird* was realized in *Roe v. Wade*, the case in which the Court first recognized that the right to privacy was a fundamental right grounded in the Fourteenth Amendment to the U.S. Constitution and that it was broad enough to encompass the decision whether to terminate a pregnancy [2].

The Supreme Court subsequently drew an explicit connection between the right to decisional privacy in the abortion context and the right to privacy in other contexts in a landmark 1992 Supreme Court case, *Planned Parenthood of Southeastern Pennsylvania v. Casey*. The *Casey* Court recognized the patient as the rightsholder and acknowledged the centrality of reproductive freedom to gender equality. In addition, the Court in *Casey* connected the right to decisional privacy identified in *Roe* and related cases involving contraception with the freedom to make other healthcare decisions, such as rejecting unwanted medical treatment. The Court thus explained that the right to decisional privacy involved more than decisions about childbearing: it is also "a rule … of personal autonomy and bodily integrity," with an "affinity to cases recognizing limits on governmental power to mandate medical treatment or to bar its rejection" [7].

Casey also introduced a unique legal standard to be applied to abortion restrictions—the "undue burden" standard—which was generally understood to be less protective of abortion rights than the standard set forth in *Roe v. Wade*. Ultimately, however, in the 2022 case *Dobbs v. Jackson Women's Health Organization*, the U.S. Supreme Court overruled both *Roe* and *Casey*, holding that there was no federal constitutional right to terminate a pregnancy [8]. The Court went to great lengths to explain why abortion was different from other decisional privacy rights in that it involved the termination of a potential life. However, the Court's reasons for rejecting a right to abortion were based primarily on the absence of any such right from the text of the Constitution and the failure of courts and governments in the U.S. to recognize such a right in the eighteenth and nineteenth centuries [8]. Thus, it is an open question whether related rights in the context of reproductive decision-making—particularly the right to use contraception—will remain protected or will similarly fall due to a lack of textual and historical support.

The line of cases recognizing the constitutional right to refuse unwanted medical treatment as an aspect of the decisional privacy right begins with *Cruzan v. Director, Missouri Department of Health* in 1990. That case centered on Nancy Cruzan, a young woman who had been in a car accident and ended up in a persistent vegetative state. The issue in the case was whether Missouri law was constitutional insofar as it prevented Cruzan's parents from terminating her life support absent "clear and convincing evidence" that Cruzan would have chosen this outcome if she were competent to make her own medical decisions. The Supreme Court held that Missouri's requirement for surrogate decisionmaking on behalf of incompetent patients was constitutional, but it also recognized that there is a "liberty interest," possessed by patients capable of autonomous choice, to refuse even life-saving medical treatment. The Court grounded this right, like the right to use contraception, in the Due Process Clause of the Fourteenth Amendment. However, in Cruzan's case, because she was incompetent to make her own decisions, her right was outweighed by the state's interest in avoiding error and abuse in such surrogate decisionmaking situations [9].

The Court applied similar reasoning in cases involving forced administration of antipsychotic drugs to detained and incarcerated individuals, deciding that such treatment was constitutionally acceptable only if "the inmate is dangerous to himself or others and the treatment is in the inmate's medical interest"—in other words, if the state's interest in protecting the individual and others outweighs the individual's interest in decisional autonomy [10]. In the case of a criminal defendant who has not yet been convicted, the Court has held, less intrusive alternatives to forced medication must also be considered, and the treatment generally may be administered involuntarily only if *necessary* to ensure the safety of the defendant or others [11]. In addition, medication may be administered involuntarily to render a defendant competent to stand trial, "but only if the treatment is medically appropriate, is substantially unlikely to have side effects that may undermine the fairness of the trial, and, taking account of less intrusive alternatives, is necessary significantly to further important governmental trial-related interests," as in case of a serious crime [12].

As demonstrated by the Supreme Court's varying approaches to the right to decisional privacy in different contexts, it is not always entirely clear what rules a court should apply when an individual claims a violation of her right to decisional privacy. As a general matter, when a recognized liberty interest (such as the right to use contraception or the right of a competent person to refuse medical treatment) is infringed, the individual's right to decisional privacy must be balanced against the governmental interests that are advanced by the challenged law. In addition, the right to terminate a pregnancy is no longer a recognized liberty interest within this framework.

There are several important limitations on the right of decisional privacy in the medical treatment context. One important limitation is that, while competent adults possess a right to refuse medical treatment, they are not constitutionally entitled to seek the assistance of a physician in order to end their lives, nor do physicians possess a constitutional right to provide that assistance. In this regard, the Supreme Court has drawn a distinction between a right to refuse treatment and a right to seek it, noting that the right to act affirmatively to end one's life does not share the long historical and legal pedigree associated with the right to avoid unwanted bodily intrusions [13]. Nonetheless, although the Supreme Court has said that the Fourteenth Amendment does not *require* states to permit medically assisted suicide, nothing in the U.S. Constitution *forbids* states to grant greater autonomy rights to patients by legalizing the practice of medical assistance in dying. Indeed, nine states plus the District of Columbia have adopted laws granting to patients a right to seek medical assistance in ending their lives. In addition, some commentators have noted that the practice of administering palliative medication in an amount sufficient to ease the patient's suffering, even if it hastens the patient's death, may be protected at the federal level by the Supreme Court's decisions [14].

Another important limit on decisional privacy is that the right to refuse unwanted medical treatment may be overridden by important public health needs. Indeed, a 1905 Supreme Court case called *Jacobson v. Commonwealth of Massachusetts* upheld a state law providing for compulsory smallpox vaccination [15]. The scope of the public-health limitation on the decisional privacy right is not well delineated, however. Early in the COVID-19 pandemic and prior to the *Dobbs* decision, for example, while courts still recognized a constitutional right to abortion, they struggled to articulate the line between individual liberty rights and the state's imperative to protect the public health in cases involving state orders mandating the closure of abortion clinics, along with other businesses and gathering places [16].

Finally, those who are not mentally competent to make their own decisions are not entitled to the same degree of autonomy as fully competent adults. Children and intellectually disabled individuals may fall into this category. Indeed, the Supreme Court has often noted that minors' constitutional rights are not necessarily coextensive with those of adults [17]. The Supreme Court has not set forth clear rules regarding the reproductive rights of intellectually disabled individuals, and state laws on this subject vary widely. In addition, *Cruzan* strongly suggests that states may limit the autonomy rights of such in dividuals when asserting an interest in protecting them from harm or abuse.

Beyond the contexts described here, it is not clear to what extent the right to decisional privacy applies to medical decisionmaking outside the realms of reproductive decisions and the right to refuse unwanted treatment. Courts have occasionally entertained claims from individuals—particularly, those seeking access to unapproved treatments—but have by and large rejected those claims [18].

3   Application

Clinical vignette: Jane Doe is a pregnant woman who suddenly collapsed at home, likely due to a pulmonary embolism. In an unconscious state, she is rushed to the hospital and connected to a ventilator. It soon becomes apparent that Doe will not be able to breathe on her own and requires the ventilator for life support. It also becomes apparent that she will not regain consciousness and her medical condition will not improve. Doe has an advance directive, which states that she would like to be removed from life support in such a situation.

However, Doe lives in one of the thirty-six states that have a so-called "pregnancy exclusion" law, which overrides or limits advance directives when a patient is pregnant and requires that the pregnant woman's life be sustained in order to attempt to save the fetus. If she were mentally competent and not pregnant, Jane would have a constitutional right to refuse life-sustaining treatment. Because she is not mentally competent, however, the state's interests in preserving the life of the fetus and in ensuring the decision reflects Doe's wishes must be balanced against Doe's decisional privacy interest. In addition, since Doe no longer has a constitutional right to terminate her pregnancy, the state's interest in the potential life of the fetus might be considered to be more compelling than her own interest in refusing unwanted care. The outcome would be uncertain if the case were to be litigated.

Case 1: Abigail Alliance for Better Access to Developmental Drugs v. von Eschenbach, 495 F.3d 695 (D.C. Cir. 2007) The Abigail Alliance for Better Access to Developmental Drugs is an organization that advocates on behalf of terminally ill patients for broader access to experimental drugs, not yet approved by the Food & Drug Administration (FDA).

On behalf of terminally ill, mentally competent patients who had no other treatment options, the Abigail Alliance sought access to experimental, unapproved drugs outside the fairly limited regulatory channels by which some patients can acquire those drugs. The plaintiffs' theory was that the risk-benefit calculation is different for terminally ill patients. For them, the delay caused by the lengthy approval process could be a "death sentence." At the same time, they were less concerned about the potential safety risks of the drugs, because they were facing death regardless and had no suitable treatment alternatives. They thus asserted a fundamental constitutional right of terminally ill individuals to make an informed decision to access potentially life-saving drugs that had passed only the first phase of clinical testing, when no approved alternative treatment is available.

The trial court decided that there was no such constitutional right. It noted that no Supreme Court case had recognized a right of access in these circumstances. It also rejected the Abigail Alliance's argument that the Supreme Court's recognition of a right to refuse life-saving treatment in *Cruzan* necessarily implied a "complementary right" to access potentially life-saving treatment [19]. On appeal, a panel of the D.C. Circuit Court of Appeals reversed the trial court and held that both the long-standing right of self-preservation and the line of cases vindicating an individual's liberty interest in decisional privacy supported a "right of terminally ill patients to make an informed decision that may prolong life, specifically by use of potentially life-saving new drugs that the FDA has yet to approve for commercial marketing but that the FDA has determined, after Phase I clinical human trials, are safe enough for further testing on a substantial number of human beings" [19].

After rehearing by the full D.C. Circuit Court of Appeals, however, the court changed course and upheld the trial court's rejection of the Abigail Alliance's claim [18]. That court concluded, like the trial court, that no such right had been recognized, explicitly or implicitly, as an aspect of the right to decisional privacy found in *Cruzan* or any other case. Rather, the court noted that the U.S. had long regulated drugs to ensure their safety and efficacy, and that the principle identified in *Cruzan* "protecting individual *freedom* from life-saving, but forced, medical treatment does not evidence a constitutional tradition of providing affirmative *access* to a potentially harmful, and even fatal, commercial good."

If the court had recognized a constitutional right to have pre-approval access to potentially life-saving experimental drugs, it would have still been possible for the court to decide that the right was outweighed by important governmental interests. For example, the court might have concluded the terminally ill patients' right was outweighed by the need to retain an incentive for individuals to participate in clinical trials; if there were an alternate route to accessing unapproved drugs, it would eliminate any incentive for participation in clinical trials. Because the court did not find that the right to decisional privacy was implicated in this case, however, it did not need to decide whether the government's interests in drug regulation outweighed the individuals' rights.

Case 2: United States v. Cannabis Cultivators Club, 5 F. Supp. 2d 1086, 1102 (N.D. Cal. 1998) In *Cannabis Cultivators Club*—a case decided well before widespread legalization of marijuana in various states across the United States—the federal government had sought to enforce the Controlled Substances Act's prohibition on distributing or manufacturing marijuana against dispensaries in California that provided cannabis to patients whose doctors recommended it in compliance with a California state law (the Compassionate Use Act of 1996). Among other claims, the dispensaries asserted that patients had a constitutional right "to be free from unnecessary pain, to receive palliative treatment for a painful medical condition, to care for oneself, and to preserve one's own life" by using medical marijuana [20]. The District Court for the Northern District of California ultimately decided that no such "fundamental right to obtain the medication of choice" existed and that the federal law could be enforced [21]. In reaching this decision, the court distinguished between the established right to refuse medical treatment recognized in *Cruzan* and the non-existent right to require the government "to set aside its regulations in order to provide dying patients access to experimental medical treatments" [21, 22].

4   Summary

The constitutional right to decisional privacy is well-established in U.S. law, but its contours are not well-defined. It encompasses the freedom to make certain personal decisions—particularly, certain reproductive health decisions and decisions about refusing medical care—autonomously and without undue interference from the government. It is not clear whether the right has applications beyond those already identified by the Supreme Court, however, and it is not always evident how courts will evaluate new claims asserting a right to decisional privacy. It is certain, however, that the right is more limited in the case of minors and mentally incompetent adults. In addition, the right does not encompass a right to medical assistance in ending one's life, and it may be overridden by important government interests, particularly in public health.

References

1. U.S. Constitution amdt. XIV.
2. Roe v. Wade, 410 U.S. 113 (1973).
3. Greene J. The So-Called Right to Privacy. U.C. Davis L. Rev. 2010;43:715–47.
4. Griswold v. Connecticut, 381 U.S. 479 (1965).

5. Eisenstadt v. Baird, 405 U.S. 438 (1972).
6. Skinner v. Oklahoma, 316 U.S. 535 (1942).
7. Planned Parenthood v. Casey, 505 U.S. 833 (1992).
8. Dobbs v. Jackson Women's Health Organization, No. 19-1392, U.S. (2022).
9. Cruzan by Cruzan v. Dir., Missouri Dep't of Health, 497 U.S. 261 (1990).
10. Washington v. Harper, 494 U.S. 210 (1990).
11. Riggins v. Nevada, 504 U.S. 127 (1992).
12. Sell v US, 539 U.S. 166 (2003).
13. Washington v. Glucksberg, 521 U.S. 702 (1997).
14. Kamisar Y. On the Meaning and Impact of the Physician-Assisted Suicide Cases. Minn. L. Rev. 1998;82:895-922.
15. Jacobson v. Commonwealth of Massachusetts, 197 U.S. 11 (1905).
16. Parmet WE. Rediscovering *Jacobson* in the Era of Covid-19. B.U. L. Rev. Online. 2020;100:117–33. https://www.bu.edu/bulawreview/files/2020/07/PARMET.pdf. Accessed 7 June 2021.
17. Dailey AC. Children's Constitutional Rights. Minn. L. Rev. 2011;95:2099–79.
18. Abigail Alliance for Better Access to Developmental Drugs v. von Eschenbach, 495 F.3d 695 (D.C. Cir. 2007).
19. Abigail Alliance For Better Access to Developmental Drugs v. von Eschenbach, 445 F.3d 470 (D.C. Cir. 2006).
20. United States v. Cannabis Cultivators Club, 5 F. Supp. 2d 1086 (N.D. Cal. 1998).
21. United States v. Cannabis Cultivator's Club, No. C 98-00085 CRB, 1999 WL 111893 (N.D. Cal. Feb. 25, 1999).
22. Preterm v. McCloud, 994 F.3d 512 (6th Cir. 2021).

Human Subjects Research

Richard S. Saver

Contents

Key Points
- Human subjects research is regulated more strictly than regular clinical care, generally requiring prior IRB approval and specialized informed consent.
- Most clinical trials are subject to the core federal research regulations, the so-called Common Rule.
- Financial relationships between investigators and the commercial entities sponsoring the research may be subject to conflicts of interest rules, which generally require disclosure of the financial ties.
- Inadequate legal compliance can lead to clinical trial shutdowns, professional discipline, and malpractice liability.

Legal Concepts
- Research definition: "a systematic investigation, including research development, testing and evaluation, designed to develop or contribute to generalizable knowledge."
- Human subject definition: "a living individual about whom an investigator…obtains information or biospecimens through intervention or interaction with the individual…or obtains, uses, studies, analyzes, or generates identifiable private information or identifiable biospecimens."

R. S. Saver (✉)
UNC School of Law and UNC School of Medicine, Chapel Hill, NC, USA
e-mail: saver@email.unc.edu

© The Author(s), under exclusive license to Springer Nature Switzerland AG 2022   479
A. S. Pasha (ed.), *Laws of Medicine*,
https://doi.org/10.1007/978-3-031-08162-0_32

- Institutional Review Board: most studies involving human subjects must obtain prior approval from this special review committee, usually housed at a medical institution and comprised of a mix of persons with scientific expertise and other backgrounds.
- Common Rule: the core federal regulations applicable to research with human subjects conducted or funded by the federal agencies and to trials testing products for FDA approval, even if privately funded.

1   Introduction

Research with human subjects is vitally important for the advancement of medical knowledge and development of new therapies. But medical experimentation also raises numerous concerns, ranging from protecting subjects from harm and exploitation to safeguarding the integrity of the research itself from undue influence by financial and other secondary interests. Thus, human subjects research has evolved from a largely peripheral domain of medicine to become a highly regulated activity. Generally, clinical trials involving human subjects require prior approval by a research review committee, known as the Institutional Review Board (IRB). In addition, specialized consent must be obtained with volunteer subjects.

This two-pronged approach—prior IRB approval plus detailed consent—generally applies to clinical trials funded by the federal agencies as well as to privately sponsored studies testing drugs, devices, and other products for approval by the Food and Drug Administration (FDA). As for potential conflicts of interest, disclosure is required, and possible other safeguards imposed, for certain financial relationships between the investigator and the commercial entity sponsoring the research, such as when the investigator also serves as a paid consultant to the drug company whose product is being tested.

2   Discussion

2.1   Historical Background

Law was initially rather hands-off with regard to medical research involving human subjects, especially before World War II. In an era before the complex protocols and high volume of clinical trials common to the research enterprise today, experimentation was not regulated very differently from ordinary medicine. Research was more often considered an ad hoc departure from the clinicians' traditional standard of care. Although deviations from the standard of care could be considered a form of medical malpractice, there were nonetheless few reported cases subjecting clinicians to liability for research activity. On the one hand, evidence that an intervention was experimental tended to suggest that the clinician was careless or reckless. On the other hand, courts at this time seemed to tolerate some level of innovation so long as it was introduced in an incremental manner [1].

A great deal changed in the aftermath of World War II. Revelations of atrocious actions by Nazi physicians, performed under the guise of scientific research, drew public attention to the dangers of unregulated medical experimentation. The Nuremburg Military Tribunal formulated the influential Nuremberg Code to provide guidance and the Code emphasized the need for voluntary consent from subjects and minimization of suffering and injury [2]. Because the Nuremberg Code addressed research with healthy subjects, as had been performed in a large number of the Nazi experiments, many U.S. clinicians perceived it as inapplicable to their work with ill subjects who might benefit from research interventions [3]. As an alternative, the World Medical Association adopted the Declaration of Helsinki in 1964 [4]. The Declaration addressed therapeutic research with already ill subjects in addition to healthy volunteers. Similar to the Nuremberg Code, the Declaration emphasized the need for informed consent and harm minimization. Also, in 1953 the National Institutes of Health (NIH) issued the first set of guidelines calling for some form of peer review of internal NIH research studies. Meanwhile, several medical centers developed their own research review committees for clinical trials conducted within their own institutions, even though the then-current legal and ethical guidelines did not require it [5].

However, egregious research actions continued, such as the Tuskegee Syphilis Study in which poor African-American men suffering from syphilis were deliberately left untreated as part of an investigation into the natural history of the disease [6]. In reaction to ongoing public outrage, Congress in 1974 enacted the National Research Act (NRA) [7], which called for additional regulations on medical experimentation and created the National Commission for the Protection of Human Subjects of Biomedical and Behavioral Research. The Commission issued the influential Belmont Report [8], a set of ethical guidelines, still widely followed, that emphasized the importance of informed consent, systematic and prospective evaluation of risks and benefits in a study, and the equitable selection of subjects.

In 1981, both FDA and the Department of Health and Human Services (HHS) relied upon the Belmont Report when extensively revising their regulations [9]. The HHS and FDA rules were substantially similar, including following the two-pronged approach of requiring (1) specialized consent from subjects and (2) that studies obtain prior approval from a research review committee, the IRB, that carefully evaluated a proposed investigation's potential risks and benefits. Eventually, multiple other federal agencies codified rules for their own sponsored research and followed the same general framework.

This core set of regulations, known as the "Common Rule," is broadly applicable to research conducted or funded by most federal agencies, even when undertaken in private settings, and to clinical trials testing products for FDA approval. After a lengthy process, the federal agencies updated the Common Rule, effective in 2018 [10], with the revisions principally addressing improvements to the research consent process and easing the regulatory burden on IRBs.

2.2   Federal Regulations and Related Laws Today

The current HHS and FDA regulations, which include the core Common Rule, have broad sweep, affecting most research activities conducted with human subjects. The HHS regulations apply to investigations that it conducts or partially funds [11]. As a good deal of biomedical research is funded by grants from NIH, a division within HHS, this means that the HHS regulations extend to a large volume of clinical trials. Meanwhile, the FDA regulations apply to clinical trials that are testing drugs, devices, and biologics for FDA approval, including industry-funded studies and in private settings such as community physicians' offices [12]. Academic institutions that conduct research funded by an agency that has adopted the Common Rule must provide a "Federal-wide Assurance" in which they agree to follow the Common Rule for such studies. Many institutions, because of the logistical difficulty in treating research protocols within the same facility differently, voluntarily elect in these assurances to follow the Common Rule for *all* research conducted at their facilities [13]. In addition, many foundations and other private funding organizations require compliance with the research regulations for any clinical trials that they support. Thus, the Common Rule requirements extend widely.

Apart from the federal regulations, state common law further imposes certain obligations on research participants. A clinician has a broad common law duty to secure the patient's informed consent to treatment. This informed consent obligation likely applies with even more rigour in the research setting, as experimentation presents unknown risks and involves activities outside the ordinary doctor-patient relationship. For example, the investigator must balance dual loyalty, with fidelity to follow the study protocol in addition to caring for the subject. The common law malpractice system serves as additional backstop regulation of human subjects research. Experimentation conducted negligently, or with insufficient consent procedures, can expose the investigator, host institution, and even the study's sponsor to tort liability. Nonetheless, even in the modern era there have been only a relatively modest, although increasing, number of "malresearch" cases where experimental subjects have recovered for negligent conduct. This is likely because consent forms properly advised subjects about the risks involved. Also, physical harm is still a required element in the typical tort action alleging negligent research. Yet many research subjects, already ill when enrolling in a clinical trial, face difficulty demonstrating that study participation made them physically worse off.

Additionally, several states have enacted statutes addressing human experimentation. Such laws generally prohibit enrollment of certain individuals, such as involuntary patients in state mental hospitals, in clinical trials [14]. In addition, an increasing number of states have laws that require special informed consent before commencement of a genetic test and/or disclosure of the test results, which can affect how research involving genetic testing is conducted [15]. Finally, every state has a professional licensing statute governing the conduct of clinicians. Under many of these statutes, conducting medical experiments without proper subject consent or prior IRB approval could be deemed an act of "unprofessional conduct" and subject the clinician to discipline, including loss of license [16].

The consequences of noncompliance with research laws can extend beyond the investigator and more broadly to the institution as a whole where the experimentation is conducted. The agencies have authority to suspend *all* federally funded research at an institution if deficiencies identified with one study indicate larger institutional-wide problems with policies and procedures for protecting human subjects. In the past, leading academic medical centers such as Johns Hopkins University and Duke University have had all federally funded research with human subjects temporarily suspended because of subjects' injuries and IRB monitoring problems [17]. In addition, investigators and institutions face potential fraud and abuse liability for billing governmental programs like Medicare for research-related interventions. Medicare generally does not pay for experimental services except for routine costs arising in clinical trials and numerous conditions must be satisfied in order to qualify for reimbursement [18].

2.3   Key Features of the Federal Regulations

The HHS and FDA regulations, which incorporate the Common Rule, have the following key features.

2.4   What Counts as "Research"?

First, and most important, is understanding how "research" is envisioned. The Common Rule defines "research" as "a systematic investigation, including research development, testing and evaluation, designed to develop or contribute to generalizable knowledge [19]." There has been understandable confusion, given this definition, about how to classify certain activities. Clinicians regularly try new therapeutic approaches with their patients, even without following a formal protocol or plans to publish their results. Much of this "innovative" therapy falls in a gray area between standard therapy and formal experimentation [20]. Concentrating further on the regulations' use of the terms (1) *systematic evaluation* and (2) *designed to develop or contribute to generalizable knowledge* offers useful guidance, however.

For example, consideration of the clinician's intent is at least partly relevant, even if not conclusive, because it helps indicate whether the clinician designed the activity with the purpose of generating generalizable information for the larger medical community.

Application

Clinical Vignette # 1
Dr. Jones practices within an academic medical center. Following an educated hunch, he prescribes in alternating sequence two different FDA approved medications for a few of his diabetes patients. Each medication is considered standard therapy but each is usually offered alone. He is noting efficacy and side effects for the few patients receiving the sequenced medications to better inform his treatment plans with future diabetes patients as to when to consider drug sequencing.

Dr. Green, practicing at the same academic medical center, collects blood sugar levels and other outcomes data for all the hospital's diabetes patients regarding their experiences with the two different FDA approved medications. She hopes to develop a better evidence base for deciding front-line therapies and will present her findings at a medical conference.

Dr. Jones is likely not engaged in research per the Common Rule. The data gathering seems more anecdotal than systematic investigation. Dr. Jones is trying innovative prescription approaches, rather than analyzing and sharing the results to contribute to the community of medical knowledge. Dr. Green, however, is likely engaged in research. Dr. Green has a more organized plan for gathering the data and the project is designed with the goal of disseminating the knowledge to the broader medical community. If Dr. Green's work is funded in part by a federal agency, the Common Rule likely will apply. Moreover, assuming Dr. Green receives no agency funding, if the medical center where Dr. Green works has filed a general assurance to submit all research at the institution to the Common Rule, her work will need to comply with its requirements.

Certain activities are further defined in regulations as *not* research, including public health surveillance [19]. One additional area of ongoing confusion, however, is how to classify "quality improvement activities."

Application

Clinical Vignette # 2

A medical center collects data on its surgery patients, looking at how length of hospitalization and readmissions correlates with the frequency of inpatient physical therapy services offered post-surgery. It does so in the hopes of improving surgical recovery procedures and quality of care for patients post-surgery.

This is probably not research subject to the Common Rule. Each quality improvement activity must be analyzed case by case. HHS has interpreted the research regulations still to apply when a quality improvement activity is also designed to accomplish a research purpose [21]. Here, the collection of data is not designed to contribute to general medical knowledge. Further, the aim is to analyze information for enhancing clinical and administrative operations, while a secondary research purpose is lacking.

2.5   Who Is a "Human Subject?"

It might seem straightforward to know when an investigative activity involves human subjects, but the issue is actually complicated. A large volume of research involves looking at medical records or discarded specimens and does not create direct interactions between the research team and live patients. The Common Rule defines a "human subject" as "a living individual about whom an investigator...obtains information or biospecimens through intervention or interaction with the individual...or obtains, uses, studies, analyzes, or generates identifiable private information or identifiable biospecimens." [19].

Thus, interventions with live patients, such as administering an experimental drug, are easily considered research with human subjects. Likewise, chart review of identifiable medical records, testing of identifiable biospecimen, and otherwise generating patient identifiable information involves human subjects.

Application

Clinical Vignette # 3
Tissue biopsies of the lung were previously obtained from several patients for clinical diagnostic purposes. Dr. Jones is interested in performing research assays on the remaining portions of the biospecimens as part of a study about lung cancer biomarkers. The specimens will be coded so that Dr. Jones will not be able to ascertain the identity of individuals.

Patient consent for the research assays is not required. The activity is not considered research with a human subject. This vignette touches on one highly controversial issue that arose during the 2018 revisions to the Common Rule. The federal agencies initially proposed extending "human subject" to apply to investigations with *unidentified* biospecimens (such as left-over portions of blood samples) and to require each subject's consent for such research. Among other reasons, regulators were trying to prevent recurrence of troubling episodes such as the infamous story of Henrietta Lacks, an African American cervical cancer patient at Johns Hopkins Hospital, who in the 1950s unknowingly had her cells extracted by researchers for starting an indefinitely reproduceable and very profitable human cell line that was used in many subsequent research studies [22]. Requiring consent in these situations would bring greater transparency, and perhaps an equitable share in potential commercial profits, for otherwise unknowing patients. At the same time, institutions, investigators, and other stakeholders expressed concern that requiring consent from every subject for research with unidentifiable biospecimens would be extremely burdensome and impede the ability to conduct important studies without significant off-setting benefits. In response, the agencies dropped the proposal and the Common Rule still provides that investigation with unidentifiable biospecimens is generally not considered research with a human subject [23].

2.6   What Research Activity Is "Exempt" from the Regulations?

The Common Rule specifically exempts certain types of research from its requirements. The common thread to these activities is that they generally involve non-physical interactions with subjects that present minimal risk and where confidentiality remains protected. Exempt research includes the review of existing medical records if the information is recorded in a manner so that subjects cannot be identified [24].

2.7   IRB Review and Specialized Consent

Human subjects research that is not exempted must generally be approved in advance by an IRB. The IRB will review the study protocol, proposed consent forms, and subject recruitment plans. The next two sections explain the vitally important IRB review and informed consent requirements in more detail.

2.8   The Critical Role of the IRB

The IRB serves as the centerpiece in the research oversight system. Modeled after the peer review committees required by the early NIH regulations in the 1950s, IRBs initially looked very much like peer review bodies. However, the regulatory requirements have evolved so that IRBs have now have more complicated structures and monitoring functions. The IRB defies easy categorization. It has at times been compared to the jury system, to a quasi-regulatory agency, and even to a deputy sheriff of the government [5].

1. **Composition.** Institutions generally establish a "home" IRB to review clinical trials conducted at their own facilities. Each IRB must include at least five members of diverse backgrounds, with consideration given to dimensions of race, gender, and culture as well as sensitivity to community attitudes [25]. The IRB is akin to a peer review body in that its membership must include at least one person with scientific expertise. However, it also must include at least one member whose primary concerns are in nonscientific areas, such as an ethicist or lawyer. Most IRB members are drawn from the institution itself, typically medical school faculty or members of a hospital's medical staff. The regulations further provide, however, that at least one IRB member must be someone who is not affiliated with the institution in order to offer an "outsider" perspective and perhaps check institutional bias [5, 25]. Despite the requirements for one community member and at least one non-scientist, the membership of most IRBs is dominated by individuals with clinical expertise who work for the home institution. A particular research study may be beyond the expertise of the current IRB members. In such instances, the regulations permit the IRB to invite additional individuals with relevant background to assist in the review, although they may not vote as regular IRB members [25].
2. **Study Approval Criteria.** Investigators submit research protocols and proposed consent forms to the IRB for review. In order to approve, the IRB must find that the study adheres to several key requirements, including that (1) risks to subjects are minimized; (2) risks to subjects are reasonable in relation to anticipated benefits; (3) selection of subject is equitable; (4) informed consent will be sought from each subject and documented; (5) data monitoring will take place to ensure subject safety; and (6) adequate protections are imposed for confidentiality of data and privacy [26]. IRBs often condition approval on requested modifications to the proposed study plans and consent forms. The regulatory review criteria, while well intentioned, are also necessarily vague, such as the general requirements that risks be "reasonable in relation to anticipated benefits" or that subject selection be "equitable." The general lofty ends of protecting the rights and welfare of human subjects cannot be neatly spelled out in rule-like specificity. As a

result, IRBs end up making many discretionary judgments, calling upon considerations of ethics, community attitudes, and institutional policies in addition to the basic legal requirements. Accordingly, it should not surprise that there can be frequent variability between IRBs in decisions reached on similar research studies [27]. Moreover, this means that the quality of the IRB's work depends to large degree on the conscience and commitment of its members, who typically serve on a volunteer, part-time basis.

3. **Expedited Review**. The Common Rule allows for "expedited review" in limited instances for minimal risk studies. Expedited review means review only by the IRB Chair or an experienced IRB member. Examples of such studies include collection of blood samples from healthy populations and obtaining biological specimens by noninvasive means, such as clipping nails or cutting hair [28, 29].

4. **Continuing Review and Monitoring**. IRB responsibilities do not end with the initial review. They generally must also conduct continuing review of already approved research studies at least once a year, looking at ongoing issues of enrollment, adverse event reports, and other matters [30]. The 2018 revisions to the Common Rule tried to minimize IRB workload burden by eliminating the continuing review requirement for studies where new risks to subjects are less likely to develop, such as when the remaining research involves only data analysis [23, 30]. In addition, an IRB has legal authority at any time to suspend or terminate research projects that fail to adhere to its initial review requirements or where unexpected, serious harm to subjects occurs [31].

5. **Special populations.** The federal regulations generally impose additional review criteria and procedures for research involving special populations, specifically pregnant women, human fetuses, neonates, prisoners, and children [32]. In these instances, the risk/benefit ratios for an IRB to approve a study may be more stringent than the "risks are reasonable in relation to anticipated benefits" standard for experimentation with the general population. As an illustration, for studies involving children, an IRB has additional review criteria to consider when the risks involved are more than minimal. To approve the study, apart from ensuring that parental consent and child assent will be obtained, the IRB must further find that the risks are justified by the anticipated benefits to the child subjects and the risk/benefit ratio is at least as favorable as available alternative approaches [33]. Similarly, the required composition of the reviewing IRB may be different when special populations are involved. For example, IRBs reviewing research with prisoners must have at least one member who is a prisoner or prison representative and the majority of IRB members cannot have affiliations with the prisons involved [34].

6. **Independent Review Boards**. A smaller subset of clinical trials occur outside of academic medical centers, such as drug company sponsored trials conducted in community physicians' private clinics. Because there is no home IRB to conduct business with, investigators can request review from a "non-institutional review board," sometimes referred to as an "independent review board." Independent review boards, typically created by private, for-profit companies, offer review services to investigators who may not have access to a home IRB. Other than the

fact that they are formally independent of any one institution, the board membership requirements and general oversight responsibilities of independent review boards remain much the same as for local IRBs [35].

2.9   Informed Consent Requirements

One of the most important roles of the IRB is to ensure that investigators obtain adequate informed consent from subjects. This typically involves review of the study's proposed consent form and recruitment procedures.

1. **General Requirements**. The Common Rule requires that prospective subjects be given sufficient opportunity to consider whether to participate in research and in a manner which minimizes the possibility of undue influence or coercion (such as conditioning access to other healthcare on research participation). The information conveyed must be in a language understandable to the subject and include, at a minimum, what a reasonable person would want to have in order to decide whether to participate. Moreover, the informed consent document cannot use exculpatory language that makes subjects waive their legal rights. In particular, subjects cannot be asked to release the research team from any liability for negligence [36].

2. **Basic Elements** For regular clinical care, informed consent doctrine generally requires that the clinician disclose to the patient a description of the proposed treatment, its likely benefits, material risks, and the medically viable alternatives. However, as previously noted, the informed consent requirements for medical experimentation are more stringent. First, the Common Rule requires that subjects should clearly be told that the intervention is experimental and the purposes of the research study [36]. Such express emphasis on the experimental nature of the activity is important because of long-standing concerns with therapeutic misconception, where subjects are prone to conflate experimental interventions presenting unknown risks and benefits with "cutting-edge" standard therapy. Additionally, subjects must be told the (1) foreseeable risks of the research; (2) benefits to the subject or others that may result; (3) alternative courses of treatment; (4) the extent to which records will be kept confidential; (5) whether compensation will be offered or additional treatment made available if the subject experiences harm from the research; and (6) a statement that the research is voluntary and that refusal to participate, or withdrawal at any time, will not lead to penalty or loss of benefits. The 2018 revisions to the Common Rule also clarified important required disclosures if the research involves collection of identifiable private information or identifiable biospecimens. In these instances, subjects must be told whether there is the possibility that the information or biospecimens, once identifiable details are removed, may be used for future research studies, and that additional informed consent will not be required for such secondary uses [36].

There has been long-standing concern about the growth and complexity of consent forms, with important information often buried in very long documents. In partial response, the 2018 revisions to the Common Rule require that consent forms begin with a "concise and focused" presentation of the key information most likely to help someone decide whether to participate in the study, such as discussion of the most important and frequent risks [36].

3. **Additional Elements** When applicable to particular studies, the Common Rule requires additional information disclosure, including (1) that the study may involve risks to the subject that are unforeseeable; (2) circumstances where the subject may be involuntarily terminated from the study (3) any additional costs that may be borne by the subject from participation (such as charges for services and products provided through the trial); (4) a statement that new findings arising during the study that may influence subjects' willingness to continue will be shared; and (5) whether the research may involve whole genome sequencing of a subject's biospecimens. Additionally, to avoid recurrence of situations like the Henrietta Lacks affair, investigators must disclose whether a subject's biospecimens may be used for profit and whether the subject will have a right to share in these profits. Note, the regulations only require disclosure *whether* a subject may share in these profits, they do *not require* that the research team share such profits with subjects [36].

4. **Broad Consent for Storage/Subsequent Use** The Common Rule allows investigators to obtain a subject's consent for uses of private identifiable information or identifiable biospecimens collected in a current study in subsequent investigations unrelated to the current clinical trial. Known as "broad consent," this precludes the need to recontact and obtain each subject's consent in the future for each additional use. In order to be valid, broad consent must, among other criteria, apprise subjects of the types of additional research that might be conducted, the approximate timeframe in which their information or biospecimens may be used, and that individuals who provide broad consent will not have to be told the details of any future research studies using their information or biospecimens [36].

5. **Emergency Research** Requiring prospective informed consent from subjects is a significant impediment to research on therapies provided in emergency and critical care situations. Certain experimental interventions, such as new cardiac resuscitation devices, can only be applied effectively during limited time windows when the subject, still unconscious or otherwise in distress, may be unable to communicate or provide consent and legal representatives are not readily available. The regulations try to balance the need for systematic testing of emergency interventions with subject protection by allowing for waiver of consent for this type of emergency research but only if an extensive number of conditions are satisfied. According to the FDA rules, IRBs may waive the requirement for prospective consent if subjects are in a "life-threatening" condition and, among other factors, obtaining consent is not feasible because (1) subjects will not be able to give consent due to their medical condition; (2) the intervention must be administered before consent can be feasibly obtained from the subjects' legal

representatives; (3) research participation holds out the prospects of direct benefit to subjects; and (4) risks are reasonable in light of what is known about the medical condition, available standard therapies, and the experimental intervention. Further, because potential research subjects may be drawn from the community without consent, investigators must in advance of the trial engage in public disclosure and a consultation process with community stakeholders about the proposed emergency research. Also, following study completion, investigators must provide follow-up disclosure to the community about the research results and demographic characteristics of the participants [37, 38].

2.10   Conflicts of Interest

The medical experimentation enterprise features increasing commercialization. Potential conflicts of interest (COIs) abound that threaten to undermine the integrity of the research and the protection of subjects. Two parallel federal regulatory regimes address COIs in research: (1) the rules for research funded by the Public Health Service (PHS) including by the all important NIH [39]; and (2) research on products regulated by the FDA, even if privately funded [40]. The PHS regulations attempt to ensure that PHS-supported clinical trials are free from bias resulting from the investigator's financial arrangements. Investigators must disclose financial interests above certain monetary thresholds to a designated conflicts official or committee within their home institution. For example, disclosure is likely required if the investigator also serves as a paid consultant to the drug company sponsoring the study. Reportable interests include payments over $5000, equity interests over $5000 in publicly traded companies, any equity interests in a non-publicly traded entity, and intellectual property rights and interests from which the investigator receives income, such as royalty and license fees [41] Institutions receiving these reportable disclosures must then evaluate the arrangement and report to PHS any interests identified as conflicting and the institution's management plan for addressing the potential conflicts [42].

The FDA regulations differ somewhat in that the disclosure obligation is on sponsors of the research, such as drug companies, not the investigator, the monetary thresholds triggering disclosure are for a narrower range of financial relationships, and disclosure must be made directly to the agency. If a clinical trial will support a party's application for FDA approval of the product, the entity generally must disclose to the FDA its payments to an investigator over $25,000, equity interests of the investigator over $50,000, proprietary interests held by the investigator in the tested product (such as patent rights), and other financial relationships in which the value of the investigator's compensation could be affected by the study outcome. Further, the party seeking FDA approval must also disclose the steps it will take to minimize the potential for bias resulting from the disclosed arrangements [40].

Apart from the FDA and PHS rules, public disclosure of research-related financial ties may also arise because of the federal Physicians Payments Sunshine Act [43]. The Sunshine Act generally requires manufacturers of drug and device products covered by the governmental healthcare programs to disclose to the Centers for Medicare and Medicaid Services (CMS) a broad range of their financial relationships (above very small threshold amounts) with physicians and teaching hospitals. For example, research-related payments (such as clinical trial sponsorship payments to investigators and institutions) are reportable, as well as ownership interests an investigator may have in the company whose product is being tested. CMS then posts this information in a publicly searchable platform, known as the "Open Payments" database [44]. The Sunshine Act only requires public disclosure of the financial relationships. It does not direct that any additional safeguards be imposed to address potential conflicts.

When a financial interest disclosure may additionally require some form of follow-up corrective action, such as with the PHS rules, institutions have tremendous discretion in coming up with management plans. For example, conflicts of interest committees (COICs) can choose to require the investigator to disclose the conflict in all published studies, move the research to another site, or ban the conflicted investigator from directly enrolling subjects for the trial [45]. Unfortunately, the failure to subject conflicts oversight to rigorous assessment means that COICs operate without a strong evidence base as to which management plans prove effective and there can be considerable variability in how different COICs address identified conflicts [46]. Moreover, the FDA and PHS rules focus exclusively on financial COIs. Yet non-financial secondary interests, such as an investigator's desire for enhanced reputation, investigative zeal, and intellectual passion, can also introduce bias concerns. Here, the principal safeguard is through IRB review of each study to ensure that it is scientifically sound, has acceptable risk/benefit ratios, and follows an adequate consent process as required by the regulations. But this still leaves much to IRB discretion and they vary considerably in how to address non-financial secondary interests, to the extent addressing them directly at all.

Application
Case 1: Estate of Gelsinger v. Trustees of University of Pennsylvania (Pa. Ct. Com. Pl. 2000) In 1999, Jesse Gelsinger, age 18, died while enrolled in a gene therapy study at the University of Pennsylvania. The clinical trial evaluated a risky procedure for infusing genetically altered viruses as treatment for a rare liver function disorder. The infamous Gelsinger episode involved alleged noncompliance with multiple provisions of the federal research regulations, ranging from inadequate IRB review of the consent form and protocol to insufficient management of COIs. The purported failings included that Gelsinger was allowed to enroll when he did not meet the protocol's narrow eligibility criteria, the consent form did not mention adverse events in prior animal studies, and the researchers failed to alert the IRB or federal agencies about serious adverse event complications with previous subjects. One of the co-investigators, James Wilson, had patents on some aspects of the procedure and Wilson and the University had equity interests in the company that stood to profit from development of treatments that might result from the research, yet no special conflicts safeguards were imposed, including no restrictions on Wilson's participation in the study [47].

Critics questioned whether the lure of financial gain, and zealous desire to offer cutting-edge technology to patients suffering from the liver condition, led to lax oversight and a rush to proceed with dangerous experimentation. Gelsinger's estate filed suit against Wilson, the University, and other related defendants alleging negligence, battery, and related counts. The suit was quickly settled after payment of a confidential sum. In addition, the FDA temporarily suspended all gene therapy trials at the University's Institute for Human Gene Therapy [47]. Also, as a condition of settlement, Wilson published a "lesson learned" article in which he acknowledged the possibility of bias due to the financial ties and non-financial interests, such as the desire for professional acclaim [48].

Case 2: Robertson v. McGee (N.D. Oklahoma 2002) Subjects enrolled in a clinical trial of an investigational melanoma vaccine at the University of Oklahoma Health Science Center [49]. Many of the individuals had advanced disease, were unresponsive to standard therapies, and had very poor prognoses. Allegedly 26 participants died during the study, although the deaths were not attributed to the vaccine itself [50]. In 2001, several of the subjects sued the investigator, hospital, University IRB, pharmaceutical company sponsor, and a University bioethicist who consulted with the IRB [49, 50]. Among other troubling allegations, the plaintiffs contended that the investigators enrolled subjects at remote sites without local IRB approval, pregnant subjects were allowed to continue in the trial even though the protocol called for their exclusion and the IRB had not reviewed the study under the special criteria applicable to pregnant women, other subjects did not meet protocol eligibility criteria, and the investigators changed the protocol without IRB approval [32, 51]. Study nurse Cherlynn Mathias turned whistleblower and helped uncover many of the improprieties [52].

Concerned about lack of oversight, HHS regulators temporarily suspended all government-sponsored studies involving human subjects at the University [53]. The University eventually agreed to settle the case, and it also fired the lead researcher, Dr. J. Michael McGee [54]. While the subjects obtained a partial settlement, the federal district court rejected their attempt to recover for affronts to dignity, concluding that federal law did not provide clear support for such intangible harm claims [49]. Nonetheless, the case illustrates how, regardless of good intent, disorganized, careless research practices can generate negative consequences beyond the amount of liability payments and beyond the single study involved. Here, the inattention to regulatory requirements, including alleged failure to follow subject eligibility criteria and making protocol changes unapproved by the IRB, led to a temporary ban on all federally sponsored human subjects trials at the University, ongoing regulatory scrutiny, and the investigator's loss of University employment.

3   Summary

Medical research with human subjects is strictly regulated and should not be conflated with regular clinical care. Most clinical trials with human subjects are governed by the core federal research regulations, which require that each study obtain prior approval by an IRB. In addition, heightened informed consent requirements apply to the decision to participate in research. Investigators' and institutions' financial relationships with the entities sponsoring the research or manufacturing the tested products may also trigger conflict of interest rules, which generally require disclosure of the financial ties and possible imposition of additional safeguards. Noncompliance with the complex set of laws governing medical experimentation can lead to clinical trial shutdowns, professional discipline, and malpractice liability.

References

1. Goldner, J. An overview of legal controls on human experimentation and the regulatory implications of taking Professor Katz seriously. Saint Louis University Law J. 1993; 38: 63- 134.
2. Nuremberg Code (1947), https://history.nih.gov/display/history/Nuremberg+Code
3. National Bioethics Advisory Commission, Ethical and Policy Issues in Research Involving Human Subjects, Appendix C: The Current Oversight System: History and Description, 2001, https://bioethicsarchive.georgetown.edu/nbac/human/overvol1.pdf
4. World Medical Association, Declaration of Helsinki (1964), https://www.wma.net/what-we-do/medical-ethics/declaration-of-helsinki/doh-jun1964/
5. Saver, R.S. Medical research oversight from the corporate governance perspective: comparing institutional review boards and corporate boards. William and Mary Law Review. 2004; 46: 619-730.
6. Jones, J.H. Bad blood: the tuskegee syphilis experiment. 2d. ed. New York: The Free Press; 1993.
7. National Research Act of 1974, Pub. L. No. 93-348, 88 Stat. 342 (codified as amended at 42 U.S.C. §§ 210 to 3000aaa-13).
8. The National Commission for the Protection of Human Subjects of Research, The Belmont Report: Ethical Principles and Guidelines for the Protection of Human Subjects of Research. 1979, https://www.hhs.gov/ohrp/regulations-and-policy/belmont-report/read-the-belmont-report/index.html
9. Department of Health and Human Services, Final Regulations Amending Basic HHS Policy for the Protection of Human Subjects, 46 Fed. Reg. 8366 (January 26, 1981); 46 Fed. Reg. 8942 (January 27, 1981) (FDA Rules).
10. Federal Policy for the Protection of Human Subjects, 82 Fed. Reg. 7149 (January 19, 2017).
11. 45 C.F.R.§46.101
12. 21 C.F.R. §50.1
13. Office of the Secretary and Food and Drug Administration, Department of Health and Human Services, Human Subject Research Protections: Enhancing Protections for Research Subjects and Reducing Burden, Delay, and Ambiguity for Investigators, 76 Fed. Reg. 44512-44531 (July 26, 2011).
14. *See, e.g.,* Mo. Ann. Stat. § 630.115
15. *See, e.g.* ILCS 410, § 513/30
16. Federation of State Medical Boards, Essentials of a State Medical and Osteopathic Practice Act, § IX.D (2015), https://www.fsmb.org/siteassets/advocacy/policies/essentials-of-a-state-medical-and-osteopathic-practice-act.pdf
17. Manier, J. Federal research funding cut over death at hopkins, Chicago Tribune. 2001 July 20.
18. Centers for Medicare and Medicaid Services, Medicare Clinical Trial Policies, https://www.cms.gov/Medicare/Coverage/ClinicalTrialPolicies
19. 45 C.F.R. § 46.102
20. Noah, L. Informed consent and the elusive dichotomy between standard and experimental therapy. Amer. J. Law and Medicine. 2002; 28: 361-408.
21. Office for Human Research Protections, Quality Improvement Activities FAQs, https://www.hhs.gov/ohrp/regulations-and-policy/guidance/faq/quality-improvement-activities/index.html

22. Skloot, R. The immortal life of henrietta lacks. Crown; 2010.
23. Menikoff, J., Kaneshiro, J., & Pritchard, I. The common rule, updated. New. Eng. J. Med. 2017; 367: 613-615.
24. 45 C.F.R. §46.104
25. 45 C.F.R. §46.107
26. 45 C.F.R. §46.111
27. Henry, S.G., Romano, P.S., & Yarborough, M. Building trust between institutional review boards and researchers. J. Gen. Intern. Med. 2016; 31(9): 987-989.
28. 45 C.F.R. §46.110
29. Office for Human Research Protections, Expedited Review Categories, https://www.hhs.gov/ohrp/regulations-and-policy/guidance/categories-of-research-expedited-review-procedure-1998/index.html
30. 45 C.F.R. §§46.109
31. 45 C.F.R. §46.113
32. 45 C.F.R. Parts B, C, and D.
33. 45 C.F.R. § 46.405
34. 45 C.F.R. §46.304
35. Food and Drug Administration, Non-Local IRB Review: Guidance for Institutional Review Boards and Clinical Investigators, Sept. 6, 2018, https://www.fda.gov/regulatory-information/search-fda-guidance-documents/non-local-irb-review
36. 45 C.F.R. § 46.116
37. 21 C.F.R. § 50.24
38. Department of Health and Human Services, Waiver of Informed Consent Requirements in Certain Emergency Research, 61 Fed. Reg. 51531 (Oct. 2, 1996).
39. 42 C.F.R. §50.604
40. 21 C.F.R. Part 54
41. 42 C.F.R. §50.603
42. 42 C.F.R. §50.605
43. 42 U.S.C. § 1320a-7h (2012)
44. Centers for Medicare and Medicaid Services, Open Payments Data, https://openpaymentsdata.cms.gov/
45. Department of Health and Human Services, Responsibility of Applicants for Promoting Objectivity in Research for Which Public Health Services Funding is Sought, 76 Fed. Reg. 53256 (2011).
46. Taylor, P.L. Innovation incentives or corrupt conflicts of interest? moving beyond jekyll and hyde in regulating biomedical academic-industry relationships. Yale J. Health Policy Law Ethics. 2013:13(1): 135-197.
47. Wilson, R.F. Estate of gelsinger v. trustees of the university of pennsylvania: money, prestige, and conflicts of interest in human subjects research, in Health Law and Bioethics: Cases in Context. Johnson, S., et al., eds.; Aspen; 2009.
48. Wilson, J.M. Lessons learned from the gene therapy trial for ornithine transcarbamylase deficiency. Molecular Genetics and Metabolism. 2009; 96(4): 151-157.
49. Robertson v. McGee, 2002 WL 535045 (N.D. Okla. 2002).
50. Shaul, R.Z., Birenbaum, S., & Evans, M. Legal liabilities in research: early lessons from north america. BMC Medical Ethics 2005; 6(4): 1-7.
51. Complaint, Robertson v. McGee (N.D. Okla. 2002), available at https://www.sskrplaw.com/files/robertson_complaint.pdf
52. Gillham, O. Cancer study case still alive, Tulsa World, Mar 16, 2003, at A17
53. Weiss, R. & Nelson, D., U.S. halts cancer tests in oklahoma, Wash. Post, July 11, 2000, at A1.
54. Gillham, O. OU settles melanoma research suit, Tulsa World, Aug. 28, 2002, at A13.

Next-Generation Genomic Sequencing and Clinician Liability

Teneille Brown and Lesley Ramey

Contents

Key Points

- In the last decade, genetic testing has evolved from simple single-gene tests to complex genome-wide associations and full genomic sequencing.
- The volume and complexity of genetic information for multifactoral traits is much more difficult to interpret.
- Clinicians with very little genetics training may be asked to interpret direct-to-consumer genetic tests that are ordered online by the patient.
- Clinicians will need to be more mindful of the risks for negligence (medical malpractice) liability from the return of genetic testing results. To avoid negligence liability for the failure to obtain informed consent, clinicians who do not plan to (1) provide warnings to family members or (2) update the warnings they provide to patients as additional clinically useful information becomes available, should make this explicit in the informed consent process.

T. Brown (✉)
S.J. Quinney College of Law, Center for Law and the Biomedical Sciences (LABS), Center for Health Ethics, Arts and Humanities (CHeEtAH), University of Utah, Salt Lake City, UT, USA
e-mail: teneille.brown@law.utah.edu

L. Ramey
S.J. Quinney College of Law, University of Utah, Salt Lake City, UT, USA
e-mail: lesley.ramey@law.utah.edu

Legal Concepts

- Duty of Ccare: Because clinicians have "special relationships" with their patients (that is, fiduciary obligations, expert knowledge, and capacity to control someone's body) courts have been more willing to extend broad affirmative duties to protect or warn them that might even extend to non-patients.
- Informed Consent: This term is both an ethical concept and a legal cause of action. Here, we use it to mean a civil negligence claim arising out of a clinician's duty of care to their patients, which has been interpreted to mean that they must disclose sufficient information about the material benefits and risks of proposed medical treatment for the patient to make a well-informed decision about their healthcare.
- Foreseeability: The primary factor that is used to determine whether clinicians owe legal duties of care. Clinicians should only be found to have a duty of care where the general type of harm is reasonably foreseeable to a clinician in a similar situation.
- Polygenic Risk Score (PRS): A number that measures how a person's entire make-up of genetic variants impacts their risk for developing particular traits or diseases.
- Wrongful Life and Wrongful Birth: Informed consent negligence claims that may be brought against a clinician for failure to perform or negligent performance of genetic testing during pregnancy that results in the child being born with a severe genetic deformity.
- Genetic Information Nondiscrimination Act (GINA): A federal statute passed by Congress in 2008 barring health insurers and employers from using genetic information in certain ways to discriminate.
- Clinical Utility: The ability of a screening or diagnostic genetic test to prevent or ameliorate adverse health outcomes such as mortality, morbidity, or disability through the adoption of efficacious treatments or interventions.

1   Introduction

While the first generation of genetic tests revealed whether someone had a simple Mendelian trait that would almost certainly be expressed, the next-generation provides results that are more subtle and complex. Indeed, improvements in genome sequencing and sampling methodologies (such as genome-wide associations or whole-exome sequencing) now mean that we can know very little about a whole lot. We can study a larger percentage of each patient's full genome, but we are still piecing together what all of it means, and how the various sequences interact with each other and the environment.

The volume and variety of genetic results that can be returned puts clinicians in a precarious situation. Laboratories now return genetic results that are of unknown significance—meaning that the lab has only modest confidence in classifying a particular mutation as pathogenic, or disease-causing. Additionally, the patient's results might reveal mutations that predict likely disease, but for which there are no available treatments. In some cases, patients might prefer to maintain their "right to an open future"—free of information about diseases that may appear late in life, like Alzheimer's disease, and for which little can be done preventatively [1]. Moreover, the interpretation of the genome, if not the genome itself, changes over time. Together, these ambiguous sorts of genetic results may create new kinds of legal duties for clinicians.

For now, what is clear is that clinicians have a duty to their patients to reasonably translate the results of genetic tests that they order. Thus, they must separate out the results that have present clinical utility from the results that do not. While genetic counselors can provide interpretive guidance, their availability and the patient's insurance coverage for them may not meet demand.

In addition to the many diagnostic and screening tests that are ordered, patients might also receive an assessment that aggregates the relative weight of thousands or even millions of genetic variants with tiny effect sizes. These are called "polygenic risk scores." Some are already being offered directly to consumers (DTC) for diseases like glaucoma, depression, prostate cancer, and type-2 diabetes [2]. Polygenic risk scores provide a glimpse into someone's comparative risk for a particular disease or group of diseases. Their discriminatory accuracy, judged by their ability to identify people without a family history of disease as being at risk, is low. However, this is improving quickly. For example, one study found that using machine-learning and genome-wide risk prediction, a particular polygenic risk score was able to identify 30% of individuals without a family history of colorectal cancer as having risks similar to those with family history [3]. This potential to accurately predict risk would be enormously helpful for fulfilling the lofty goals of precision medicine. Individuals at high-risk could be offered targeted screening and interventions, while those at low-risk could avoid these costly and cumbersome procedures. But we are not there yet.

At present, most polygenic risk scores provide little by way of individual predictive power. In many cases the aggregated risk score only explains a tiny percentage of the risk variance. For example, some of the most predictive polygenic risk scores in psychiatry are for schizophrenia. In this "best case" scenario, current scores can only explain 7% of the total risk for developing the disease [4]. However, this has not stopped some companies from claiming much better predictive values for their tests. Many of these tests are considered lab-developed tests (LDTs). These are tests that are designed, manufactured, and run within a single facility. Previously, the Food and Drug Administration (FDA) has declined to exercise significant regulatory oversight LDTs. However, the FDA recently announced that LDTs must still obtain premarket approval, which suggests that they may exercise more authority over LDTs in the future.

The minimal FDA oversight of LDTs to date is troubling. Research suggests that DTC genetic testing companies often oversimplify the results of their own tests for marketing purposes [5]. They may claim that their tests are x% accurate, which tells the user very little about the risks for false negatives and false positives. This is particularly concerning as risk for complex traits is often considered to be deterministic and binary—either present or absent. In reality, risk information is much more nuanced, and can vary based on subtle interactions between one's unique environment and thousands of variants. Explaining this to patients is difficult, especially if they struggle generally with numeracy [6].

Given that patients will seek guidance from their primary care doctor on how to interpret these results, this next generation of genomics may lead to unexpected types of liability [7]. This chapter is meant to raise some of the legal issues general practitioners should be aware of when ordering or interpreting genetic tests. Specifically, we will address the current and future landscape of liability for the negligent (non)disclosure of genetic testing results and some unique privacy concerns related to genetic data.

2   Discussion

2.1   Informed Consent and Genetic Testing

Genetic testing comes in many forms. It can be used for screening or diagnostic purposes. This distinction matters a great deal as false positives are expected when screening people for diagnostic testing. Our expectation and tolerance for false positives is lower when tests are used to diagnose particular disorders. Either way, the results can reveal information about the patient's likelihood of developing a particular condition, the likely progression of a condition, or how a patient may react to treatment.

Before ordering genetic tests, clinicians must obtain the patient's informed consent. Clinicians who fail to provide patients with an "opportunity to evaluate knowledgeably the options available" and the attendant risks can be liable for negligence, or medical malpractice, in U.S. courts [8]. While it is often difficult in practice to obtain true, meaningful informed consent, clinicians are expected to share all "material" information with their patients or be subject to paying damages [8]. Material information is information that may cause the patient to make a different choice. States use either a "prudent patient" or "reasonable clinician" standard for determining what should have been disclosed. The former standard recognizes that in informed consent cases it is the patients' perspective that ought to be prioritized. Thus, materiality is measured by referring to what a prudent patient would want to know. In contrast, the reasonable clinician standard sets the standard for disclosure at what a reasonable physician would likely share.

An informed consent claim presupposes that there was a choice to be made—usually in the form of an intervention that was available. If there is nothing clinically different a patient could have done with this information, then the plaintiff typically cannot prove that the defendant caused her cognizable harm. Typically, the risk information must exceed a certain threshold of probability and magnitude to be shared, as clinicians need not disclose risks that are quite unlikely or that indicate either an obvious or minimal harm.

It is critical that juries and judges put themselves in the shoes of the clinician at the time the informed consent took place and not superimpose what we know now on what was known then. However, because "materiality" can be very difficult to determine, it may be subject to significant hindsight bias. There are four reasons for this, and each will be described below.

First, whether the disease would foreseeably develop depends on something called genetic penetrance. Penetrance is a measure of the percentage of people with a given mutation (or genotype) that will exhibit symptoms or signs of a related disorder (or phenotype). If the penetrance for a genotype is quite low, then possessing that mutation might not be very predictive for developing the disease. In many cases, we do not know what the true penetrance of a mutation is. Most of the people who are tested for mutations already have symptoms. This means we do not know the true population base rate. Researchers must test many more asymptomatic people to understand the actual risk of developing disease. As population research develops, estimates of penetrance will likely decrease.

Next, even if penetrance is reliably known, and the condition might have a modest likelihood of being expressed, this still must be evaluated in the context of the patient's unique situation. Clinicians must know whether there are other clinical factors at play that might to a patient's risk. This requires them to interrogate the patient's lifestyle, family history, environmental risk, or protective factors that may change their individual likelihood of developing this disease. Given that most people assume genetic risk is more deterministic than it is, sharing information about the potentially large role of the environment is critical to meaningful informed consent [9].

Another wrinkle in assessing materiality is that the clinical utility of genetic results will change over time. Put simply, our ability to identify mutations has outpaced our ability to interpret them. Many mutations are not yet classified as likely pathogenic, or disease-causing. Instead, they are reported as variants of "unknown significance." Researchers might have good guesses, based on the location of the mutation and its correlation with particular metabolic processes. However, as the connections between mutations and disease are better mapped out in larger populations, clinicians will be better able to say what any given mutation actually means. Given this, some have argued that clinicians may have a duty to update their patients with genetic information that becomes clinically significant some time after the initial results were returned. To avoid negligence liability for failure to warn of genetic risk, clinicians who do not plan to update the warnings they provide to patients as more precise information becomes available should make this explicit in the informed consent process.

Depending on the clinical utility of the information, clinicians might have a duty to return incidental genetic findings. Incidental findings are those that are generated from a genetic test, but which were not the primary purpose of the test. For example, a patient might undergo exome sequencing to determine possible causes of severe developmental delays. The results may reveal mutations indicating a high risk of an unrelated disorder, such as cancer. Or, a patient might be part of a clinical research study that secondarily reveals a risk for a genetic disorder that is unrelated to the research. Whether a clinician has a duty to search for these findings and/or report them back to the patient depends on the foreseeability and magnitude of any potential harms, compared against the potential burden in requiring physicians to assume this obligation.

But as part of the informed consent, clinicians may now be required to discuss with the patient whether she can expect to be informed of incidental findings, and whether the clinician may breach their confidentiality to share the genetic results with the patient's relatives [10]. Legal scholars have speculated that cases like this are likely to increase. However, to date there are not many published opinions finding liability for clinicians' failing to share incidental genomic findings.

Below we will present five very different types of informed consent claims to help clarify some of the novel legal implications of next-generation genetic tests. The first is in the context of reproductive decision-making, the second is in the context of familial risk, the third is in the context of variants of unknown significance, the fourth is in the context of incidental findings, and the fifth is in the context of prescription drug warnings. Four are based on actual cases and one is based on a published case-study.

Case 1: Wrongful Birth Wrongful birth claims are a particular type of informed consent claim. In wrongful birth claims, the parents argue that their clinician either failed to perform necessary genetic testing or performed genetic testing negligently, which infringed on the parents' right to make an informed decision about whether to terminate the pregnancy. Testing may be negligently performed when results are reported incorrectly or the testing is performed too late to be actionable. States that bar such claims do so under the theory that life cannot be an injury [11]. These claims have been controversial and are only recognized in 24 states, despite presenting fairly typical features of an informed consent claim. So-called "wrongful birth" claims are brought by the parents when their child is born with severe genetic defects, and "wrongful life" claims are brought by the children themselves. Wrongful life claims are even more controversial and only allowed in three states, as they require the plaintiff to argue that they should never have been born [11]. Despite the philosophical problem of recognizing that life can be an injury, in many cases the parents are merely seeking money damages to help cover the extra costs associated with caring for their severely disabled child [12]. Statutes may limit the injury the parents can claim in these cases [13].

In *Clark v. Children's Memorial Hospital,* the parents had genetic testing done on their son when he began to exhibit symptoms consistent with Angelman's syndrome [14]. The ordering clinician incorrectly told the parents that the genetic testing had ruled out all possible genetic causes for the disease. Comforted that their son's disorder was not genetic, they decided to conceive another child. However, after the second son was born, they learned that the earlier genetic tests indicated that any children they later conceived would have a 50% chance of being born with Angelman syndrome. The parents were able to demonstrate that the hospital and clinicians incorrectly reported the results of the genetic tests. They also ostensibly proved that if they had known about the risk of their children having Angelman's they would not have had another child. As a result, an Illinois court permitted them to bring a "wrongful birth" informed consent claim. The damages went toward paying for the extraordinary medical expenses required for the second child.

Case 2: Duty to Warn Family Members A couple of published cases have held that the clinician's duty to warn their patient of actionable genetic risk may extend to non-patients. Here, the informed consent claim is brought not by the clinician's patient, but by their patient's relatives. The relatives claim they were owed a duty of care and the clinician was negligent in failing to disclose the risk of heritable disease to *them*.

In *Safer v. Estate of Pack*, a patient who suffered from colon cancer brought an informed consent claim against the clinician who had treated her father years earlier for colon cancer [15]. She alleged that the clinician should have known that the disease was hereditary and he therefore should have also warned her. Had she been aware of her risk, the plaintiff argued she would have been more vigilant in monitoring herself for colon cancer. The trial court dismissed the action and the daughter appealed. The Superior Court of New Jersey analogized imposing a duty to warn of genetic risks to a duty to warn of serious infectious diseases. Because the harm to the patient's close relatives was foreseeable, the duty "extend[s] beyond the interests of a patient to members of the immediate family of the patient who may be adversely affected by [non-disclosure] [15]." *Safer* held that while imposing broad duties to non-patients might not be wise, in the case of warning of "avertible risk from genetic causes, by definition a matter of familial concern" imposing a duty to warn is fair and feasible [15].

The court further held that in *Safer*, the duty to the relative was considered discharged by informing the patient of the risk, and assumed that the patient would share this information with his relative. However, the court reserved the possibility of one day breaching a patient's confidential health information to warn their relatives, if there were reason to believe they did not communicate. If other states were to follow this precedent, it could create liability for clinicians for failing to warn the relatives of their patients of foreseeable health risks. This could impose new burdens on clinicians to identify and contact their patient's relatives.

Case 3: Failure to Update Disclosure of Variants of Unknown Significance In the case of *Williams v. Quest Diagnostics*, the plaintiff sued a clinical laboratory for ostensibly misclassifying her son's genetic variant [16]. The patient was a young boy with seizures. His clinician ordered genetic tests to try to determine whether he might have a seizure disorder called Dravet's. The results came back with a SCN1A mutation, that the laboratory had labeled a variant of unknown significance [16]. While some research had indicated that the SCN1A mutation was linked with Dravet's, at that time the laboratory claimed that there was not enough data to label it pathogenic, or "disease-causing." However, it has since been linked with Dravet's.

The mother argued that the misclassification resulted in the clinician prescribing her son a seizure medication that was contraindicated with Dravet's, and which caused his premature death. The outcome of the published appeal focused on whether the laboratory could be considered a healthcare provider under South Carolina's statute of limitations. However, the decision recognized that medical providers could be liable for misclassifying genetic variants if this caused their patients harm. Laboratories do not have uniform reporting standards, and there is disagreement in the field as to when there is sufficient data to link a mutation to causing a particular disease. To be sure, the standards in this area are in flux. Given this, it is possible jurors might engage in a form of hindsight bias—superimposing what is known at trial with what should have been known when the results were shared. To avoid liability for misclassified variants, clinicians should involve genetic counselors as much as possible and try to stay abreast of changes in a variant's classification. They should also use caution when making clinical decisions based on genetic variants of unknown significance. This might require doing personal research into the link between mutations and disease, or disclosing to patients what is and what is not known at the time.

Case 4: Duty to Disclose Incidental Findings of Suspected Incest In a fictionalized case report, based on an actual case, a 14-year-old gave birth to a child with severe skin lesions. The baby's clinical presentation suggested a genetic disease called epidermolysis bullosa (EB). The medical team ordered a multigene panel, which confirmed that the baby had EB and would likely not live more than a year. The gene panel showed that the baby was homozygous for the rare disorder, which can also mean that his parents were close relatives. Given the birth mother's age, and how involved her father was in the medical discussions, this raised a suspicion of incest in the medical team. In part to investigate this suspicion, a clinician then ordered a microarray analysis of both the birth mother and the baby, without informing the birth mother that the results could indicate consanguinity, or incest. When the results were returned, they revealed that the birth father and the birth mother were likely second-degree relatives (such as a parent and child). Before discussing the results with the birth mother, the social worker reported the suspected incest to the state's Child Protective Services. Some members of the medical team were worried that they had violated the birth mother's privacy and unfairly turned the medical team into forensic investigators.

Despite these compelling ethical concerns, laws in each state require clinicians to report suspected child abuse when they have reasonable suspicion it has occurred [17]. To whom and what they must report varies between states. In some states, this obligation can include requiring clinicians to run follow-up tests when a child appears to have been abused. While mandatory reporting laws were written when suspected abuse could be investigated through x-rays of broken bones or inspection of bruises, genetic testing is not different in kind from these methods [17]. Mandatory reporting laws allow clinicians, and in some cases the standard of care requires them, to order genetic testing to determine whether abuse has occurred [18]. Even so, the doctrine of informed consent must be respected, even when the patient is a minor. If the state's laws permit testing the birth mother without her consent, to protect her autonomy she should still have been told that one of the reasons for the microarray analysis was to determine whether she had been sexually abused by her father. Genetic counseling should also have been offered, along with counseling support. In states that require patients to either consent to the test or be served with a judicially-approved warrant, the failure of the clinicians to obtain the birth mother's consent to the microarray might give rise to an informed consent cause of action. At the same time, failing to follow-up on suspected abuse can also lead to a medical malpractice claim. Thus, to thread the needle of potential liability, clinicians should perform the tests necessary to confirm suspected abuse while also ensuring they either obtain a warrant or the patient's consent to the test.

Case 5: Failure to Warn of Drug Defects and the Learned Intermediary Doctrine A drug can be considered defective if it is distributed without an adequate warning on its label or FDA-approved package insert. Once a drug manufacturer has included a warning on the approved label, the manufacturer's duty to warn is satisfied. It is presumed that the clinician will share these risks with patients as part of the informed consent process. The duty to warn now passes to the clinician, who as a "learned intermediary," is expected to share any material risks with the patient. Clinicians are expected to read the fine-print in the drug's package insert, and to choose when it is appropriate to share these warnings with patients. Informing patients that certain genetic traits may limit or eliminate the effectiveness of the drug may be considered part of this duty. Failing to fulfill this duty to warn could count as medical malpractice. As precision medicine develops, this will require clinicians to either recommend genetic testing before prescribing medications, or to inquire into the patient's known genetic risks.

Recently, this sort of failure to warn claim came up in Hawaii. Drug manufacturers Sanofi and Bristol-Myers Squibb were sued for product liability over their marketing of Plavix a prescription antiplatelet agent they produced (that is also available in generic form as clopidogrel) [19]. The drug did not work well for many people of Asian or Pacific Islander descent because of a genetic mutation that blocked proper metabolization of the drug [19]. From 1998 to 2010, the drug manufacturers failed to warn clinicians that their patients of Asian or Pacific Islander descent would not see a reduced risk of heart attack or stroke, the drug's primary approved indication. In Hawaii, as much as 30% of users may not have benefited from the drug [19]. In 2010, the Food and Drug Administration (FDA) required the manufacturers to add a black box warning on the drug's label [19]. The Hawaii court found the drug manufacturers were liable for misleading marketing and for a failure to warn clinicians and consumers for the period from 1998 to 2010. The court ordered the defendants to pay $834 million [19]. This landmark case is likely to be appealed.

2.2   Genetic Privacy

Americans report being quite concerned about the possibility of their genetic privacy being breached. This may be because we have a perfect storm for potential discrimination in our country—namely, significant investment in clinical genetics testing and commercial biobanks, with no universal, publicly-funded health insurance. Many types of genomic information are not that different from other types of sensitive health information, like blood pressure or cholesterol levels. However, there are some aspects of our genes that are unique [20]. For one, our genomes are immutable. At the same time, what they mean for our future health is still being decoded, as we place the complicated interaction of genetics and the environment in context. In short, if we decide to share information with third-parties that we assume is "junk DNA", it might not be junk in a few years. Or, information about our cancer risk might later reveal correlated information about our risk for mental illnesses for which there exists no effective treatment. Our genetic information also has the potential to reveal risks for future diseases that our close genetic relatives might prefer not to know. However, if the disease is often expressed like sickle cell or Huntington's, the close family members likely already know it is heritable from their lived experiences. Finally, if the disease is recessive and varies in expression, a relative's genetic information may not be very predictive for us. Whether it is based on genetic exceptionalism or the immutability of genetic information, violations of genetic privacy continue to cause great concern about who has access to this information and how it may be used against us. This has led to heightened expectations of privacy surrounding genetic information [21].

With the explosion of the DTC genetic testing industry, there remains great concern over the privacy protections afforded to this kind of genetic information. Consumers continue to enthusiastically provide saliva samples to companies like 23andMe or Color, either to investigate their genealogy or to learn about their risks for common cancers and heart disorders. While some of these companies purport to comply with the Health Insurance Portability and Accountability Act (HIPAA) regulations, many of them are recreational and not HIPAA-covered entities. Thus, any compliance with HIPAA may be voluntary, and could be changed unilaterally through revising the consumer terms of service. In many cases, DTC genetic testing companies could decide to sell their user's data, and the only meaningful legal protections would come through consumer protection or contract law. This is because these companies operate outside of the traditional health care model. They are thus capable of avoiding many health care regulations like HIPAA, while offering what is essentially health care information to unsophisticated consumers. Insecure data transmission and security systems lead to vulnerabilities as well. In sum, there are many concerns related to commercial DTC companies.

However, if a patient brings her DTC genetic testing results to a HIPAA-regulated clinic and this becomes part of her medical record, the DTC results now become protected health information. Clinicians are prohibited from disclosing their patient's protected health information. In addition to HIPAA, some states have passed special laws protecting the confidentiality of genetic information. In the

United States, the privacy accorded to genetic results depends on where the data are stored and whether the tests were conducted through a HIPAA-covered entity.

There are other ways the federal government supports heightened confidentiality protections for genetic information. If genetic information is included in a clinical study, the researchers can apply for a Certificate of Confidentiality through the National Institutes of Health. These certificates are meant to prohibit unconsented disclosures of identifiable, sensitive research information to anyone not connected to the research. However, these protections are not ironclad. When enforcing the certificates runs up against a criminal defendant's constitutional rights, they are not likely to be upheld. In the few cases challenging them, the judges have not afforded the sensitive data robust protection [22]. Given the perfect storm of increased sharing of genetic information with a patchwork of privacy protections that are far from airtight, scholars anticipated large spikes in litigation related to genetic discrimination and privacy violations. However, so far the deluge of claims has not yet materialized.

2.3   Genetic Information and Nondiscrimination

After the remarkable completion of the Human Genome Project, the promise of genetic research was not being fully realized. Policymakers believed this was because Americans were afraid of participating in genetic research. In addition to the privacy breaches described in the previous section, people were worried that health insurance companies and employers could use their genetic information to discriminate against them—and either deny them health insurance or employment altogether or offer it on worse terms. The Genetic Information Nondiscrimination Act (GINA) of 2008 was passed by Congress to respond to this fear, by prohibiting the acquisition and use of genetic information to discriminate in healthcare insurance and employment [23]. The federal law does not apply to discrimination in disability or life insurance, which has led some states to pass their own laws to prohibit discrimination in these domains [24]. GINA also contains a number of exceptions. For example, an employer can legally obtain genetic information, such as through voluntary wellness programs. Further, GINA notably does not prohibit genetic discrimination by banks, landlords, or schools [25].

GINA has not generated the high-profile litigation that scholars expected, and the cases that have been published do not involve the kind of overt genetic discrimination that motivated the passage of the statute. Most of the published cases have turned on the definition of "genetic information," with litigants claiming discrimination based on something other than their genome [25]. A few cases have dealt with the legal contours of employee wellness programs that encourage participation by reducing the cost of health insurance. The dearth of cases may mean that GINA has been effective, that proving discrimination remains very difficult, or that concerns of genetic discrimination were overblown. Here, we will canvas two cases that help explore the contours of GINA. However, these sorts of cases are unlikely to present liability for individual clinicians. Instead, they may just be useful as background information for clinicians to provide to their patients.

GINA Case #1: The Definition of "Genetic Information" The plaintiff in *Poore v. Peterbilt of Bristol* complained that he was terminated in violation of GINA when his employer discovered that his wife had multiple sclerosis, and asked the employee about her prognosis [26]. The federal district court held that the complaint failed to state a violation of GINA because the manifestation of a disease does not count as genetic information. Courts have held that genetic information means "a genetic propensity to acquire [a] disease," that is predictive in nature. It does not cover diseases that are already diagnosed, unless there is a significant heritable component that speaks to the employee's genetic risk. The plaintiff did not demonstrate that his wife's multiple sclerosis diagnosis counted as genetic information *with respect to the employee.* This same definition of "genetic information" would apply in the context of potential health insurance discrimination.

GINA Case #2: Employee Wellness Programs Another area of litigation surrounding GINA challenges the legality of employee-sponsored wellness programs. These cases have turned on whether GINA permits employers to obtain genetic information that is part of a "voluntary" wellness program [27]. The American Association for Retired Persons (AARP) challenged the Equal Employment Opportunity Commission (EEOC) regulations that permitted a 30% reduction in insurance premiums if employees participated in a wellness program. The AARP argued that these programs were too coercive to be voluntary [28]. While the regulations permit employers to obtain information about the employee's genetic information, the employer could not use the incentives to collect genetic information about the employee's spouse or children. Even so, the federal district court found that the EEOC failed to provide a reasoned explanation for its 30% incentive level, so it was deemed arbitrary. Rather than vacating the regulations entirely, however, the court remanded them to the EEOC for reconsideration.

3   Summary

The clinical use of genetic testing will likely increase, given that the tests are cheaper and widely available. However, clinicians will face new ethical and legal challenges as the kinds of genetic information being returned are multi-faceted and clinically complex. As clinician's duties to patients and non-patients hinge primarily on the foreseeability of harm, this will require clinicians to investigate the clinical utility of the various genetic tests and results, and to share any information that might be material to a patient's healthcare decisions. Of course this might be difficult to discern prospectively, and may be subject to juror's hindsight bias. Thus, documentation of the decision-making process is important.

The primary way that clinicians will need to consider liability for genetic testing results is in the context of informed consent. As a result, this chapter focused on the many different kinds of claims that might be brought in this context. The majority of the claims that are likely to arise will continue to stem from clinicians' interpretation and disclosure of complicated genetic results. To protect against informed consent medical malpractice liability, clinicians should recommend tests in appropriate situations, consult genetic counselors for interpretive guidance, ensure all material information is shared before and after testing, and check that results are accurately recorded.

References

1. Bredenoord AL, de Vries MC, van Delden JJ. Next-generation sequencing: does the next generation still have a right to an open future?. Nat Rev Genet. 2013 Mar;14:306. doi: https://doi.org/10.1038/nrg3459.
2. Petrone J. MyHeritage moves into health with new genetic testing offering, underscoring industry trend. genomeweb [Internet]. 2019 May 28 [cited 2019 Nov 18];Applied Markets:[about 3 p.]. Available from: https://www.genomeweb.com/applied-markets/myheritage-moves-health-new-genetic-testing-offering-underscoring-industry-trend
3. Thomas M, Sakoda LC, Hoffmeister M, Rosenthal EA, Lee JK, van Duijnhoven FJ, Platz EA, Wu AH, Dampier CH, de la Chapelle A, Wolk A, Joshi AD, Burnett-Hartman A, Gsur A, Lindblom A, Castells A, Win AK, Namjou B, Van Guelpen B, Tangen CM, He Q, Li CI, Schafmayer C, Joshu CE, Ulrich CM, Bishop DT, Buchanan DD, Schaid D, Drew DA, Muller DC, Duggan D, Crosslin DR, Albanes D, Giovannucci EL, Larson E, Qu F, Mentch F, Giles GG, Hakonarson H, Hampel H, Stanaway IB, Figueiredo JC, Huyge JR, Minnier J, Chang-Claude J, Hampe J, Harley JB, Visvanathan K, Curtis KR, Offit K, Li L, Le Marchand L, Vodickova L, Gunter MJ, Jenkins MA, Slattery ML, Lemire M, Woods MO, Song M, Murphy N, Lindor NM, Dikilitas O, Pharoah PD, Campbell PT, Newcomb PA, Milne RL, MacInnis RJ, Castellvi-Bel S, Ogino S, Berndt SI, Bezieau S, Thibodeau SN, Gallinger SJ, Zaidi SH, Harrison TA, Keku TO, Hudson TJ, Vymetalkova V, Moreno V, Martin V, Arndt V, Wei WQ, Chung W, Su YR, Hayes RB, White E, Vodicka P, Casey G, Gruber SB, Schoen RE, Chan AT, Potter JD, Brenner H, Jarvik GP, Corley DA, Peters U, Hsu L, Genome-wide modeling of Polygenic Risk Score in colorectal cancer risk. Am J Hum Genet. 2020 Sept;107(3):432-444. doi: https://doi.org/10.1016/j.ajhg.2020.07.006
4. Zheutlin AB, Ross DA. Polygenic Risk Scores: what are they good for? Biol. Psychiatry. 2018 Jun;83(11):e51-e53. doi: https://doi.org/10.1016/j.biopsych.2018.04.007
5. Kious BM, Docherty AR, Botkin JR, Brown TR, Francis LP, Gray DD, Keeshin BR, Stark LA, Witte B, Coon H. Ethical and public health implications of genetic testing for suicide risk: family and survivor perspectives. Genet Med. 2020 Oct; 23(2): 289–297. doi:https://doi.org/10.1038/s41436-020-00982-1
6. Lea D, H, Kaphingst K, A, Bowen D, Lipkus I, Hadley D, W: Communicating genetic and genomic information: health literacy and numeracy considerations. Public Health Genomics. 2011;14:279-289. doi: https://doi.org/10.1159/000294191
7. Clayton EW. Be ready to talk with parents about direct-to-consumer genetic testing. JAMA Pediatr. 2020 Feb;174(2):117–118. doi:https://doi.org/10.1001/jamapediatrics.2019.5006
8. Canterbury v. Spence, 464 F.2d 772 (D.C. Cir. 1972).
9. Tabery J. Beyond versus: the struggle to understand the interaction of nature and nurture. 1st ed. United Kingdom: MIT Press; c2014. 101–103, 188 pp.
10. American Medical Association. Code of Medical Ethics [Internet]. c2016. Chapter 4, Opinions on Genetics and Reproductive Medicine; [cited 2021 Oct 18][about 9 p.]. Available from: https://www.ama-assn.org/sites/ama-assn.org/files/corp/media-browser/code-of-medical-ethics-chapter-4.pdf
11. Gaiparashvili M, Wrongful birth and wrongful life cases - comparative study. Herald of Law. 2020 Dec;1(1): 24-44.
12. Robinson v. Mitchell, 323 So.3d 982 (La.App. 2 Cir. 2021).

13. *See, e.g.*, Abortion Based on Genetic Abnormality, Prohibition, LA. STAT. ANN. §40:1061.1.2 (2018); GUTTMACHER INSTITUTE, ABORTION BANS IN CASES OF SEX OR RACE SELECTION OR GENETIC ABNORMALITY (2021).
14. Clark v. Children's Mem'l Hosp., 391 Ill. App. 3d 321, 907 N.E.2d 49 (Ill. App. Ct. 1st Dist. 2009), *aff'd in part, rev'd in part,* 2011 IL 108656, 955 N.E.2d 1065 (Ill. 2011).
15. Safer v. Est. of Pack, 677 A.2d 1188 (N.J. Super. Ct. App. Div. 1996).
16. Williams v. Quest Diagnostics, Inc., 353 F. Supp. 3d 432 (D.S.C. 2018).
17. Danny Veilleux, *Validity, Construction, and Application of State Statute Requiring Doctor or Other Person to Report Child Abuse*, 73 A.L.R. 4th (2019).
18. McGowan M, Brown T, Biller AB, de Sante-Bertkau J. Genomic testing, unexpected consanguinity, and adolescent parents. Hasting Center Report. 2021 Aug;51(5):8-11. doi: https://doi.org/10.1002/hast.1276
19. Feeley J. Bristol-Myers Sanofi ordered to pay 834 million over Plavix. Bloomberg [Internet]. 2021 Feb. 15 [cited 2021 Oct 18];Business:[about 5 p.]. Available from: https://www.bloomberg.com/news/articles/2021-02-15/bristol-myers-sanofi-ordered-to-pay-834-million-over-plavix
20. Gostin LO, Hodge JG. Genetic privacy and the law: an end to genetics exceptionalism. Jurimetrics. 1999 Fall:21-58.
21. Amato v. Dist. Att'y for Cape & Islands Dist., 952 N.E.2d 400 (Mass. App. Ct. 2011).
22. Teneille Brown & Kelly Lowenberg, *Biobanks, Privacy, and the Subpoena Power*, 2009 STANFORD J.L., SCI., & POL'Y 89 (2009).
23. Genetic Information Nondiscrimination Act of 2008, Pub. L. No. 110-233, 122 Stat. 881 (May 21,2008).
24. Genetic Privacy Act, N.J. STAT. ANN. § 10:5-43 (West 1996); *see also* Discrimination on the Basis of Genetic Information or Testing, ME. REV. STAT. tit. 24-A, § 2159-C (2019).
25. Bradley A. Areheart & Jessica L. Roberts, *GINA, Big Data, and the Future of Employee Privacy*, 128 YALE L.J. 710, 728 (2019).
26. Poore v. Peterbilt of Bristol, LLC, 852 F. Supp. 2d 727 (W.D.Va. 2012).
27. 42 U.S.C.A. § 2000ff-1 (West).
28. AARP v. United States Equal Emp. Opportunity Comm'n, 267 F. Supp. 3d 14 (D.D.C. 2017), *on reconsideration,* 292 F. Supp. 3d 238 (D.D.C. 2017).

Decisions Near the End of Life

Robert S. Olick

Contents

R. S. Olick (✉)
Center for Bioethics and Humanities, State University of New York Upstate Medical
University, Syracuse, NY, USA

Department of Internal Medicine, George Washington University School of Medicine and
Health Sciences, Washington, DC, USA
e-mail: olickr@upstate.edu

A. S. Pasha (ed.), *Laws of Medicine*,
https://doi.org/10.1007/978-3-031-08162-0_34

Key Points
- Both competent and incompetent patients have the right to refuse life-sustaining treatment. This right is based on the Constitution and the common law right of self-determination.
- Competent patients may exercise this right in the process of informed consent.
- Family members or other surrogates may refuse treatment on behalf of incompetent loved ones based on the patient's previously expressed wishes and best interests.
- Competent adults may execute advance directives to control decisions near the end of life when they have lost decisional capacity due to illness, disease or injury. The proxy directive designates a trusted person to make decisions on the patient's behalf; the living will states the person's wishes for future treatment with some specificity. The two approaches can be combined in a single document.
- Both competent and incompetent patients may refuse any and all medical interventions.
- The law often limits forgoing of treatment for incompetent patients to conditions of terminal illness or permanent unconsciousness, but these limitations do not apply for competent patients.
- Clinicians are obligated to respect decisions to refuse life-sustaining treatment.

Legal Concepts

Competence (capacity) is the ability to understand and appreciate the nature and consequences of healthcare decisions, including their benefits, risks and burdens and the option of no treatment, and to make an informed decision. When patients are incompetent (lack capacity) healthcare decisions must be made on their behalf by someone else.

A surrogate decision maker (surrogate) is someone legally authorized to make decisions on the patient's behalf.

The right to refuse treatment, including life-sustaining treatment, belongs to both competent and incompetent patients. It is variously established under both the federal and state constitutions and the common law right of self-determination.

Advance directives are legal documents that allow individuals to express their wishes for healthcare at a time of future incapacity to decide for oneself. They are most often used to direct forgoing of life-sustaining treatment, but can also be used to request treatment. Advance directives may designate a family member or other trusted person to act on the patient's behalf (the proxy directive), may state the patient's wishes with respect to forgoing of life-sustaining treatment (the living will), or both (the combined directive).

Death is determined based on one of the two legal standards of death: (1) irreversible loss of cardiorespiratory function, or (2) irreversible loss of all functions of the entire brain including the brain stem.

Medical aid in dying (also known as physician-assisted suicide) is the writing of a prescription by a clinician to be self-administered by a terminally ill patient with the intention of hastening death.

Professional conscience refers to the right of clinicians to refuse or withdraw from care of a patient on grounds of a moral, religious or professional objection to the patient's wishes regarding end-of-life treatment. Withdrawal from care typically requires an appropriate transfer of the patient's care to another clinician.

1 Introduction

All 50 states and the District of Columbia recognize the rights of both competent and incompetent adult patients to refuse unwanted life-sustaining treatments. The right to refuse treatment is protected by the U.S. Constitution, by some State constitutions, and is grounded in the common law right of self-determination and bodily integrity [1–4]. Competent patients may direct withholding or withdrawal of treatment contemporaneously in the process of informed consent (see Chap. 30). The rights of incompetent patients may be exercised on their behalf by a family member or other surrogate decision maker based on the patient's previously expressed wishes and best interests. State laws nationally give competent adults the right to exercise control over life-sustaining treatment decisions at a time of future incapacity by writing advance directives for healthcare. This chapter describes the substantial legal consensus that governs end-of-life decisions, identifies some of the important ways that state laws vary in their approach to these decisions, and discusses some common dilemmas in end-of-life care. Later sections address two issues of growing importance where no legal consensus exists—medical futility and medical aid in dying. Rights of professional conscience to withdraw from patient care, and the role of ethics consultation as a resource to resolve ethical-legal dilemmas are also discussed. Notably, the legal consensus described here also substantially reflects the ethical and clinical consensus.

Most adult patients die after a decision is taken to forgo life-sustaining interventions, such as cardiopulmonary resuscitation (CPR), a ventilator or a feeding tube. Every day in hospitals and nursing homes across the country families, friends, physicians and other clinicians face the challenges of deciding for patients who have lost the capacity to decide for themselves. A central goal of this chapter is to set forth core themes and provisions of the extensive body of law governing end-of-life decisions for incompetent patients. End-of-life law is very much a matter of state law developed in the "laboratory of the states," and specific rules may differ from state to state. Clinicians should be familiar with the law of their practicing state.

We begin with the threshold question of the legal standard of death.

2 Determination of Death

There are two uniform, national standards of death: the traditional standard of irreversible loss of cardiorespiratory function, and the modern standard of neurological death, also known as "whole brain death." This doctrine, promulgated (1981) in the Uniform Determination of Death Act (UDDA) states that "An individual who has sustained either (1) irreversible cessation of circulatory and respiratory functions or (2) irreversible cessation of all functions of the entire brain, including the brain stem, is dead." [5]. This language has been adopted in law verbatim, or nearly so, in all 50 states and the District of Columbia [6]. While the law establishes the standards of death, the criteria for determining whether death has occurred (e.g., absence of brain stem reflexes) are established by the medical profession. New York directs by regulation that hospitals adopt formal policies and sets forth the clinical criteria in department of health guidelines [7]. Nevada is the only state to specify by statute that determination of neurological death be made in accordance with current guidelines from leading professional organizations [8]. It should be noted that the consent of a family member to perform the neurological death exam is not required in most states, but has been required in a few cases, though the law is typically silent on this question [6]. In most cases neurological death is declared and support is removed when neurological criteria are satisfied (sensitivity to the family's grieving process customarily involves a short delay); there is no duty to continue to treat a dead body. But there are three circumstances where there may be an obligation to continue physiological support for a period of time – organ donation, pregnancy, and accommodation of religious objection to neurological death.

At least four states (California, Illinois, New Jersey and New York) require some form of reasonable accommodation of a patient's religious or moral objection to determination of neurological death [6]. New Jersey was the first state to enact a religious exemption by statute, and represents the most rigorous commitment to respect for the patient's religious beliefs. Under New Jersey law, when there is reason to believe the patient would object to having death declared on the basis of neurological criteria, physiological support is to be continued and death is to be declared solely on the basis of cardiorespiratory criteria. The exemption applies to the patient's own expressions of religious beliefs, based on information provided by families or a religious leader or as stated in the patient's advance directive, but it is not intended to apply to the family's own religious views [9]. Other states have adopted different interpretations of reasonable accommodation that give hospitals some discretion to define what is reasonable. California and New York call for accommodation of both religious and moral objections [6].

3   Refusal of Life-Sustaining Treatment: The Legal Consensus

Since the seminal opinion from the New Jersey Supreme Court in the case of Karen Ann Quinlan (1976) a judicial and legislative consensus that governs decisions near the end of life has been firmly established. *Quinlan* involved a young woman in persistent vegetative state (PVS) whose parents sought the right to authorize withdrawal of a respirator on her behalf. The NJ court held that competent adults have a constitutionally-protected right to privacy to refuse unwanted bodily interventions, including life-sustaining treatment. The court further held that when patients cannot exercise that right due to loss of decisional capacity, under the doctrine of substituted judgment family members may decide on behalf of incompetent loved ones based on the patient's prior wishes and best interests. When clinicians implement patient and family wishes, forgoing life-sustaining treatment allows a natural dying process to take its course; it does not constitute killing, assisted suicide or suicide. The court's approach called for confirmation of Karen's PVS before removal of the respirator [10]. *Quinlan* was only binding in NJ, but courts in many other states have found the reasoning persuasive and have grounded the right to refuse treatment in either the Constitution or the common law right of self-determination and bodily integrity. Over the ensuing 15 years a judicial consensus recognizing the right to refuse treatment and the authority of family members to decide on behalf of incompetent loved ones emerged. Most of these cases involved patients who were either terminally ill or permanently unconscious [1–4].

The U.S. Supreme Court's decision in the case of Nancy Beth Cruzan (1990) affirmed the constitutional basis of the right to refuse treatment and cemented the legal consensus. *Cruzan* involved a young Missouri woman whose life in persistent vegetative state was being sustained by a gastronomy tube. The Court rejected any legal distinction between forgoing of feeding tubes and other forms of life-sustaining treatment, and held that her parents could make the decision to withdraw treatment if they could satisfy Missouri's requirement that Nancy's wishes to refuse a feeding tube under these circumstances be shown by providing "clear and convincing evidence." In upholding Missouri's evidentiary standard *Cruzan* also affirmed that states have authority to establish their own rules for end-of-life decisions, as many had already done, provided state laws do not unduly infringe upon patients' constitutionally-protected rights to control their own health care [11]. Notably the vast majority of states reject this high evidentiary standard. Rather in most states the patient's previous statements and expressions to refuse treatment must be trustworthy, shown by a preponderance of the evidence, or similar language; all easier burdens for families to meet [1–4]. When clinicians find information about the patient's treatment refusal to be trustworthy and reliable, the formal evidentiary standard is rarely a significant issue at the bedside.

In the aftermath of *Quinlan*, California became the first state to enact advance directive legislation. State legislatures responded to calls for action amidst high profile cases in their own states. Today all 50 states and DC have statutes that recognize the legal right of competent adults to put their wishes for end-of-life care in writing in order to provide direction and exercise control over treatment decisions when capacity for contemporaneous decisionmaking has been lost due to the ravages of illness, disease or injury. These laws are based on the same core principles that both competent and incompetent patients have the right to refuse life-sustaining treatment and incorporate many of the principles found in the judicial consensus [1–4].

4   Competent Patients

Courts have consistently ruled in favor of the competent patient's right to refuse life-sustaining treatment. Treatment refusal starts with the right of self-determination and bodily integrity that grounds the doctrine of informed consent.

The two core components of the doctrine are that intrusions upon the body in the form of medical diagnosis or treatment require the voluntary consent of the patient, and that patients must be provided with appropriate information to make informed decisions. The corollary of the right to consent is the right to say no to unwanted medical interventions, including life-sustaining treatment. (see Chap. 30 on informed consent).

Several well-known cases have tested the limits of treatment refusal. A California case upheld the right of Elizabeth Bouvia, a young woman with severe cerebral palsy and quadriplegia, to refuse a nasogastric tube [12]. A Georgia case upheld the right of Larry McAfee, a quadriplegic, to be disconnected from the ventilator sustaining his life [13]. A Michigan court ruled that a patient with amyotrophic lateral sclerosis (ALS) had the right to refuse a respirator [14]. Numerous courts have upheld treatment refusals made on religious grounds, for example refusal of blood transfusion by a Jehovah's Witness [15]. These cases are representative of a judicial consensus that the patient's right to refuse is not limited to any particular medical condition and is not dependent on the severity of illness or disease. The patient's view that continued medical interventions would mean an intolerable quality of life takes precedence over the clinician's contrary view of the patient's (future) quality of life. The right to refuse applies to any and all life-sustaining treatments. This basic right has been recited numerous times in cases involving incompetent patients and in advance directive statutes nationally. (See below.) The right of competent patients to refuse unwanted life-sustaining interventions is so well-established that today disagreements rarely end up in court.

The rights of competent patients are sometimes characterized as "near absolute." This means that clinicians have a clear and firm obligation to respect the competent patient's informed decision to refuse life-sustaining treatment. But there are recognized exceptions. The US Supreme Court *Jacobson* decision (1905) upheld compulsory vaccination against smallpox, establishing a public health exception to treatment refusal [16]. Isolated early cases denied the right to refuse treatment and ordered treatment to protect the interests of dependent children [17]. The phrase "near absolute" also allows for the possibility of rare cases where it may be unsettled and controversial whether the patient's refusal should be respected. An emerging but unresolved issue is refusal of food and fluids by mouth, as most end-of-life law expressly contemplates forgoing of artificially-provided fluids and nutrition. Another controversial issue is the patient's refusal of pain medication, given the professional duty to provide relief from pain and suffering that is codified in many places. A competent patient's request for medical aid in dying, discussed below, raises different legal concerns.

5   Competence and Decisional Capacity

All adults 18 and older are presumed competent to make their own decisions. When patients lose the capacity to decide, the locus of decisional authority shifts to a surrogate decisionmaker and/or an advance directive. The terms "competence" and "capacity" are often used interchangeably. The strict definition is that (in)competence is a legal concept that involves a judicial determination of inability to make decisions, and sometimes applies not just to medical decisions but also to those involving financial and other personal matters. A court determination is necessary for example when there is a court-appointed guardian for the patient. The court's judgment of incompetence can only be reversed by returning to court. "Decisional capacity" is more commonly used in medical ethics and clinical settings to refer to the assessments made by physicians as to whether patients are capable of making their own decisions. But the legal and clinical/ethical concepts substantially overlap. A common and useful definition of decisional capacity is the "ability to understand and appreciate the nature and consequences of health care decisions, including the benefits and risks of each, and alternatives to any proposed health care, and to reach an informed decision." [18]. In practice this definition captures what physicians consider when they determine patient (in)capacity. States may require confirmation of incapacity by a second physician and this determination should be documented in the medical record [19]. It is common practice to "call psych" to confirm incapacity, but states tend to require a psychiatry consult only when this expertise is important for particular patients, such as those with a diagnosed mental disorder or who are developmentally disabled [19].

Determinations of incapacity at the bedside are often decision-specific and focus on the demands and complexities of specific health care decisions (e.g., the patient may be capable of choosing a proxy, but may not understand the risks and benefits of heart surgery). Sometimes patients have fluctuating capacity and are able to make their own decisions in windows of lucidity. For example, some dementia patients exhibit fluctuating capacity. Respect for patient autonomy and informed consent calls on clinicians to be attentive to circumstances where patients regain capacity to make their own decisions. Periodic re-assessment of capacity may be warranted for some patients. The decision-specific approach rejects status-based judgments of incapacity. Patients who suffer from dementia or depression are not therefore decisionally incapable. Rather those conditions call for a closer look at the patient's ability to make specific contemporaneous decisions [20].

Physician determinations of patient incapacity are regularly made without the need for legal involvement and this is entirely consistent with the law. Because courts and statutes consistently use the language of competence (though the terms are sometimes used interchangeably in legal analyses) this chapter uses the term (in)competent most of the time.

6   When Families Decide for Incompetent Patients

Who decides for the incompetent patient who has not designated a health care proxy is a key threshold question. Statutes in many states establish a priority list of surrogate decisionmakers, often modeled on the long-standing approach to consent for organ donation. The hierarchy of decisionmakers commonly is: spouse; adult child; parent; adult sibling; close adult friend or relative [21]. A significant number of states, such as NY, recognize the rights of a domestic partner on a par with a spouse [21]. A court-appointed guardian, who could be a family member, would typically have priority over others on the surrogate list.

When called upon to bear the burdens of decision surrogates should seek, first and foremost, to determine what the patient would choose for him- or herself, and should also act in the patient's best interests (court decisions use the term substituted judgment). In the absence of advance directives, surrogates are called upon to recall and recount past conversations and statements indicative of the patient's wishes and values and to present a narrative of the kind of person the patient has been over a lifetime (e.g., fiercely independent, deeply religious). Most states require surrogates to show reliable, trustworthy evidence that it is more likely than not forgoing life-sustaining treatment is consistent with the patient's wishes. Missouri's more demanding clear and convincing evidence standard, upheld in *Cruzan*, is the law in only a few other states [22].

It is not uncommon for the choice of surrogate to raise ethical and legal concerns. Family members may disagree about the forgoing of treatment or have different accounts of their loved one's wishes and best interests. A person lower on the priority list may know the patient better and seem a preferred surrogate to someone higher on the list. On rare occasion the legal surrogate may take a position inconsistent with the law or act contrary to the patient's wishes, perhaps with ill motive. If the issue cannot be resolved it may be necessary to petition the court to remove the legally authorized surrogate from that responsibility, at least where there is a statutory list of surrogates.

The legal consensus holds that surrogate decisions to implement the patient's wishes to refuse life-sustaining treatment are to be honored when the patient is terminally ill or permanently unconscious. Following the rule for qualification for hospice care, a terminal condition is often defined as a prognosis of 6 months of life or less remaining. Some states consider likely death within a year to be terminal, while others state that a terminal condition means death "within a short time." It is commonly stated that terminal condition is determined "with or without provision of life-sustaining treatment." [1–3] Beyond the parameters of terminal condition and permanent unconsciousness the scope and limits of medical circumstances where forgoing treatment is permitted varies among the states. Some courts have been reluctant to authorize forgoing of treatment for non-terminally ill patients in a minimally conscious state unless a higher evidentiary standard of proof of the patient's wishes is met [23]. But the law in this regard is considered unsettled. Operative language in New York's Family Health Care Decisions Act authorizes forgoing of treatment when "provision of treatment would involve such pain, suffering or other burden that it would reasonably be deemed inhumane or extraordinarily burdensome," though not necessarily terminal [24]. Although patients have the same right to refuse feeding tubes as any other form of treatment, some states require a specific statement or indication that the patient would refuse artificially-provided fluids and nutrition in order to implement this decision. This rule reflects historical debate as to whether forgoing artificial fluids and nutrition raises special concerns (some hold that provision of fluids and nutrition is obligatory basic care) and follows *Cruzan's* ruling that states may adopt their own rules for end-of-life care so long as consistent with patients' constitutional rights.

7   When Patients Have Advance Directives

All 50 states and DC have laws that recognize the right of competent adults to write advance directives to control end-of-life decisions at a future time when they have lost the capacity to make their own contemporaneous decisions [25]. There are three types of advance directives. Proxy directives (also known as "healthcare powers of attorney") designate and empower a trusted family member, friend or advisor to engage with the physician and healthcare team to make decisions in accordance with the patient's wishes. Living wills (also known as "instruction directives") state the patient's wishes and directions to those responsible for the patient's care with some specificity but do not designate a proxy. A combined directive merges these two approaches into a single document. Nearly all states recognize all three of these approaches to advance care planning. New York, Massachusetts and Michigan laws formally recognize only the health care proxy. However, a patient's living will is still important evidence of the patient's wishes in these states that should not be ignored. Advance directives often provide the best and most reliable information about the patient's wishes [26].

7.1   Formal Requirements

Clinicians should be familiar with the formal requirements for executing a valid advance directive for at least two reasons. First, at the bedside only legally valid directives are entitled to respect. Second, clinicians play an important role in engaging patients with advance care planning. Consultation with the individual's chosen proxy and with the individual's physician is strongly recommended, particularly when the patient has a diagnosed, potentially life-threatening condition. Seeking to encourage use of advance directives, federal law allows physician reimbursement for advance care planning consultation [27]. Though advance directives are legal documents with requisite formalities individuals most often write advance directives without consulting an attorney.

Formalities for executing advance directive forms are highly uniform nationally, with some significant variations. Formalities for legally valid execution include that the document be signed, witnessed and dated. Individuals may choose a family member, close friend or religious advisor to serve as proxy, but most states prohibit the patient's physician or long-term care provider from simultaneously holding the role of physician and proxy. Many states have adopted and make available "standard" forms, however use of this form is typically optional. Any of a range of advance directive forms may be used and are entitled to respect so long as they are validly executed. Most states recognize out-of-state documents ("reciprocity"), provided the document complies with the formal requirements of either the practicing or neighboring state. Out-of-state directives are to be respected on the same basis as in-state directives, with the caveat that clinicians have no duty to comply with patient wishes that contravene the law where they practice [28, 29]. Rules for revoking a directive generally provide that the directive's author may revoke the document at any time (again, there is some variation across the states). When a competent individual writes a new document to update their wishes the new document replaces and invalidates an earlier one. (One exception would be when the earlier document is of a different type; for example, the patient designates a proxy today and has a living will from years ago, both may be valid.)

Pursuant to the federal Patient Self-Determination Act (PSDA), hospitals and other healthcare facilities are required to provide information about advance care planning to patients and families, to inquire about the patient's "advance directive status" and seek to obtain a copy for the patient's record. The law prohibits requiring anyone to write a directive as a condition of receiving care; nor can insurance companies require directives as a condition of coverage. The PSDA does not create or modify substantive rights to make end-of-life decisions under state law [30].

7.2   Decision Making with Advance Directives

As recited both in law and advance directive forms themselves, the proxy's fiduciary duty is to make decisions based on what is known of the patient's wishes and values, and secondarily to act in the patient's best interests. The proxy effectively "stands in the shoes of the patient" to engage in the process of informed consent with the physician. The principal authority of the proxy concerns end-of-life decisions, but proxies may also make related decisions for discharge to home, hospice or a nursing facility, and may permit or limit access of other family members to confidential information. Proxy directives may limit or otherwise expressly state the intended scope of the proxy's authority, though patients commonly use short-form documents that give the proxy broad authority without committing their specific wishes to writing. Clinicians are obligated to respect the rights and authority of proxies. When clinicians and facilities act in good faith to respect the patient's wishes (at the proxy's direction) to forgo treatment in accordance with accepted medical standards, statutes commonly confer immunity from civil and criminal liability and professional discipline. Immunity provisions are intended to encourage honoring advance directives in clinical practice, but their legal import has rarely been tested in court.

Clinicians sometimes believe that proxy authority is absolutely binding; that they must comply with whatever the proxy wants. However, this view is incorrect. Patients invariably choose as proxy someone they trust who faithfully carries out the responsibilities of decision in accordance with their wishes. A proxy appointment often serves to facilitate family dialogue and alleviate some of the burdens on loved ones. But on occasion other family may disagree with the proxy's decision, believe it does not reflect the patient's wishes, or even challenge the authority of the proxy himself. Family and friends are always important sources of information about the patient's wishes and should not be ignored simply because they are not the proxy. It is also possible that the proxy is unable or unwilling to act as required, or in rare cases presses for decisions clearly contrary to the patient's wishes. Legally the proxy's authority takes precedence but when serious disagreements and concerns persist and overriding a proxy's decision or removing the proxy altogether is warranted it is often necessary to resort to the courts [26].

Parallel to the judicial consensus, advance directive laws uniformly permit refusal of life-sustaining treatment when the patient is either terminally ill or permanently unconscious (including but not limited to PVS). "Terminal condition" may be defined as death within 1 year or "within a short time." A common guideline is 6 months (again, tracking hospice rules). That the patient is terminally ill or permanently unconscious typically requires the determination of the attending and a second concurring physician. It is understood in clinical practice that terminal prognoses can be uncertain, and criteria may be interpreted flexibly to honor the patient's wishes for comfort through the dying process. A minority of states allow forgoing of life-sustaining treatment when the patient has a progressive, irreversible condition that does not meet the definition of "terminal" but based on the patient/proxy wishes the burdens of continued intervention outweigh the benefits. For example, Florida law uses the term "end-stage condition;" [31] Oregon law authorizes forgoing treatment for "progressive illness that will be fatal and is an advanced stage" [32]. Patient advance directives may refuse any and all unwanted medical modalities. However, some states require a specific statement in the document to forgo artificially-provided fluids and nutrition or that the patient's refusal of fluids and nutrition is "reasonably known" [33, 34].

Designation of a healthcare proxy is the preferred and most common use of advance directives, but sometimes patients have only a living will. Clinicians are obligated to respect the patient's living will. These documents can provide clear guidance as to the patient's wishes under the circumstances, but they can also be difficult to interpret and implement when they are written remotely in time and express decisions about medical conditions or treatments that do not fit the patient's current circumstances. Living wills are framed as instructions to physicians and family members. They do not give family the authority of a proxy and contemplate that primary responsibility for implementing the terms of the document rests with the physician [35]. Family and friends (if available) remain important participants in determining the patient's wishes and should be consulted if the living will is ambiguous or unclear. It may be argued that a spouse or adult child is the proper decisionmaker under the state's surrogate priority list. However, the relationship between these laws and advance directive laws in the several states is unclear. That the patient has not chosen a proxy but does have a living will implies the patient did not have anyone they chose to trust with this responsibility.

8   POLST: Physician (Provider) Orders for Life-Sustaining Treatment

Many hospitals nationally use POLST forms, first developed in Oregon, to document orders for care and treatment near the end of life. All states have POLST programs; in some states they go by different names, for example Medical Orders for Life-Sustaining Treatment (MOLST) in NY and PAPOLST in Pennsylvania [36]. In some states POLST is recognized by statute, in others it is authorized by department of health regulation, elsewhere POLST is established through clinical guidelines that have consensus support of professional organizations, institutions and clinicians. POLST laws and governing regulations are typically procedural and are not a source of patients' substantive rights [37, 38]. POLST is used within the framework of legislation and case law described above, and the forms are commonly tailored to reflect and be consistent with the governing state law. It has been noted that clinicians using POLST may have stronger legal protections in states with enabling legislation compared to those who rely on consensus practice in the absence of legislation, but compliance with other extant law governing surrogate decision making and advance directives should largely obviate these concerns.

POLST forms allow clinicians to document in one place all end-of-life orders such as do-not-resuscitate (DNR), and thus provide a useful tool for communicating the treatment plan. (POLST forms are often brightly colored and stand out in the chart.) Physician orders implement decisions made by family, or pursuant to the patient's advance directive, and may also document competent patients' oral directives for care, treatment and choice of preferred surrogate. POLST is typically completed in consultation with the physician, documents contemporaneous decisions and orders for end-of-life care for patients being treated (usually) in hospital, and bears the physician's signature to end-of-life orders. POLST forms are portable and should follow the patient when transferred from one facility to another. As with advance directives, clinicians should be familiar with the content and formalities for executing POLST forms [37, 38]. When used to establish out-of-hospital DNR orders (permitted in a few states) [38] clinicians should be aware of separate legislation governing out-of-hospital DNR orders that may exist in their state of practice.

9   When the Patient's Wishes are Unknown

Sometimes there is no evidence of the incompetent patient's wishes and end-of-life decisions must be based solely on the patient's best interests. There are two types of patients for whom this applies. Some patients have left no evidence of their wishes and have no one to act as their surrogate decision maker (the "unbefriended" patient). End-of-life decisions must sometimes be made for developmentally disabled patients who have never been competent. In either situation there may be a guardian appointed to act on the patient's behalf. A state agency may have authority to make or oversee decisions for developmentally disabled patients who are wards of the state or otherwise under the state's purview. For example, NY law authorizes a "two physician DNR order" for unbefriended patients under certain circumstances, and requires involvement of the Office for People with Developmental Disabilities when end-of-life decisions are contemplated for patients within the agency's purview [39]. The law governing end-of-life decisions for unbefriended and developmentally disabled patients is complex and beyond the scope of this chapter.

10   When the Patient is Incarcerated

Clinicians sometimes provide end-of-life care for incarcerated patients. The incarcerated patient has forfeit his or her freedom, but retains the right to healthcare. In *Estelle* (1976) the Supreme Court held that the right to health care for inmates is guaranteed by the US Constitution. Central to this principle is that access to care is controlled by the prison system that must therefore provide access for its inmates who have been deprived of liberty to seek health care on their own [40]. While incarcerated patients have the right to consent to or refuse treatment including life-sustaining treatment, to execute advance directives and to have family members decide on their behalf when they have lost capacity for self-determination, these rights can be more limited than for the general population. Some courts have upheld the right of correctional facilities to impose treatment over the objection of inmates when doing so would protect institutional interests in the safety, security and health of the prison population. It has been held that a violent prisoner can be medicated without consent [41], and that a prisoner must accept routine vaccination for diphtheria-tetanus or face isolation in order to prevent spread of infection in the prison population [42]. Reportedly some correctional facilities have denied inmates the opportunity to write advance directives, though the legal basis for doing so is questionable. Unbefriended patients who perhaps have seriously damaged relationships with family or no family at all to speak on their behalf are especially vulnerable. Both law and practice vary with respect to the role of corrections officials in end-of-life decisions for unbefriended inmates and even for those with involved family [43, 44].

11   Medical Futility

Patient and/or family insistence on continued provision of treatment that clinicians consider ethically and clinically non-beneficial, harmful or otherwise inappropriate, commonly referred to as medical futility, presents challenging dilemmas. This is a slowly developing and largely unaddressed area of end-of-life law. Among the few states with legislation, Texas permits physicians to forgo futile treatment over family objection with approval of a hospital committee [45]. A number of states provide that clinicians may refuse requests for life-sustaining treatment where contrary to accepted medical standards or treatment would not provide significant benefit. These laws, known as "unilateral decision statutes," appear to support forgoing treatment despite family objection but they have not been tested and their meaning is unclear. One leading legal commentator characterizes these laws as purely enabling, and concludes they likely do not offer a safe harbor of protection from legal liability if physicians override family/proxy requests for futile treatment [46]. Uncertainty about the law often inclines clinicians to accede to proxy/family demands. When medical futility disputes go to court they are sometimes framed as "wrongful life" cases, discussed below.

12   Professional Conscience

As participants and partners in end-of-life decisions, physicians, nurses, and other clinicians also have rights to decline to participate in forgoing life-sustaining treatment if doing so would violate their sincerely held personal or professional convictions. This right of professional conscience is of long-standing, is found in both statute and case law, and applies to a range of issues. Advance directive laws are a model for how professional conscience is honored and implemented. These laws commonly provide that in the exercise of professional conscience physicians and other clinicians should act in good faith to inform patient, proxy and family of their objection and desire to withdraw from the patient's care, and should follow institutional policy for withdrawal and transfer to another clinician or facility. There should be a timely and respectful transfer of care. Pending transfer to another clinician willing to follow patient/proxy/family wishes, care and treatment is continued to assure the patient is not abandoned.

13   Ethics Committees and Ethics Consultants

Ethics committees and ethics consultants can be found in hospitals and other healthcare facilities across the country. Ethics committees typically engage in the core functions of policy review and institutional education, and may discuss select cases, usually retrospectively. Ethics consultants are available to assist at the bedside when challenging ethical issues in patient care arise. Ethics consultants may function as members of the ethics committee or independently as part of an ethics consult service. The rapid growth of ethics committees and consultants can be traced to the 1992 Joint Commission on Accreditation of Healthcare Organizations accreditation standard that facilities establish and maintain "a mechanism for the consideration of ethical issues arising in the care of patients…" [47]. There is otherwise very little law directly applicable to the conduct of ethics committees or consultants. Only a handful of states (e.g., Texas [45] and Maryland [48]) have relevant statutes.

Ethics consultants are important resources for clinicians. The ethics consultant works to identify ethical issues, clarify misunderstandings and to help resolve disagreements. But the ethics consultant's role is advisory only. Authority and responsibility for treatment decisions rests in the physician-patient-family relationship. It should be noted that requests for ethics consultation are typically not limited to physicians. In many hospitals anyone with direct involvement in the patient's care, including physicians, nurses, social workers and family members may request an ethics consult.

14   Legal Liability and "Wrongful Prolongation of Life"

Decisions to forgo life support are made every day within the privacy of the patient-family-physician-institution relationship, without involvement of the courts. Though litigation is extremely rare compared to the frequency of end-of-life decisions, clinicians may be understandably concerned that overtreatment or undertreatment in the face of adamant opposition by family or proxy can lead to lawsuits (or the threat of one). The cases that form the judicial consensus were essentially declaratory judgment actions that involved determining the rights and obligations of the parties, but did not seek to impose legal liability on clinicians. Advance directive statutes, court decisions and other laws provide immunity from legal liability for clinicians acting in good faith to implement the patient's wishes.

Families have sometimes sued for failure to respect patient and family refusal of treatment that results in unwanted prolongation of life. "Wrongful prolongation of life" cases tend to be based on a number of legal theories; courts have been more receptive to some than to others. Early wrongful life cases consistently ruled in favor of clinicians, but recently courts have been more receptive to claims that dying patients are injured when their treatment refusals are not honored, particularly if the patient has written an advance directive [49]. To illustrate, the Ohio Supreme Court (1995) denied relief to the family of an 82 year-old patient with a history of chronic heart disease and heart attacks who suffered a paralyzing stroke after being resuscitated despite a DNR order entered at his direction [50]. But the Georgia Supreme Court (2016) recognized as valid the proxy's claims for violation of her terminally ill 91 year-old grandmother's rights when she was intubated and placed on mechanical ventilation despite the proxy's refusal of ventilation in accordance with her grandmother's previously stated wishes [51].

Clinicians should be aware that poor communication with families has been associated with the risk of lawsuits. Good communication skills have been associated with decreased risk of legal entanglements [52, 53]. Families who feel respected and supported even in the face of serious disagreement are less likely to pursue lawsuits.

15   Medical Aid in Dying

The term medical aid in dying refers to requests from competent, terminally ill patients for a clinician to write a prescription to be self-administered by the patient with the purpose of hastening the dying process. In most states this practice is illegal and writing the prescription risks criminal prosecution. But the law has been slowly trending toward legalization of medical aid in dying. Two cases decided by the US

Supreme Court (1997) rejected challenges to state criminal laws in NY and Washington state and ruled that there is no constitutional right to medical aid in dying (the cases used the phrase physician-assisted suicide). The Court drew a bright line between refusal of unwanted life-sustaining treatment and requests for assistance in dying, calling them both factually and legally distinct. These cases also held that whether or not to legalize medical aid in dying, as Oregon had already done, should be determined "in the laboratory of the states" [54, 55].

As of this writing, ten states and DC have enacted laws that authorize medical aid in dying. The first such law was enacted in Oregon in 1994 (Death with Dignity Act). States have closely followed the Oregon model, and the laws all contain very similar features, if not always the exact same language. In summary, a competent adult who is terminally ill (prognosis of 6 months or less of life remaining) may request a prescription for medication to be self-administered by the patient with the purpose of hastening death. Some states limit prescribing authority to physicians, others give prescribing authority to other clinicians as well (e.g., nurse practitioners). Clinicians must ensure the patient is competent and making an informed and voluntary decision. There is a waiting period that requires the patient to re-affirm the request is voluntary and informed. Clinicians who participate in this process in good faith have immunity from liability. Clinicians with a conscientious objection to medical aid in dying are not obligated to comply with the patient's request, but there may be a duty to refer the patient to another clinician who will respect the patient's decision [56].

Medical aid in dying should not be confused with the distinct concept of "terminal sedation," the practice of sedating a dying patient into unconsciousness for the purpose of relieving pain and suffering. Terminal sedation is a common practice that occurs within the bounds of the law [57]. Legal issues surrounding pain management and palliative care for dying patients are beyond the scope of this chapter.

16   Application

Cases: *In re Farrell, 529 A.2d 404 (1987); In re Peter, 529 A.2d 419 (1987); In re Jobes, 529 A.2d 434 (1987).* This trilogy of cases decided by the NJ Supreme Court embodies many of the pillars of today's legal-ethical consensus. All three cases involved withdrawal of life-sustaining treatment from women suffering from incurable and irreversible disease.

Kathleen Farrell, 37 years old, lived at home, paralyzed and in need of 24/7 nursing care with an advanced terminal stage of ALS. Reciting the time-honored common law right of self-determination to control what happens to one's own body, the court upheld Ms. Farrell's right as a competent adult to have her voluntary informed request for removal of the respirator sustaining her life respected.

Hilda Peter was a 65-year-old PVS patient in a nursing home. When previously competent she had appointed her good friend Mr. Johanning with durable power of attorney. Based on his knowledge and conversations with Ms. Peters he believed that she would want the nasogastric tube sustaining her life removed. The court ruled that this decision was authorized as an exercise of her right when competent to choose whether or not she wanted life-sustaining treatment and to choose who should be able to make that decision on her behalf.

Nancy Ellen Jobes was a 31-year-old nursing home patient whose life in PVS was sustained by a jejunostomy tube. Her husband requested removal of the feeding tube to allow her to die. Based on testimony of family and friends, the court found there was sufficient trustworthy evidence that Ms. Jobes would not want her life prolonged in this manner and that her husband could direct withdrawal of the j-tube in the exercise of substituted judgment.

Each of these cases came before the judiciary due to uncertainty about the scope and limits of the right to refuse treatment and the obligations of clinicians and institutions. That uncertainty no longer exists. Collectively these cases rest squarely on the right of all patients, competent or incompetent, to refuse life-sustaining treatment, rights that are protected by the constitution and the common law right of self-determination and bodily integrity. They also represent other key pillars of the consensus, including that there is no distinction between withholding and withdrawing treatment and no legal difference between refusal of artificially-provided fluids and nutrition and other medical modalities. This trilogy of cases also stands for the rule that patients' rights to refuse treatment do not depend on whether they are in hospital, nursing home, or at home. To protect the interests of dying patients the court set forth procedural rules for effectuating patients' treatment refusals in these different care settings, for example confirmation of PVS for patients in nursing homes (*Peter* and *Jobes*) and of the patient's competent informed treatment refusal when at home (*Farrell*). Importantly, the court also granted judicial immunity from civil and criminal liability to all those who act in good faith and in accordance with the procedures set forth in the opinions to effectuate the patient's wishes to refuse treatment.

16.1 Clinical Vignette 1

Mr. B is a 67-year-old man with bladder cancer, multiple metastases to the lung, liver and spleen, and sepsis. He has been deemed to lack decisional capacity by two physicians. Mrs. B was appointed as healthcare proxy several years ago. The medical team believes that continuing chemotherapy and other aggressive measures would be "futile" and recommends a palliative care plan. Mrs. B has agreed to a DNR/DNI order, but insists on other aggressive measures including chemotherapy. The patient's two adult children agree with the physician's palliative care recommendations and insist on making their father comfortable through the dying process. The team has called for an ethics consult hoping to resolve this disagreement. But in the family meeting Mrs. B remains steadfast, and the children threaten to contact their attorney.

This case presents several challenging issues some of which fall within and others outside the parameters of the extant legal consensus. Clinicians believe palliative care for Mr. B is the best ethical and clinical path, but the law does not support simply overriding the proxy's insistence on aggressive measures. Seeking court removal of the proxy is an option but is unlikely to succeed on futility grounds alone. The case against the proxy would be stronger if the children recount trustworthy statements their father made showing he would not want to prolong the dying process in this way, contrary to Mrs. B's decision. (The children also have the option to seek removal of their mother as proxy.) The physician or other clinician with strong objection could seek to withdraw from the case in accordance with institutional policy, provided there is an appropriate transfer of care and the patient is not abandoned.

Family conflict is a common reason to call for an ethics consult and this often serves to facilitate an acceptable resolution. Deterioration of the patient's condition with time often brings greater clarity. But clinicians are understandably reluctant to pursue legal recourse to overrule or remove the proxy. Given Mrs. B's clear legal authority as proxy it is unlikely a lawsuit for wrongful prolongation of their father's life would be successful here. Continued dialogue respectful of all family members may resolve the issues and decreases the likelihood of such a lawsuit.

16.2   Clinical Vignette 2

Mr. C is a 75 year-old man who has struggled with advancing Parkinson disease for many years. This morning the home health aide found Mr. C unconscious on the floor of his apartment. She quickly called 911, and then called Mr. C's daughter. Within minutes of the ambulance arrival at the ER Mr. C's daughter rushes into the ER asking for her father. The patient is unconscious and unresponsive, and needs immediate intubation, IV fluids and pressors to sustain his life. He is highly unlikely to regain consciousness even with a "full court press." The EMS technician hands the physician a handwritten note and an empty bottle of prescription pain pills that the home health aide found next to Mr. C's body. The note recites his history of struggles with this progressive, irreversible disease and states his wish to refuse life-sustaining interventions should he be hospitalized. The note also says in part: "I would even take my own life if my suffering becomes too much to bear." It is addressed to "my daughter and my doctors." Mr. C's daughter confirms that they have discussed her father's wishes for the end of life and that he was adamant in his refusal of life-sustaining treatment. She insists that he not be intubated.

On its face this case asks whether Mr. C's apparent attempted suicide vitiates the obligation to forgo life-sustaining interventions. Clinicians may ask whether complying with the daughter's decision involves assisting the patient's suicide attempt. Initially, the emergency team has a duty to diagnose and stabilize the patient's condition under the Emergency Medical Treatment and Active Labor Act (EMTALA) [58]. (see Chap. 6.) They should not delay rescue efforts to explore and resolve any uncertainties about the patient's wishes if life hangs in the balance. The related question upon admission is whether the daughter is the appropriate surrogate for the patient. It is discovered that Mrs. C died 2 years earlier and that the daughter is the only adult child. As the sole family member and surrogate Mr. C's daughter has legal authority to refuse continued interventions in accord with her father's wishes and best interests. The patient's history and suicide note invite inquiry into whether Mr. C's wishes were influenced by depression or perhaps Parkinson dementia. Concerns about the patient's prior mental status may cast doubt on the daughter's validation of his wishes, but the patient's possible incapacity should not be presumed nor should his wishes be discounted based on suspicions that cannot be verified and are contradicted by his daughter. In this case the patient's prior competent wishes not to have his compromised quality of life continued by life-sustaining interventions, committed to writing and affirmed by his daughter based on her reports of conversations between father and daughter should be honored. Though Mr. C's hospitalization and critical situation was precipitated by his apparent suicide attempt, not to honor his surrogate's decision would violate the patient's considered intentions to take charge of how he would die in the face of a progressive debilitating disease. Understood in this way, to withdraw intubation and make the patient comfortable is to respect his wishes and his daughter's authority, not assisting in suicide. (Note that here Mr. C's daughter is still the proper surrogate and his wishes still matter even though set forth in a handwritten note without the formalities of recognized advance directives.)

[This vignette is modeled on "The Suicide Note" and case commentary [59]. The facts of the vignette and some of the issues presented have been changed for purposes of this chapter.]

Note: This would be a different case if the patient were admitted following a suicide attempt with no prior underlying life-threatening condition. Compared to Mr. C's case, clinicians may feel strongly that withdrawal of life-sustaining treatment means assisting in suicide, and may be more likely to object and seek not to participate in this decision.

If the patient's condition is reversible and continued interventions could reasonably lead to discharge there is a strong argument for following this path. However, if despite aggressive measures the patient is likely to remain in serious condition and dependent on life-sustaining treatment, patients and families have the right to forgo continued treatment based on the patient's wishes and best interests pursuant to the consensus described here.

17   Summary

All 50 states and DC recognize the rights of both competent and incompetent patients to refuse unwanted life-sustaining treatments. The right to refuse treatment is protected by the US constitution, by some State constitutions, and is also grounded in the common law right of self-determination and bodily integrity. Refusal of life-sustaining treatment has been the subject of numerous court cases and extensive legislation. For competent patients the right to direct withholding or withdrawal of treatment is near absolute and is based on contemporaneous decisions made in the process of informed consent. For incompetent patients, this right may be exercised on their behalf by family members acting in accordance with the patient's previously expressed wishes and/or best interests. End-of-life decisions for now incompetent patients are also made in accordance with patients' advance directives for healthcare. All 50 states and DC recognize the right to plan ahead for a time when the ravages of illness and disease have taken the capacity to decide by putting our wishes regarding life-sustaining treatment in writing in the form of an advance directive.

The core principles of the end-of-life consensus can be summarized as follows:

- Competent patients have the right to refuse unwanted treatment, including life-sustaining treatment and to have their treatment refusals respected. This right is based on the Constitution and the common law right of self-determination and is protected by the doctrine of informed consent.
- Incompetent patients have the same right to refuse life-sustaining treatment as competent patients, but the rights of incompetent patients are exercised in a different manner. Family members or other surrogate decisionmakers should act first and foremost in accordance with the patient's previously expressed wishes, and in the patient's best interests.
- Competent adults may execute advance directives to control decisions near the end of life at a future time of decisional incapacity. The patient's previously expressed wishes should be respected by proxies, families, physicians and other clinicians.
- Patients may refuse any and all forms of life-sustaining treatment. There is no legal distinction between respirators, feeding tubes, CPR, antibiotics or other interventions.
- There is no legal distinction between withholding and withdrawal of treatment. When treatment is withheld or withdrawn from a dying patient this allows the patient's underlying disease process to take its course. Implementing this decision does not constitute killing, assisted suicide or suicide.
- State laws and judicial decisions impose a number of limitations on the rights of incompetent patients to refuse life-sustaining treatment that do not apply for competent patients, such as the requirement in many (but not all) states that the patient have a terminal condition or a state of permanent unconsciousness before forgoing treatment is authorized.
- Medical aid in dying is illegal in most states. There is a slowly growing trend toward legalization.

When legal-ethical dilemmas arise in patient care, ethics consultation can help to clarify ethical issues and to resolve disagreements. Clinicians should be aware of their rights of professional conscience to withdraw from patient care, provided there is an appropriate transfer of care. Current law does not offer strong support for overriding patient/family insistence on continued treatment in medical futility cases, but this area of law continues to develop.

End-of-life law is very much a matter of state law. Clinicians should be familiar with the law of their practicing state.

References

1. Meisel A, Cerminara KL, Pope TM, editors. The Right to Die: The Law of End-of-Life Decisionmaking. 3rd ed. Wolters Kluwer; 2016.
2. Olick RS.Taking Advance Directives Seriously: Prospective Autonomy and Decisions Near the End of Life. Wash., D.C.: Georgetown University Press; 2001.
3. Cantor NL. Advance Directives and the Pursuit of Death with Dignity. Bloomington, IN: Indiana University Press; 1993.
4. Meisel A.The legal consensus about forgoing life-sustaining treatment: its status and prospects. Kennedy Institute of Ethics Journal 1992;2(4):309-345.
5. Uniform Determination of Death Act (1981), 12A Uniform Laws Annotated 777 (West 2008).
6. Lewis A, Bonnie RJ, Pope TM et al. Determination of Death by Neurologic Criteria in the United States: The Case for Revising the Uniform Determination of Death Act. J Law, Medicine & Ethics 2019;47(4):9-24.
7. New York State Department of Health and New York Task Force on Life and the Law. Guidelines for Determining Brain Death (Nov. 2011). 2011. https://www.health.ny.gov/professionals/hospital_administrator/letters/2011/brain_death_guidelines.htm. Accessed 21 Nov 2021.
8. Russell JA, Epstein LG, Greer DM, Kirschen M, Rubin MA, Lewis A. Brain death, the determination of brain death, and member guidance for brain death accommodation requests: AAN position statement. Neurology 2019;92:228-232.
9. Olick, RS. Brain Death, Religious Freedom, and Public Policy: New Jersey's Landmark Legislative Initiative. Kennedy Institute of Ethics Journal 1991;1:275-292.
10. In re Quinlan, 355 A.2d 647, cert. denied sub nom. Garger v. New Jersey, 429 U.S. 922 (1976).
11. Cruzan v. Director, Missouri Dept. of Health, 497 U.S. 261 (1990).
12. Bouvia v. Superior Court, 179 Cal. App. 3d 1127 (1986).
13. State v. McAfee, 259 Ga. 579 (1989).
14. In re Culham, No. 87-340537-AZ, slip op. (Cir. Ct. Mich., Dec. 15, 1987).
15. In re Osborne, 294 A.2d 372 (D.C. 1972).
16. Jacobson v. Massachusetts, 197 U.S. 11 (1905).
17. Application of President & Directors of Georgetown College, 331 F.2d 1000 (D.C. Cir.), cert. denied, 377 U.S. 978 (1964).
18. NJ Rev Stat 26:2H-55 (2013).
19. NY Public Health Law §2994-C (2015).
20. Ganzini L, Volicer LL, Nelson WA, Fox E, Derse AR.Ten myths about decision-making capacity. JAMDA 2005;6:S100-S104.
21. American Bar Association Commission on Law and Aging. Default Surrogate Consent Statutes (Sept. 2019). https://www.americanbar.org/content/dam/aba/administrative/law_aging/2019-sept-default-surrogate-consent-statutes.pdf. Accessed 12 Nov 2021.
22. Meisel A, Snyder L, Quill T. Seven Legal Barriers to End-of-Life Care: Myths, Realities, and Grains of Truth. JAMA 2000;284(19):2495-2501.
23. In re Martin, 538 N.W.2d 399 (Mich. 1995).
24. NY Public Health Law §2994-d(5) (2015).
25. American Bar Association Commission on Law and Aging. State HealthCare Power of Attorney Statutes: Selected Characteristics (Sept. 2019). https://www.americanbar.org/content/dam/aba/administrative/law_aging/2019-sept-state-health-care-power-of-attorney-statutes.pdf. Accessed 21 Nov 2021.
26. Olick RS. Defining features of advance directives in law and clinical practice. Chest 2012; 141:232-238.
27. Barwise AK, Wilson ME, Sharp RR, DeMartino ES. Ethical Considerations About Clinical Reimbursement for Advance Care Planning. Mayo Clinic Proceedings 2020;95(4):653-657.
28. National Hospice and Palliative Care Organization. Caring Info:Advance Directives. https://www.caringinfo.org/planning/advance-directives/. Accessed 21 Nov 2021.

29. Sabatino CP. Overcoming the Balkanization of State Advance Directive Laws. Journal of Law, Medicine & Ethics 2018;46(4):978-987.
30. Ulrich LP. The Patient Self-Determination Act: Meeting the Challenges in Patient Care. Wash., D.C.: Georgetown University Press; 1999.
31. Fla Stat Ann §765.302 (West 2020).
32. Or Rev Stat §127.540 (Or 2009).
33. NY Public Health Law §2982 (2014).
34. 20 Pa CSA §5456 (West suppl 2011).
35. NJ Rev Stat 26:2H-64 (2013).
36. National POLST. POLST Program Names. https://polst.org/program-names/. Accessed 12 Nov 2021.
37. Hickman SE, Sabatino CP, Moss AH, Nester JW. The POLST (Physician Orders for Life-Sustaining Treatment) paradigm to improve end-of-life care: potential state legal barriers to implementation. J Law Med Ethics 2008;36(1):119-140.
38. Pope TM, Hexum M. Legal Briefing: POLST: Physician Orders for Life-Sustaining Treatment. The Journal of Clinical Ethics 2012; 23(4):353-376.
39. NY Public Health Law §2994, et seq (2015).
40. Estelle v. Gamble, 429 US 97 (1976).
41. Washington v. Harper, 494 US 210 (1990).
42. Zaire v. Dalsheim, 698 F.Supp. 57 (S.D.N.Y. 1988), aff'd 904 F.2d 33 (2d Cir. 1990).
43. Tobey M, Simon L. Who Should Make Decisions for Unrepresented Patients Who are Incarcerated? AMA Journal of Ethics. https://journalofethics.ama-assn.org/article/who-should-make-decisions-unrepresented-patients-who-are-incarcerated/2019-07. Accessed 12 Nov 2021.
44. Natterman J, Rayne P. The Prisoner in a Private Hospital Setting: What Providers Should Know. J Health Care L & Pol'y 2017;19(1):119-147.
45. Tex. Health & Safety Code §166.046.
46. Pope TM. Medical Futility Statutes: No Safe Harbor to Unilaterally Refuse Life-Sustaining Treatment. Tennessee L. Rev. 2007;71(1):1-81.
47. Joint Commission on Accreditation of Healthcare Organizations, 1994. Accreditation Manual for Hospitals, Vol. 1, Standards (Oakbrook Terrace, Ill.: Joint Commission, 1994), R1.1.1.6.1.
48. Md. Code Ann., Health—Gen. §§19-370 et seq. (Michie 1996).
49. Hodge SD. Wrongful Prolongation of Life – A Cause of Action That May Have Finally Moved into the Mainstream. Quinnipiac L Rev 2019;37:167-198.
50. Donohue J. "Wrongful Living": Recovery for a Physician's Infringement on an Individual's Right to Die. Journal of Contemporary Health Law & Policy 1998;14(2):391-419.
51. Doctors Hosp. of Augusta, LLC v. Alicea, 788 S.E.2d 392 (Ga. 2016).
52. Levinson W, Roter DL, Mullooly JP, Dull VT, Frankel RM. Physician-patient communication: the relationship with malpractice claims among primary care physicians and surgeons. JAMA. 1997;277:553-559.
53. Lo B, Quill T, Tulsky J, for the ACP-ASIM End-of-Life Care Consensus Panel. Ann Intern Med 1999; 130: 744-749.
54. Washington v. Glucksberg, 521 U.S. 702 (1997).
55. Vacco v. Quill, 521 U.S. 793 (1997).
56. Compassion & Choices. Understanding Medical Aid in Dying. https://compassionandchoices.org/end-of-life-planning/learn/understanding-medical-aid-dying/. Accessed 21 Nov 2021.
57. Cantor NL, Thomas III GC. The Legal Bounds of Physician Conduct Hastening Death. Buffalo L. Rev. 2000;48(1):83-173.
58. Emergency Medical Treatment and Active Labor Act, 42 U.S.C.A. §1395dd (2011).
59. d'Oronzio JC. The Suicide Note. Cambridge Quarterly of Healthcare Ethics 2002;11:422-431.

Part XIII

Right to Healthcare

Is There a Right to Health in U.S.?

Nicholas J. Diamond and Alice Hall-Partyka

Contents

Key Points
- The right to health exists in international instruments and governments have the obligation to progressively realize the right to health over time.
- The U.S. does not generally recognize positive rights and the right to health does not exist under U.S. law.
- Several judicial decisions, primarily from the U.S. Supreme Court, have defined new rights that resemble the right to health.
- Over the last century, Congress has enacted various laws that increase coverage for and access to healthcare, which achieve some of the goals envisioned by a right to health.

N. J. Diamond (✉)
Georgetown University Law Center, Washington, DC, USA
e-mail: njd9@georgetown.edu

A. Hall-Partyka
Crowell & Moring LLP, Los Angeles, CA, USA
e-mail: ahallpartyka@crowell.com

Legal Concepts
- Rights are entitlements to perform, or not perform, certain actions.
- Under international law, the right to health provides for the enjoyment of the highest attainable standard of physical and mental well-being.
- U.S. law does not recognize a right to healthcare.
- Various U.S. laws guarantee healthcare coverage for certain U.S. citizens and residents and implement other protections relating to the availability, accessibility, acceptability, and quality of healthcare.

1 Introduction

What does it mean for "health" or "healthcare" to be recognized as a right? Under the law, rights can be understood as entitlements. They might be entitlements to perform (or not perform) certain actions. They might also be entitlements that others perform (or not perform) certain actions. A right to "health" acknowledges that all individuals are entitled to enjoy the highest attainable standard of health.

Domestically, rights may be provided for in a country's founding documents, such as the fundamental civil and political liberties found in the U.S. Constitution, or created through legislative or judicial processes. Internationally, rights may be provided for in treaties entered into between countries and their supporting documents.

In the U.S., a right to health is not formally acknowledged, and there are no protections for such a right in the Constitution or through the U.S.'s participation in international treaties. Therefore, under the law, U.S. citizens are not granted a general entitlement to be healthy or to receive healthcare services. However, there are protections created through legislative and judicial processes that resemble a right to health in the U.S.

Clinicians may encounter these issues in caring for patients—whether through a patient's eligibility for coverage for a particular healthcare service, the cost-sharing that the patient is expected to pay for that service, or the specific requirements that a clinician must comply with when providing services to a patient covered under a government-funded program, such as Medicare or Medicaid.

2 Health as a Human Right

2.1 The Right to Health Under International Law

Under international law, the right to health can be traced back to the Constitution of the World Health Organization, signed in 1946 under the auspices of the United Nations (UN). The Constitution provides that "[t]he enjoyment of the highest attainable standard of health is one of the fundamental rights of every human being without distinction of race, religion, political belief, economic or social condition" [1]. It likewise broadly defines health as "a state of complete physical, mental and social well-being and not merely the absence of disease or infirmity."

The Universal Declaration of Human Rights, adopted under the auspices of the UN in 1948, provides that "[e]veryone has the right to a standard of living adequate for the health and well-being of himself and of his family, including food, clothing, housing and medical care and necessary social services" [2]. A more specific commitment to the right to health appears in the International Covenant on Economic, Social and Cultural Rights (ICESCR), which "recognize[s] the right of everyone to the enjoyment of the highest attainable standard of physical and mental health" [3].

General Comment No. 14 clarifies the content of the right to health under the ICESCR, expressing that this right encompasses four components—availability, accessibility, acceptability, and quality [4]. Availability refers to a "sufficient quantity" of healthcare facilities, goods, services, and programs, with the acknowledgement that such a quantity will vary by country. Accessibility emphasizes that such healthcare facilities, goods, services, and programs should be accessible without discrimination. It likewise recognizes the physical dimension of accessibility, especially for vulnerable or rural populations, as well as the economic dimension of accessibility, which encompasses affordability with a specific emphasis on equity. Acceptability refers to the goal of providing healthcare facilities, goods, services, and programs in both an ethically and culturally appropriate manner. Finally, quality underscores the need for healthcare facilities, goods, services, and programs to be "scientifically and medically appropriate and of good quality," such as employing appropriately trained healthcare personnel and supplying unexpired medicines.

Governments have the obligation to progressively realize the right to health over time. The right to health follows the traditional tripartite human rights obligations to respect, protect, and fulfill. "Respect" includes refraining from, among other practices, denying or limiting equal access to health services of all types, enforcing discriminatory practices, and imposing discriminatory practices relating specifically to women's health status and needs. "Protect" includes, but is not limited to, adopting measures ensuring equal access to health services, ensuring that privatized health services remain available, accessible, acceptable, and of good quality, and guaranteeing the appropriate training of healthcare personnel. "Fulfill" requires governments to adopt legislative, administrative, budgetary, judicial, promotional, and other measures towards the full realization of the right to health.

2.2 Is There a Right to Health Under U.S. Law?

The U.S. Constitution enshrines in U.S. law a system of primarily negative, rather than positive, rights. Negative rights are prohibitory constraints on governmental power, such as not interfering with the free exercise of religion, while positive rights are affirmative duties with which the government must comply, such as guaranteeing a defendant's right to an attorney in a criminal trial. Therefore, most fundamental civil and political liberties provided for under the U.S. Constitution are prohibitions on governmental power, as opposed to affirmative duties.

Under international law, the right to health is a positive right because it establishes several affirmative obligations on governmental power. No such right to health exists in the U.S. Constitution, which neither establishes a right to *healthcare* (i.e., services related to health) nor health *insurance* (i.e., financial coverage of services related to health and medical care). The U.S. could decide, under its constitutional power to enter into treaties, to ratify the ICESCR, which would require a right to health to then be created under U.S. law. However, the U.S. has only signed, not ratified, the ICESCR [5].

As discussed below, there are several legislative and judicial examples of what resembles a right to healthcare or health insurance, or perhaps even a right to health, under U.S. law. However, no such right to health—in the broad sense of the right to health under international law—exists as a fundamental liberty under the U.S. Constitution.

3   Judicial Approach

The judiciary is constitutionally tasked with interpreting U.S. law. As a result, the judiciary can define and shape the content of existing rights and, in certain instances, define new rights. Several judicial decisions regarding health, primarily from the U.S. Supreme Court, have defined new rights that resemble a right to health.

For example, in *Griswold v. Connecticut* (1965), the Court announced a newly defined right, the so-called right to privacy, which *Roe v. Wade* (1973) and its progeny subsequently applied to establish a woman's right to make reproductive decisions, including the right to decide to have an abortion. As another example, in *Estelle v. Gamble* (1976), the Court established the right to adequate medical care for prisoners. Later, in *Cruzan v. Director, Missouri Department of Health* (1990), the Court established the right to refuse medical treatment, based on a right to privacy and bodily integrity.

Each of these examples demonstrates the establishment of a right that relates to health and, more narrowly, to *healthcare*. While patients may benefit from these protections, these judicial decisions do not equate to a fundamental right to health. In contrast to the right to health under international law, they also do not establish positive obligations on the U.S. government to take steps to respect, protect, and fulfill these judicially created rights.

4   Legislative Approach

Even if there is no "right" to healthcare in the U.S., Congress has enacted legislation that increases access and availability of healthcare services and guarantees health coverage for some of the country's most vulnerable populations, such as children and older adults. Over the past 60 years, Congress has gradually expanded access to government-funded healthcare programs to millions of U.S. citizens and residents who may not otherwise have health coverage. For an even broader number of U.S. residents, it has taken numerous steps to further increase access to healthcare and expand protections for individuals who may have trouble obtaining and affording it. By breaking down access, availability, and coverage barriers, Congressional actions may provide some of the same benefits as a (positive) right to healthcare.

4.1   Government-Funded Programs to Guarantee Health Coverage

One of the most significant steps taken by Congress to expand access to health coverage was the passage of the Social Security Amendments in 1965, which created the Medicare and Medicaid programs [6]. These programs were designed to provide basic insurance to qualifying individuals who did not have health insurance, but they have evolved and expanded over the years.

The Medicare program was designed as a healthcare program to ensure that elderly individuals had access to healthcare, regardless of income or medical history. Today, the Medicare program provides health coverage to U.S. citizens and legal permanent residents over the age of 65, along with people of all ages with certain disabilities or with End-Stage Renal Disease.

Over the years, the Medicare program has expanded and changed. Notably, the Balanced Budget Act of 1997 formalized the use of contracted private health organizations (called Medicare Advantage organizations) to administer benefits and manage care for certain beneficiaries (Medicare Part C) [7]. The Medicare Prescription Drug, Improvement, and Modernization Act of 2003 expanded prescription drug benefits (Medicare Part D) [8]. In 2020, over 62 million Medicare beneficiaries in the U.S. were enrolled in hospital and/or medical benefits, with almost forty percent of those individuals enrolled through Medicare Advantage organizations [9].

While Medicare eligibility is largely based on age, Medicaid is a guarantee of health coverage for certain low-income individuals and families. Unlike Medicare, which is a federal program administered by the Centers of Medicare and Medicaid Services (CMS) within the U.S. Department of Health and Human Services, Medicaid is jointly administered by state agencies who receive some oversight and guidance from CMS. Originally, Medicaid eligibility included low-income families, pregnant women, people with disabilities, and people who need long-term care.

As with Medicare, however, Medicaid has significantly expanded beyond its original roots. For instance, various states expanded Medicaid to provide coverage for developmentally disabled children. In 1997, Congress buttressed this trend by establishing another program—the Children's Health Insurance Program (CHIP)—which provides health insurance for children in families with income too high to qualify for Medicaid but too low for private coverage and is often operated by state Medicaid agencies [10]. As of November 2020, there were almost 80 million individuals enrolled in either the Medicaid program or CHIP [11].

More recently, in 2010, the Patient Protection and Affordable Care Act (ACA) sought to expand Medicaid to other low-income adults through "Medicaid Expansion" [12]. As designed, the ACA expanded eligibility for Medicaid to all adults with incomes up to 138% of the federal poverty level, with the federal government committing to cover the majority of expenses associated with this new expansion population. However, in a 2012 decision [13], the U.S. Supreme Court held that states could not be required to expand Medicaid. As a result, a dozen states have not expanded coverage, leaving many low-income adults in these states without access to affordable health coverage [14].

The ACA also aimed to provide affordable coverage to individuals through the development of marketplaces to purchase health insurance [15]. The marketplaces were intended to make the purchase of health insurance easier and more affordable. Further, low-income households with incomes above the Medicaid eligibility threshold may receive financial assistance from the government to defray some of the costs of coverage.

Though not an absolute guarantee, these programs demonstrate how Congress has incrementally ensured that vulnerable populations who might not otherwise have access to affordable and quality coverage have options for receiving care.

4.2   Other Guarantees of Basic Health Coverage

Beyond creating programs that directly provide coverage to vulnerable and low-income individuals, Congress has enacted other protections that make health insurance and healthcare accessible and affordable to a broad range of individuals. Many of these protections apply to all individuals, regardless of whether they fall into the specific groups covered by the coverage programs described above.

After a few state supreme courts recognized a right to healthcare in emergencies [16], Congress passed the Emergency Medical Treatment and Active Labor Act (EMTALA) in 1986 [17]. EMTALA requires hospitals participating in Medicare to screen and stabilize all patients who use their emergency rooms, regardless of their ability to pay. Despite its limited scope, EMTALA creates a widely accessible right because it extends to everyone at these emergency rooms, including undocumented immigrants who are not eligible for other government healthcare programs [18].

The ACA also made several changes intended to ensure that individuals can keep their existing private insurance and continue to receive comprehensive and quality health services. For instance, the ACA guaranteed that most health plans cover "essential health benefits" including hospitalization, mental health, prescription drugs, and birth control, and set certain annual cost-sharing limitations [19]. Further, the ACA created a tax penalty for large employers that do not offer employees coverage that meets ACA specifications and prohibited insurance companies from denying coverage to those with preexisting conditions [20]. While these protections do not provide any entitlement to health coverage, the ACA made health insurance that adequately covers needed medical expenses affordable and accessible for many Americans.

Most recently, Congress, along with the executive branch, has expanded further into guaranteeing access to healthcare services in response to the COVID-19 pandemic. For instance, in addition to requiring that health plans cover certain

COVID-19 testing and treatment for their members, Congress funded the creation of a COVID-19 Uninsured Program to reimburse claims related to COVID-19 testing, treatment, and vaccine administration for uninsured individuals [21]. Lastly, the federal government provided access to the COVID-19 vaccine free of charge for all U.S. residents, regardless of insurance or immigration status.

5 What's Next?

The right to health is unlikely to be recognized in the U.S. in the near future. It is not expected that the government will ratify any treaties or otherwise take steps to recognize such a right. Further, based on the current conservative composition of the U.S. Supreme Court, the Court is more likely in the near future to take steps to limit the previously recognized rights that resemble the right to health than to take any steps that would increase such rights.

Still, public support has increased for single-payer and public option approaches to health insurance, through which more or even all U.S. citizens and residents could receive access to care [22]. While it is not clear that Congress and the executive branch will pursue these options before there is broader support for these larger measures, Congress in the meantime may continue expanding the subsets of U.S. population eligible for coverage under government programs, such as Medicaid, or for financial assistance to purchase coverage through the marketplaces.

Further, Congress can be expected to continue its current piecemeal trajectory of implementing further protections that increase access to and availability of healthcare. The steps taken in light of the COVID-19 pandemic may prove to be a model for the ability of the federal government to take more significant steps to ensure coverage of healthcare, particularly where there are public health concerns, for all U.S. residents.

Because of this steady movement, clinicians can expect to see continual change in how patients pay for and access healthcare.

References

1. Constitution of the World Health Organization, preamble, *signed* July 22, 1946, *entered into force* Apr. 7, 1948.
2. Universal Declaration of Human Rights, art. 25(1), proclaimed by the General Assembly, resolution 217 A (III), A/RES/3/217 A, *ratified* Dec. 10, 1948.
3. International Covenant on Economic, Social and Cultural Rights, art. 12(1), Dec. 16, 1966, S. Treaty Doc. 95-19, 6 I.L.M. 360 (1967), 993 U.N.T.S. 3.
4. Committee on Economic, Social and Cultural Rights, Substantive Issues Arising in the Implementation of the International Covenant on Economic, Social and Cultural Rights: General Comment 14, para. 12, 30–31, 33–37, U.N. Doc. E/C.12/2000/4 (Aug. 11, 2000).
5. Under U.S. law, the President may form, negotiate, and sign a treaty, but the treaty is not binding until the Senate ratifies it with a two-thirds vote.
6. Social Security Amendments 1965, Pub. L. 89-97, tit. I.
7. *See* Balanced Budget Act 1997, Pub. L. 105-33 (U.S.), tit. IV.

8. *See* Medicare Prescription Drug, Improvement, and Modernization Act 2003, Pub. L. 108-173, tit. I.
9. Ctrs. for Medicare and Medicaid Services: Medicare Enrollment Dashboard. https://www.cms.gov/Research-Statistics-Data-and-Systems/Statistics-Trends-and-Reports/CMSProgramStatistics/Dashboard (2021). Accessed 13 June 2021.
10. Balanced Budget Act 1997, Pub. L. 105-33, tit. IV, § 4901.
11. Ctrs. for Medicare and Medicaid Services: November 2020 Medicaid & CHIP Enrollment Data Highlights. https://www.medicaid.gov/medicaid/program-information/medicaid-and-chip-enrollment-data/report-highlights/index.html (2020). Accessed June 13, 2021.
12. Patient Protection and Affordable Care Act 2010, Pub. L. 111-148, tit. II, § 2001.
13. Nat'l Fed'n of Indep. Bus. v. Sebelius, 567 U.S. 519, 588 (2012).
14. As of the date that this chapter was written, twelve states had not expanded Medicaid under the ACA. Henry J. Kaiser Family Foundation: Status of State Medicaid Expansion Decisions. https://www.kff.org/medicaid/issue-brief/status-of-state-medicaid-expansion-decisions-interactive-map/ (2021). Accessed June 13, 2021.
15. *See* Patient Protection and Affordable Care Act 2010, Pub. L. 111-148, tit. I, § 2001.
16. *See e.g.,* Wilmington Gen. Hosp. v. Manlove, 174 A.2d 135, 139 (Del. 1961); Guerrero v. Copper Queen Hosp., 537 P.2d 1329, 1331 (Ariz. 1975).
17. Consolidated Omnibus Budget Reconciliation Act 1985, Pub. L. 99-272, tit. IX, § 9121.
18. Some municipalities offer basic health coverage for undocumented individuals who are ineligible for other government healthcare programs. *See, e.g.,* Los Angeles County, Department of Health Services: My Health LA. https://dhs.lacounty.gov/my-health-la/. Accessed June 28, 2021; San Francisco Department of Public Health: HealthySF. https://healthysanfrancisco.org/. Accessed June 28, 2021.
19. Patient Protection and Affordable Care Act 2010, Pub. L. 111-148, tit. I, § 1302.
20. *See, e.g., id.* tit. I, §§ 1201, 1513.
21. *See, e.g.,* Families First Coronavirus Response Act 2020, Pub. L. 116-127; Paycheck Protection Program and Health Care Enhancement Act 2020, Pub. L. 116-139; Coronavirus Aid, Relief, and Economic Security (CARES) Act 2020, Pub. L. 116-136; American Rescue Plan Act 2021, Pub. L. 117-2.
22. *See, e.g.,* Pew Research Center: Majority of Democrats Favor a Single National Government Program to Provide Health Care Coverage. https://www.pewresearch.org/fact-tank/2020/09/29/increasing-share-of-americans-favor-a-single-government-program-to-provide-health-care-coverage/ft_2020-09-29_healthcare_01/ (2020). Access October 2, 2021.

Healthcare Should Be a Right

Laura D. Hermer

Contents

Key Points
- Racial, ethnic, socioeconomic, and healthcare disparities lead to disparities in health status.
- The current means of financially accessing healthcare services in the Unites States is fragmented, inconsistent, opaque, and non-universal.
- Creating a legally enforceable right to healthcare would help ensure that all legal U.S. residents could get the healthcare services they need and would help equalize health status among different groups.

Legal Concepts
- Enforceable Right to Healthcare: A right that gives the holder a remedy at law. In other words, the holder may bring a claim, whether administrative, judicial, or both, against a party for the alleged denial or abridgement of that right, where the holder has suffered a legal injury as a result of such denial or abridgement.

L. D. Hermer (✉)
Mitchell Hamline School of Law, Saint Paul, MN, USA
e-mail: Laura.hermer@mitchellhamline.edu

© The Author(s), under exclusive license to Springer Nature Switzerland AG 2022   541
A. S. Pasha (ed.), *Laws of Medicine*,
https://doi.org/10.1007/978-3-031-08162-0_36

1   Introduction

Some Americans believe we enjoy the best healthcare in the world, but this belief is not correct. While many of our acute care services excel for those able to access them, the United States performs poorly in most population health outcomes in comparison to other wealthy nations [1].

Causes of our poor performance on these measures are multifactorial: poverty; racism; educational and environmental disparities; and fragmented, inconsistent, and often barrier-laden access to healthcare, among others [2]. By treating healthcare as a right and reforming our systems of health coverage and care accordingly, we would make access to care more equitable and thereby help improve the health status of all Americans.

2   Discussion

In the United States, evidence supports a strong and consistent correlation between low socioeconomic status (SES) and poorer health. For example, the average life expectancy of a 40 year-old American man in the lowest 1% of income distribution is 14.6 years less than that of a 40 year-old man in the highest 1% [2]. At the same time, strong and consistent correlations exist between Black, Latinx, or Native American race or ethnicity and poorer health in comparison with whites [3]. SES plays into these correlations, but even among Black and white Americans in the same income group, substantial health disparities still exist. For example, as income increases, self-reported health for white adults improves at a faster rate than for Black adults, and Black adults at the highest education levels reported significantly poorer health than white adults at the same education level [4, 5]. These divergent racial outcomes have little to do with "inherent" racial disparities. For example, one 2005 study found that while 44% of Black Americans had hypertension (defined as having a blood pressure equal to or greater than 140/90 mm Hg, or taking antihypertension medications), as compared to 26.8% of white Americans, only 13.5% of Nigerians had hypertension [6]. Instead, substantial evidence suggests that these disparities in health are founded on structural socioeconomic and educational inequities, and thus are impacted substantially by the neighborhood, family, and other circumstances into which a person is born [5, 7, 8].

Evidence finds a strong correlation between generous, universal health coverage and superior health [9]. As discussed in the previous chapter, the U.S. population does not currently enjoy such coverage. Medicare, the federal health insurance program for the elderly and certain disabled individuals, comes the closest to meeting these qualifications, but does not meet the mark. Medicare covers only about 80% of an average beneficiary's healthcare expenses in the absence of any secondary insurance program, premiums for parts of the traditional program have become means-tested over time, and the program covers less than 20% of the U.S. population [10, 11].

Most Americans instead rely on a patchwork of other coverage options: private and usually employer-sponsored coverage, Medicaid, or one or more smaller programs. Approximately 67% of Americans have private coverage: 56.4% through employment and 10.2% through the individual market [11]. These percentages hide disparities. For example, fewer than half of full-time workers earning less than 250% of the federal poverty level had employer-sponsored health insurance in 2018, as compared to 85% of those earning more than 400% of the federal poverty level, a gap that has widened over the last 20 years [12].

If a person is fortunate enough to have employer-sponsored coverage but loses their job, coverage is usually lost as well. Fluctuations in coverage can be common, and are associated with uninsurance, exposure to financial risk, and reduced access to care [13]. In 2019, 29.6 million Americans were uninsured [11]. Cost is a substantial contributor to this state of affairs: The average annual premium for an employer-sponsored family plan in 2019 was $20,486, with the employee responsible, on average, for nearly 28% of the share [14, 15]. Accordingly, one survey found that 73.7% of non-elderly, uninsured respondents cited cost as a reason they lacked coverage. Of the non-elderly, uninsured respondents who worked, 72.5% reported that their employer did not offer health insurance benefits [16]. Obtaining and keeping health insurance can be difficult or confusing due to a complex labyrinth of rules and opacity regarding program existence and requirements, especially for those who do not have access to coverage through their employment. This may partially account for the 13.1 million people who are eligible either for a free individual market plan through the ACA Marketplace or a Medicaid plan, but who nevertheless are uninsured [17].

Because of Medicaid's problems, some argued before the 2014 implementation of the ACA that Medicaid coverage was worse for beneficiaries than being uninsured [18]. Yet growing evidence compiled since the ACA's implementation suggests the contrary: Lower-income populations experience better health and economic outcomes when they have improved access to coverage, even when that coverage is Medicaid. Individuals living in states that expanded Medicaid were far more likely to have a usual source of care, receive preventive care, pay less money out of pocket for healthcare, present earlier and be more likely to receive optimal management for certain surgical conditions, be diagnosed and treated earlier for various cancers, have fewer avoidable hospitalizations for chronic conditions, and have better self-reported mental and physical health, among other improvements [19–25]. Some studies also found reduced racial and ethnic disparities in coverage, health, and/or economic status following Medicaid expansion [26–29].

The ACA's expansion of Medicaid coverage in participating states and its extension of both subsidies and group market protections to individuals with non-group private health plans have made coverage more accessible and affordable for millions of Americans. However, much more can be done. An important first step would be to institute a legally enforceable right for all legal U.S. residents to a particular set of healthcare services. Such a right would presumably not guarantee access to any and all healthcare, but rather would ensure access to all reasonable and necessary healthcare ordered or prescribed by a clinician, or to a specific subset of such healthcare. To make the right enforceable, individuals would be able to bring an administrative claim or lawsuit if they were denied such access in violation of the law.

As a practical matter, we would need to make our healthcare system more equitably accessible and affordable for everyone. Guaranteeing and simplifying access to care would help ensure that everyone can get the preventive, acute, and chronic care they need, without delays due to cost or lack of insurance. We would also need to reduce wasteful care to avoid unreasonable expense. Instituting a transparent and communal way of funding services would provide additional impetus to avoid providing low-value or unnecessary care, and to encourage integrated forms of healthcare delivery.

Instituting such a right would not, in itself, determine the structure of access. Theoretically, such a guarantee could be fulfilled by extending private coverage as it presently exists today to everyone, with wrap-around benefits where needed. Such a system would be inordinately costly, and therefore unlikely to be instituted. Much more likely would be, for example, a heavily regulated all-payer system, in which people had several means of accessing coverage and where all providers of each type in the same geographic region accepted the same, negotiated reimbursement rate from all payers. Alternatively, there could be a statewide or nationwide single-payer system that reimbursed providers at rates exceeding Medicare but less than most private payers, or a state or federal public option providing an automatic default, should a legal resident lose other coverage.

Germany provides one example of a heavily-regulated all-payer system on which the United States could draw for ideas. There, multiple private payers negotiate rates annually with health care facilities and clinicians, and providers overall are subject to global, top-down budgeting [30, 31]. While the German government mandates coverage for all Germans and closely regulates the coverage, payment, and delivery of health care, it does not generally employ clinicians and other health care providers and does not offer public coverage. Rather, competing, not-for-profit, non-governmental health insurance plans financed through compulsory wage contributions and supplemental payments cover most Germans and private plans make up the remainder. For those who are unemployed, the government makes their contributions. All insurers must cover a specified set of services. Clinicians are paid on a fee-for-service basis, up to a quarterly threshold based on the number of patients per practice and reimbursement points per patient. After that threshold, payment amounts decline substantially, incentivizing clinicians to keep within their budget [31].

As a single-payer alternative, Medicare offers an obvious, home-grown example, though as mentioned earlier it is not without fault. It covers many necessary medical services with low administrative expenses and wide coverage of the eligible population [32]. It has the advantage of widespread familiarity and acceptance among both the populace and providers alike. If considered as a template for more universal coverage, it would require modifying or discarding some of its more antiquated features, such as payment methods and regulations that encourage fragmented healthcare delivery and overutilization. Another possibility would be state-based single-payer systems, such as those proposed in recent years in Vermont, California, and New York [33–35]. Because those work within our currently-existing framework of private and public coverage, however, they would require multiple waivers of federal law, significant shifts in revenue sources, and total price tags that have given the public pause, even though those sums do not exceed present health care expenditures.

3   Application

3.1   Clinical Vignette

Mr. Smith is a 53 year-old male who is diagnosed with stage 4 metastatic prostate cancer. His clinicians start him on leuprorelin, abiraterone, and prednisone. The treatment works well, reducing Mr. Smith's testosterone level to <0.01 ng/dL. Based on Mr. Smith's response, his clinicians believe Mr. Smith has a good chance of keeping his cancer in remission for some time, perhaps even for several years.

Mr. Smith works as a software developer at a small firm, which provides him with health insurance as one benefit of employment. Several months after Mr. Smith's treatment for his prostate cancer commenced, the firm lets him go. While the firm cites difficult economic conditions and Mr. Smith's distracted and lackluster performance over the past several months, Mr. Smith wonders if the cost of his care may have factored into the decision. According to the insurance statements he receives, he knows the cost of his ongoing medication and monitoring tests, which he will need to receive for the remainder of his life, will total over $120,000/year. This sum substantially exceeds the gross annual salary he earned at the firm. Mr. Smith is not aware of this, but the firm self-insures (i.e., pays claims directly rather than purchasing coverage from a health insurer) despite its small size. The sharp increase in utilization costs from Mr. Smith's care caused a shock to the firm's finances. If he were able to prove that the firm terminated him based on his health status, Mr. Smith might ultimately be able to prevail in a wrongful termination claim against his now-former employer. However, there are many ways to defend against such claims in the absence of clear proof, and in any event, Mr. Smith might not survive long enough to see his claim succeed.

When Mr. Smith's employer terminates him, it also terminates his health insurance.

Mr. Smith is now uninsured and unemployed, with a terminal illness. As long as he remains sufficiently healthy to work, Mr. Smith might be able to find new employment that might offer health insurance as a benefit. Until that time, he will need to pay for his treatment out-of-pocket, or else try to find coverage. He has a variety of different options available to him. However, federal law requires that he be notified only of one of them, and he may not know about the existence of the others. Moreover, even if he is aware of them, he may or may not be able to take advantage of any of them, depending on their cost.

1. Federal law requires Mr. Smith's former employer to notify him that he can opt to maintain his health insurance through the firm for up to 18 months. However, he will be responsible for the full cost of the premium plus administrative expenses, for a total of $836/month, in addition to his $2500 annual deductible and substantial out-of-pocket expenses.

2. Mr. Smith may not be aware of it, but he could alternatively seek coverage through his state's ACA Marketplace or exchange. If he seeks coverage through the Marketplace, he might qualify for federal subsidies to help him pay his premiums for private coverage and possibly also his out-of-pocket costs, depending on his income (including unemployment benefits). The subsidies will not cover all his costs. Moreover, no federal subsidies are available to him for a private plan if his income declines below 100% of the federal poverty level. Additionally, he will need to determine whether any of the different health insurance plans available to him through the Marketplace cover all his different providers and drugs. If none does, he may need to investigate different coverage options, if any, or different treatment options, if any.

3. If Mr. Smith seeks coverage through his state's ACA Marketplace, the financial information he provides in the process will also be evaluated to see if he qualifies for his state's Medicaid program. If his income is low enough, and if he lives in a state that expanded Medicaid coverage to working-age adults under the ACA, then he should be able to obtain Medicaid. If he lives in a non-expansion state, however, he will not, unless he is able to qualify on the basis of disability, is sufficiently well-informed to be aware of that possibility, and is able to successfully apply to his state's Medicaid agency and appeal any denial. Medicaid will cost Mr. Smith much less than any subsidized private coverage he could otherwise obtain. Moreover, as long as he retains his current clinicians, he will likely be able to continue with them under Medicaid. If he needs to change providers, however, he might have difficulty finding new ones. Also, if he needs to move to a different state and continues to need Medicaid, he will need to verify that he will qualify for coverage there, and he will then need to reapply.

4. Mr. Smith might not look for coverage on his state's ACA Marketplace, but might instead consult with a private insurance broker. The insurance broker may or may not inform Mr. Smith of public or potentially-subsidized private options through his state's ACA Marketplace. Any private coverage purchased outside of the state's ACA Marketplace will not qualify for federal subsidies.

5. Mr. Smith might not be able to afford any private coverage options available to him. He might not live in a state that expanded Medicaid coverage. He might try to qualify for Medicaid on the basis of disability and fail. He might eventually obtain coverage after a lapse in both coverage and care, with negative health consequences. He might not have any option for coverage, liquidate his assets and spend his remaining savings on treatment, and then end treatment when the money runs out and any charity care he can cobble together ends.

If there were an enforceable right to healthcare in the United States, Mr. Smith's health and economic outcomes likely would be different. Regardless of the financing system used, Mr. Smith would likely be able to both continue his treatment and keep his job for as long as he remained healthy enough to do so. Under most likely formulations, his firm would not be responsible, whether via direct payment or via premiums paid for health insurance, for the cost of its employees' health care. Instead, both the costs and risks would be spread more broadly.

4 Summary

Despite the United States' wealth and its vaunted healthcare system, Americans have inequitable and inconsistent access to care. Access is strongly contingent on socioeconomic status, age, employment status, and geographic vagaries, among other factors. Guaranteeing and simplifying access to healthcare would help address socioeconomic disparities in health and ensure that individuals would have reliable access to medical services throughout their lives. While we could theoretically extend coverage to all Americans under our present system, moving either to a more heavily-regulated multi-payer system or to a single-payer option or system would make better financial and organizational sense.

References

1. Tikkanen R, Abrams MK. U.S. health care from a global perspective, 2019: higher spending, worse outcomes? 2020 [cited 2021 June 30]. Available from: https://www.commonwealth-fund.org/publications/issue-briefs/2020/jan/us-health-care-global-perspective-2019.
2. Adler NE, Cutler DM, Fielding JE, Galea S, Glymour M, Koh HD et al., Addressing Social Determinants of Health and Health Disparities. In Dzau VJ, McClellan MB, McGinnis JM, Finkelman EM, eds. Vital Directions for Health & Health Care. Washington D.C.: National Academy of Medicine. 2017 [cited 2021 June 28]; p. 71–95. Available from: https://nam.edu/wp-content/uploads/2018/02/Vital-Directions-for-Health-and-Health-Care-Final-Publication-022718.pdf.
3. Introduction and literature review. In Smedley BD, Stith AY, Nelson AR, eds. Unequal Treatment: Confronting Racial and Ethnic Disparities in Healthcare. Washington D.C.: Institute of Medicine. 2003; p. 29–79.
4. Farmer MM, Ferraro KF. Are racial disparities in health conditional on socioeconomic status? Soc. Sci. Med. 2005;60:1 191–204. doi: https://doi.org/10.1016/j.socscimed.2004.04.026.
5. Braveman PA, Cubbin C, Egerter S, Williams DR, Pamuk E. Socioeconomic disparities in health in the United States: What the patterns tell us. Am. J. Pub. Health. 2010;100:S1. p. S186–S196. doi: https://doi.org/10.2105/AJPH.2009.1660-82.
6. Cooper RS, Wolf-Maier K, Luke A, Adeyemo A, Banegas JR, Forrester T, et al. An international comparative study of blood pressure in populations of European vs. African descent. BMC Med. 2005; 3:2. doi: https://doi.org/10.1186/1741-7015-3-2.
7. Kennedy BP, Kawachi I, Glass R, Prothow-Stith D. Income distribution, socioeconomic status, and self rated health in the United States: multilevel analysis. BMJ. 1998;317: p. 917-921. doi: https://doi.org/10.1136/bmj.317.7163.917.
8. Glymour MM, Avendano M, Kawachi I. Socioeconomic status and health. In Berkman LF, Kawachi I, Glymour MM, eds. Social Epidemiology. 2nd ed. Oxford University Press; 2014.

9. Bergqvist K, Yngwe MA, Lundberg O. Understanding the role of welfare state characteristics for health and inequalities – an analytical review. BMC Pub. Health. 2013;13:1234. doi: https://doi.org/10.1186/1471-2458-13-1234.

10. McArdle F, Stark I, Levinson Z, Newman T. How does the benefit value of Medicare compare to the benefit value of typical large employer plans? A 2012 update. 2012. [cited 2021 June 30]. Available from: https://www.kff.org/wp-content/uploads/2013/01/7768-02.pdf.

11. Keisler-Starkey K, Bunch LN. Health Insurance Coverage in the United States: 2019. Washington D.C.: U.S. Census Bureau. 2020 [cited 2021 June 28]. Available from: https://www.census.gov/content/dam/Census/library/publications/2020/demo/p60-271.pdf.

12. Rae M, McDermott D, Levitt L, Claxton G. Long-Term Trends in Employer-Based Coverage. Peterson-KFF Health System Tracker. 2020 [cited 2021 June 29]. Available from: https://www.healthsystemtracker.org/brief/long-term-trends-in-employer-based-coverage/.

13. Sommers BD, Gourevitch R, Maylone B, Blendon RJ, Epstein AM. Insurance churning rates for low-income adults under health reform: lower than expected but still harmful for many. Health Affairs. 2016;35:10. doi: https://doi.org/10.1377/hlthaff.2016.0455.

14. Kaiser State Health Facts. Average annual family premium per enrolled employee for employer-based health insurance. 2019 [cited 2021 June 29]. Available from: https://www.kff.org/other/state-indicator/family-coverage/?currentTimeframe=0&sortModel=%7B%22colId%22:%22Location%22,%22sort%22:%22asc%22%7D.

15. Claxton G, Rae M, Young G, McDermott D. Employer Health Benefits: 2019 Annual Survey. San Francisco: Kaiser Family Foundation. 2019 [2021 June 29]. Available from: https://www.kff.org/report-section/ehbs-2019-section-1-cost-of-health-insurance/.

16. Tolbert J, Orgera K, Damico A. Key facts about the uninsured population. 2020 [cited 2021 June 29]. Available from: https://www.kff.org/uninsured/issue-brief/key-facts-about-the-uninsured-population/.

17. Rae M, Cox C, Claxton G, McDermott D, Damico A. How the American Rescue Plan Act affects subsidies for Marketplace shoppers and people who are uninsured. 2021 [cited 2021 June 29]. Available from: https://www.kff.org/health-reform/issue-brief/how-the-american-rescue-plan-act-affects-subsidies-for-marketplace-shoppers-and-people-who-are-uninsured/.

18. Roy A. Why Medicaid is a humanitarian catastrophe. Forbes. 2011 [cited 2021 June 29]. Available from: https://www.forbes.com/sites/theapothecary/2011/03/02/why-medicaid-is-a-humanitarian-catastrophe/?sh=1f0dae8644bf.

19. Sommers BD, Maylone B, Blendon RJ, Orav EJ, Epstein AM. Three-year impacts of the Affordable Care Act: Improved medical care and health among low-income adults. Health Affairs. 2017;36:6 1119–1128. doi: https://doi.org/10.1377/hlthaff.2017.0293.

20. Lin S, Brasel KJ, Chakraborty O, Glied SA. Association between Medicaid expansion and the use of outpatient general surgical care among US adults in multiple states. JAMA Surgery. 2020;155(11):1058-1066. doi: https://doi.org/10.1001/jamasurg.2020.2959.

21. Eguia E, Cobb AN, Kothari AN, Molefe A, Afshar M, Aranha GV et al. Impact of the Affordable Care Act (ACA) Medicaid expansion on cancer admissions and surgeries. Ann. Surg. 2018;268:4 584–590. doi: https://doi.org/10.1097/SLA.0000000000002952.

22. Mondesir FL, Kilgore ML, Shelley JP, Levitan EB, Huang L, Riggs KR et al. Medicaid expansion and hospitalization for ambulatory care-sensitive conditions among nonelderly adults with diabetes. J. Ambulatory Care Mgmt. 2019;42:4 pp 312–320. doi: https://doi.org/10.1097/JAC.0000000000000280.

23. Takvorian SU, Oganisian A, Mamtani R, Mitra N, Shulman LN, Bekelman JE et al. Association of Medicaid expansion under the Affordable Care act with insurance status, cancer stage, and timely treatment among patients with breast, colon, and lung cancer. JAMA Network Open. 2020;3:2. doi: https://doi.org/10.1001/jamanetworkopen.2019.21653.

24. Soni A, Wherry LR, Simon KI. How have ACA insurance expansions affected health outcomes? Findings from the literature. Health Affairs. 2020;39:3 pp. 371–378. doi: https://doi.org/10.1377/hlthaff.2019.01436.

25. Loehrer AP, Chang DC, Scott JW, Hutter MM, Patel VI, Lee JE et al. Association of the Affordable Care Act Medicaid expansion with access to and quality of care for surgical conditions. JAMA Surgery. 2018;153(3):e175568. doi: https://doi.org/10.1001/jamasurg.2017.5568.

26. Patel MR, Tipimeni R, Kieffer EC, Kullgren JT, Ayanian JZ, Chang T et al. Examination of changes in health status among Michigan Medicaid expansion enrollees from 2016 to 2017. JAMA Network Open. 2020;3:7. doi: https://doi.org/10.1001/jamanetworkopen.2020.8776.

27. Brown CC, Moore JE, Felix HC, Stewart K, Bird TM, Lowery CL et al. Association of state Medicaid expansion status with low birth weight and preterm birth. JAMA. 2019;321(16):1598-1609. doi: https://doi.org/10.1001/jama.2019.3678.

28. Hayes SL, Riley P, Radley DC, McCarthy D, Reducing racial and ethnic disparities in access to care: Has the Affordable Care Act made a difference? The Commonwealth Fund [Internet]. 2017 [cited 2021 June 28]. Available from: https://www.commonwealthfund.org/publications/issue-briefs/2017/aug/reducing-racial-and-ethnic-disparities-access-care-has.

29. Han X, Jemal A, Zheng Z, Sauer AG, Fedewa S, Yabroff KR. Changes in noninsurance and care unaffordability among cancer survivors following the Affordable Care Act. J. Natl Cancer Inst. 2020;112:7 pp. 688–697. doi: https://doi.org/10.1093/jnci/djz218.

30. Reinhardt, UE. Operating under a global budget: Perspectives from the United States and abroad. In Institute of Medicine. Changing the Health Care System: Models from Here and Abroad. National Academies Press; 1994.

31. Blumel M, Busse R. International Health Care System Profiles: Germany. Commonwealth Fund; 2020. [cited 2021 October 26]. Available from: https://www.commonwealthfund.org/international-health-policy-center/countries/germany.

32. Henry J. Kaiser Family Foundation. An Overview of Medicare. 2019. [cited 2021 October 28]. Available from: https://files.kff.org/attachment/issue-brief-an-overview-of-medicare.

33. Hunter K, Kendall D. Single-payer health care: A tale of three states. Third Way; 2019. [cited 2021 October 28]. Available from: https://www.thirdway.org/report/single-payer-health-care-a-tale-of-3-states.

34. Hart A, Bluth R. New single-payer bill intensifies Newsom's political peril. Kaiser Health News; 2021. [cited 2021 October 28]. Available from: https://khn.org/news/article/new-single-payer-bill-intensifies-newsoms-political-peril/.

35. Young, S. Lawmakers face an uphill climb on single-payer in 2020. Politico; 2019. [cited 2021 October 28]. Available from: https://www.politico.com/states/new-york/albany/story/2019/12/30/lawmakers-face-an-uphill-climb-on-single-payer-in-2020-1235558.

Index